RELATIONS FOR CONVERSIONS OF UNITS[†]

Acceleration:	1 ft/s^2	= 0.3048 m/s^{2*}
	1 in./s^2	= 25.4 mm/s^{2*}
Area:	1 ft^2	= 0.092903 m^2
	1 in^2	= 645.16 mm^{2*}
Density:	1 lbm/ft^3	= 16.018 kg/m^3
Energy:	1 Btu	= 1055.1 J
	1 ft·lb	= 1.3558 J
Force:	1 kip	= 4.4482 kN
	1 lb	= 4.4482 N
Impulse:	1 lb·s	= 4.4482 N·s
Length:	1 ft	= 0.3048 m*
	1 in.	= 25.4 mm*
	1 mi (= 5280 ft)	= 1.6093 km
	1 NM	= 1852 m*
Mass:	1 lbm	= 0.45359237 kg*
	1 slug	= 14.594 kg
	1 ton (= 2000 lbm)	= 907.18 kg
Moment of Force:	1 lb·ft	= 1.3558 N·m
	1 lb·in.	= 0.11298 N·m
Moment of Inertia, area:	1 in^4	= 416231 mm^4
Moment of Inertia, mass:	1 slug·ft^2 (= 1 lb·ft·s^2)	= 1.3558 kg·m^2
Momentum, linear:	1 lb·s	= 4.4482 N·s
Momentum, angular:	1 lb·ft·s	= 1.3558 kg·m^2/s
Power:	1 ft·lb/s	= 1.3558 W
	1 hp (= 550 ft·lb/s)	= 745.70 W
Pressure, stress:	1 lb/ft^2	= 47.880 Pa
	1 psi (= 1 lb/in^2)	= 6.8948 kPa
Velocity:	1 ft/s	= 0.3048 m/s*
	1 mph (= 1 mi/h)	= 0.44704 m/s*
	1 kn (= 1NM/h)	= 0.51444 m/s
Volume:	1 ft^3	= 0.028317 m^3
	1 in^3	= 16.387 cm^3
	1 gal (= 231 in^3)	= 3.7854 L
Work:	1 ft·lb	= 1.3558 J

[†]Refer to Table 1.1 on p. 9.

*Exact value

Engineering Mechanics STATICS

Engineering Mechanics
STATICS

I. C. JONG

Professor of Mechanical Engineering
University of Arkansas
Fayetteville, Arkansas

B. G. ROGERS

Professor of Civil Engineering
Lamar University
Beaumont, Texas

New York   Oxford
OXFORD UNIVERSITY PRESS

Oxford University Press

Oxford   New York
Auckland Bangkok Buenos Aires Cape Town Chennai
Dar es Salaam Delhi Hong Kong Istanbul Karachi Kolkata
Kuala Lumpur Madrid Melbourne Mexico City Mumbai
Nairobi São Paulo Shanghai Taipei Tokyo Toronto

Published by Oxford University Press, Inc.
198 Madison Avenue, New York, New York, 10016
http://www.oup-usa.org

Oxford is a registered trademark of Oxford University Press

Cover: © COMSTOCK INC./ Brooklyn Bridge

ISBN: 0-19-517139-X

Printing number: 9 8 7 6 5 4 3 2 1

Printed in the United States of America
on acid-free paper

Preface

This book and the following volume, *Engineering Mechanics: Dynamics*, are designed for the fundamental courses in statics and dynamics, which are usually offered in the sophomore or junior year in engineering curricula. In the development of the text, students are assumed to have a background in algebra, geometry, trigonometry, and basic differential and integral calculus. Nevertheless, students with a prior knowledge in college physics normally have a higher degree of readiness for learning engineering mechanics.

Mechanics has long been recognized as mainly a deductive science; however, the learning process is largely inductive. In the text, simple topics and problems precede those that are more complex and advanced. For example, the computation of moments of forces using scalar algebra is presented before that using vector algebra, explanations of the rudiments of vector analysis are given before their applications in mechanics, and exercise problems are arranged in order of increasing difficulty. Overall, the text is written to give a clear and up-to-date presentation of the theory and application of engineering mechanics; it is aimed at helping engineering students develop an ability to apply well-established principles to analyze and solve problems in a logical and effective manner.

Organization and Features

As shown in the contents overview, the first ten chapters cover statics, and the remaining ten chapters cover dynamics. The text has been especially organized to help students understand and retain the concepts presented. A more definitive approach is used in developing the text. Each chapter is divided into two or three definitive parts with descriptive headings, and each part contains an appropriate number of sections. The *key concepts* to be covered are listed at the beginning of each chapter, and the list is followed by an introductory paragraph. In addition to sets of traditional problems, concept-reinforcing *developmental exercises* are provided, and concluding remarks as well as a set of review problems are given at the end of each chapter.

Appendices A and B include materials for a *review of basic mathematics* from algebra through calculus, Appendix C verifies the vector characteristics of infinitesimal angular displacements, while Appendices D and E provide general *computer programs* for solving linear and nonlinear equations in mechanics. Yet, omission of computer programs will not prejudice the understanding of the text.

The list of *key concepts* at the beginning of each chapter serves as a summary of major topics presented. By identifying the key concepts in a given chapter, this list helps students focus their efforts and increase their efficiency of study. The introductory paragraph, which follows the list, gives an overview of the subject matter and the relevance of the topics to be covered.

The numerous *developmental exercises* interspersed throughout the text are intended for use as learning exercises. Users may either use or skip them as they see fit. These exercises are designed to (a) prompt students to describe and reinforce their understanding of the concepts, laws, and principles they are learning; and (b) provide opportunities for students to practice problem-solving skills by working through exercises that contain cue questions or similarities to preceding examples.

It is believed that *examples* depicting real-life situations are a useful means of providing motivation for students to learn and to develop the skills needed to reduce actual physical situations to models so that basic principles can be applied. Thus, the text includes many examples that illustrate real-life, as well as theoretical, situations. Each example contains a sample problem and a solution to that problem. The solution stresses the interpretation of the situation and the setting up of equations. Whenever intermediate steps in the solution are routine in nature, they are omitted to avoid distraction from the main solution.

The sets of *problems* provided in each chapter are also designed to create real-life, as well as theoretical, situations. Students can practice applying the principles presented in the text to these problems. Further, the text offers an abundance of problems, which provide the instructor with choices of assignments. Approximately two thirds of the problems have answers given in the back of the book. To alert users to a problem without a given answer, an asterisk (*) is placed after the problem number.

The last section in each chapter is set aside to give *concluding remarks* on the subject matter in that chapter. Generally, these remarks cover the highlights and special features of the chapter. At the end of each chapter, a set of *review problems* is given. These problems generally illustrate the key concepts listed at the beginning of the chapter. Besides providing additional homework problems, they may be used by students as a test of the knowledge and skills they are expected to have acquired upon completing the study of that chapter.

Systems of Units

In recognition of current trends and practices in the United States, both the SI units *(Le Système International d'Unités)* and the U.S. customary units are used. Chapter 1 offers a discussion of the conversion between the two systems of units. Approximately 60 percent of the problems are stated in SI units, and the remainder retain U.S. customary units.

Optional Sections and Order of Coverage

To allow choices of coverage and to serve a wider group of users, the text includes more than a few optional sections. Each of these sections is marked with a star (★) before the section number for distinction from the other sections that form the core of the basic mechanics course. Optional sections may be skipped without affecting the understanding of the core of the course.

At the discretion of the instructor, some of the materials may be covered in a different sequence without losing continuity. One may cover the concepts of forces and moments in Chapters 2 and 4 prior to covering the concepts of equilibrium in Chapters 3 and 5; and moments of inertia in Chapter 7 may be covered toward the end of the course, such as after Chapter 9.

Computer Programs and Diskettes

Although students may use a *calculator* as a tool to solve linear simultaneous equations and find roots of nonlinear equations, they will often benefit further by using a *computer*, which is nowadays utilized in a wide variety of situations. Therefore, some computer programs are included. These programs are written in BASIC and will run on IBM PC and compatibles. Students using a computer should understand the essential concepts involved in solving the problem and be prepared to exercise good judgement in interpreting the output from the computer. The short programs appearing in some sections of the text are simple application programs, which serve as illustrations. The program in Appendix D is a general program for solving a set of n linear simultaneous equations in n unknowns, while the program in Appendix E is a general program for finding roots of a nonlinear or transcendental equation. Several problems that are amenable to computer solutions are identified in the text by footnotes, hints, or suggestions.

All programs listed in the text and the supplementary software, as described in Appendix F, are stored on computer diskettes, which may be obtained from bookstores or the publisher. The authors wish to thank Professor Lynn D. Wills of the University of Arkansas for his contribution in preparing and programming the supplementary software.

Unique Topic Coverage

This text includes a number of presentations, illustrations, and emphases not usually found in similar texts. In particular, note the following:

1. Contrasting *relativistic viewpoints* on length, time, and mass are presented in Sec. 1.14. These viewpoints provide glimpses of the difference between engineering mechanics and modern physics.

2. The *computer solution of simultaneous equations*, encountered in Example 3.8, is illustrated in Appendix D. Opportunities to practice this skill are provided in some problems (e.g., Probs. 3.43, 3.44, 3.45, 3.57, 8.24, 8.25, etc.).

3. The *principle of moments*, synthesized from the concept of equivalent systems of forces, is identified as the key principle for locating the resultant of parallel forces, population center, centroid, geographic center, center of mass, and center of gravity in Sec. 6.2. It is a unifying principle which simplifies the study of the topics mentioned.

4. *Useful formulas* are established in Example 6.7 for locating the centroid of any plane area using triangular finite elements (cf. Probs. 6.40, 6.41, 6.42, and 6.96). Their application to any composite plane area furnishes an illustration of the finite element approach.

5. *Useful formulas* are established in Example 7.11 and the associated developmental exercises D7.28 and D7.29 for finding the moments of inertia and the product of inertia of any plane area using triangular finite elements (cf. Probs. 7.67 through 7.78, and 7.161). Their application to any composite plane area furnishes another illustration of the finite element approach.

6. *Effects of distributed moments* are taken into account in Eq. 8.9 and illustrated in Example 8.21. Without that equation, the slope in the bending-moment diagram for a beam with distributed moments would be inexplicable.

7. The *computer solution of a transcendental equation*, encountered in Example 8.24, is illustrated in Appendix E. Opportunities to practice using a computer to solve transcendental and nonlinear equations are provided in some problems (e.g., Probs. 8.113, 9.114, 10.24, 10.25, 10.26, 10.34, etc.).

8. The distinction between *compatible virtual displacement* and *constrained virtual displacement* is made in Sec. 10.4 and illustrated in Examples 10.7 through 10.10 for both frames and machines.

9. *Infinitesimal angular displacements* are shown to be vectors in Appendix C. It establishes the background against which angular velocities and angular accelerations can readily be discerned as vectors.

10. The *displacement center* for finding virtual displacements is introduced in Sec. 10.6 and illustrated in Examples 10.6 and 10.10. This concept provides an alternative to the traditional approach using variational calculus.

11. A *quantitative definition for potential energy* is stated in Sec. 10.9. This definition makes potential energy as easy to understand and compute as work of a force.

12. *Taylor's series expansion* is employed to establish a *higher-order criterion for stability* in Sec. 10.14. It shows explicitly why the system is stable when the first nonzero derivative of the potential energy is of *even* order and has a *positive* value.

Acknowledgments

The authors thank their colleagues Jim H. Akin, Charles W. Crook, and Charles R. Fagg for their thoughtful suggestions, and department heads, William F. Schmidt (University of Arkansas) and Enno Koehn (Lamar University), for their encouragement. They appreciate the numerous helpful suggestions and comments made on drafts of the text by the reviewers:

Paul Chan, New Jersey Institute of Technology
Sahib Chehl, Southern University
Rick Gill, University of Idaho
Richard Golembiewski, Milwaukee School of Engineering
Reese Goodwin, Brigham Young University
David Hartman, Northern Arizona University
Robert Howland, Notre Dame University
Richard Longman, Columbia University
Scott Martin, Youngstown State University
Luis Clemente Mesquita, University of Nebraska
Saeed Niku, California State Polytechnic University
Semih Oktay, University of Maryland
Rick Reimer, University of California, Berkeley
Eugene Stolarik, University of Minnesota
Robert Swaim, Oklahoma State University
William Van Arsdale, University of Houston
Carl Vilmann, Michigan Technological University

Special thanks are extended to Barbara Gingery (Acquisitions Editor), Marc Sherman (Project Editor), and Saunders College Publishing for their effort and support during the various stages of the project.

I. C. Jong
B. G. Rogers

June 1990

A Note for Students

As students, your goal here is learning mechanics as well as developing a high level of analytical capability. It is going to require some of your *time* and *effort*. However, many people find that it may be more efficiently attained by noting the following hints:

- **Study methodically.** As you begin each new chapter, try first to gain an overview. This may be achieved by doing the following: (a) thumb through the chapter to look it over, (b) go over the key concepts at the beginning of the chapter, (c) read the introduction following the key concepts, and (d) turn to the last section to read the concluding remarks. Then, read rapidly the sections as assigned or desired, and later return for an in-depth study.

- **Draw diagrams.** It is important to realize early on that engineering mechanics is a physical science which heavily involves *geometry*. Pertinent diagrams or sketches are valuable aids in studying the state of rest or motion of physical systems. There is no substitute for the perception and insight gained from the diagrams or sketches drawn.

- **Work problems.** Work the problems assigned or suggested by your instructor. This is not only an effective way to learn the subject but also a way to show to yourself how well you have learned it.

- **Ask questions.** Maintain your inquisitive mind in the study. Most instructors and teaching assistants have a sincere interest in helping students deepen their understanding of the new material.

The authors sincerely hope that the above hints will prove to be helpful, and that you will see the logic, beauty, and relevance of mechanics as you progress in your learning.

List of Symbols

a	Acceleration
a	Constant; radius; distance; axis; semimajor axis
A, B, C, . . .	Vectors; forces; reactions from supports and connections
$A, B, C,$ . . .	Points; magnitudes
$A', B', C',$ . . .	Points; magnitudes
A	Area
$\overline{AB}, \overline{AC},$ . . .	Line segments; distances
b	Width; distance; axis; semiminor axis
c	Constant
C	Centroid; constant; compression
d	Distance
d_s	Shortest distance
e	Base of natural logarithms; coefficient of restitution
F	Force; friction force
$\mathbf{F}_k$	Kinetic friction force
$\mathbf{F}_m$	Maximum friction force
$\mathbf{F}_{\parallel}$	Component of **F** parallel to a specific axis
$\mathbf{F}_{\perp}$	Component of **F** perpendicular to a specific axis
g	Gravitational acceleration
G	Constant of gravitation; center of gravity; mass center
h	height; sag of cable
i, j, k	Unit vectors along Cartesian coordinate axes
$I, I_x, I_y,$ . . .	Moments of inertia
$I_{xy}, I_{uv},$ . . .	Products of inertia
$\bar{I}$	Centroidal moment of inertia of an area; central moment of inertia of a body
J	Polar moment of inertia
k	Spring modulus; constant
k_x, k_y, k_O	Radii of gyration
$\bar{k}$	Centroidal radius of gyration of an area; central radius of gyration of a body
K	Torsional spring modulus; constant
L	Length; span
m	Mass of a body
$\tilde{m}$	Distributed torque; moment per unit length

$\mathbf{M}$	Moment; torque
M	Magnitude of moment or torque; mass of earth
M_{AB}	Moment about axis AB
$\mathbf{M}_O$	Moment about point O
$\mathbf{M}_O^R$	Resultant moment about point O
$\mathbf{M}_\parallel$	Component of $\mathbf{M}$ parallel to a specific axis
$\mathbf{M}_\perp$	Component of $\mathbf{M}$ perpendicular to a specific axis
n	Normal axis; normal direction
$\mathbf{N}$	Normal component of reaction
O	Origin of coordinates; point
p	Pressure; distance; pitch of a screw
P_{cr}	Critical load for buckling
$\mathbf{q}$	Displacement
$\mathbf{q}_\parallel$	Component of $\mathbf{q}$ parallel to a specific force
$\mathbf{r}$	Position vector
$\mathbf{r}_{B/A}$	Position vector of B relative to A
$\mathbf{r}_\parallel$	Component of $\mathbf{r}$ parallel to a specific axis
$\mathbf{r}_\perp$	Component of $\mathbf{r}$ perpendicular to a specific axis
r	Radius; distance; polar coordinate
r_f	Radius of circle of friction
$\mathbf{R}$	Resultant force; resultant vector; reaction
R	Radius; radius of earth; magnitude
s	Length of arc; length of cable; axis
t	Thickness; time; tangential axis; tangential direction
T	Tension; concentrated moment
u	Rectangular coordinate or axis; polar coordinate
U	Work
$U_{1\rightarrow2}$	Work of a force acting on a body from position 1 to position 2
$\mathbf{v}$	Velocity
v	Rectangular coordinate or axis; speed
V	Volume; shear; potential energy
V_a, V_e, V_g	Applied, elastic, and gravitational potential energies
w	Distributed load; force per unit length
$\mathbf{W}$	Weight force
W	Weight
x, y, z	Rectangular coordinates or axes; distances
$\bar{x}, \bar{y}, \bar{z}$	Rectangular coordinates of centroid or center of mass
x', y', z'	Centroidal axes; central axes; auxiliary rectangular coordinate axes
x'', y'', z''	Auxiliary rectangular coordinate axes
x''', y''', z'''	Auxiliary rectangular coordinate axes
α, β, γ	Angles
β	Angle of contact in belt friction
γ	Specific weight
γ_A, γ_L	Weight per unit area; weight per unit length
δ	Deflection
$\delta\mathbf{r}$	Virtual displacement vector
$\delta x, \delta y$	Virtual displacements
$\delta\theta$	Angular virtual displacement
$\Delta\theta$	Angular displacement

δU	Virtual work
ε	Eccentricity of conic section or of orbit
η	Variable analogous to y; axis
θ	Angle; angular coordinate; polar coordinate
$\theta_1, \theta_2, \ldots$	Angles; generalized coordinates
θ_A	Directional angle of $\mathbf{A}$; angle at point A
$\theta_x, \theta_y, \theta_z$	Direction angles of a vector
$\boldsymbol{\lambda}$	Unit vector along a line
μ_k	Coefficient of kinetic friction
μ_r	Coefficient of rolling resistance
μ_s	Coefficient of static friction
ξ	Variable analogous to x; axis
ρ	Mass density; radius of curvature
ρ_A, ρ_L	Mass per unit area, mass per unit length
ϕ	Angle
ϕ_k	Angle of kinetic friction
ϕ_s	Angle of static friction
ψ	Angle
*	Asterisk, marking a problem with no answer given
★	Star, marking an optional section
†	Dagger, marking a footnote
⇧	Hollow arrow, representing a moment or torque
⇒	Equipollent to
◄	Answer pointer

Contents Overview

[†]Chapter 7 may be covered after Chap. 9 without affecting continuity.

Contents

†Also called *second moments of areas*.

Fundamentals

Chapter 1

KEY CONCEPTS

- Scalar, vector, force, mechanics, statics, equilibrium, dynamics, kinematics, kinetics, particle, rigid body, continuum, concentrated force, dimension, length, mass, time, weight, apparent weight, unit, slug, and newton.
- Characteristics of a force and idealizations in mechanics.
- Dimensions and units in the SI and U.S. customary system.
- Rounding rule, numerical accuracy, and conversion of units.
- Newton's laws of motion and law of gravitation, the parallelogram law, and the rigid-body principle.

To form a basis for the study of elementary mechanics, this chapter presents fundamental concepts, definitions, laws, and principles. The fundamental importance of the concept of force is obvious since the study of effects of forces on physical bodies at rest or in motion is, in truth, the central effort of mechanicists. Particles, rigid bodies, continua, and concentrated forces are all idealized concepts which have practical significance in the solution of mechanics problems. Dimensions and units are important to engineers since they work with numerical quantities which are usually used in conjunction with certain units. The distinction between *mass* and *weight* is emphasized because some beginners in mechanics frequently confuse the mass with the weight of a body. As a guide, some rules and concepts related to numerical accuracy are given and discussed. Furthermore, problems are provided to let students practice converting units from the SI to the U.S. customary system, and vice versa.

1.1 Prior Basic Mathematical Skills[†]

The effective study of mechanics as presented in this text requires certain mathematical skills. A set of problems reflecting some of such skills is presented in App. A, which may be used for self-assessment. The mathematical

---

[†]Skills in computer usage and programming are not required for a basic course of study; however, such skills will be helpful if the use of computers is incorporated to enrich the course of study. As mentioned in the Preface, some computer programs are included in the text.

background needed in answering the problems in App. A is tacitly assumed and will generally not be explained in the development of the text material. For convenience, a short review of basic mathematics is presented in App. B, which includes basic concepts and useful topics in algebra, geometry, trigonometry, analytic geometry, and calculus.

FUNDAMENTAL CONCEPTS

1.2　Scalar and Vector Quantities

A *scalar* may be defined as a quantity which has magnitude and possibly sense, but no orientation. The sense of a scalar is denoted by a positive or a negative sign before the magnitude of the scalar. Examples of scalar quantities include length, time, mass, temperature, cost, humidity, work, energy, and power.

A *vector* may be defined as a directed quantity which has magnitude, sense, and orientation, and which obeys the laws for addition of vectors. The sense and orientation of a vector constitute its direction. The result of the addition of two vectors must be independent of the order of addition of these two vectors. Concepts and operations involving vectors will be defined and illustrated later. Typical vector quantities are force, moment of force, linear displacement, velocity, acceleration, impulse, and momentum.

In tensor analysis, whose general theory is not needed in the present study, both scalars and vectors are regarded as special cases of tensors: A scalar is defined as a tensor of the zeroth order, and a vector is defined as a tensor of the first order. Furthermore, specially defined moments of inertia[†] as well as other quantities are classified as tensors of the second order.[††] Operations involving scalars and vectors will be employed for the effective study of the topics in this text.

Developmental Exercises

D1.1[†††]　Define a scalar and give three examples of scalars.

D1.2　Define a vector and give three examples of vectors.

D1.3　A book is shown in Fig. D1.3(a). First, rotate the book about the x axis through 90° as shown in Fig. D1.3(b). This finite rotation is said to have a direction parallel to the x axis and a magnitude of 90°. From this new position, rotate the book about the y axis through 90° and note the orientation of the book. (i) If you first rotated the book in Fig. D1.3(a) about the y axis through 90°, then rotated it about the x axis through 90°, would you obtain the same orientation for

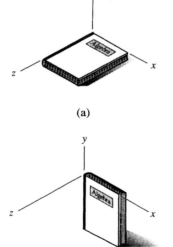

(a)

(b)

Fig. D1.3

[†]Cf. the first footnote on p. 237.

[††]Second-order tensors are sometimes called *bisors*. There are higher-order tensors defined and used in advanced mechanics.

[†††]The developmental exercises (D's) may be *used* or *skipped* as explained in the Preface. Whenever a developmental exercise asks for the definition of a term or description of a law or principle, students are encouraged to answer in their *own words*, which reflect their understanding of the concept involved. Answers to selected developmental exercises are given at the end of the text.

the book as that obtained earlier? (ii) Can we classify finite rotations (which have directions and magnitudes) of a body as vectors? (Cf. App. C.)

D1.4 Is it proper to say that all physical quantities can be classified as either scalar or vector quantities?

1.3 Representation of a Vector

In a typeset text, a vector quantity is often represented by a **boldface** letter, while its magnitude is simply represented by the *italic* version of the same letter. If not typeset, a vector is often represented by a letter with a *short arrow* drawn above it (e.g., $\vec{F}$), while its magnitude is simply represented by that letter (e.g., F).

At times, a vector is graphically represented by a directed line segment. The *magnitude* of the vector is usually denoted by a lightface letter or by a numerical value near the directed line segment; alternatively, the length of the directed line segment drawn to scale may be used to denote the magnitude. The *direction* of the vector is specified by the orientation of the line segment and by the arrowhead indicating the sense at the tip of the segment. For instance, a vector **V** in the *xy* plane may be represented as shown in Fig. 1.1, where the sense indicates that **V** is directed from the point *A* toward the point *B*. If the sense of **V** is reversed, we obtain $-\mathbf{V}$. Furthermore, a vector may be represented by its analytical expression, which will be covered in Chap. 2.

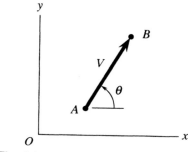

Fig. 1.1

Developmental Exercises

D1.5 Which are vector quantities in the equation written as $\mathbf{F} = m\mathbf{a}$?

D1.6 In Fig. 1.1, what part of the vector **V** does each of the following represent? (a) The letter *V*, (b) the angle θ, (c) the arrowhead at the point *B*.

1.4 Characteristics of a Force

A *force* may be defined as an action in the form of a push, a pull, or a shear exerted by one body on another body where the action changes or tends to change the state of motion of the body on which it acts. A force is a vector. The force exerted by one body on another body *by contact* is called a *contact force* or *surface force*. The force exerted by one body on another body *at a distance* is called a *body force*. Gravitational forces and magnetic forces are examples of body forces. In some cases, a force applied to a body will cause the body to change its direction of motion, or its speed, or both. In other cases, the application of a force to a body at rest will cause the other forces (e.g., the reaction forces from its supports) acting on the same body to change without causing the body to move.

The properties which distinguish one force from a different force are called the characteristics of a force. The *characteristics of a force* are (a) its point of application, (b) its magnitude, and (c) its direction (orientation and sense). The rotational effect produced by a given force acting at a point on

Fig. 1.2

a body is generally different from that produced by another force with the same magnitude and direction (as the given force) but acting at a different point on the same body. This may be illustrated in Fig. 1.2, where the curved arrow indicates the rotational effect of the force **F** on the rigid body.

In mathematics, any two vectors which have the same magnitude and direction are considered to be equal, regardless of whether or not they have the same point of application or the same line of action. Force vectors are not so. Efforts to differentiate such situations have led to the practice of referring to vectors in mathematics as *free vectors* and to force vectors as *bound* (or *localized*) *vectors* in some circumstances and as *sliding vectors* in other circumstances (cf. Sec. 1.12). Note that there are some free vectors in mechanics (cf. Secs. 4.4 and 4.12).

(a)   (b)

(c)   (d)

Fig. D1.9

Fig. D1.10

Developmental Exercises

D1.7   Define a force and state the two ways in which a body may exert a force on another body.

D1.8   What are the three characteristics of a force?

D1.9   Which of the following forces are contact forces? (a) The force exerted by a toolbox on a workbench, (b) the force exerted by the workbench on the toolbox, (c) the force exerted by the earth on an apple as it falls to the ground, (d) the force exerted by a hammer on a nail.

D1.10   A force **F** is applied to the point A as shown. (a) What are the coordinates of the point of application? (b) What is the direction of the force? (c) Is a force vector a free vector?

1.5   Mechanics

Mechanics may be defined as the physical science which describes and predicts the conditions of rest or motion of bodies under the action of force systems. In other words, where there is motion or force, there is mechanics. In engineering, mechanics is generally based on Newton's laws (cf. Sec. 1.11) and is often called *newtonian mechanics,* after the English physical scientist and mathematician Sir Isaac Newton (1642–1727). Conditions involving speeds of bodies close to the speed of light (about 300×10^6 meters per second) and conditions requiring consideration of bodies with extremely small mass and size (such as subatomic particles with sizes in the order of 10^{-12} meter and smaller) cannot be adequately described by Newton's laws; these extreme conditions are treated in *relativistic mechanics* and *quantum mechanics*. However, for a vast range of problems between these extremes, Newton's laws give accurate results and are far simpler to apply.

A force system acting on a body is said to be *balanced* if it has no tendency to change the state of rest or motion of the body in any way. If a body is in *equilibrium,* the force system acting on it must be *balanced.* Furthermore, a body in a state of *equilibrium* must be at rest or moving at a constant speed in a straight line. *Statics* is the branch of mechanics that describes and predicts the conditions of bodies under the action of balanced force systems. Most problems in statics concern bodies at rest.

Dynamics is the branch of mechanics that describes and predicts the conditions of bodies under the action of unbalanced force systems. For the purpose of study, dynamics is often divided into kinematics and kinetics. *Kinematics* deals with the study of motion of bodies without considering the cause of motion. *Kinetics* relates the motion of a body to the force system causing the motion; it usually contains some kinematics.

Developmental Exercises

D1.11   Define the terms: (a) mechanics, (b) newtonian mechanics, (c) statics, (d) dynamics.

D1.12   Give an example of a condition which cannot be adequately described by Newton's laws.

D1.13   State the condition in which a body may be said to be in equilibrium.

D1.14   Name the two parts into which dynamics is often divided for study.

D1.15   How does kinetics differ from kinematics?

1.6   Idealizations in Mechanics

A *particle* is an idealized body which may have finite or negligible mass[†] and whose size and shape can be disregarded without introducing appreciable errors in the description and prediction of its state of rest or motion. It is sometimes viewed as a mass point or a corpuscle.

In *statics,* if the lines of action of all the forces acting on a body intersect at a common point, the body (regardless of the amount and distribution of its mass) may be treated as a particle located at that common point. In *kinematics,* if only the translational motion (in which the body maintains the same orientation) of a body is of concern, the body is treated as a particle. In *kinetics,* if the motion of a body caused by the action of an unbalanced force system is exclusively translational, the body is often treated as a particle, which must have a finite mass.

[†]The equilibrium of a particle with negligible mass (e.g., a knot or small connector) is studied in statics.

A *rigid body* is an idealized body composed of a large number of particles, all of which always remain at fixed distances from each other. In reality, no physical body under the action of forces can remain absolutely rigid. In practice, a body may be regarded as rigid when its deformation is too small to affect the desired accuracy of the analysis. Most bodies in this text are assumed to be rigid bodies. Elastic springs and cordlike elements, such as cords, strings, cables, wires, ropes, belts, and chains, are, of course, nonrigid bodies.

A *continuum* is an idealized body whose matter is assumed to be totally continuous, not porous. The continuum theory is especially practical when one is interested only in the macroscopic manifestations of the matter and nothing else. It can be said that, for many practical purposes in engineering, we are not considering what happens at the molecular level. In this text, each body is assumed to be a continuum; i.e., the molecular structure of matter is disregarded. Note that a continuum may be in the form of a rigid body or a deformable body.

A *concentrated force* (also called a *point force*) is an idealized force assumed to act at a point of a continuum. Although the force exerted by the earth on a body is actually distributed throughout the body, we frequently idealize this situation by assuming the total force exerted by the earth on a body to be a single concentrated force acting at a specific point, which is called the *center of gravity* of the body. It is clear that a contact force exerted on a body by another body is actually distributed over the area of contact between the two bodies. If the area of contact is relatively small, the contact force between the two bodies may be considered as a concentrated force. However, a distributed force system acting on a rigid body is often replaced by a concentrated force to simplify the analysis and solution of a problem.

The assumptions and simplifications leading to the formulation and frequent utilization of the concepts of *particle, rigid body, continuum,* and *concentrated force* are some of the most important *idealizations in mechanics.* Without them, the study of mechanics problems in engineering would be unnecessarily difficult in many cases.

Developmental Exercises

D1.16   Define a particle.

D1.17   Which of the following bodies in the circumstances described may be considered as particles? (a) The determination of the tensile forces holding a *crate and its connecting ring* in equilibrium, (b) the determination of the unknown forces acting on a *beam* in equilibrium, (c) the determination of the maximum altitude of a *satellite,* (d) the determination of the safe speed for a *car* traveling over a vertical curve, (e) the determination of the force holding up one end of a *car.*

D1.18   Define (a) a rigid body, (b) a continuum, (c) a concentrated force.

D1.19   Describe when a body may be regarded as a rigid body.

D1.20   A load of sandbags is uniformly placed on a beam. The reactions from

Fig. D1.17

the supports at A and B are to be determined. Can the distributed weight force of the sandbags be replaced by a single concentrated force?

D1.21   What concepts represent some of the most important idealizations in mechanics?

Fig. D1.20

1.7   Dimensions: Physical Quantities

A *dimension* is a qualitative description of a physical quantity which can be quantified by a certain standard of measure, called *unit*.[†] Examples of dimensions include length, mass, time, force, velocity, area, volume, and others. In 1971, the 14th General Conference of Weights and Measures, an international body, selected *length, mass, time, electric current, thermodynamic temperature, amount of matter,* and *luminous intensity* as the *seven base dimensions* for practical measurement in the International System of Units, abbreviated SI from the French *Le Système International d'Unités,* which is an *absolute* system of units. However, we need to use only *length, mass,* and *time* out of the foregoing seven as the base dimensions in elementary mechanics. A dimension which is developed in terms of base dimensions is called a *derived dimension*; e.g., velocity and volume are derived dimensions. The U.S. customary system of units is a *gravitational* system of units, which uses *length, force,* and *time* as base dimensions in elementary mechanics. Since the United States is in a transition period to

---

[†]Units are discussed in Sec. 1.8. The term *dimension* is often referred to as *quantity* in other literature. Elsewhere, *dimension* is used to indicate the measurement in width, length, thickness, or others.

adopt the SI, U.S. engineers will need to be familiar with both systems of units, which are therefore used in this text.

The base dimensions length, mass, and time in the SI are all scalars and may be described as follows, where the definitions given were adopted by the General Conference of Weights and Measures in the years indicated:[†]

- The dimension *length* is a concept for measuring quantitatively the linear extent of the size or position of an object. The standard of length is defined in terms of the distance traveled by light during a specified fraction of a second. Precisely, *one meter* is defined as the length of the path traveled by light in vacuum in 1/299 792 458 of a second.[††] (Adopted in 1983.)[†††] This means that the speed of light is now *defined* to be exactly 299 792 458 meters per second. In the United States, *one foot* is legally equal to 0.3048 meter.

- The dimension *mass* is a concept for describing quantitatively the translational inertia of a body. Besides, the mass of a body characterizes the mutual gravitational attraction with another body. The standard of mass is defined in terms of the mass of a certain cylinder. Precisely, *one kilogram* is defined as the amount of mass equal to the mass of the *international prototype kilogram,* which is a platinum-iridium cylinder kept at the International Bureau of Weights and Measures at Sèvres, near Paris, France. (Adopted in 1889.) Note that the mass of 1 liter of water is approximately equal to 1 kilogram. In the United States, *one pound-mass is* legally equal to 0.453 592 37 kilogram.

- The dimension *time* is a concept for ordering the flow, or for measuring the duration, of events. The standard of time is defined in terms of a certain number of periods of atomic radiation. Precisely, *one second* is defined as the duration of 9 192 631 770 periods of the radiation corresponding to the transition between the two hyperfine levels of the ground state of the cesium-133 atom. (Adopted in 1967.)

Developmental Exercises

D1.22 Define (a) dimension, (b) derived dimension.

D1.23 How many physical quantities have been selected as base dimensions for practical measurement in the SI?

D1.24 Which SI base dimensions are used in elementary mechanics?

D1.25 Explain briefly the dimensions length, mass, and time.

D1.26 Give three examples of derived dimensions in the SI.

1.8 Units and Prefixes

A *unit* of a dimension is a standard of measure for the quantitative description of the dimension. The primary unit of a base dimension is called a *base*

[†]The relativistic viewpoints of length, mass, and time are briefly described in Sec. 1.14.

[††]When it is desirable to separate the digits in a number into groups of three, the SI practice is to use spaces, not commas, to do so.

[†††]*One meter* was previously defined in 1960 to be equal to 1 650 763.73 vacuum wavelengths of the orange-red radiation emitted under specified conditions by the krypton-86 atom. This definition is now superseded by the 1983 definition of the speed-of-light meter.

unit. The SI uses the *meter* (m), *kilogram* (kg), and *second* (s) as the base units for the base dimensions length, mass, and time, respectively. Furthermore, the base units for the base dimensions electric current, thermodynamic temperature, amount of matter, and luminous intensity are ampere (A), kelvin (K), mole (mol), and candela (cd), respectively. In the U.S. customary system of units, the *foot* (ft), *pound* (lb), and *second* (s) are the base units for length, force, and time, respectively. The unit of a derived dimension is called a *derived unit* and is generally a compound unit, unless it is given a special name; e.g.,[†]

$$1 \text{ kg} \cdot \text{m/s}^2 = 1 \text{ newton} = 1 \text{ N} \quad \text{(for force)}$$

$$1 \text{ N/m}^2 = 1 \text{ pascal} = 1 \text{ Pa} \quad \text{(for pressure or stress)}$$

$$1000 \text{ cm}^3 = 10^{-3} \text{ m}^3 = 1 \text{ liter} = 1 \text{ L} \quad \text{(for volume)}$$

The abbreviations of SI units and other units commonly used in mechanics are listed in Table 1.1. As a rule, symbols for the units are not pluralized, the practice with respect to upper and lower case modes is to be followed, and a period is used only with the abbreviation for inch. However, when the inch is raised to a power, the period may be omitted.

Table 1.1   Units and Their Abbreviations

Name	Symbol	Name	Symbol	Name	Symbol
meter	m	liter	L	mile	mi
kilogram	kg	minute	min	pound	lb
second	s	hour	h	pound-mass	lbm
newton	N	day	d	gallon	gal
joule	J	radian	rad	barrel	bbl
hertz	Hz	degree	°	nautical mile	NM
watt	W	inch	in.	knot	kn
pascal	Pa	foot	ft	acre	a

Table 1.2   SI Prefixes

Factor	Prefix	Symbol	Factor	Prefix	Symbol
10^{18}	exa	E	10^{-1}	deci*	d
10^{15}	peta	P	10^{-2}	centi*	c
10^{12}	tera	T	10^{-3}	milli	m
10^{9}	giga	G	10^{-6}	micro	μ
10^{6}	mega	M	10^{-9}	nano	n
10^{3}	kilo	k	10^{-12}	pico	p
10^{2}	hecto*	h	10^{-15}	femto	f
10^{1}	deka*	da	10^{-18}	atto	a

*The use of these prefixes should be avoided, except for measuring area, volume, body and clothing, and others where numbers would be otherwise awkward.

Generally, each dimension has a *primary unit* and several *secondary units.* In forming secondary units in the SI, the *prefixes* in Table 1.2 are used. For example, 1 gigapascal = 10^9 pascals (or 1 GPa = 10^9 Pa), 1 kilonewton = 10^3 newtons (or 1 kN = 10^3 N), and 1 millimeter = 10^{-3}

[†]The primary unit of a derived dimension is *not* called a base unit. Cf. *ASME Guide SI-1,* 9th ed., 1982.

meter (or 1 mm $= 10^{-3}$ m). Note that *gram* (g) is to be considered a *secondary unit* because kilogram (kg) is selected as a base unit, where 1 g $= 10^{-3}$ kg.

Developmental Exercises

D1.27 (a) Define the terms: unit, base unit, derived unit. (b) Write the abbreviations for newton, kilogram, pound, and pound-mass.

D1.28 Which of the following units are base units (a) in the SI, (b) in the U.S. customary system? Units: N, lb, lbm, slug, g, kg, m, mm, km, Pa, ft, mi, s, h, gal, and L.

D1.29 Express the following quantities in base units in the SI or U.S. customary system: (a) 322 lbm, (b) 0.00524 kN, (c) 1452 g, (d) 30 μs.

1.9 Significant Digits and Rounding Rule

Any nonzero digit in a number is a significant digit; and the digit 0 is also a significant digit except when it is used to fix the decimal point or to fill the places of discarded digits. To avoid ambiguity, we define the *significant digits* of a number as the digits beginning with the first nonzero digit on the left and ending with the last digit on the right.[†]

The *scientific notation* and the *engineering notation* for numerical values are powers-of-ten notations in which each number is expressed as a product consisting of a number multiplied by 10 raised to an appropriate power. For instance, 1.9312128×10^7 and 3.048×10^{-1} are scientific notations, while 19.312128×10^6 and 304.8×10^{-3} (where each exponent is a multiple of ± 3) are engineering notations for the numbers 19312128 and 0.3048, respectively. These notations are particularly useful in expressing very large and very small numbers which have just a few significant digits. (Note that 19.312128 Mm $= 12000$ mi and 304.8 mm $= 1$ ft.)

From time to time, it is necessary to *round off* or simply *round* a number. The rule for rounding a number is summarized in the following table:

Table 1.3 Rounding Rule*

When the First Digit Dropped Is:	The Last Digit Retained Is:
Smaller than 5	Unchanged
Greater than 5 or 5 followed by one or more nonzero digits	Increased by 1
5 which is not followed by any nonzero digit	Unchanged if it is even Increased by 1 if it is odd

*For discussion of the rule summarized here, see J. B. Scarborough, *Numerical Mathematical Analysis,* 6th ed. (Baltimore, Md: The Johns Hopkins Press, 1966), pp. 2–3. Note that when the first digit dropped is 5 which is not followed by any nonzero digit, the last digit retained is such that it becomes, or remains, an *even* digit. The errors due to rounding are generally minimized when the rule is followed consistently.

[†]The number of significant digits of a number has also been defined as the number of digits of the number beginning with the first nonzero digit on the left and ending with the last nonzero digit on the right.

Developmental Exercises

D1.30 State the number of significant digits in each of the following numbers:
(a) 41.500, (b) 41.5021, (c) 42.50, (d) 42.501, (e) 0.006245, (f) 1360000, (g)
5.218×10^6.

D1.31 Round the numbers in D1.30 to *three* significant digits.

D1.32 Round the numbers in D1.30 to *two* significant digits.

1.10 Numerical Accuracy and Conversion of Units

The accuracy of the result in an answer should be closely related to the
accuracy of the given data. Let us here assume that the last significant digit
of each given number is an estimated digit. When a given number is *added
to* or *subtracted from* another given number, the best possible value of the
resulting number will contain no significant digits after any of the decimal
places where the estimated digits of these two given numbers are located.
For instance, we write

$$1.2 + 9 = 10 \qquad 9.81 - 20.3 = -10.5$$

$$12.34 \times 10^3 + 3.41 = 12.34 \times 10^3$$

$$12.34 \times 10^3 - 341 \times 10^6 = -341 \times 10^6$$

When a given number is *multiplied* or *divided by* another given number, the
best possible value of the resulting number may contain one more digit than,
but at least as many digits as, the number of significant digits of the less
accurate given number. For instance, we write

$$1.2(9) = 11 \qquad 9.81/2.34 = 4.19$$

$$(12.34 \times 10^3)(3.41) = 42.1 \times 10^3$$

$$(12.34 \times 10^3)/(3.41 \times 10^6) = 3.62 \times 10^{-3}$$

In the practice of engineering, the data are seldom known with an accuracy
greater than 0.2%. We note that the relative error between 9.97 and 9.99 is
about the same as that between 1.001 and 1.003 (about 0.2% in both cases).
Unless otherwise specified, it will generally be assumed and understood in the
given data and in the *reporting of numerical final answers* in this text that all
numbers whose first significant digit is 1 have a degree of accuracy of at least
four significant digits, and other numbers have a degree of accuracy of at least
three significant digits. For example, a force of 12 N is to be read 12.00 N, a
length of 1.7 m is to be read 1.700 m, and a force of 50 N is to be read 50.0
N. Nevertheless, it is suggested that, if possible, *all intermediate computations be carried through with more significant digits than wanted or justified
in the final results*. As a rule of thumb, *a numerical final answer is generally
reported to four significant digits if its first significant digit is 1; otherwise, it
is generally reported to three significant digits.*[†]

[†]An exception is given to the case where a computed number is added to or subtracted from an
exact number (e.g., 360°). The accuracy in that rule of thumb is known as the *slide rule
accuracy*. A slide rule is an analog calculator, which was widely used by engineers before the
advent of electronic digital calculators in the early 1970s.

The practice of reporting final results according to the number of significant digits may have no more practical value than according to the number of digits after the decimal point. For instance, reporting the reaction components from the support at A as $A_x = 1.234$ kN and $A_y = 78.9$ kN has no more practical value than reporting them as $A_x = 1.2$ kN and $A_y = 78.9$ kN for design and general purposes. However, the former practice allows readers to check the accuracy of numerical results more closely.

From time to time, engineers may need to convert units from the U.S. customary system to the SI, and vice versa. The following conversions are basic and well-known to engineers:

1 mi = 5280 ft	1 yd = 3 ft	1 ft = 12 in.
1 ft = 0.3048 m	1 d = 24 h	1 h = 60 min
1 min = 60 s	π rad = 180°	1 ton = 2000 lbm
1 kip = 1000 lb	1 lb = 16 oz	1 hp = 550 ft · lb/s

Developmental Exercises

D1.33 Assuming that the last significant digit of each number is a doubtful digit, compute the values of the following: (a) 32.2 + 1.234, (b) 32.2 − 1.234, (c) 5.35(3.14), (d) 12.67/25.4, (e) $26.7 \times 10^3 - 21.9$, (f) $3.92/(322 \times 10^3)$, (g) $\sqrt{32.2}$, (h) ln 9.81, (i) $e^{0.234}$, (j) sin 125.5°, (k) cos 125.5°.

D1.34 When you use an electronic calculator in your solution of a problem, should you report your final numerical answer to as many significant digits as shown on the calculator?

EXAMPLE 1.1

The angle of contact between a belt and a drum is $\beta = 36.9°$. Express β in radians.

Solution. We know that π rad = 180°. Thus, we write

$$\beta = 36.9° = 36.9° \left(\frac{\pi \text{ rad}}{180°} \right)$$

$$\beta = 0.644 \text{ rad} \blacktriangleleft$$

REMARK. Note that the final answer is given to only three significant digits because the first significant digit of β is other than 1.

Developmental Exercise

D1.35 Express the following quantities in SI base units: (a) 2 in., (b) 6 yd, (c) 1000 ft³.

EXAMPLE 1.2

The efficiency of an automobile is often measured by its fuel economy (or gas mileage). Making use of 1 gal = 231 in^3 and 1 m^3 = 1000 L, determine the fuel economy of 40 mi/gal of a compact car in km/L.

Solution. Note that the given gas mileage of 40 mi/gal is to be read 40.0 mi/gal. Using the basic conversions (cf. p. 12) as well as those given in the problem, we write

$$40 \text{ mi/gal} = \frac{40 \text{ mi}}{\text{gal}} \cdot \frac{5280 \text{ ft}}{1 \text{ mi}} \cdot \frac{0.3048 \text{ m}}{1 \text{ ft}} \cdot \frac{1 \text{ km}}{1000 \text{ m}} \cdot \frac{1 \text{ gal}}{231 \text{ in}^3}$$

$$\cdot \frac{(12)^3 \text{ in}^3}{1 \text{ ft}^3} \cdot \frac{1 \text{ ft}^3}{(0.3048)^3 \text{ m}^3} \cdot \frac{1 \text{ m}^3}{1000 \text{ L}}$$

$$= \frac{40(5280)(0.3048)(12)^3}{231(0.3048)^3(10)^6} \frac{\text{km}}{\text{L}}$$

$$40 \text{ mi/gal} = 17.01 \text{ km/L} \blacktriangleleft$$

REMARK. Note that the final answer is given to four significant digits in this case because its first significant digit is 1.

PROBLEMS[†]

1.1   A motorist traveling on a highway notes that the posted speed limit is 90 km/h. Determine the speed limit in (a) mi/h, (b) ft/s.

1.2   The rate of flow of oil in a pipe is 90 bbl/min. Making use of 1 bbl = 42 gal, 1 gal = 231 in^3 and 1 m^3 = 1000 L, determine the rate of flow of oil in L/s.[††]

1.3*   A meteorologist reports a rainfall of 1 in. Making use of 1 a = 43560 ft^2, 1 gal = 231 in^3, and 1 m^3 = 1000 L, determine the rainfall on a residential lot of one half acre in (a) liters, (b) gallons.

1.4   A submarine is cruising at a speed of 45 kn. Making use of 1 kn = 1 NM/h and 1 NM = 1852 m, determine the speed of the submarine in ft/s.

FUNDAMENTAL LAWS

1.11   Newton's Laws of Motion and Law of Gravitation[†††]

Newton's laws of motion are Newton's first law, Newton's second law, and Newton's third law. *Newton's first law* states that every particle remains at

---

[†]Answers are provided at the end of the text for all problems except those marked with an asterisk.

[††]It may be noted that 1 barrel = 1 bbl = 42 gal of petroleum or 31 gal of fermented beverage.

[†††]Newton's work, entitled *Philosophiæ Naturalis Principia Mathematica,* was written in Latin and published in 1687; it was revised in 1713 and 1726. For a modern translation, see Florian Cajori, trans., *Sir Isaac Newton's Mathematical Principles of Natural Philosophy and His System of the World* (Berkeley: University of California Press, 1934).

rest, or in motion with constant velocity, unless an unbalanced force acts on it. In other words, when the resultant force acting on the particle is zero, the particle will remain at rest if it is originally at rest, and the particle will move with constant speed in a straight line if it is originally in motion. Thus, Newton's first law is also called the *law of inertia*. Note that the *inertia* of a particle is an *intrinsic property* of the particle to preserve its status quo. The *mass* of a particle is the quantitative measure of the *inertia* of the particle.

In modern terminology, *Newton's second law* states that the derivative of the momentum of a particle with respect to the time t is proportional to the resultant force **F** acting on the particle, where the momentum is equal to the product of the mass m and the velocity **v** of the particle. The constant of proportionality can be made unity by appropriate choices of units so that Newton's second law becomes

$$\mathbf{F} = \frac{d}{dt}(m\mathbf{v}) \tag{1.1}$$

Fig. 1.3

We know that the acceleration **a** of the particle is the time derivative of its velocity **v**; i.e., $\mathbf{a} = d\mathbf{v}/dt$. For a particle with constant mass m as shown in Fig. 1.3, Eq. (1.1) yields

$$\mathbf{F} = m\mathbf{a} \tag{1.2}$$

Besides being applicable to a body with a constant mass m, Newton's second law in the form of Eq. (1.1) is found to be directly applicable to a body gaining or losing mass due to the change in its velocity such as those encountered in relativistic mechanics (cf. Sec. 1.14). However, Newton's second law in the form of either Eq. (1.1) or Eq. (1.2) is *not* directly applicable to a variable body gaining or losing mass through absorbing or expelling particles such as those encountered in the motion of a chain being pulled up from a pile or the motion of a rocket (cf. Sec. 14.13). Thus, it is a *misconception* to think that Newton's second law in the form of Eq. (1.1) is intended for direct application to a body where the particles it contains are continuously changing.

By Newton's second law, we know that when the resultant of all forces acting on the particle is zero, the acceleration of the particle must also be zero. Since zero acceleration means constant velocity and constant velocity includes zero velocity (i.e., at rest), we see that Newton's first law can be regarded as a *special case* of Newton's second law where the acceleration of the particle is zero. On the other hand, Newton's first law may well be considered as the *definition of a reference frame* in space (called the *inertial reference frame*) for which the second law is then valid.[†] In statics, Newton's second law is mainly used to relate weight to mass (cf. Sec. 1.13). In dynamics, Newton's second law is extensively used.

Fig. 1.4

Newton's third law states that to every action there is a collinear reaction having the same magnitude and being opposite in sense. This is to say that the mutual actions of two bodies upon each other are always equal in magnitude and directed in opposite senses along a common line, as shown in Figs. 1.4 and 1.5. From this law, we note that any action must always be

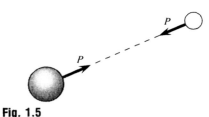

Fig. 1.5

[†]Cf. Sec. 12.1.

accompanied by a corresponding reaction. Newton's third law is very important in both statics and dynamics.

Newton's law of gravitation states that two particles are mutually attracted by a pair of opposite forces that are directed along the line joining the particles and have the same magnitude which is directly proportional to the product of their masses and inversely proportional to the square of the distance between them.[†] For two particles with masses M and m separated by a distance r, as shown in Fig. 1.6, the magnitude of the attractive force may be written as

$$F = G\frac{Mm}{r^2} \tag{1.3}$$

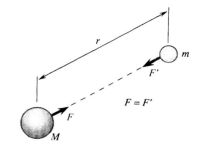

Fig. 1.6

where G is the *constant of gravitation*. Experiments show that

$$G = 66.726 \times 10^{-12} \frac{m^3}{kg \cdot s^2} \tag{1.4}[††]$$

This law has little application in statics; it is mainly used in dynamics in studying the motions of spacecraft, satellites, and planets.

Developmental Exercises

D1.36 Describe (a) Newton's first law, (b) Newton's second law, (c) Newton's third law, (d) Newton's law of gravitation.

D1.37 Is it correct to infer from Newton's third law that "forces always occur in pairs"?

1.12 Parallelogram Law and Rigid-Body Principle

The *parallelogram law* states that the combined effect of two forces acting on a particle is equivalent to the effect of a single force, called their *resultant,* which acts on the particle and is equal to the diagonal of the parallelogram whose consecutive sides intersecting at that diagonal are equal to those two forces.[†††] This is the basic law which defines the addition of two vectors in physical space. The parallelogram law may be illustrated with a graphical equation as shown in Fig. 1.7, which means that

Fig. 1.7

$$\mathbf{P} + \mathbf{Q} = \mathbf{R} \tag{1.5}$$

where $\mathbf{R}$ is the *resultant* of $\mathbf{P}$ and $\mathbf{Q}$. The *resultant* of more than two forces may be obtained by repeated applications of Eq. (1.5).

[†]The statement of this law is a synthesis from *Book Three, System of the World* in Newton's *Principia* (Cf. Cajori, trans. *op. cit.*). In certain astronomical studies, this law is replaced by the law of gravitation as covered in the *general* theory of relativity (1915) of Albert Einstein (1879–1955). Cf. Sec. 1.14.

[††]The accepted value of G as of 1982 is $G = (66.726 \pm 0.005) \times 10^{-12}$ m³/(kg · s²).

[†††]Although fundamental in nature, the parallelogram law was treated by Newton as *Corollary II* to his laws of motion. (Cf. Cajori, trans. *op. cit.*, pp. 13–15.) For Varignon's contribution to this law, cf. Sec. 4.18.

Fig. 1.8

The *rigid-body principle* states that any two opposite forces acting on a rigid body will have no effect on the condition of rest or motion of the rigid body if they have the same magnitude and the same line of action. This principle is illustrated in Fig. 1.8 and is employed in deriving the so-called *principle of transmissibility* in Fig. 1.9, which may be stated as follows: The effect of a force on the condition of rest or motion of a rigid body will remain unchanged if the force is moved to act at another point on its line of action. Thus, a force acting on a rigid body is sometimes called a *sliding vector*.

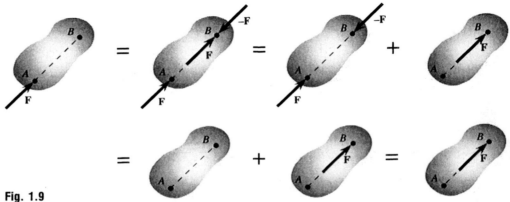

Fig. 1.9

Developmental Exercises

D1.38   Describe (a) the parallelogram law, (b) the rigid-body principle, (c) the principle of transmissibility.

D1.39   Three forces **A**, **B**, and **C** form the sides and a diagonal of a parallelogram as shown. (a) Is **A** + **B** = **C**? (b) Is **B** + **C** = **A**? (c) Is **C** + **A** = **B**? (d) Is $B = A + C$?

Fig. D1.39

1.13   Mass Versus Weight and Force

The *mass* of a body is a quantitative measure of the *inertia of the body* with respect to translational motion. In newtonian mechanics, the mass of a body is a positive scalar quantity and does not vary with the orientation, location, or state of motion of the body. However, the inertia of a body with respect to rotational motion is dependent on not only the amount of mass but also on the way the mass is distributed with respect to the axis of rotation of the body.[†] The *weight force* **W** of a body is the *gravitational force* exerted by the earth on the body and is always directed vertically downward toward the earth.[††] A weight force is a body force. The *weight W* of a body is the magnitude of the weight force **W** of the body. Such a weight is a *gravitational weight*. By Newton's second law, the weight of a body with mass *m* may be written as

$$W = mg \tag{1.6}$$

where g is the gravitational acceleration of the body. The value of g is known to vary with the location of the body.[†††] The *standard gravitational acceleration* is defined to be exactly equal to 9.80665 m/s². However, for practical purposes, we will use

$$g = 9.81 \text{ m/s}^2 \quad \text{or} \quad g = 32.2 \text{ ft/s}^2 \tag{1.7}$$

as the gravitational acceleration of all bodies near the surface of the earth. These approximate values of g are well-known to engineers.

On the other hand, people feel their "weight" by the contact forces exerted on them from their supports (e.g., seats, floors, and the ground). If their supports exert no contact forces on them, they feel "weightless." Such a "weight" felt by people should be classified as *apparent weight* (or contact weight). Obviously, the apparent weight of a body in motion is generally different from the gravitational weight of the body at the same location. In mechanics, the term *weight* is to be understood as the *gravitational weight*, not the apparent weight. Note that the mass of a one-kilogram gold bullion remains the same, no matter if we place it on the surfaces of the Moon, Mars, or Jupiter. However, the weight of this bullion will be different at these different locations. Remember that we use kilogram (kg) as a unit of mass, not of weight or force, in the SI.

Newton's second law and the concept of weight force as described earlier are useful in defining the units of force and mass. *One newton* (N) is defined as the net force required to give a body with a mass of 1 kg an acceleration of 1 m/s²; i.e.,

$$1 \text{ N} = 1 \text{ kg} \cdot \text{m/s}^2 \tag{1.8}$$

The SI is termed an *absolute system* since mass is taken as an absolute or base quantity. The U.S. customary system is termed a *gravitational system* since force as measured from gravity is taken as a base quantity.

One pound (lb) of force is defined as the weight of a body with one pound-mass (1 lbm) measured under the standard gravitational acceleration, which is 9.80665 m/s². Thus, by Newton's second law, we write

$$1 \text{ lb} = 1 \text{ lbm} \cdot 9.80665 \text{ m/s}^2 \qquad 1 \text{ lb} = \frac{9.80665}{0.3048} \text{ lbm} \cdot \text{ft/s}^2 \tag{1.9}$$

[†]Cf. Sec. 7.11.

[††]The weight force of a body on the Moon or Mars is, of course, taken to be the gravitational force exerted by the Moon or Mars on the body, as appropriate.

[†††]Considering that the earth is a rotating oblate spheroid with flattening at the poles, the International Gravity Formula gives $g = 9.78049 (1 + 0.0052884 \sin^2 \phi - 0.0000059 \sin^2 2\phi)$, where ϕ is the latitude and g is measured in m/s², at the sea level. Thus, g varies from 9.780 m/s² at the equator to 9.832 m/s² at the poles.

Notice that the abbreviations for *pound-mass* and *pound* are lbm and lb, respectively; the former is a unit of mass, while the latter is a unit of force. *One slug* is defined as the amount of mass of a body which will be accelerated at the rate of 1 ft/s² by a net force of 1 lb; i.e.,

$$1 \text{ lb} = 1 \text{ slug} \cdot \text{ft/s}^2 \tag{1.10}$$

In the U.S. customary system, the *slug* is the *primary unit* of mass, while the *pound-mass* is a *secondary unit*. We see from Eqs. (1.9) and (1.10) that

$$1 \text{ slug} = 9.80665/0.3048 \text{ lbm} \qquad 1 \text{ slug} \approx 32.2 \text{ lbm} \tag{1.11}$$

Developmental Exercises

D1.40 How does the *weight* of a body differ from its *mass*?

D1.41 Astronauts in a space shuttle orbiting around the earth experience no contact force from their surroundings and are reported as being "weightless." Which of the following is zero: (a) their gravitational weight or (b) their apparent weight?

EXAMPLE 1.3

Express 1 slug in terms of the base units ft, lb, and s.

Solution. From Newton's second law and the definition of 1 slug, we write

$$1 \text{ lb} = 1 \text{ slug} \cdot (1 \text{ ft/s}^2)$$

$$1 \text{ slug} = 1 \frac{\text{lb} \cdot \text{s}^2}{\text{ft}} \blacktriangleleft$$

Developmental Exercises

D1.42 Define the units: (a) 1 N, (b) 1 lbm, (c) 1 lb, (d) 1 slug.

D1.43 Which of the units in D1.42 are for (i) force, (ii) mass?

EXAMPLE 1.4

Making use of the exact relations that 1 lbm = 0.45359237 kg and 1 slug = 9.80665/0.3048 lbm, express 1 newton in terms of pounds to four significant digits.

Solution. From Newton's second law and the definitions of units, we write

$$1 \text{ N} = 1 \text{ kg} \cdot (1 \text{ m/s}^2)$$

$$= 1 \text{ kg} \cdot \frac{1 \text{ m}}{s^2} \cdot \frac{1 \text{ ft}}{0.3048 \text{ m}} \cdot \frac{1 \text{ lbm}}{0.45359237 \text{ kg}} \cdot \frac{1 \text{ slug}}{9.80665/0.3048 \text{ lbm}}$$

$$= \frac{1}{0.45359237(9.80665)} \frac{\text{slug} \cdot \text{ft}}{s^2} = 0.224809 \text{ slug} \cdot \text{ft/s}^2$$

$$1 \text{ N} = 0.2248 \text{ lb} \quad \blacktriangleleft$$

Developmental Exercises

D1.44 Express the value of G in Eq. (1.4) in base units in the U.S. customary system.

D1.45 A cart weighing 981 N is being held on a ramp as shown. Which of the following will give the correct mass of the cart?

(a) $\dfrac{981 \sin 30°}{9.81}$ kg. (b) $\dfrac{981 \cos 30°}{9.81}$ kg. (c) $\dfrac{981}{9.81}$ kg. (d) $981(9.81)$ kg.

Fig. D1.45

PROBLEMS†

1.5 A newborn baby in a hospital weighs 7 pounds and 5 ounces. Making use of 1 lbm = 0.4536 kg, determine the baby's mass in kg.

1.6 Making use of 1 lbm = 0.4536 kg, express a gas pressure of 120 kPa in psi (pounds per square inch).

1.7 The allowable compressive stress for a structural member is 20 ksi (kips per square inch). Making use of 1 lbm = 0.4536 kg, determine the allowable compressive stress in MPa.

1.8* Making use of 1 lbm = 0.4536 kg, determine the acceleration in ft/s² of a 10-Mg space vehicle under the action of a 10-lb thrust force exerted by a course correction rocket.

1.9 The gravitational acceleration on the surface of the moon is approximately 5.35 ft/s². Making use of 1 lbm = 0.4536 kg, determine the weight of an instrument of 10 slugs on the surface of the moon in (a) pounds, (b) newtons.

1.10 The deflection δ of the free end of a cantilever beam, as shown, is given by $\delta = PL^3/(3EI)$, where P is the weight of the 10-kg crate, $L = 2$ m, $E = 200$ GPa, and $I = 5 \times 10^4$ mm⁴. Determine the value of δ in inches.

1.11* The angle of twist (in radians) of a circular shaft, as shown, is given by $\theta = TL/(JG)$, where $T = Fd$, $F = 300$ lb, $d = 2$ in., $L = 300$ mm, $J = 10^4$ mm⁴, and $G = 11.5 \times 10^6$ psi. Determine the value of θ in degrees.

Fig. P1.10

Fig. P1.11*

---

†Recall that 1 pascal = 1 Pa = 1 N/m² in solving some of the problems.

1.14   Concluding Remarks

The concepts of force, equilibrium, and idealization, as well as the parallelogram law, the rigid-body principle, and Newton's first and third laws, are extensively used in the study of statics. In dynamics, Newton's second law and law of gravitation are additionally used in the study. Note that the applicability of Newton's third law is *not* limited to bodies in contact. In explaining the third law in his *Principia*, Newton stated that "this law takes place also in attractions . . . ."[†] Thus, Newton's third law should be understood to be applicable to any two bodies interacting with each other either *by contact* or *at a distance*.

In relativistic mechanics, the measurements of dimensions are affected by the relative motion of the observer with respect to the object. For example, the concepts of length, time, and mass are not absolute in relativistic mechanics as they are in newtonian mechanics. For enrichment, a brief glance at some of the contrasting relativistic viewpoints would perhaps be of interest, although they are not needed for understanding the remainder of this text. To describe them, we first let the speed of light be denoted by c and consider two coordinate systems S and S' where S' moves with a constant velocity $\mathbf{v}$ with respect to S. By Einstein's special theory of relativity (1905), we have the following:

1. A rigid body of length L_0 as measured in the direction of $\mathbf{v}$ in the system S, in which the rigid body is at rest, measures only L as observed from the system S' where $L = L_0(1 - v^2/c^2)^{1/2}$; and reciprocally an identical *length contraction* occurs for a rigid body at rest in S' as observed from S.
2. The time interval t_0 measured between two events that occur at the same place in the system S is found to be t as observed from the system S' where $t = t_0/(1 - v^2/c^2)^{1/2}$; and reciprocally an identical *time dilation* (or retardation) occurs for a time interval measured in S' as observed from S.
3. A body of mass m_0 as measured in the system S, in which the body is at rest, manifests a mass m as observed from the system S' where $m = m_0/(1 - v^2/c^2)^{1/2}$; and reciprocally an identical *mass increase* occurs for a body at rest in S' as observed from S. Thus, Newton's second law is to be written as $\mathbf{F} = d(m\mathbf{v})/dt$ and the total energy of the particle is $E = mc^2$. It can be shown that $mc^2 = m_0c^2 + \frac{1}{2}m_0v^2$ when $v \ll c$.

By Einstein's general theory of relativity (1915), which deals with the law of gravitation in the large scale structure of the universe, gravity is looked upon not merely as action at a distance but rather as some modification of space. According to Einstein, mass is the cause of the *curvature* of the four-dimensional space-time, which is not flat. In the universe, everything (including light) must experience gravitation. Einstein's law of gravitation controls a geometrical quantity called *curvature* in contrast to Newton's law of gravitation, which controls a mechanical quantity called *force*.

[†]Cf. Cajori, trans. *op. cit.*, pp. 13–14.

From the relativistic viewpoints, we see that the fundamental principles formulated by Newton do have limitations. However, the relativistic effects are unobservably small in most engineering practice where $v \ll c$, and newtonian mechanics still remains the basis of today's engineering sciences.

REVIEW PROBLEMS

1.12 The critical load P_{cr} for the buckling of a column fixed at one end as shown is given by $P_{cr} = (\pi^2 EI)/(4L^2)$. For $E = 10 \times 10^6$ lb/in², $I = 10^3$ mm⁴, and $L = 1.5$ m, determine the largest mass m in kg of a block which may be placed on the top of the column without causing the column to buckle.

1.13 The magnitude v_{esc} of the escape velocity of a space vehicle at a distance r from the center of the earth is equal to $[(2gR^2)/r]^{1/2}$ where g is the standard gravitational acceleration and R is the radius of the earth. Knowing that the circumference of the earth is 40 Mm and the space vehicle is 300 miles above the surface of the earth, determine the value of v_{esc} in miles per hour.

1.14* It is known that $GM = gR^2$ where G is the constant of gravitation, M is the mass of the earth, and g and R are the same as those defined in Prob. 1.13. Using the value of G as given in Eq. (1.4), determine the mass of the earth in pounds-mass.

Fig. P1.12

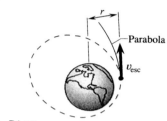

Fig. P1.13

Choose the correct or best item to complete the sentence in each of the following:

1.15 A scalar may be defined as a quantity which has (a) magnitude only, (b) magnitude and possibly sense, (c) magnitude and direction, (d) sense, but no orientation, (e) magnitude and possibly orientation.

1.16 A vector may be defined as a directed quantity which has (a) sense and orientation, and which obeys the laws for addition of vectors; (b) magnitude and orientation, and which obeys the principle of addition; (c) magnitude and direction, and which obeys the laws for addition of vectors; (d) magnitude and sense, and which obeys the principle of addition; (e) magnitude, orientation, and sense.

1.17 In statics, if the lines of action of the forces acting on a body intersect at a common point, the body may be treated as a (a) rigid body, (b) particle, (c) beam, (d) elastic body, (e) continuum.

1.18 The dimension which is a concept for measuring quantitatively the linear extent of the size or position of an object is (a) a base dimension, (b) mass, (c) time, (d) length, (e) a unit.

1.19 The base unit for mass in the SI is (a) kilogram, (b) slug, (c) gram, (d) newton, (e) pound-mass.

1.20 Rounding the number 0.03065 to three significant digits, we write (a) 0.0306, (b) 0.03, (c) 0.0307, (d) 3.065×10^{-2}, (e) 0.031.

1.21 It can be shown that 1 slug is equivalent to (a) 1 ft · lb/s², (b) 1 lb · s²/ft, (c) 1 ft · s²/lb, (d) 32.2 lb, (e) 9.81 kg.

1.22 One newton is defined as (a) the weight of a 1-kg mass, (b) the weight of a 1-slug mass, (c) the net force required to accelerate a mass of 1 kg at the rate of 9.80665 m/s², (d) the net force required to accelerate a mass of 1 slug at the rate of 1 ft/s², (e) the net force required to accelerate a mass of 1 kg at the rate of 1 m/s².

1.23   The mass of a body is (a) equal to the weight of the body, (b) the magnitude of the attractive force exerted by the earth on the body, (c) the magnitude of the inertia force of the body, (d) a qualitative measure of the rotational inertia of the body, (e) a quantitative measure of the translational inertia of the body.

1.24   When a net force of 1 lb acts on a body of mass m, the acceleration of the body is found to be 1 ft/s^2. The value of m is (a) 1 newton, (b) 1 kilogram, (c) 1 pascal, (d) 1 slug, (e) 1 pound-mass.

Forces   # Chapter 2

KEY CONCEPTS

- Directional angle, slope triangle, zero vector, couple, triangle rule, polygon rule, unit vector, vector components, scalar components, magnitude of a vector, analytical expression of a vector, direction angles, and position vector.
- Addition and resolution of forces in a plane using the parallelogram law.
- Addition and resolution of forces in a plane using vector algebra.
- Addition and resolution of forces in space using vector algebra.

A knowledge of how forces are added and resolved is basic to the study of the conditions of bodies which are acted on by forces. This chapter presents fundamental studies on the addition and resolution of forces in a plane and of forces in space. Two basic ways are used in representing forces. The *graphical* representation of forces is a basic step in applying the parallelogram law to the addition and resolution of forces, while the *analytical* representation of forces is a basic step in using vector algebra for adding and resolving forces. The method using vector algebra is shown to be consistent with the parallelogram law. The distinction between the vector components and the scalar components of a vector is stressed. Furthermore, the generation of a set of simultaneous scalar equations from a vector equation is illustrated in the method using vector algebra. The representations and operations pertaining to force vectors are applicable to other vector quantities.

FORCES IN A PLANE[†]

2.1 Graphical Representation

A force in a plane may be represented graphically by a segment of its line of action plus an arrowhead at the tip of the line segment to indicate the sense of the force. Using an appropriate scale, we may let the magnitude of

---

[†]A plane is a *two-dimensional space*. Thus, *forces in a plane* are forces in a two-dimensional space, and are, at times, referred to as *two-dimensional forces*. This seems to be an appropriate terminology. However, some engineers and mathematicians object to attaching the term "dimension" to a single force or vector in space because dimension is a property of space, not of a force or vector. (Some suggest the use of an *n*-component force or vector when it contains *n* components.)

Fig. 2.1

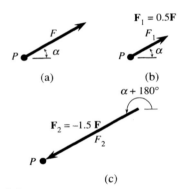

Fig. 2.2

Fig. 2.3

the force be represented by the length of that line segment. Alternatively, we may denote the magnitude of a force with an *italic* letter or a numerical value near the directed line segment used to represent the force. When a force lies in the xy plane, the counterclockwise angle measured from the positive direction of the x axis to the positive direction of the force is used to specify the orientation and sense of the force and is called the *directional angle* of the force. Figure 2.1 illustrates the graphical representations of a 400-N force **P** at C and a 200-N force **Q** at B, where θ_P and θ_Q, as indicated, are the *directional angles* of **P** and **Q**.

The *orientation* of a force in the xy plane may also be indicated with a *slope triangle*, which is a small right triangle having sides a, b, and hypotenuse c, as shown in Fig. 2.2. Since $c = (a^2 + b^2)^{1/2}$, the value of c is often not indicated on the slope triangle. By proportion, we may write the projections of **F** as follows:

$$F_x = \frac{b}{c} F \qquad F_y = \frac{a}{c} F \qquad (2.1)$$

We see that the *directional angle* θ_F of **F** can be found from

$$\tan \theta_F = \frac{a}{b} \qquad (2.2)$$

Depending on the direction of **F**, the directional angle θ_F may have a value ranging from 0 to 360° (or 0 to 2π rad).

The *product* $c\mathbf{F}$ of a scalar c and a force **F** represents another force which has a magnitude equal to the product of the magnitude of **F** and the absolute value of c, and which points in the direction of **F** if c is positive, but in the opposite direction of **F** if c is negative. These concepts are illustrated in Fig. 2.3. If $c = 0$, then $c\mathbf{F}$ is a *zero* (or *null*) vector **0**. A *zero vector* has a magnitude of zero and an indeterminate direction. If $c = -1$, then $c\mathbf{F} = (-1)\mathbf{F} = -\mathbf{F}$, which is the *negative vector* of **F**. When **F** and $-\mathbf{F}$ represent forces at different points of a rigid body and have different lines of action, they are said to form a *couple*. The sum of the forces in a couple is

$$\mathbf{F} + (-\mathbf{F}) = \mathbf{0} \qquad (2.3)$$

However, the effect of a couple is not simply zero, but is a twisting or turning action, which is discussed in Chap. 4.

Developmental Exercises

D2.1 Define (a) the directional angle of a force in the xy plane, (b) a zero vector, (c) a couple.

D2.2 Represent graphically (a) a 50-N force **F** at the point $A(1, 2)$ m with $\theta_F = 210°$, (b) a 30-N force **T** at the point $B(2, 1)$ m with $\theta_T = 60°$.

D2.3 A 50-lb force **F** with $\theta_F = 60°$ acts at the point $(2, 0)$ ft. Sketch the following: (a) **F**, (b) $\mathbf{F}_1 = 2\mathbf{F}$, (c) $\mathbf{F}_2 = -0.5\mathbf{F}$.

D2.4 Two forces **P** and **Q** are applied to an anchor as shown, where slope triangles are indicated. Determine the directional angles θ_P and θ_Q of **P** and **Q**, respectively.

Fig. D2.4

2.2 Addition of Forces: Parallelogram Law

*Forces must be added in a way consistent with the parallelogram law, or
its extended rules*: the *triangle rule* and the *polygon rule*. The *triangle rule*
states that when two vectors are drawn to scale and in tip-to-tail fashion, the
vector connecting, and directed from, the tail of the first vector to the tip of
the second one gives the resultant **R** of those two vectors. This rule may be
dubbed the "half-of-a-parallelogram law." The triangle rule, as illustrated
in Fig. 2.4, shows that

$$\mathbf{P} + \mathbf{Q} = \mathbf{Q} + \mathbf{P} \qquad (2.4)$$

Thus, *vector addition is commutative*.

The *polygon rule* states that when several vectors are drawn to scale and
in tip-to-tail fashion, the vector connecting, and directed from, the tail of
the first vector to the tip of the last one gives the resultant of those several
vectors. This rule can be shown to be valid through repeated applications of
the triangle rule. The polygon rule, as illustrated in Fig. 2.5, shows that

$$\mathbf{P} + \mathbf{Q} + \mathbf{S} = (\mathbf{P} + \mathbf{Q}) + \mathbf{S} = \mathbf{P} + (\mathbf{Q} + \mathbf{S}) \qquad (2.5)$$

Thus, *vector addition is associative*.

Generally, the parallelogram law alone gives the magnitude and direction, but not the point of application, of the sum of the forces. Figure 2.6
shows that the sum of the two forces **P** and **Q** may graphically be obtained
by placing the tails of the two force vectors at the same point, say *A*, and
forming a parallelogram with these two force vectors as consecutive sides.

Fig. 2.4

Fig. 2.5

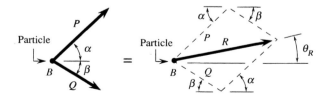

Fig. 2.6

The sum is denoted by **R** as shown, and is equal to the *diagonal vector*
whose tail is located at the point *A*. This sum is the *resultant force*, or
simply the *resultant*, of the two forces.[†] When the two forces **P** and **Q** act
on a particle, it is evident that the particle itself is also the *point of application* of the resultant force **R** as shown in Fig. 2.7, where R and θ_R of **R**
may analytically be determined.

Fig. 2.7

[†]Note that a *resultant force* is not an actually applied force. Nevertheless, it is a vector representing an *equivalent force*.

Fig. 2.8

When the two forces **P** and **Q** act on a rigid body, we see from the principle of transmissibility that the line of action of the resultant force **R** must contain the *point of intersection C* of the lines of action of **P** and **Q** as shown in Fig. 2.8. To locate the resultant force of two or more parallel forces, we need the concepts of *moment* and *equivalence*, which are presented in Chaps. 4 and 5.

The subtraction of the force **Q** from the force **P** may be written as

$$\mathbf{P} - \mathbf{Q} = \mathbf{P} + (-\mathbf{Q}) \tag{2.6}$$

Thus, the *subtraction* of **Q** from **P** is the same as the *addition* of **P** and the negative vector of **Q**. The parallelogram law may, therefore, be used to find the *difference of two forces* as shown in Fig. 2.9.

Fig. 2.9

Developmental Exercises

D2.5 For the three vectors shown, their relation is given by
(a) **F** + **Q** = **P**.   (b) **P** + **Q** = **F**.   (c) **P** + **F** = **Q**.
(d) **F** − **Q** = **P**.   (e) **F** − **P** = **Q**.

D2.6 The three vectors shown are drawn to scale. Their resultant **R** is given by

(a)   (b)   (c)   (d)

Fig. D2.5

Fig. D2.6

EXAMPLE 2.1

A 200-N force **P** and a 400-N force **Q** are applied to a small boat stuck in a shallow stream as shown. The resultant force of **P** and **Q** is **R**. Determine the magnitude R and the directional angle θ_R of **R**.

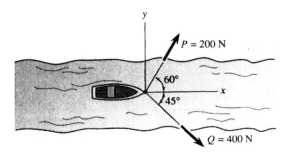

Solution.   The parallelogram for the forces and the relevant angles are shown. Using the laws of cosines and sines, we write

$$R^2 = (200)^2 + (400)^2 - 2(200)(400) \cos 75°$$ $R = 398$ N ◀

$$\frac{200}{\sin \alpha} = \frac{R}{\sin 75°}$$ $\alpha = 29.0°$

$$\theta_R = 270° + 45° + \alpha$$ $\theta_R = 344.0°$ ◀

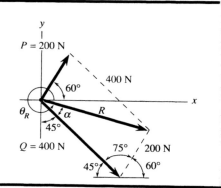

EXAMPLE 2.2

The boat in Example 2.1 remains stuck. A decision is made to get the boat out by reversing the sense of **Q** such that the resultant is **D** = **P** − **Q**. Determine the magnitude D and the directional angle θ_D of **D**.

Solution.   The parallelogram for finding **D** is shown. Using the laws of cosines and sines, we write

$$D^2 = (200)^2 + (400)^2 - 2(200)(400) \cos 105°$$ $D = 491$ N ◀

$$\frac{200}{\sin \alpha} = \frac{D}{\sin 105°}$$ $\alpha = 23.2°$

$$\theta_D = 180° - 45° - \alpha$$ $\theta_D = 111.8°$ ◀

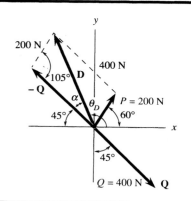

Developmental Exercise

D2.7   Describe the procedure for (a) adding two given forces, (b) subtracting a given force from another given force.

2.3   Resolution of Forces: Parallelogram Law

The reverse of the process of adding two forces to obtain an equivalent single force is the process of resolving a single force into two component forces which, together, have the same effect as that of the single force. The *resolution* as well as the *addition* of forces must conform to the *parallelogram law*, or its extended rules: the *triangle rule* and the *polygon rule*, described in Sec. 2.2.

As illustrations, let us consider the following three different specifications and problems where *a given single force* **F** *is resolved into two component forces* **P** *and* **Q**:

(1) *The magnitudes of* **P** *and* **Q** *are specified and* $P + Q > F$, *the directions of* **P** *and* **Q** *are to be determined.*

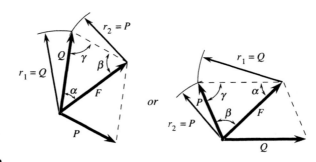

Fig. 2.10

The force triangle can be constructed by drawing two arcs of radii P and Q with centers at the tail and tip of F as shown in Fig. 2.10. In this case, F, P, and Q are known quantities; the senses of **P** and **Q** are readily determined from the triangle rule or the parallelogram law. The main task is the determination of the angles α and β. To solve the problem analytically, we may first use the law of cosines to find the angle α:

$$P^2 = F^2 + Q^2 - 2FQ \cos\alpha$$
$$\alpha = \cos^{-1}[(F^2 + Q^2 - P^2)/(2FQ)] \tag{2.7}$$

Then, using the law of sines, we have

$$\frac{P}{\sin\alpha} = \frac{Q}{\sin\beta}$$
$$\beta = \sin^{-1}[(Q \sin\alpha)/P] \tag{2.8}$$

The angles α and β define the directions of **P** and **Q** with respect to the given force **F** as shown in Fig. 2.10.[†] Note that it is possible to have *two different solutions* for the directions of **P** and **Q** relative to the direction of **F** if the magnitudes only are specified.

(2) *The orientations of* **P** *and* **Q** *are specified, the magnitudes and senses of* **P** *and* **Q** *are to be determined.*

The parallelogram may be constructed by drawing lines from the tip of **F**, parallel to the specified orientations of **P** and **Q**, as shown in Fig. 2.11. In this case, F, α, and β are known quantities; the senses of **P** and **Q** are readily determined from the parallelogram law once the parallelogram is constructed as shown. The main task is the determination of the magnitudes P and Q. Since the sum of α, β, and γ must be $180°$, we have $\gamma = 180° - \alpha - \beta$. By the law of sines, we write

$$\frac{F}{\sin\gamma} = \frac{P}{\sin\alpha} = \frac{Q}{\sin\beta}$$

$$P = \frac{F \sin\alpha}{\sin\gamma} = \frac{F \sin\alpha}{\sin(\alpha + \beta)} \qquad Q = \frac{F \sin\beta}{\sin\gamma} = \frac{F \sin\beta}{\sin(\alpha + \beta)} \tag{2.9}$$

Fig. 2.11

[†]The same solutions for α and β may alternatively be obtained by constructing right triangles without using the law of cosines and the law of sines. This is illustrated in Example 2.8.

(3) *The force* **P** *is specified, the force* **Q** *is to be determined.*
Drawing **F** and **P** from the same point, we immediately see from the relation
F = **P** + **Q** and the triangle rule that the closing side of the force triangle
must be the force **Q** as shown in Fig. 2.12. In this case, F, P, and β are
known quantities; the sense of **Q** is readily determined from the triangle rule
as shown. The main task is the determination of Q and α (or γ). Using the
laws of cosines and sines, we write

Fig. 2.12

$$Q^2 = F^2 + P^2 - 2FP \cos\beta$$

$$Q = (F^2 + P^2 - 2FP \cos\beta)^{1/2} \qquad (2.10)$$

$$\frac{Q}{\sin\beta} = \frac{P}{\sin\alpha} = \frac{F}{\sin\gamma}$$

$$\alpha = \sin^{-1}\left(\frac{P \sin\beta}{Q}\right) \qquad \gamma = \sin^{-1}\left(\frac{F \sin\beta}{Q}\right) \qquad (2.11)$$

EXAMPLE 2.3

A man and his son attach a piece of rope to their disabled car in the garage
and pull with a force **F** in the x direction, but the rope breaks at a tension
of 400 N just as the car starts to move. The man then attaches one piece of
the rope to the car and pulls with a force **Q** of 300 N, while his son uses
the other piece of rope and pulls with a force **P** of 200 N as shown. It is
known that $\alpha < \beta$, and the resultant of **P** and **Q** is to be the same as the
previous 400-N force **F**. Determine the directional angles θ_P and θ_Q of **P**
and **Q**.

Solution. Since $\alpha < \beta$ is specified, there is only one solution. The parallelogram for the forces is constructed as shown, where $\theta_P = \beta$. By the
law of cosines, we write

$$(300)^2 = (400)^2 + (200)^2 - 2(400)(200) \cos\beta$$

$$\beta = 46.6° \qquad\qquad \theta_P = 46.6° \blacktriangleleft$$

Applying the law of sines, we have

$$\frac{200}{\sin\alpha} = \frac{300}{\sin\beta} \qquad \alpha = 29.0°$$

$$\theta_Q = 360° - \alpha \qquad \theta_Q = 331.0° \blacktriangleleft$$

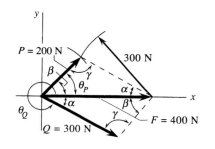

Developmental Exercise

D2.8 What fundamental law must we observe in resolving a force into two
component forces?

EXAMPLE 2.4

A crate is to be moved in the direction of $\overrightarrow{OA}$ as shown. It is known that a 50-kip force **F** in this direction is required but obstructions prevent a direct application of such a force. Consequently, the forces **P** and **Q** are applied instead as shown. It is desired that the resultant of **P** and **Q** be equivalent to **F**. Determine the magnitudes of **P** and **Q**.

Solution. The parallelogram may be constructed by drawing lines parallel to the specified orientations of **P** and **Q** from the tip of **F** as shown, where the angles indicated are determined from geometry. Referring to the figure and using the law of sines, we write

$$\frac{50}{\sin 45°} = \frac{P}{\sin 60°} = \frac{Q}{\sin 75°}$$

$$P = 61.2 \text{ kips} \qquad Q = 68.3 \text{ kips} \blacktriangleleft$$

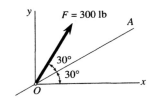

Fig. D2.9

Developmental Exercise

D2.9 A 100-lb force **F** is to be resolved into two forces **P** and **Q** in the directions as shown. Determine their magnitudes P and Q.

EXAMPLE 2.5

A 300-lb force **F** is to be resolved into a 400-lb force **P** along the direction of the line OA and another force **Q**. Determine the magnitude Q and the directional angle θ_Q of **Q**.

Solution. The force **Q** may be obtained from the parallelogram law as shown. By the laws of cosines and sines, we write

$$Q^2 = (300)^2 + (400)^2 - 2(300)(400)\cos 30°$$

$$Q = 205.31 \qquad Q = 205 \text{ lb} \blacktriangleleft$$

$$\frac{Q}{\sin 30°} = \frac{400}{\sin \alpha} \qquad \alpha = 76.9° \text{ or } 103.1°$$

Since the projection of **P** onto **F** is

$$(400 \text{ lb}) \cos 30° = 346 \text{ lb} > 300 \text{ lb} = F$$

we conclude that $\alpha > 90°$. We choose $\alpha = 103.1°$ and write

$$\theta_Q = 60° + \alpha \qquad \theta_Q = 163.1° \blacktriangleleft$$

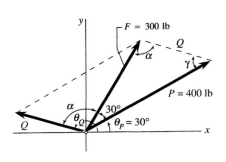

Developmental Exercise

D2.10 A 30-lb force **F** is to be resolved into a 20-lb force **P** and another force **Q** as shown. Determine the magnitude Q and the directional angle θ_Q of **Q**.

Fig. D2.10

PROBLEMS

2.1 A 300-lb force **A** and a 280-lb force **B** act on a hook as shown. If their resultant is **C**, determine the magnitude C and the directional angle θ_C of **C**.

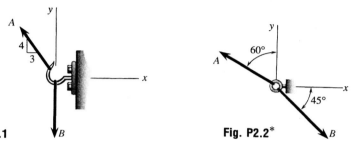

Fig. P2.1   **Fig. P2.2****

2.2* A 400-N force **A** and a 600-N force **B** act on an eyebolt as shown. If their resultant is **C**, determine the magnitude C and the directional angle θ_C of **C**.

2.3 The hydraulic cylinder AB exerts at B a force **F** of 600 N in the direction of $\overrightarrow{AB}$ against the adjustable platform BC as shown. Determine the magnitudes P and Q and the directional angles θ_P and θ_Q of **P** and **Q**, which are the components of **F** parallel and normal to BC, respectively.

Fig. P2.3

2.4 A 200-lb force **A** and a 600-lb force **B** act on the top of a post as shown. If **D** = **A** − **B**, determine the magnitude D and the directional angle θ_D of **D**.

Fig. P2.4   **Fig. P2.5****

Fig. P2.6

2.5* A 200-N force **A** and a 300-N force **B** act on a bracket as shown. If **D** = **A** − **B**, determine the magnitude D and the directional angle θ_D of **D**.

2.6 A man exerts a force **F** of 150 N on the handles of a wheelbarrow as shown. Resolve **F** into two forces **P** and **Q**, which are parallel and normal to the incline, respectively, and determine the values of P, Q, θ_P, and θ_Q.

2.7 The 400-N force **F** as shown is to be resolved into two forces **A** and **B** where A = 350 N, B = 600 N, and $90° < \theta_B < 150°$. Determine the directional angles θ_A and θ_B.

2.8* Solve Prob. 2.7 if A = 300 N, B = 600 N, and $0 < \theta_A < 150°$.

Fig. P2.7

$Q = 200$ N

Fig. P2.9

2.9 The force **Q** as shown is to be resolved into two forces **A** and **B** where **A** is parallel to the x axis and **B** is parallel to a line whose slope is -1. Determine A, B, and θ_B.

2.10* Solve Prob. 2.9 if **A** is parallel to the y axis.

2.11 A tractor is being transported by a helicopter to a remote location. If the tensile force $\mathbf{T}_1$ in the cable AB as shown is 4 kips and the resultant of the tensile forces $\mathbf{T}_1$ and $\mathbf{T}_2$ is vertical, determine T_2.

Fig. P2.11

2.12 A 400-N force $\mathbf{F}_1$ and a 500-N force $\mathbf{F}_2$ act on a hook as shown. If $\theta_1 = 45°$ and the resultant of $\mathbf{F}_1$ and $\mathbf{F}_2$ is vertical, determine θ_2.

Fig. P2.12 and P2.13*

2.13* Two forces $\mathbf{F}_1$ and $\mathbf{F}_2$ act on a hook as shown, where $\theta_1 = 30°$ and $F_1 = 200$ lb. If the resultant of $\mathbf{F}_1$ and $\mathbf{F}_2$ acts vertically downward and has a magnitude of 500 lb, determine F_2 and θ_2.

2.14 A 600-N force **S** acts on an anchor as shown. If $\mathbf{S} = \mathbf{A} + \mathbf{B}$, where $A = 500$ N and $\theta_A = 350°$, determine B and θ_B.

Fig. P2.14

2.15* Solve Prob. 2.14 if $A = 500$ N and $\theta_A = 170°$.

2.16 A beam *BC* is being lowered into position by a motor crane. At the instant shown, $\theta_1 = 35°$ and the tensions in the cables *AB* and *AC* are 800 lb and 1000 lb, respectively. Furthermore, the resultant **R** of the tensile forces in *AB* and *AC* is known to act vertically. Determine θ_2 and R.

Fig. P2.16

2.4 Analytical Representation

Unit vectors are vectors of magnitude equal to 1. The unit vectors directed along or parallel to the *x* and *y* axes, as shown in Fig. 2.13, are denoted by **i** and **j**, respectively, and are referred to as *cartesian unit vectors*,† which have fixed directions in the *xy* plane. A *general unit vector* is a unit vector which is not parallel to any of the coordinate axes. In Fig. 2.13, the line *AB* is not parallel to any of the coordinate axes. Therefore, the unit vector **λ** directed along the line *AB* is a general unit vector.

When a vector is resolved into two vectors which are perpendicular to each other, the two vectors are called the *rectangular vector components*, or

Fig. 2.13

---

†René Du Perron Descartes (born 1596 in France, died 1650 in Sweden) published in 1637 his magnum opus, *Géométrie*, in which he applied algebra to geometry and introduced the rectangular coordinate system for studying geometric problems.

Fig. 2.14

more briefly the *vector components*, of the original vector. A vector in the *xy* plane is often resolved into two components parallel to the *x* and *y* axes. We may denote the vector components of **F** parallel to the *x* and *y* axes as **F**$_x$ and **F**$_y$, respectively, as shown in Fig. 2.14. We write

$$\mathbf{F} = \mathbf{F}_x + \mathbf{F}_y \qquad (2.12)$$

Furthermore, the vector components **F**$_x$ and **F**$_y$ of **F** may be expressed in terms of scalars and the unit vectors **i** and **j** as follows:

$$\mathbf{F}_x = F_x\mathbf{i} \qquad \mathbf{F}_y = F_y\mathbf{j} \qquad (2.13)$$

where the scalar quantities F_x and F_y are called the *rectangular scalar components*, or more briefly the *scalar components*, of **F**. The scalar components F_x and F_y are also simply called the *x component* and the *y component* of **F**, respectively; they may be positive or negative, depending on whether or not the senses of the vector components **F**$_x$ and **F**$_y$ are the same as the senses of **i** and **j**, respectively. Of course, either of the values of F_x and F_y may be zero.

In our study, the term *analytical expression* of a vector will be used to refer to the expression of the vector written in terms of scalars and cartesian unit vectors. The analytical expression of a vector is the so-called *cartesian vector form* of the vector. We may write the *analytical expression* of a force **F** in the *xy* plane as

$$\mathbf{F} = F_x\mathbf{i} + F_y\mathbf{j} \qquad (2.14)$$

The *magnitude F* of **F** is equal to the *absolute value* |**F**|, a nonnegative scalar. Applying the Pythagorean theorem, we write

$$F = (F_x^2 + F_y^2)^{1/2} \qquad (2.15)$$

Letting $\boldsymbol{\lambda}_F$ be the unit vector in the direction of **F**, we write

$$\mathbf{F} = F\boldsymbol{\lambda}_F \qquad (2.16)$$

$$\boldsymbol{\lambda}_F = \cos\theta_F\mathbf{i} + \sin\theta_F\mathbf{j} \qquad (2.17)$$

Note that the magnitude of $\boldsymbol{\lambda}_F$ is

$$|\boldsymbol{\lambda}_F| = (\cos^2\theta_F + \sin^2\theta_F)^{1/2} = 1 \qquad (2.18)$$

Moreover, the scalar components F_x and F_y of **F** may be expressed in terms of F and θ_F as follows:

$$F_x = F\cos\theta_F \qquad F_y = F\sin\theta_F \qquad (2.19)$$

The signs of F_x and F_y are, of course, dependent on the quadrant of θ_F.

Developmental Exercises

D2.11 Discuss the differences and similarities among the terms: (a) unit vector, (b) cartesian unit vector, (c) general unit vector.

D2.12 Define the terms: (a) vector components, (b) scalar components, (c) analytical expression of a vector.

EXAMPLE 2.6

For the forces **P** and **Q** in Example 2.1, write (a) the analytical expressions of the unit vectors λ_P and λ_Q in the directions of **P** and **Q**, (b) the analytical expressions of **P** and **Q**.

Solution. The directional angles of **P**, **Q**, λ_P, and λ_Q may be indicated as shown. By Eq. (2.17) and then Eq. (2.16), we write

$$\lambda_P = \cos\theta_P \mathbf{i} + \sin\theta_P \mathbf{j} = \cos 60°\mathbf{i} + \sin 60°\mathbf{j}$$

$$\lambda_Q = \cos\theta_Q \mathbf{i} + \sin\theta_Q \mathbf{j} = \cos 315°\mathbf{i} + \sin 315°\mathbf{j}$$

$$\lambda_P = 0.5\mathbf{i} + 0.866\mathbf{j} \qquad \lambda_Q = 0.707\mathbf{i} - 0.707\mathbf{j} \blacktriangleleft$$

$$\mathbf{P} = P\lambda_P = 200(0.5\mathbf{i} + 0.886\mathbf{j})$$

$$\mathbf{Q} = Q\lambda_Q = 400(0.707\mathbf{i} - 0.707\mathbf{j})$$

$$\mathbf{P} = 100\mathbf{i} + 173.2\mathbf{j} \text{ N} \qquad \mathbf{Q} = 283\mathbf{i} - 283\mathbf{j} \text{ N} \blacktriangleleft$$

Developmental Exercises

D2.13 A 100-N force **G** and a 130-N force **H** are applied to an eyebolt as shown. Determine (a) the vector components of **G** and **H**, (b) the scalar components of **G** and **H**, (c) the magnitude of $\mathbf{H}_x$.

D2.14 A stump is pulled with a 250-N force **P** as shown. (a) Write the analytical expression of **P**. (b) What is the directional angle of **P**?

Fig. D2.13

2.5 Addition and Resolution of Forces: Vector Algebra

If two forces **P** and **Q** in a plane are analytically expressed as

$$\mathbf{P} = P_x\mathbf{i} + P_y\mathbf{j} \qquad \mathbf{Q} = Q_x\mathbf{i} + Q_y\mathbf{j}$$

then, by the associative property of the vector addition, we may write the *addition* of these two forces as

$$\mathbf{P} + \mathbf{Q} = (P_x + Q_x)\mathbf{i} + (P_y + Q_y)\mathbf{j} \qquad (2.20)$$

Note that this result is consistent with the parallelogram law for the addition of two forces as sketched in Fig. 2.15. The *subtraction* of **Q** from **P** may

Fig. D2.14

Fig. 2.15

Fig. 2.16

similarly be written as

$$\mathbf{P} - \mathbf{Q} = (P_x - Q_x)\mathbf{i} + (P_y - Q_y)\mathbf{j} \qquad (2.21)$$

The analytical expressions of forces may likewise be used to perform the addition or subtraction of more than two forces.

The *resolution* of a given force **F** in a plane into two component forces **P** and **Q** may *analytically* be formulated in terms of their magnitudes F, P, and Q, and their directional angles θ_F, θ_P, and θ_Q as shown in Fig. 2.16, where

$$\mathbf{F} = \mathbf{P} + \mathbf{Q} \qquad (2.22)$$

In terms of magnitudes and directional angles, we write

$$F \cos\theta_F \mathbf{i} + F \sin\theta_F \mathbf{j} = (P \cos\theta_P + Q \cos\theta_Q)\mathbf{i}$$
$$+ (P \sin\theta_P + Q \sin\theta_Q)\mathbf{j}$$

For this vector equation to be true, we must require that the coefficients of the **i**'s and **j**'s on both sides of the equal sign be equal; i.e., the ensuing set of simultaneous scalar equations is

$$F \cos\theta_F = P \cos\theta_P + Q \cos\theta_Q$$
$$F \sin\theta_F = P \sin\theta_P + Q \sin\theta_Q \qquad (2.23)$$

In these two simultaneous equations, F and θ_F are given or known; therefore, if any two of the four quantities P, Q, θ_P, and θ_Q are specified or known, the other two may be determined by solving the two simultaneous equations in Eqs. (2.23). Furthermore, if any two vectors in Eq. (2.22) are known, the third vector can be found using vector algebra.

EXAMPLE 2.7

Determine the resultant force **R** of the five forces $\mathbf{F}_1$, $\mathbf{F}_2$, $\mathbf{F}_3$, $\mathbf{F}_4$, and $\mathbf{F}_5$ acting on an eyebolt as shown.

Solution. This problem can be solved by using *vector algebra* as follows:

$$\mathbf{F}_1 = 500\boldsymbol{\lambda}_1 = 500\mathbf{i} \qquad \mathbf{F}_2 = 200\boldsymbol{\lambda}_2 = 200(\cos 30°\mathbf{i} + \sin 30°\mathbf{j})$$

$$\mathbf{F}_3 = 100\boldsymbol{\lambda}_3 = 100\mathbf{j} \qquad \mathbf{F}_4 = 340\boldsymbol{\lambda}_4 = 340\left(-\frac{8}{17}\mathbf{i} - \frac{15}{17}\mathbf{j}\right)$$

$$\mathbf{F}_5 = 350\boldsymbol{\lambda}_5 = 350\left(\frac{3}{5}\mathbf{i} - \frac{4}{5}\mathbf{j}\right) \qquad \mathbf{R} = \mathbf{F}_1 + \mathbf{F}_2 + \mathbf{F}_3 + \mathbf{F}_4 + \mathbf{F}_5$$

$$\mathbf{R} = \left[500 + 200\cos 30° + 340\left(-\frac{8}{17}\right) + 350\left(\frac{3}{5}\right)\right]\mathbf{i}$$

$$+ \left[200\sin 30° + 100 + 340\left(-\frac{15}{17}\right) + 350\left(-\frac{4}{5}\right)\right]\mathbf{j}$$

$$\mathbf{R} = 723\mathbf{i} - 380\mathbf{j} \text{ lb} \quad \blacktriangleleft$$

Developmental Exercise

D2.15   Refer to Example 2.7. Determine the resultant force **P** of the forces **F**₃,
F₄, and **F**₅.

EXAMPLE 2.8

Work Example 2.3 by constructing right triangles and using the *Pythagorean theorem*, instead of the laws of cosines and sines.

Solution.   The parallelogram for resolving **F** into **P** and **Q** was shown in
Example 2.3. The top portion of this parallelogram may be divided into two
right triangles where their perpendicular sides are related to the scalar components of **P** and **Q** as indicated.

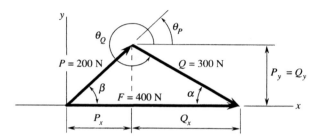

We are to find the directional angles θ_P and θ_Q. By the Pythagorean theorem
and the figure shown, we write

$$(200)^2 = P_x^2 + P_y^2 \qquad (300)^2 = Q_x^2 + Q_y^2$$

$$P_y = Q_y \qquad P_x + Q_x = 400$$

To solve the above four simultaneous equations, we write

$$(200)^2 - (300)^2 = (P_x^2 + P_y^2) - (Q_x^2 + Q_y^2) = P_x^2 - Q_x^2$$

$$= (P_x + Q_x)(P_x - Q_x) = 400(P_x - Q_x)$$

Thus, $P_x - Q_x = -125$. This and the preceding equations readily yield

$$P_x = 137.50 \qquad P_y = 145.24 \qquad Q_x = 262.50 \qquad Q_y = 145.24$$

Referring to the figure shown, we write

$$\tan \beta = \frac{145.24}{137.50} \qquad \beta = 46.6° = \theta_P$$

$$\tan \alpha = \frac{145.24}{262.50} \qquad \alpha = 29.0°$$

$$\theta_Q = 360° - \alpha \qquad \theta_P = 46.6° \qquad \theta_Q = 331.0° \quad \blacktriangleleft$$

Developmental Exercise

D2.16   A 10-kip force **F** is to be resolved into an 8-kip force **P** and a 9-kip force
Q. Determine the angle γ as shown by constructing right triangles and using the
Pythagorean theorem, instead of the laws of cosines and sines.

Fig. D2.16

EXAMPLE 2.9

Using *vector algebra*, work Example 2.4.

Solution. The parallelogram for resolving **F** into **P** and **Q** is shown, where the magnitudes P and Q are to be determined. Referring to the figure, we write

$$\mathbf{F} = 50\lambda_F = 50(\cos 120°\mathbf{i} + \sin 120°\mathbf{j})$$

$$\mathbf{P} = P\lambda_P = P(\cos 195°\mathbf{i} + \sin 195°\mathbf{j})$$

$$\mathbf{Q} = Q\lambda_Q = Q(\cos 60°\mathbf{i} + \sin 60°\mathbf{j})$$

Since $\mathbf{F} = \mathbf{P} + \mathbf{Q}$, we write

$$50 \cos 120°\mathbf{i} + 50 \sin 120°\mathbf{j}$$
$$= (P \cos 195° + Q \cos 60°)\mathbf{i} + (P \sin 195° + Q \sin 60°)\mathbf{j}$$

Equating the coefficients of **i**'s and **j**'s, we have

$$50 \cos 120° = (\cos 195°)P + (\cos 60°)Q$$

$$50 \sin 120° = (\sin 195°)P + (\sin 60°)Q$$

Solution of the above two equations yields

$$P = 61.2 \text{ kips} \qquad Q = 68.3 \text{ kips} \blacktriangleleft$$

EXAMPLE 2.10

Using *vector algebra*, work Example 2.5.

Solution. The parallelogram for resolving **F** into **P** and **Q**, as specified, is shown, where the magnitude Q and the directional angle θ_Q are to be determined. Referring to the figure, we write

$$\mathbf{F} = 300\lambda_F = 300 \cos 60°\mathbf{i} + 300 \sin 60°\mathbf{j}$$

$$\mathbf{P} = 400\lambda_P = 400 \cos 30°\mathbf{i} + 400 \sin 30°\mathbf{j}$$

Since $\mathbf{F} = \mathbf{P} + \mathbf{Q}$, we write

$$\mathbf{Q} = \mathbf{F} - \mathbf{P} = (300 \cos 60° - 400 \cos 30°)\mathbf{i}$$
$$+ (300 \sin 60° - 400 \sin 30°)\mathbf{j}$$
$$= -196.4\mathbf{i} + 59.8\mathbf{j}$$

$$Q^2 = (-196.4)^2 + (59.8)^2 \qquad Q = 205 \text{ lb} \blacktriangleleft$$

$$\tan \theta_Q = 59.8/(-196.4) \qquad \theta_Q = 163.1° \blacktriangleleft$$

Developmental Exercise

D2.17   It is known that $(2 + 3x)\mathbf{i} + 37\mathbf{j} = -4\mathbf{i} + (2x + y)\mathbf{j}$. Determine the values of x and y.

PROBLEMS

2.17   Knowing that $\mathbf{F} = 80\mathbf{i} + 150\mathbf{j}$ N and $\mathbf{S} = 90\mathbf{i} - 400\mathbf{j}$ N, determine their magnitudes F and S and directional angles θ_F and θ_S.

2.18*   Knowing that $\mathbf{P} = -240\mathbf{i} + 70\mathbf{j}$ lb and $\mathbf{Q} = -22\mathbf{i} - 120\mathbf{j}$ lb, determine their magnitudes P and Q and directional angles θ_P and θ_Q.

2.19   The cable of a motor crane is loaded with a 3.4-kip force $\mathbf{F}$ as shown. Determine the magnitudes of the forces $\mathbf{P}$ and $\mathbf{Q}$ which are the components of $\mathbf{F}$ parallel and normal to the boom AB, respectively.

Fig. P2.19

2.20   A heavy stump is being dislodged with three tractors as shown, where the forces exerted by the tractors are $P = 6.5$ kN, $Q = 5$ kN, and $S = 8.5$ kN. Determine the resultant $\mathbf{R}$ of these three forces.

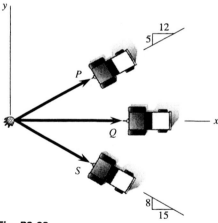

Fig. P2.20

2.21*   The two forces $\mathbf{P}$ and $\mathbf{Q}$ applied to a hook as shown have a resultant $\mathbf{R} = -100\mathbf{i} - 600\mathbf{j}$ N. If $\mathbf{P}$ has an x component $\mathbf{P}_x = -200\mathbf{i}$ N and $\mathbf{Q}$ has a y component $\mathbf{Q}_y = -300\mathbf{j}$ N, determine $\mathbf{P}$ and $\mathbf{Q}$.

2.22   Determine the resultant $\mathbf{R}$ of the three forces which act on the riveted bracket as shown.

Fig. P2.22

Fig. P2.21*

2.23* Determine the resultant **R** of the four axial forces in the truss members that are fastened to a gusset plate at a hinge support as shown.

Fig. P2.25 and P2.26*

Fig. P2.23*   **Fig. P2.24**

2.24 The resultant of the four forces acting on the anchor shown is known to be **R** = 559**i** + 788**j** N. Determine the force **Q**$_3$.

2.25 The horizontal component of the resultant **R** of the four forces acting on the crate shown is **R**$_x$ = 1160**i** N. If F_1 = 260 N, F_2 = 500 N, F_4 = 760 N, and $\theta_3 = \tan^{-1}(5/12)$, determine F_3.

2.26* The crate shown is acted on by four forces, where F_1 = 260 N, F_2 = 500 N, and F_3 = 50 N. If the resultant of the four forces is **R** = 160**i** + 750**j** N, determine θ_3 and F_4.

2.27 A small boat at A is being towed with two forces **F**$_1$ and **F**$_2$ as shown. If F_1 = 60 lb and the resultant of **F**$_1$ and **F**$_2$ is directed from A to B, determine F_2.

Fig. P2.27   **Fig. P2.28**

2.28 When a 330-N force **P** is applied to the structure at the joint C as shown, the forces **F**$_{AC}$ and **F**$_{BC}$ developed in the members AC and BC are known to direct along the axes of the respective members, and the resultant of **F**$_{AC}$ and **F**$_{BC}$ is equal to −**P**. Determine the magnitudes of **F**$_{AC}$ and **F**$_{BC}$.

2.29* through 2.42* Using *vector algebra*, solve the following problems:

2.29*	Prob. 2.1.	**2.30***	Prob. 2.2*.
2.31*	Prob. 2.3.	**2.32***	Prob. 2.4.
2.33*	Prob. 2.5*.	**2.34***	Prob. 2.6.
2.35*	Prob. 2.9.	**2.36***	Prob. 2.10*.
2.37*	Prob. 2.11.	**2.38***	Prob. 2.12.
2.39*	Prob. 2.13*.	**2.40***	Prob. 2.14.
2.41*	Prob. 2.15*.	**2.42***	Prob. 2.16.

FORCES IN SPACE†

2.6 Representation of a Force in Space

The unit vectors directed along or parallel to the x, y, and z axes of a fixed rectangular coordinate system are denoted by **i**, **j**, and **k**, respectively, which form a set of *cartesian unit vectors* in space. Each of these three unit vectors is a *constant vector* and is perpendicular to the other two as shown in Fig. 2.17.

A force **F** in space is often resolved into three components $\mathbf{F}_x$, $\mathbf{F}_y$, and $\mathbf{F}_z$ parallel respectively to the x, y, and z rectangular coordinate axes as shown in Fig. 2.18, where

$$\mathbf{F} = \mathbf{F}_y + \mathbf{F}_h = \mathbf{F}_y + (\mathbf{F}_x + \mathbf{F}_z)$$

Fig. 2.17

Fig. 2.18

Since vector addition is associative and commutative, we write

$$\mathbf{F} = \mathbf{F}_x + \mathbf{F}_y + \mathbf{F}_z \tag{2.24}$$

where $\mathbf{F}_x$, $\mathbf{F}_y$, and $\mathbf{F}_z$ are called the *rectangular vector components*, or more briefly the *vector components*, of **F**. Expressing them as products of appropriate scalars and the unit vectors **i**, **j**, and **k**, we write

$$\mathbf{F}_x = F_x\mathbf{i} \qquad \mathbf{F}_y = F_y\mathbf{j} \qquad \mathbf{F}_z = F_z\mathbf{k}$$

$$\mathbf{F} = F_x\mathbf{i} + F_y\mathbf{j} + F_z\mathbf{k} \tag{2.25}$$

where F_x, F_y, and F_z are called the *rectangular scalar components*, or more briefly the *scalar components*, of **F**. Furthermore, F_x, F_y, and F_z are also called the *x component*, the *y component*, and the *z component* of **F**, respectively. Note that Eq. (2.25) shows the *analytical expression of the vector* **F** *in space*. Referring to Fig. 2.18 and applying the Pythagorean theorem, we see that the *magnitude* (or *absolute value*) of **F** is

$$F = (F_y^2 + F_h^2)^{1/2} = [F_y^2 + (F_x^2 + F_z^2)]^{1/2}$$

$$F = (F_x^2 + F_y^2 + F_z^2)^{1/2} \tag{2.26}$$

The nonnegative angles measured from the positive directions of the x, y, and z coordinate axes to the positive direction of **F** are referred to as the *direction angles* of **F** and are denoted by θ_x, θ_y, and θ_z, respectively, as

---

†Unless otherwise stated, the term space will be understood to imply *three-dimensional space*. *Forces in space* are, at times, called *three-dimensional forces*. However, some object to the latter terminology. Cf. the footnote on p. 23.

Fig. 2.19

shown in Fig. 2.19.[†] Depending on the signs of the scalar components, the direction angles may have values ranging from 0 to 180° (or from 0 to π rad).

The cosines of the direction angles are called the *direction cosines*. Referring to Fig. 2.19, we write

$$F_x = F \cos\theta_x \qquad F_y = F \cos\theta_y \qquad F_z = F \cos\theta_z \quad (2.27)$$

Thus, the scalar components F_x, F_y, and F_z of **F** are, in fact, the *projections* of **F** onto the x, y, and z axes, respectively. If a scalar component or a projection of the vector onto an axis is positive, zero, or negative, then the corresponding direction angle must be less than, equal to, or greater than 90°, respectively. Using Eqs. (2.26) and (2.27), we write

$$F^2 = F_x^2 + F_y^2 + F_z^2 = F^2(\cos^2\theta_x + \cos^2\theta_y + \cos^2\theta_z)$$

$$\cos^2\theta_x + \cos^2\theta_y + \cos^2\theta_z = 1 \qquad (2.28)$$

Equation (2.28) is called the *constraint equation* on the direction angles. We see from this equation that only two of the three direction angles θ_x, θ_y, and θ_z are independent; the third one must satisfy the constraint equation, which states that the sum of the squares of the three direction cosines equals 1.

Developmental Exercises

D2.18 What are the scalar components of the vectors (a) **F** = 3**i** − 4**j** − 12**k** kN, (b) **r** = 40**i** − 9**k** m?

D2.19 What is the magnitude of the force **F** = 4**i** + 4**j** − 7**k** kips?

D2.20 A force vector **F** is shown. (a) Which direction angles are less than 90°? (b) Is there any direction angle which is greater than 180°? (c) Is there any direction angle which is negative? (d) Is there any direction cosine which is negative?

D2.21 For **F** = 8**j** − 15**k** lb, what are (a) the projections of **F** onto the coordinate axes, (b) the direction angles of **F**?

Fig. D2.20

2.7 Position Vector and Unit Vector in Space

The *position vector* of a point P is defined as the vector drawn from the origin O of the coordinate system to the point P. It has a magnitude equal to the distance between O and P and is directed from O to P.

As shown in Fig. 2.20, the position vector **r** of the point $P(x, y, z)$ is

$$\mathbf{r} = \overrightarrow{OP} \qquad (2.29)$$

The projections of **r** onto the coordinate axes are equal to x, y, and z, respectively. Thus, we may express **r** as

$$\mathbf{r} = x\mathbf{i} + y\mathbf{j} + z\mathbf{k} \qquad (2.30)$$

Fig. 2.20

[†]Note that the term *direction angle* is associated with a vector in three-dimensional space, while the term *directional angle*, as defined in Sec. 2.1, is associated with a vector in a plane.

The name *position vector* of the point P for $\mathbf{r}$ is quite fitting because $\mathbf{r}$ contains the information which completely defines the position of P.

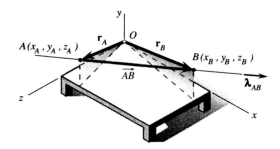

Fig. 2.21

The position vectors $\mathbf{r}_A$ and $\mathbf{r}_B$ of the points $A(x_A, y_A, z_A)$ and $B(x_B, y_B, z_B)$ in Fig. 2.21 may be expressed as

$$\mathbf{r}_A = x_A\mathbf{i} + y_A\mathbf{j} + z_A\mathbf{k} \qquad \mathbf{r}_B = x_B\mathbf{i} + y_B\mathbf{j} + z_B\mathbf{k}$$

The vector which is directed from A to B is denoted by $\overrightarrow{AB}$. The vector $\overrightarrow{AB}$ is called *the position vector of B relative to A*. As shown in Fig. 2.21, the vectors $\mathbf{r}_A$, $\overrightarrow{AB}$, and $\mathbf{r}_B$ form a vector triangle OAB. Applying the *triangle rule* in Sec. 2.2, we write

$$\mathbf{r}_A + \overrightarrow{AB} = \mathbf{r}_B$$
$$\overrightarrow{AB} = \mathbf{r}_B - \mathbf{r}_A = (x_B\mathbf{i} + y_B\mathbf{j} + z_B\mathbf{k}) - (x_A\mathbf{i} + y_A\mathbf{j} + z_A\mathbf{k})$$
$$\overrightarrow{AB} = (x_B - x_A)\mathbf{i} + (y_B - y_A)\mathbf{j} + (z_B - z_A)\mathbf{k} \qquad (2.31)$$

The magnitude of $\overrightarrow{AB}$ is given by the absolute value

$$|\overrightarrow{AB}| = [(x_B - x_A)^2 + (y_B - y_A)^2 + (z_B - z_A)^2]^{1/2} \qquad (2.32)$$

Thus, the unit vector $\boldsymbol{\lambda}_{AB}$ directed from $A(x_A, y_A, z_A)$ to $B(x_B, y_B, z_B)$ may be computed from the formula

$$\boldsymbol{\lambda}_{AB} = \frac{\overrightarrow{AB}}{|\overrightarrow{AB}|} \qquad (2.33)$$

where $\overrightarrow{AB}$ and $|\overrightarrow{AB}|$ are given by Eqs. (2.31) and (2.32).

In Fig. 2.22, a unit vector $\boldsymbol{\lambda}$ is shown, where the direction angles θ_x and θ_y are known, but θ_z is unknown. In this case, the cosine of θ_z may first be

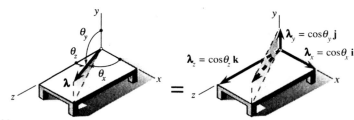

Fig. 2.22

obtained from Eq. (2.28) as

$$\cos\theta_z = \pm(1 - \cos^2\theta_x - \cos^2\theta_y)^{1/2} \qquad (2.34)$$

where the upper sign corresponds to $0 \le \theta_z \le 90°$, and the lower sign corresponds to $90° < \theta_z \le 180°$. If the range of values of θ_z is not specified, there will be two possible values for θ_z. In any case, the analytical expression of the unit vector $\boldsymbol{\lambda}$ may be written as

$$\boldsymbol{\lambda} = \cos\theta_x\mathbf{i} + \cos\theta_y\mathbf{j} + \cos\theta_z\mathbf{k} \qquad (2.35)$$

Furthermore, we may express any vector $\mathbf{F}$ in space as the product of its magnitude F and the unit vector $\boldsymbol{\lambda}_F$ which is directed along $\mathbf{F}$; i.e.,

$$\mathbf{F} = F\boldsymbol{\lambda}_F \qquad (2.36)$$

Equation (2.36) is a useful *formula* for obtaining the analytical expression of any vector $\mathbf{F}$ when F and $\boldsymbol{\lambda}_F$ are known. On the other hand, if the analytical expression of $\mathbf{F}$ is known, we may obtain the analytical expression of $\boldsymbol{\lambda}_F$ by dividing F into $\mathbf{F}$.

Developmental Exercises

D2.22 A point P is located at $(6, -4, 12)$ m. Determine (a) the analytical expression of the position vector of P, (b) the distance from the origin to P.

D2.23 A vector directed from A to B is shown. (a) Write the analytical expression of $\overrightarrow{AB}$. (b) Determine the length of $\overrightarrow{AB}$. (c) Write the analytical expression of the unit vector $\boldsymbol{\lambda}_{AB}$.

D2.24 A 100-lb force $\mathbf{F}$ has direction angles $\theta_x = 45°$, $90° < \theta_y \le 180°$, and $\theta_z = 60°$. Determine the analytical expressions of (a) the unit vector $\boldsymbol{\lambda}_F$, (b) the force $\mathbf{F}$.

Fig. D2.23

2.8   Addition and Resolution of Forces in Space

Forces in space are seldom added or resolved by direct application of the parallelogram law. Such forces are more effectively handled by using vector algebra. Suppose that the analytical expressions of the forces $\mathbf{P}$ and $\mathbf{Q}$ are

$$\mathbf{P} = P_x\mathbf{i} + P_y\mathbf{j} + P_z\mathbf{k} \qquad \mathbf{Q} = Q_x\mathbf{i} + Q_y\mathbf{j} + Q_z\mathbf{k}$$

Then, by the associative property of the vector addition, we write

$$\mathbf{P} + \mathbf{Q} = (P_x + Q_x)\mathbf{i} + (P_y + Q_y)\mathbf{j} + (P_z + Q_z)\mathbf{k} \qquad (2.37)$$

The subtraction of $\mathbf{Q}$ from $\mathbf{P}$ may similarly be written as

$$\mathbf{P} - \mathbf{Q} = (P_x - Q_x)\mathbf{i} + (P_y - Q_y)\mathbf{j} + (P_z - Q_z)\mathbf{k} \qquad (2.38)$$

The addition or subtraction of three or more forces in space may likewise be performed by using vector algebra.

Suppose that a given force $\mathbf{F}$ is to be resolved into three noncoplanar forces **A**, **B**, and **C** with *unknown magnitudes* but *known directions*. We write

$$\mathbf{F} = \mathbf{A} + \mathbf{B} + \mathbf{C} \qquad (2.39)$$

$$F_x\mathbf{i} + F_y\mathbf{j} + F_z\mathbf{k} = A\boldsymbol{\lambda}_A + B\boldsymbol{\lambda}_B + C\boldsymbol{\lambda}_C \qquad (2.40)$$

where $\boldsymbol{\lambda}_A$, $\boldsymbol{\lambda}_B$, and $\boldsymbol{\lambda}_C$ are known and may be expressed in terms of the unit vectors **i**, **j**, and **k**. We may rearrange and equate the coefficients of **i**'s, **j**'s, and **k**'s on both sides of Eq. (2.40) to yield a set of three simultaneous scalar equations with the magnitudes A, B, and C as the three unknowns, which may then be solved. (Cf. Example 2.13.)

EXAMPLE 2.11

The guy wire AB, as shown, has a tension of 2 kips. Determine the analytical expressions of (a) the unit vector $\boldsymbol{\lambda}_{AB}$ directed along AB, (b) the tensile force $\mathbf{T}_{AB}$ exerted by the guy wire on the utility pole.

Solution. Let the unit vector $\boldsymbol{\lambda}_{AB}$ as shown be first resolved into a vertical component $\boldsymbol{\lambda}_y$ and a horizontal component $\boldsymbol{\lambda}_h$. The horizontal component $\boldsymbol{\lambda}_h$ is then projected onto the zx plane and resolved into two rectangular components $\boldsymbol{\lambda}_x$ and $\boldsymbol{\lambda}_z$ which lie along the x and z axes, respectively. We see from the figure that

$$|\boldsymbol{\lambda}_y| = |\boldsymbol{\lambda}_{AB}|\,(\cos 60°) = (1)(\cos 60°) = \cos 60°$$

$$|\boldsymbol{\lambda}_h| = |\boldsymbol{\lambda}_{AB}|\,(\sin 60°) = (1)(\sin 60°) = \sin 60°$$

$$|\boldsymbol{\lambda}_x| = |\boldsymbol{\lambda}_h|\,(\cos 50°) = \sin 60°\,\cos 50°$$

$$|\boldsymbol{\lambda}_z| = |\boldsymbol{\lambda}_h|\,(\sin 50°) = \sin 60°\,\sin 50°$$

By Eq. (2.36), we write

$$\boldsymbol{\lambda}_y = |\boldsymbol{\lambda}_y|(-\mathbf{j}) = -\cos 60°\mathbf{j}$$

$$\boldsymbol{\lambda}_h = \boldsymbol{\lambda}_x + \boldsymbol{\lambda}_z = |\boldsymbol{\lambda}_x|\mathbf{i} + |\boldsymbol{\lambda}_z|(-\mathbf{k}) = \sin 60°\,\cos 50°\mathbf{i} - \sin 60°\,\sin 50°\mathbf{k}$$

$$\boldsymbol{\lambda}_{AB} = \boldsymbol{\lambda}_y + \boldsymbol{\lambda}_h = -\cos 60°\mathbf{j} + (\sin 60°\,\cos 50°\mathbf{i} - \sin 60°\,\sin 50°\mathbf{k})$$

$$\boldsymbol{\lambda}_{AB} = 0.557\mathbf{i} - 0.5\mathbf{j} - 0.663\mathbf{k} \quad \blacktriangleleft$$

Check. $\qquad (0.557)^2 + (0.500)^2 + (0.663)^2 \approx 1.000$

Since $T_{AB} = 2$ kips, we apply Eq. (2.36) to write

$$\mathbf{T}_{AB} = T_{AB}\boldsymbol{\lambda}_{AB} = 2(0.5567\mathbf{i} - 0.5\mathbf{j} - 0.6634\mathbf{k})$$

$$\mathbf{T}_{AB} = 1.113\mathbf{i} - \mathbf{j} - 1.327\mathbf{k} \text{ kips} \quad \blacktriangleleft$$

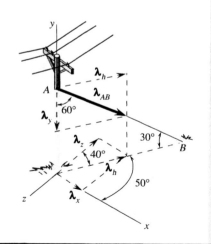

Developmental Exercise

D2.25 Write the analytical expressions of (a) $\boldsymbol{\lambda}_F$ if $\mathbf{F} = 400\mathbf{i} + 700\mathbf{j} - 400\mathbf{k}$ lb, (b) the 260-lb force $\mathbf{P}$ if $\boldsymbol{\lambda}_P = \frac{1}{13}(3\mathbf{i} - 12\mathbf{j} + 4\mathbf{k})$.

2.4 m F_{HC}

F_{HB}

F_{HA}

0.6 m

C

B

0.8 m

1.8 m

z 1.2 m A 0.8 m x

EXAMPLE 2.12

Three forces $\mathbf{F}_{HA}$, $\mathbf{F}_{HB}$, and $\mathbf{F}_{HC}$ act on a hook at H as shown. If the magnitudes of these forces are $F_{HA} = 420$ N, $F_{HB} = 500$ N, and $F_{HC} = 390$ N, determine the resultant force $\mathbf{R}$ acting on the hook.

Solution. From the geometry and the data given, we write

$$\overrightarrow{HA} = 1.2\mathbf{i} - 2.4\mathbf{j} + 0.8\mathbf{k} \qquad \lambda_{HA} = \tfrac{1}{7}(3\mathbf{i} - 6\mathbf{j} + 2\mathbf{k})$$

$$\overrightarrow{HB} = -2.4\mathbf{j} - 1.8\mathbf{k} \qquad \lambda_{HB} = -\tfrac{1}{5}(4\mathbf{j} + 3\mathbf{k})$$

$$\overrightarrow{HC} = -0.8\mathbf{i} - 2.4\mathbf{j} + 0.6\mathbf{k} \qquad \lambda_{HC} = \tfrac{1}{13}(-4\mathbf{i} - 12\mathbf{j} + 3\mathbf{k})$$

$$\mathbf{F}_{HA} = F_{HA}\lambda_{HA} = 420\lambda_{HA} \qquad \mathbf{F}_{HA} = 60(3\mathbf{i} - 6\mathbf{j} + 2\mathbf{k})$$

$$\mathbf{F}_{HB} = F_{HB}\lambda_{HB} = 500\lambda_{HB} \qquad \mathbf{F}_{HB} = -100(4\mathbf{j} + 3\mathbf{k})$$

$$\mathbf{F}_{HC} = F_{HC}\lambda_{HC} = 390\lambda_{HC} \qquad \mathbf{F}_{HC} = 30(-4\mathbf{i} - 12\mathbf{j} + 3\mathbf{k})$$

Thus, we have

$$\mathbf{R} = \mathbf{F}_{HA} + \mathbf{F}_{HB} + \mathbf{F}_{HC} \qquad \mathbf{R} = 60\mathbf{i} - 1120\mathbf{j} - 90\mathbf{k} \text{ N} \blacktriangleleft$$

Developmental Exercise

D2.26 If the forces $\mathbf{F}$ and $\mathbf{P}$ in D2.25 act at the origin O, determine their resultant $\mathbf{R}$.

EXAMPLE 2.13

If the resultant of the three forces $\mathbf{F}_{HA}$, $\mathbf{F}_{HB}$, and $\mathbf{F}_{HC}$ in Example 2.12 is $\mathbf{R} = -1000\mathbf{j}$ N, determine the magnitude of each force.

Solution. From the solution in Example 2.12, we have

$$\mathbf{F}_{HA} = F_{HA}\lambda_{HA} = \frac{F_{HA}}{7}(3\mathbf{i} - 6\mathbf{j} + 2\mathbf{k})$$

$$\mathbf{F}_{HB} = F_{HB}\lambda_{HB} = -\frac{F_{HB}}{5}(4\mathbf{j} + 3\mathbf{k})$$

$$\mathbf{F}_{HC} = F_{HC}\lambda_{HC} = \frac{F_{HC}}{13}(-4\mathbf{i} - 12\mathbf{j} + 3\mathbf{k})$$

Since $\mathbf{F}_{HA} + \mathbf{F}_{HB} + \mathbf{F}_{HC} = \mathbf{R} = -1000\mathbf{j}$, we write

$$\left(\frac{3}{7}F_{HA} - \frac{4}{13}F_{HC}\right)\mathbf{i} + \left(-\frac{6}{7}F_{HA} - \frac{4}{5}F_{HB} - \frac{12}{13}F_{HC}\right)\mathbf{j}$$

$$+ \left(\frac{2}{7}F_{HA} - \frac{3}{5}F_{HB} + \frac{3}{13}F_{HC}\right)\mathbf{k} = -1000\mathbf{j}$$

Equating the coefficients of **i**'s, **j**'s, and **k**'s, we obtain

$$\frac{3}{7} F_{HA} - \frac{4}{13} F_{HC} = 0 \tag{1}$$

$$-\frac{6}{7} F_{HA} - \frac{4}{5} F_{HB} - \frac{12}{13} F_{HC} = -1000 \tag{2}$$

$$\frac{2}{7} F_{HA} - \frac{3}{5} F_{HB} + \frac{3}{13} F_{HC} = 0 \tag{3}$$

which may be solved to yield[†]

$$F_{HA} = 339 \text{ N} \qquad F_{HB} = 343 \text{ N} \qquad F_{HC} = 472 \text{ N} \quad \blacktriangleleft$$

PROBLEMS

2.43 A force $\mathbf{F} = 3\mathbf{i} - 12\mathbf{j} + 4\mathbf{k}$ kN acts at the origin O. Determine (a) the scalar components of $\mathbf{F}$, (b) the magnitude of the horizontal component $\mathbf{F}_h$, (c) the unit vector $\boldsymbol{\lambda}_F$ directed along $\mathbf{F}$, (d) the direction angles of $\mathbf{F}$.

2.44[*] through 2.49 Let $\boldsymbol{\lambda}_{AB}$ be a unit vector directed along $\overrightarrow{AB}$ as shown. Determine (a) $\boldsymbol{\lambda}_{AB}$, (b) the magnitude of the horizontal projection of $\boldsymbol{\lambda}_{AB}$.

Fig. P2.44[*]

Fig. P2.45[*]

Fig. P2.46

Fig. P2.47

Fig. P2.49

Fig. P2.48[*]

[†]Cf. Sec. B1.3 of App. B.

Fig. P2.50*

2.50* Two guy wires are attached to a pole supporting a dish antenna as shown. If the tensions in the guy wires are $T_{PA} = 400$ N and $T_{PB} = 600$ N, determine their resultant force **R** at P.

2.51 Three forces $\mathbf{F}_{PA}$, $\mathbf{F}_{PB}$, and $\mathbf{F}_{PC}$ are applied to a crate at P as shown. If $F_{PA} = 100$ lb, $F_{PB} = 150$ lb, and $F_{PC} = 195$ lb, determine their resultant force **R** at P.

2.52 Three forces $\mathbf{F}_{PA}$, $\mathbf{F}_{PB}$, and $\mathbf{F}_{PC}$ are applied to a crate at P as shown. If their resultant is 340**j** lb, determine the magnitude of each force.

2.53 A car is suspended by four cables HA, HB, HC, and HD as shown. If the tension in each cable is 4.9 kN, determine their resultant force **R** at H.

Fig. P2.51 and P2.52   **Fig. P2.53**

2.9   Concluding Remarks

We saw in this chapter that force vectors may be represented *graphically* or *analytically*. These two basic representations are put to use in presenting the two traditional methods for adding and resolving forces. The first method uses the *parallelogram law*, while the second uses *vector algebra*. In the first method, the laws of cosines and sines along with the parallelogram law are extensively utilized. In the second method, the skill to write the analytical expressions of vectors is a basic requirement. We have chosen to use the vector algebra in handling forces in space because it is generally rather awkward to add or resolve forces in space by the direct application of the parallelogram law.

The knowledge of how forces are added and resolved is fundamental in mechanics. It is a prerequisite for studying the conditions of any physical bodies under the action of forces. All representations and operations of force vectors discussed in this chapter are also applicable to other vector quantities, such as those in dynamics.

Fig. P2.55

$F = 300$ lb
$A = 200$ lb
$B = 250$ lb

$\mathbf{F} = \mathbf{A} + \mathbf{B}$ $90° < \theta_B < 180°$

REVIEW PROBLEMS

2.54 If $\mathbf{P} = -10\mathbf{i} + 4\mathbf{j}$ N, $\mathbf{Q} = 6\mathbf{i} - 2\mathbf{j}$ N, and $\mathbf{D} = \mathbf{P} - \mathbf{Q}$, determine P, Q, D, θ_P, θ_Q, and θ_D.

2.55 The force **F** is to be resolved into the forces **A** and **B** as indicated. Using the *parallelogram law*, determine θ_A and θ_B.

2.56   The force **Q** is to be resolved into the forces **S** and **T** where **S** is parallel to the *y* axis and **T** is parallel to the line *AB* as indicated. Using *vector algebra*, determine the magnitudes *S* and *T*.

2.57   Three mooring lines are attached to the bow of a ship as shown. If the tensions in the lines are $F_{AB} = 900$ N, $F_{AC} = 1200$ N, and $F_{AD} = 1400$ N, determine the magnitude of their resultant force **R** at *A*.

Fig. P2.56

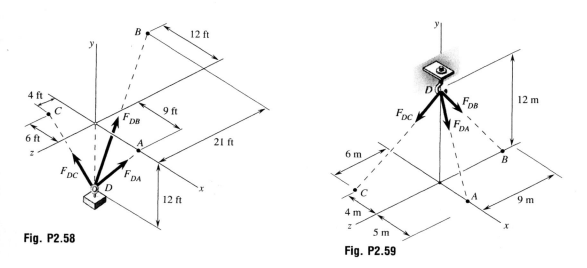

Fig. P2.57

2.58   Three forces act on an eyebolt at *D* as shown, where $F_{DA} = 150$ lb, $F_{DB} = 180$ lb, and $F_{DC} = 140$ lb. Determine their resultant force **R** at *D*.

Fig. P2.58

Fig. P2.59

2.59   Three forces act on a hook at *D* as shown. If their resultant force at *D* is **R** $= -370$**j** N, determine the magnitude of each force.

Fig. P2.60*

2.60* The pole for an antenna is held by three guy wires as shown. If the resultant force exerted by these wires on the pole at D is $\mathbf{R} = -168\mathbf{j}$ lb, determine the tension in each wire.

2.61 A transmission tower OD is guyed with three cables as shown. If the resultant force exerted by these cables on the tower at D is $\mathbf{R} = -22.9\mathbf{j}$ kN, determine the tension in each cable.

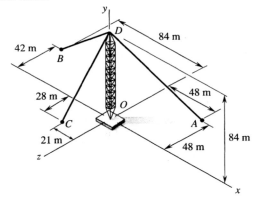

Fig. P2.61

Equilibrium of Particles

Chapter 3

KEY CONCEPTS

- Inertial reference frame, concurrent force system, reaction, internal force, external force, linear spring, space diagram, free-body diagram, and equilibrium of a particle.
- Usual assumptions and properties related to cordlike elements, pulleys, linear springs, rollers, and smooth supports.
- Drawing a good free-body diagram.
- Scalar method, analytical vector method, and semigraphical method for the solution of equilibrium problems of particles in a plane, where the force systems are concurrent and coplanar.
- Method of solution for equilibrium problems of particles in space, where the force systems are concurrent but noncoplanar.

A system of particles in equilibrium in a plane or in space is characterized by the feature that each particle in the system is subjected to a balanced system of forces whose lines of action all intersect at a common point. This chapter covers the equilibrium of systems with a single particle and systems with two or more connected particles. The fundamental concepts of inertial reference frame, reaction, internal force, external force, and free-body diagram are discussed in detail. In particular, helpful guides for drawing a good free-body diagram are given and emphasized. The scalar method, the analytical vector method, and the semigraphical method for solving equilibrium problems of particles in a plane are described and illustrated. Furthermore, a general procedure for solving equilibrium problems of particles in space is presented.

FUNDAMENTALS AND FREE-BODY DIAGRAM
3.1   Inertial Reference Frame

To describe accurately the state of rest or motion of particles or rigid bodies, a proper reference frame must be used. An *inertial reference frame* is a frame of reference in which Newton's laws of motion are valid or highly

Fig. 3.1

Fig. 3.2

Fig. 3.3

Fig. 3.4

accurate.[†] Therefore, an inertial reference frame is also called a *newtonian reference frame*. Experiments indicate that a nonaccelerating reference frame, which is either stationary or translating uniformly relative to the sun,[††] is generally valid as an inertial reference frame.

A frame of reference in which Newton's laws of motion are only accurate enough for practical purposes is called an *approximate inertial reference frame*. Since the earth rotates about its own axis and revolves around the sun, any reference frame fixed to the earth is in an accelerated motion relative to the sun. However, in most situations, a reference frame fixed on the earth is an approximate inertial reference frame because the errors introduced are generally negligible.

In situations such as the studies of motion of satellites, spacecraft, intercontinental missiles, ocean currents, and hurricanes, where the duration and distance of motion are relatively large, a reference frame fixed on the surface of the earth will become a *poor* approximate inertial reference frame because the errors introduced can be significant. In such situations, *a nonrotating reference frame attached to the center of the earth* serves better as an approximate inertial reference frame.[†††]

For practical purposes, a satisfactory approximate inertial reference frame will simply be referred to as an *inertial reference frame*. In situations under consideration, if *no* coordinate system is shown, it will be assumed that fixed rectangular coordinate systems, oriented as shown in Fig. 3.1 for systems in a plane and in Fig. 3.2 for systems in space, are being used. Note that a set of x, y, and z axes, as shown in Fig. 3.2, is said to form a *right-handed rectangular coordinate system* if when the thumb of the right hand extends along the z axis the other four fingers of the right hand can curl only in the direction in which the x axis could be rotated through 90° to coincide with the y axis.

Developmental Exercise

D3.1   What is an inertial reference frame?

3.2   Concurrent Force Systems

A *concurrent force system* is a system of forces whose lines of action intersect at a common point, which is called the *point of concurrence*. For instance, the point C in Fig. 3.3 is the point of concurrence of the concurrent forces $\mathbf{F}_1$, $\mathbf{F}_2$, $\mathbf{F}_3$, and $\mathbf{F}_4$. A *coplanar concurrent force system* is a concurrent force system where the lines of action of the forces all lie in the same plane. A *collinear force system* is a concurrent force system consisting of forces which have the same line of action, as illustrated in Fig. 3.4.

[†]A *reference frame* is a frame (or body) of reference in which one or more coordinate systems may be embedded. For conciseness, the phrase "in a reference frame" is, from time to time, used to mean "with respect to a reference frame."

[††]More accurately, to the center of mass of the solar system.

[†††]The magnitude of the acceleration of the center of the earth relative to the sun is about 6 mm/s[2].

Developmental Exercise

D3.2   Define (a) a concurrent force system, (b) a coplanar concurrent force system.

3.3   Reaction, Internal Force, and External Force

We know that an *action* of a body is an exertion of the body on another body. A *reaction* is an induced opposing action; it is a counteraction. Thus, we may regard the action as the cause and the reaction as the effect. *Reactions* from the supports or connections of a body are the *constraining forces* acting on the body to oppose the movements of the body relative to the supports or connections. In both statics and dynamics, an action exerted at a location is always accompanied, at the same location, by a collinear reaction, which is equal in magnitude but opposite in direction to the action.

The relation between action and reaction is described by *Newton's third law* and is illustrated in Fig. 3.5 for a cart of weight W_C supported by a rope and an incline, and in Fig. 3.6 for a girl of weight W_G in a swing. Note that the *weight force* of a body is the attractive force exerted at a distance by the earth on the body. At the same time, the *weight force* of a

A cart
being held

Fig. 3.5

Actions exerted by the
cart on its supports
and the earth

Reactions from the
supports and the
earth to the cart

A swinging girl

Fig. 3.6

Actions exerted by
the girl on her supports
and the earth

Reactions from the
supports and the
earth to the girl

body may be viewed as a *reaction* from the earth to the attractive force exerted on it by the body.[†]

An *internal force* to a body (which may be a system of connected bodies) is a force of interaction existing *within* the body under consideration. By Newton's third law, internal forces must occur in collinear pairs which are equal in magnitude and opposite in direction; i.e., the sum of internal forces in a system must be zero. An *external force* to a body is a force exerted on the body by another body, which is *outside* the body. Notice that the weight force of a body is an external force to the body.

Developmental Exercises

D3.3 Ascertain whether each of the following sentences gives a true or false statement about the *weight force* of a body. (a) It is the attractive force exerted by the earth on the body. (b) It is equal in magnitude but opposite in direction to the attractive force exerted by the body on the earth. (c) It is the action exerted by the body on the earth. (d) It is the reaction from the earth to the body. (e) It is equal to the mass of the body.

D3.4 Is the coupling force in a two-car train an internal force or an external force to (a) the train, (b) either car?

D3.5 Define the terms: (a) reaction, (b) internal force, (c) external force.

3.4 Cordlike Elements, Pulleys, and Connections

In engineering applications, we frequently use cordlike elements to connect different bodies and exert pulling forces along straight lines or over curved surfaces. Examples of *cordlike elements* include *cords, strings, ropes, wires, cables, belts,* and *chains.* Unless otherwise specified, cordlike elements will be assumed to be *perfectly flexible, inextensible,* and *of negligible mass.* In other words, the internal forces developed in cordlike elements under the action of applied loads are always *tensile forces,* which are collinear with or tangent to their respective cordlike elements.

In statics, the letter T with one or two appropriate subscripts (e.g., T_1, T_{AB}) is often used to represent the *magnitude* of the tensile force developed in a cordlike element. This magnitude of the tensile force is usually referred to as the *tension* in the cordlike element. Although the order of two subscripts may indicate the sense of the direction of a vector, it should be remembered that the magnitude of a vector is a scalar quantity. Thus, $\mathbf{T}_{AB} = -\mathbf{T}_{BA}$ but $T_{AB} = T_{BA}$.

From the assumptions stated earlier, it is clear that the *magnitude* of the tensile force is the *same* everywhere in the *same* cordlike element which passes around smooth pegs or pulleys (or sheaves) with frictionless bearings. This property, which is illustrated in Fig. 3.7 for a smooth peg and in Fig. 3.8 for a pulley with frictionless bearing, may be proved using the concepts of rigid-body equilibrium presented in Chap. 5. Unless otherwise specified, *pegs and pulleys will be assumed to be frictionless and the weights of pulleys will be assumed to be negligible.* The words "smooth" and "friction-

Fig. 3.7

Fig. 3.8

[†]Cf. Newton's law of gravitation in Sec. 1.11.

less'' will be assumed to have the same meaning and hence may be used interchangeably.

Two or more cordlike elements may be fastened together to form a *knot* or be connected to a *small ring* or *connector* (whose weight is taken as negligible) in supporting a load. The tensile forces in *different* cordlike elements, which are connected by a knot or a ring, are generally *different* in magnitude as well as direction. This property is illustrated in Fig. 3.9 for a *knot connection* and in Fig. 3.10 for a *ring connection*.

Fig. 3.9

Fig. 3.10

Developmental Exercises

D3.6   What are the usual assumptions and properties related to (a) cordlike elements, (b) pegs, (c) pulleys?

D3.7   A tow truck A is pulling a car B out of a ditch as shown. What is the sense of the reaction (a) from the cable to the car B, (b) from the cable to the truck A?

Fig. D3.7

D3.8   A crate suspended by a system of pulleys is held with a force **P** as shown. (a) Is the tension T_{AB} in the cable from A to B equal to P? (b) Is T_{AB} equal to the tension T_{CD} in the cable from C to D? (c) Is $T_{CD} = T_{EF}$? (d) Is $T_{AB} = T_{GH}$?

D3.9   In the system shown, which of the following pairs of sections of string have tensile forces which are equal in magnitude? (a) AB and BC, (b) BC and CD, (c) CD and DE, (d) CD and DF, (e) DE and DF.

Fig. D3.9

Fig. D3.8

3.5   Linear Springs

Springs are common elastic elements in mechanical systems. In a broader sense, any material which is capable of developing elastic restoring forces may be considered as a spring. A *linear spring* develops a tensile or compressive force whose magnitude is proportional to the magnitude of the change in length (i.e., elongation or contraction) of the spring. Otherwise, the spring is called a *nonlinear spring*. Unless otherwise noted, all springs are taken as linear and have negligible mass.

The *free length* (or *natural length*) of a spring is the length of a spring in its undeformed state.[†] For a spring of free length L experiencing an elongation x as shown in Fig. 3.11(a), the magnitude F of the tensile force developed is proportional to x and is depicted in Fig. 3.11(b). Mathematically, we write

$$F = kx \qquad (3.1)$$

where the constant k is called the *spring modulus* (or *spring constant*). The dimensions of k are *force/length*. Note in Eq. (3.1) that F represents a compression if x is the contraction (or shortening) of the spring.

(a)   (b)

Fig. 3.11

Developmental Exercise

D3.10   The spring AB has a spring modulus of 2 kN/m. In the position shown, determine the magnitude F of the spring force if the free length of the spring is (a) 0.4 m, (b) 0.3 m.

3.6   Rollers and Smooth Supports

A *roller* is a support which allows the point of support to move freely parallel to the surface on which the roller rests. For instance, a caster or an unpowered wheel is a roller support. Unless otherwise stated, rollers are assumed to have negligible mass. When *rollers* are used to support a body, the reaction from each roller to the body must be normal (i.e., perpendicular) to the surface on which the roller rests. The normal reactions at the rollers are sometimes combined into a single resultant force to allow the treatment of the body as a particle, as shown in Fig. 3.12, where a cart of weight W is held in equilibrium in (a), and the two reactions from the rollers in (b) are combined into one in (c). However, in some situations (e.g., beam problems in Chap. 5), this treatment may cease to be appropriate.

A friction force on a surface is a tangential force acting on the surface. When the effect of the friction force between a body and its support is negligibly small compared with the effects of other forces, the support may be taken as smooth. If a body rests on a *smooth support*, the reactions from the support to the body must be normal to the surface of the support and are generally distributed over the surface of contact. However, the distributed

Fig. D3.10

[†]This term is also called the *unstretched length* (or *undeformed length*) of the spring by others.

Fig. 3.12
(a)   (b)   (c)

Fig. 3.12

reactions are sometimes combined into a single resultant force to allow the treatment of the body as a particle, as shown in Fig. 3.13, where a block of weight W resting on a smooth incline is held in equilibrium in (a), and the distributed reactions from the incline in (b) are combined into one in (c). Note that $\mathbf{R}$, $\mathbf{T}$, and $\mathbf{W}$ form a concurrent force system as indicated.

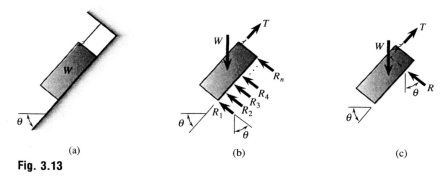

(a)   (b)   (c)

Fig. 3.13

Developmental Exercises

D3.11 How is the reaction from a roller support directed?

D3.12 When may a support be taken as smooth?

3.7 Free-Body Diagram

A *space diagram* is a *sketch* showing the *physical and geometrical conditions* of a system. In other words, a space diagram is a graphical description of the system. It generally shows the shape and size of the system, the weights, the externally applied loads, the connections, and the supports of the system.

The *free-body diagram* of a body is a *sketch* in which we *free* (or isolate) the *body* by removing its supports and connections in the space diagram and then *show* on this freed body all the pertinent weight forces, the externally applied loads, and the reactions from its supports and connections existing just before their removal. The "freed body" is often simply called the "free body." A *free-body diagram* may be viewed as an *engineer's ledger* which accounts the system of forces that bring about the state of rest or motion of the body under consideration. A simplified but practical definition of the free-body diagram may be written in the form of an equation as follows:

$$\boxed{\text{Free-body diagram}} = \boxed{\begin{array}{l}\text{Space diagram of the body with}\\ \text{pertinent weight forces and}\\ \text{applied loads indicated}\end{array}}$$
$$- \boxed{\begin{array}{l}\text{Supports and connections of}\\ \text{the body}\end{array}}$$
$$+ \boxed{\begin{array}{l}\text{Reactions from the supports}\\ \text{and connections existing}\\ \text{just before their removal}\end{array}} \qquad (3.2)$$

The importance of *drawing free-body diagrams* cannot be overemphasized for mechanics students. Those who develop the habit of drawing appropriate free-body diagrams usually find their study of mechanics considerably better organized, more at their command, and easier.

EXAMPLE 3.1

A crate of weight W is supported by three ropes in a ring connection as shown. Draw the free-body diagrams of (a) the crate and the ring combination, (b) the crate, (c) the ring.

Solution. The free-body diagrams are drawn as shown.

REMARK. Note that (1) all forces developed in the ropes are tensile forces, (2) the tensile force in the rope CD is not shown in case (a) because it is an internal force to the crate and the ring combination, (3) the tensile force in the rope CD becomes an external force to the free body of just the crate or just the ring and is shown in cases (b) and (c), (4) when ropes are cut, portions of them are usually shown on the free body to help us identify the reactions from the connections existing just before their removal.

Space diagram

(a) Free-body diagram of *the crate and the ring*

(b) Free-body diagram of *the crate*

(c) Free-body diagram of *the ring*

Developmental Exercises

D3.13 Define the terms: (a) space diagram, (b) free-body diagram.

D3.14 Refer to the free-body diagrams drawn in Example 3.1. Is each of the "freed particles" shown as a *point* or as a *body with a shape* as depicted in the space diagram?

3.8 Guides for Drawing a Good Free-Body Diagram

A *good* free-body diagram is a sketch which *correctly* and *effectively* exhibits the system of forces that brings about the state of rest or motion of the body under consideration. The following are helpful guides for drawing a good free-body diagram:

1. The *body to be freed* (or isolated) for consideration may be the entire system or any portion of the system; however, it is important to make a clear decision as to which portion of the system is to be freed.
2. The *free body* drawn should have no external supports or connections.
3. Any *adopted coordinate system* whose axes are not in the horizontal and vertical directions should be shown.
4. Appropriate *dimensions* (including slopes or angles) or the *coordinates* of certain key points, which are needed in defining the configuration of the force system, should be indicated.
5. Each *applied load* is externally applied to act on the body and should be indicated with an arrow and labeled either with its known magnitude or with a letter when it is not known.
6. The *weight force* of the free body is an external force exerted by the earth on the body and should be indicated with a *vertical* downward arrow and labeled like an applied load if the weight is not negligible.
7. The *actions exerted by the body* on its supports and connections existing just before their removal *should not be indicated* on the free body in the diagram. (It should be noted that such actions are exerted *by*, *not on*, the body. Only those forces which are exerted *on* the body by other bodies are to be indicated in the free-body diagram.)
8. The *reactions* (or components of the reactions) from the supports and connections existing just before their removal should be indicated with arrows and labeled like the applied loads. (This is obviously essential in making the free body exhibit the same condition of rest or motion as that of the body shown in the space diagram.)
9. The *sense* of an unknown force, when not reasonably obvious, may be assumed and corrected later if the value of the variable representing the magnitude of the force is subsequently found to be negative.[†]
10. The *forces in the cords or members which are uncut* in the free body are internal forces to the free body and *should not be shown*.

[†]If the value of an unknown quantity is found to be negative in the solution, the magnitude of this unknown quantity is simply equal to the absolute value of the solved value, but the true sense of this unknown quantity is opposite to that initially assumed. However, this practice may cease to be valid when the unknown quantity is a *friction force* at a location where slipping *impends* or *occurs*. The sense of a friction force is explained in Sec. 9.2.

Developmental Exercises

D3.15 · Is it true that the weight force of a body is an internal force to the body and should be excluded from the free-body diagram of the body?

D3.16   Which of the following should be indicated in the free-body diagram of a body? (a) The actions exerted by this body on its surrounding bodies, (b) the reactions from the surrounding bodies to the actions exerted by this body, (c) the forces in the cords or members which are uncut in the free body.

EXAMPLE 3.2

Three cylinders A, B, and C suspended by a system of cables and pulleys are in equilibrium as shown. The weights of the cylinders A and C are indicated. Denoting the weight of the cylinder B by W_B, draw the free-body diagrams of (a) the cylinder A, (b) the cylinder C, (c) the cylinder B and the pulley E combination.

Solution.   By the usual assumptions, we note that (1) the weights of the pulleys are negligible, (2) the magnitude of the tensile force is the same everywhere in the same cable which passes around pulleys, (3) we may use T_1, T_2, and T_3 to denote the tensions in the cables $ADEFB$, HJC, and $BGHI$, respectively. Thus, we may draw the required free-body diagrams as shown.

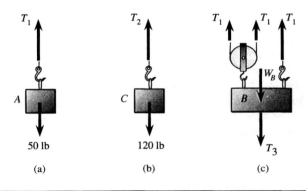

Developmental Exercises

D3.17   Draw the free-body diagram of the pulley H in Example 3.2.
D3.18   Draw the free-body diagram of the ring A in Example 3.3.

EXAMPLE 3.3

A 10-kg block is hung with a system of wires and small rings, as shown. Draw the free-body diagrams of (a) the block, (b) the ring B, (c) the ring C.

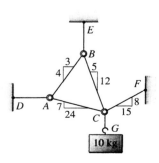

Solution.   Since all forces in the wires are tensile forces, we may draw the required free-body diagrams as shown.[†]

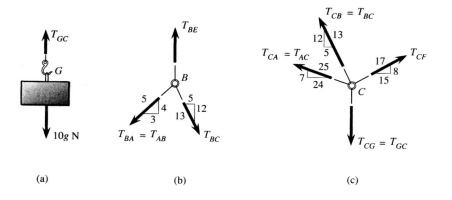

(a)   (b)   (c)

PROBLEMS

3.1 through 3.6* A crate is supported by three cables in a ring connection as shown. Draw the free-body diagrams of (a) the crate, (b) the ring, (c) the crate and the ring combination.

Fig. P3.1 and P3.17

Fig. P3.2* and P3.18

Fig. P3.3* and P3.19*

Fig. P3.4 and P3.20

Fig. P3.5* and P3.21

Fig. P3.6* and P3.22*

[†]Note in this text that g which is not separated from a number by a space, nor stands alone, is *not* the abbreviation for the unit *gram*, but represents the value of 9.81 (m/s^2) in problems with SI units.

Fig. P3.8*

Fig. P3.11*, P3.14*, and P3.27*

3.7* A man exerts a horizontal force **P** to hold a 60-lb ice block in equilibrium on a ramp as shown. Neglecting the friction force between the ice block and the ramp, draw the free-body diagram of the ice block.

3.8* A boy standing on a skateboard on a steep hill is held in equilibrium by a force **P** as shown. The boy and his skateboard have a mass of 40 kg. Draw the free-body diagram of the boy and his skateboard.

3.9 through 3.11* A block of mass m_B is being held in equilibrium by the force **P** through a system of ropes and pulleys as shown. (a) How many different ropes are there in the system? (b) Using T_i ($i = 1, 2, \ldots, n$) to represent the tensions in different ropes, draw the free-body diagrams of (i) the block B, (ii) each pulley, separately.

Fig. P3.9, P3.12, and P3.25

Fig. P3.10*, P3.13*, and P3.26

3.12 Draw the free-body diagram of the block B and the pulley E combination.

3.13* Draw the free-body diagram of the block B and the pulleys E and G combination.

3.14* Draw the free-body diagram of the block B and the pulleys E, F, and H combination.

3.15* and **3.16*** The package at G is supported by wires and rings as shown. Draw the free-body diagrams of (a) the package, (b) each ring, separately.

Fig. P3.15* and P3.43

Fig. P3.16* and P3.44*

EQUILIBRIUM IN A PLANE[†]

3.9 Equilibrium of Particles in a Plane

If a body is originally in equilibrium in the space diagram, the *free body* of this body must also be in equilibrium under the action of the forces indicated in the free-body diagram. Since all pairs of collinear forces having equal

---

[†]*Equilibrium in a plane* is equilibrium in a two-dimensional space, and is also termed *equilibrium in two dimensions*.

magnitudes and opposite senses are balanced themselves, they can have no
net external effect on a particle or a rigid body on which they act. Such
forces may be considered as *trivial forces*. The *nontrivial forces* acting on a
particle must be *concurrent forces*, regardless of the amount and distribution
of mass of the body which is being treated as a particle.[†]

The *equilibrium of a particle* is the state of the particle ·in which the
particle is at rest or moving with a constant velocity in an inertial reference
frame. We know from Newton's second law that the acceleration of the
particle is zero if the resultant force acting on it is zero. Since zero acceleration means constant (including zero) velocity, which is a manifestation
of equilibrium, we see that the *necessary and sufficient condition* for the
equilibrium of a single particle is that the resultant force **R** of all known and
unknown forces $\mathbf{F}_1$, $\mathbf{F}_2$, . . . , $\mathbf{F}_n$ indicated in the free-body diagram of the
particle equals zero. Since $\mathbf{R} = \Sigma\mathbf{F}$, the *necessary and sufficient condition*
for equilibrium of a particle is

$$\Sigma\mathbf{F} = 0 \qquad (3.3)$$

which is also called the *vector equation of equilibrium* of a particle.

An *equilibrium problem of particles in a plane* is characterized by the
feature that at least one set of coplanar, concurrent, but noncollinear unknown forces acting on a particle will be encountered in the solution; however, no set of noncoplanar unknown forces will be encountered in that
solution. For each particle in equilibrium in the *xy* plane, a set of *scalar
equations of equilibrium* may be written from Eq. (3.3) as

$$\Sigma F_x = 0 \qquad \Sigma F_y = 0 \qquad (3.4)$$

It is to be noted that the *x* and *y* axes for Eqs. (3.4) may be chosen to be
any two orthogonal axes in the plane of the forces. This is illustrated in
Example 3.4, which serves as a forerunner of the scalar method of solution
presented in Sec. 3.10.

Forces in some equilibrium problems involving pulleys are parallel,
rather than concurrent, forces. However, many of such problems may simply be solved by applying one of the two equations in Eqs. (3.4). For this
reason, some of them are included in this chapter.

EXAMPLE 3.4

A girl is sitting in an old automobile tire which is suspended as shown. If
the girl and the tire together have a mass of 60 kg, determine the tension
T_{CA} in the rope *CA*.

[†]Cf. Example 3.1 and the concept of particle discussed in Sec. 1.6.

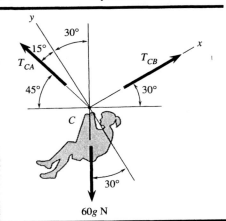

Solution. By choosing the x axis to be along the unknown tensile force $\mathbf{T}_{CB}$ in the free-body diagram shown, we can solve for T_{CA} from a single equilibrium equation as follows:

$$+\nwarrow \; \Sigma F_y = 0: \quad T_{CA}\cos 15° - 60g\cos 30° = 0$$

$$T_{CA} = 528 \text{ N} \quad \blacktriangleleft$$

Developmental Exercises

D3.19   Solve for the tension T_{CB} in Example 3.4 in a similar manner.

D3.20   Is the resultant force acting on the free body necessarily equal to zero if the body in the space diagram is in equilibrium?

D3.21   What is the characterizing feature in an equilibrium problem of particles in a plane?

D3.22   What is the necessary and sufficient condition for the equilibrium of a particle?

3.10   Scalar Method of Solution[†]

A *scalar method of solution* places emphasis on the scalar components of vectors and makes use of scalar equations, rather than vector equations. Although certain problems can best be solved by employing a special set of steps, the following procedure is generally helpful in *solving equilibrium problems of particles in a plane*:

1. *Analyze the problem situation* carefully and make preliminary computations from the given geometric and material properties if needed.
2. *Indicate the chosen xy rectangular coordinate axes* if they are not in the horizontal and vertical directions.
3. *Draw correctly the relevant free-body diagram* of a particle where no more than two unknowns will be involved. [This is desirable because no more than two independent equilibrium equations can be obtained in step (4) below.] If this step is impossible, go to step (7) below.
4. *Sum to zero the scalar components of all forces* in the x and y directions, respectively, to obtain a set of two scalar equations.
5. *Solve for the unknowns* in the equations obtained in step (4).
6. *Repeat steps (2) through (5)* for other particles until the desired unknown or unknowns have been determined.

[†]Students are encouraged to be creative and to treat *any* method outlined in this text as only a "guide" in solving a class of problems. The guide is intended to help students get started. There is no need to follow the suggested procedure in a rigid fashion.

7. If the system consists of *n* particles (some of which may have negligible weight) and *the free-body diagram drawn for any of the n particles in the system involves more than two unknowns*, then we draw the free-body diagram of each particle in the system. We may write two scalar equilibrium equations for each particle as described in step (4) and obtain a total of 2*n* simultaneous equations. If the total number of unknowns equals 2*n*, the unknowns may be determined.

8. *Add appropriate units to the final answers.*

EXAMPLE 3.5

Two identical springs *AB* and *BC* are connected by a hook at *B* and stretched horizontally between two posts with a tension of 200 N. A 35-kg boy grabs the hook at *B* and hangs from it as shown. When equilibrium exists, the deflection $\overline{BB'}$ of the hook is found to be 41 mm. Determine the spring modulus *k* of each spring.

Solution. By virtue of symmetry (or by applying $\Sigma F_x = 0$), we know that the tensions in the two springs must be the same. For simplicity in the solution, let the spring modulus of each spring be *k* N/mm, the free length of each spring be *L* mm, the elongation of each spring in the horizontal position be x_1 mm, the elongation of each spring in the inclined position be x_2 mm, and the magnitude of the tension in each spring in the inclined position be T_2. Applying the Pythagorean theorem, we write

$$\overline{AB'} = [(840)^2 + (41)^2]^{1/2} = 841$$

The initial tension in each spring is $T_1 = 200$ N. By the property of linear springs, we write

$$T_1 = kx_1 = 200 \qquad L = \overline{AB} - x_1 = 840 - \frac{200}{k}$$

$$x_2 = \overline{AB'} - L = 1 + \frac{200}{k} \qquad T_2 = kx_2 = k + 200$$

From the free-body diagram of the boy and the hook combination, we write

$$+\uparrow \ \Sigma F_y = 0: \qquad 2T_2\left(\frac{41}{841}\right) - 35g = 0$$

Substituting the expression of T_2 into this equilibrium equation and noting that $g = 9.81$ (m/s²), we write

$$2(k + 200)\left(\frac{41}{841}\right) - 35g = 0 \qquad k = 3321 \text{ N/mm}$$

$$k = 3.32 \text{ MN/m} \blacktriangleleft^\dagger$$

Developmental Exercise

D3.23 Solve for *k* in Example 3.5 if the mass of the boy is 40 kg.

---

†In writing the final answer, it is recommended that SI prefixes be used only in the numerator, except the base unit kilogram.

EXAMPLE 3.6

A cart and a block are connected and supported by a system of cables, pulleys, rollers, and a small ring as shown. The weight of the block at G is 150 lb. Determine the weight W of the cart at A consistent with equilibrium.

Solution. We will not start with the free-body diagram of the cart at A because it involves three unknowns: the weight W, the magnitude T_{AB} of the tension in the cable AB, and the magnitude R of the resultant of the roller reactions. Since the cable $BDEFG$ supports the 150-lb block at G and passes around pulleys D, E, and F, it is clear that the tension in the cable BD must be 150 lb. Thus, the free-body diagram of the ring at B involves only two unknown magnitudes T_{AB} and T_{BC} of the tensile forces in the cables AB and BC as shown. For equilibrium, we write

$$\xrightarrow{+} \Sigma F_x = 0: \quad \frac{1}{\sqrt{2}}(150) - \frac{15}{17}T_{BA} = 0 \qquad T_{BA} = 85\sqrt{2} \text{ lb}$$

Having determined T_{BA}, we are now ready to consider the cart at A as shown. Note in the free-body diagram of the cart at A that the angle between the vertical line of action of the weight force $\mathbf{W}$ and the y axis is θ and from the slope triangle that $\sin\theta = 8/17$. For equilibrium, we write

$$+\nearrow \Sigma F_x = 0: \quad 85\sqrt{2} - W\sin\theta = 0$$

$$W = 255 \text{ lb} \quad \blacktriangleleft$$

Developmental Exercise

D3.24 Solve for the tension T_{BC} in Example 3.6.

EXAMPLE 3.7

Refer to Example 3.2. Determine the weight W_B of the block B necessary to hold the system in equilibrium.

Solution. The relevant free-body diagrams for the solution of the problem are first drawn as shown. For equilibrium of *the block A*, we write

$$+\uparrow \Sigma F_y = 0: \quad T_1 - 50 = 0 \qquad T_1 = 50 \text{ lb}$$

For equilibrium of *the block C*, we write

$$+\uparrow \Sigma F_y = 0: \quad T_2 - 120 = 0 \qquad T_2 = 120 \text{ lb}$$

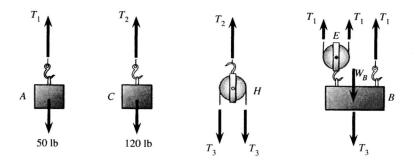

For equilibrium of *the pulley H*, we write

$$+\uparrow \Sigma F_y = 0: \quad T_2 - T_3 - T_3 = 0 \qquad T_3 = \tfrac{1}{2} T_2 = 60 \text{ lb}$$

For equilibrium of *the block B and the pulley E*, we write

$$+\uparrow \ \Sigma F_y = 0: \quad T_1 + T_1 + T_1 - W_B - T_3 = 0$$

$$W_B = 3T_1 - T_3 = 90 \qquad W_B = 90 \text{ lb} \blacktriangleleft$$

EXAMPLE 3.8

Refer to Example 3.3. Determine the tension in each wire of the system.

Solution.   The relevant free-body diagrams for the solution of the problem are first drawn as shown.

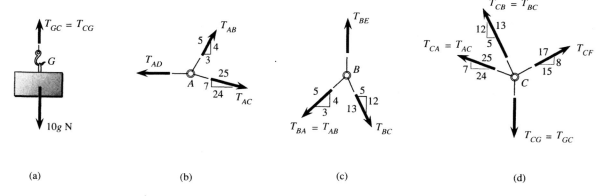

(a)   (b)   (c)   (d)

For equilibrium of *the block* at *G*, we write

$$+\uparrow \Sigma F_y = 0: \quad T_{CG} - 10g = 0 \qquad T_{CG} = 98.1 \text{ N} \blacktriangleleft$$

Since the free-body diagram drawn for any of the three small rings involves three (not two) unknowns, we should solve the rest of this problem according to step (7) in Sec. 3.10. From the three free-body diagrams of the rings *A*, *B*, and *C*, we see a total of six unknowns: T_{AB}, T_{AC}, T_{AD}, T_{BC}, T_{BE}, and T_{CF}. Moreover, from each of these three free-body diagrams, we may write two independent force equilibrium equations and obtain a total of six independent equations in six unknowns. Thus, the six unknowns may be determined. For equilibrium of *the rings* as shown in their free-body diagrams, we write

Ring A

$$\xrightarrow{+} \Sigma F_x = 0: \qquad \frac{3}{5} T_{AB} + \frac{24}{25} T_{AC} - T_{AD} = 0 \qquad (1)$$

$$+\uparrow \Sigma F_y = 0: \qquad \frac{4}{5} T_{AB} - \frac{7}{25} T_{AC} = 0 \qquad (2)$$

Ring B

$$\xrightarrow{+} \Sigma F_x = 0: \qquad -\frac{3}{5} T_{AB} + \frac{5}{13} T_{BC} = 0 \qquad (3)$$

$$+\uparrow \Sigma F_y = 0: \qquad -\frac{4}{5} T_{AB} - \frac{12}{13} T_{BC} + T_{BE} = 0 \qquad (4)$$

Ring C

$$\xrightarrow{+} \Sigma F_x = 0: \qquad -\frac{24}{25} T_{AC} - \frac{5}{13} T_{BC} + \frac{15}{17} T_{CF} = 0 \qquad (5)$$

$$+\uparrow \Sigma F_y = 0: \qquad \frac{7}{25} T_{AC} + \frac{12}{13} T_{BC} + \frac{8}{17} T_{CF} - 10g = 0 \quad (6)$$

The preceding six equations may be written in the form of a matrix equation as follows:[†]

$$
\begin{bmatrix}
\frac{3}{5} & \frac{24}{25} & -1 & 0 & 0 & 0 \\
\frac{4}{5} & -\frac{7}{25} & 0 & 0 & 0 & 0 \\
-\frac{3}{5} & 0 & 0 & \frac{5}{13} & 0 & 0 \\
-\frac{4}{5} & 0 & 0 & -\frac{12}{13} & 1 & 0 \\
0 & -\frac{24}{25} & 0 & -\frac{5}{13} & 0 & \frac{15}{17} \\
0 & \frac{7}{25} & 0 & \frac{12}{13} & 0 & \frac{8}{17}
\end{bmatrix}
\begin{bmatrix}
T_{AB} \\ T_{AC} \\ T_{AD} \\ T_{BC} \\ T_{BE} \\ T_{CF}
\end{bmatrix}
=
\begin{bmatrix}
0 \\ 0 \\ 0 \\ 0 \\ 0 \\ 10g
\end{bmatrix}
$$

Note that *the elements in the rows of the coefficient matrix correspond to coefficients of the unknowns as arranged in the unknown column matrix.* The solution of these simultaneous equations yields:

$$T_{AB} = 24.4 \text{ N} \qquad T_{AD} = 81.5 \text{ N} \qquad T_{BE} = 54.6 \text{ N} \quad \blacktriangleleft$$

$$T_{AC} = 69.7 \text{ N} \qquad T_{BC} = 38.0 \text{ N} \qquad T_{CF} = 92.4 \text{ N} \quad \blacktriangleleft$$

[†]The matrix representation of simultaneous equations is employed to facilitate the solution with a computer as demonstrated in App. D. No other background in matrix algebra is required for understanding the rest of this text.

Developmental Exercises

D3.25 What is the maximum number of independent equilibrium equations we can write for a particle in equilibrium in a plane?

D3.26 If a system of particles in equilibrium in a plane is to be analyzed and we cannot find a single particle whose free-body diagram involves no more than two unknowns, what should we do?

3.11 Analytical Vector Method of Solution

The *analytical vector method of solution* for equilibrium problems of particles in a plane may be formulated by amending the procedure given in Sec. 3.10 in such a way that step (4) is replaced by the following two *alternative steps*:

4a. *Add the analytical expressions of all forces* shown in the free-body diagram of the particle to obtain the resultant force in the *xy* plane in terms of known and unknown quantities.

4b. *Set each of the two scalar components* of the resultant force, obtained in step (4a), *equal to zero* to obtain a set of two scalar equations. (This step follows from the requirement that each of the scalar components of a zero vector must be equal to zero. For a body in equilibrium, the resultant force acting on it is, of course, a zero vector.)

The use of vector analysis frequently leads to more concise formulations and derivations of the fundamental principles of mechanics. This method is illustrated in Example 3.9.

Developmental Exercise

D3.27 In solving an equilibrium problem of particles in a plane by the analytical vector method, why do we set the two scalar components of the resultant force equal to zero?

3.12 Semigraphical Method of Solution

When a particle is in equilibrium in a plane, the resultant force of all coplanar concurrent forces shown in the free-body diagram of the particle must be equal to zero. By the polygon rule in Sec. 2.2, we can readily conclude that a *closed figure* or *polygon* will always be obtained when all the forces shown in the free-body diagram of the particle in equilibrium are drawn to scale and arranged in tip-to-tail fashion.

By graphics, the unknowns associated with an equilibrium problem may, of course, be determined by direct measurement from the force polygon drawn. However, the *unknowns* may also be *determined from the force polygon by using geometry and trigonometry*. This latter approach is called a *semigraphical method of solution*; it is useful in solving simple equilibrium problems of particles in a plane, for which the force polygons are in the form of triangles.

3g N

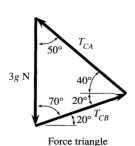

Force triangle

EXAMPLE 3.9

A 3-kg flowerpot is suspended as shown. Determine the tensions T_{CA} and T_{CB} in the wires CA and CB by (a) the scalar method, (b) the analytical vector method, (c) the semigraphical method.

Solution. Regardless of the method to be used, the free-body diagram of the flowerpot and the ring combination is drawn as shown. Note that the x axis is chosen to be collinear with the unknown force $\mathbf{T}_{CB}$ to simplify the algebraic work in the solution.

(a) *Scalar method.* For equilibrium of the free body as shown, we write

$$+\nwarrow \Sigma F_y = 0: \quad T_{CA} \cos 30° - 3g \cos 20° = 0$$

$$+\nearrow \Sigma F_x = 0: \quad T_{CB} - T_{CA} \sin 30° - 3g \sin 20° = 0$$

Solving these equations, we obtain

$$T_{CA} = 31.9 \text{ N} \qquad T_{CB} = 26.0 \text{ N} \blacktriangleleft$$

(Note the *algebraic advantage* in applying $\Sigma F_y = 0$ before $\Sigma F_x = 0$ in the above solution.)

(b) *Analytical vector method.* From the free-body diagram, we write

$$\mathbf{T}_{CB} = T_{CB}\boldsymbol{\lambda}_{CB} = T_{CB}\mathbf{i}$$

$$\mathbf{T}_{CA} = T_{CA}\boldsymbol{\lambda}_{CA} = T_{CA}(-\sin 30°\mathbf{i} + \cos 30°\mathbf{j})$$

$$\mathbf{W} = W\boldsymbol{\lambda}_W = 3g(-\sin 20°\mathbf{i} - \cos 20°\mathbf{j})$$

For equilibrium, we write

$$\Sigma \mathbf{F} = \mathbf{T}_{CB} + \mathbf{T}_{CA} + \mathbf{W} = \mathbf{0}:$$

$$(T_{CB} - T_{CA} \sin 30° - 3g \sin 20°)\mathbf{i} + (T_{CA} \cos 30° - 3g \cos 20°)\mathbf{j} = \mathbf{0}$$

Equating the coefficients of $\mathbf{i}$ and $\mathbf{j}$ to zero, we have

$$T_{CB} - T_{CA} \sin 30° - 3g \sin 20° = 0$$

$$T_{CA} \cos 30° - 3g \cos 20° = 0$$

Solving these equations, we obtain

$$T_{CA} = 31.9 \text{ N} \qquad T_{CB} = 26.0 \text{ N} \blacktriangleleft$$

(c) *Semigraphical method.* Referring to the free-body diagram in (a), we first draw the 3g-N weight force. The tensile forces $\mathbf{T}_{CA}$ and $\mathbf{T}_{CB}$ have known directions and are, therefore, drawn as shown to form a force triangle. Indicating the angles in the triangle and applying the law of sines, we write

$$\frac{3g}{\sin 60°} = \frac{T_{CA}}{\sin 70°} = \frac{T_{CB}}{\sin 50°}$$

$$T_{CA} = 31.9 \text{ N} \qquad T_{CB} = 26.0 \text{ N} \blacktriangleleft$$

Finally, we may check the answers by summing horizontal and vertical scalar components of the resultant force as follows:

$$\Sigma F_{horiz} = T_{CB} \cos 20° - T_{CA} \cos 40°$$

$$= 26.0 \cos 20° - 31.9 \cos 40° \approx 0$$

$$\Sigma F_{vert} = T_{CB} \sin 20° + T_{CA} \sin 40° - 3g$$

$$= 26.0 \sin 20° + 31.9 \sin 40° - 3g \approx 0$$

PROBLEMS

3.17 through 3.22* *(The figures for these problems are specified along with those for Probs. 3.1 through 3.6* on p. 61.)* For the system shown, determine the tension in the cables CA and CB.

3.23 Determine the magnitude of **P** in Prob. 3.7*.

3.24* Determine the tension in the rope in Prob. 3.8*.

3.25 through 3.27* *(The figures for these problems are specified along with those for Probs. 3.9 through 3.11* on p. 62.)* For the system shown, determine the mass m_B of the block B consistent with equilibrium.

3.28 A welded joint is in equilibrium as shown. Determine the magnitudes of the forces **P** and **Q**.

Fig. P3.28

Fig. P3.29

3.29 An instrument on a base is to be isolated from floor vibrations as shown. If the instrument and base together have a mass of 2 kg and $\theta = 60°$, find the magnitudes S_1 and S_2 of the forces in the springs when the instrument is in equilibrium.

3.30 A 30-kg boy grabs a wire and hangs from it as shown. If $\theta = 8°$ when the boy is in equilibrium, determine the tension T in the wire.

3.31 A 40-kg cart rests on an incline and is held by a cable which passes over a smooth pulley and has one end fastened to a scale as shown. Determine the angle θ which the incline makes with the horizontal if the scale indicates that the tension in the cable is 300 N.

Fig. P3.30

Fig. P3.31

3.32   A crate of newly cured concrete cylinders in a laboratory is lifted with a force **P** as shown. Determine the magnitude of **P** when the crate is held in equilibrium.

Fig. P3.32   (a)   (b)   (c)   (d)

60 kg   60 kg   60 kg   60 kg

3.33*   A collar C sliding on a smooth horizontal rod is connected to an anchor at A by a spring which has a modulus of 5 kN/m. When the collar is directly above the anchor, the tension in the spring is 100 N. Determine the magnitude of the force **P** required to hold the collar in equilibrium as shown if $a = 50$ mm.

3.34   A 140-kg crate is hung on the hook of a crane as shown. If the length of the cable sling is 1.98 m and the crate is in equilibrium, determine the tension in the cable sling.

Fig. P3.33*

Fig. P3.35

Fig. P3.34   **Fig. P3.36***

3.35 and 3.36*   The breaking strength of each of the two wires AC and BC is F_b as indicated. Determine the maximum magnitude of **P** and its directional angle θ_P for which neither wire will break.

3.37   Determine the tension in each cable if $F = 250$ lb and the crate shown is in equilibrium.

Fig. P3.37 and P3.38

3.38 Determine the range of values of F for which no cable will become slack and the crate shown remains in equilibrium.

3.39 through 3.41* Determine the mass of the body B necessary to hold the system shown in equilibrium.

Fig. P3.39

Fig. P3.40

Fig. P3.41*

3.51 slugs

3.42 The tension in the cable CE as indicated by the scale is 630 N. Determine the masses of the carts A and B consistent with equilibrium.

3.43 and 3.44* *(The figures for these problems are specified along with those for Probs. 3.15* and 3.16* on p. 62.)* Determine the tension in each wire.[†]

3.45 Two bodies are hung as shown. Determine the tension in each wire.[†]

Fig. P3.42

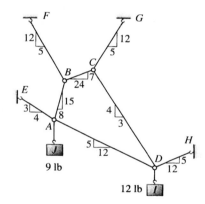

Fig. P3.45

EQUILIBRIUM IN SPACE[††]

3.13 Equilibrium of Particles in Space

The equilibrium of a particle has been discussed in Sec. 3.9. An *equilibrium problem of particles in space* is characterized by the feature that at least one set of *noncoplanar but concurrent* unknown forces acting on a particle will be encountered in the solution. The *necessary and sufficient condition* for equilibrium of a particle in space is still given by the *vector equation of*

[†]Cf. App. D for computer solution of simultaneous equations.

[††]*Equilibrium in space* is equilibrium in three-dimensional space, and is also termed *equilibrium in three dimensions.*

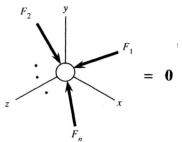

Fig. 3.14

equilibrium in Eq. (3.3); i.e.,

$$\Sigma \mathbf{F} = 0 \qquad (3.3)$$
(repeated)

where $\Sigma \mathbf{F}$ represents the summation (or the resultant) of all the known and unknown noncoplanar but concurrent forces $\mathbf{F}_1$, $\mathbf{F}_2$, . . . , $\mathbf{F}_n$ which appear in the free-body diagram of the particle as illustrated in Fig. 3.14.

For each particle in equilibrium in space, a set of *scalar equations of equilibrium* may be written from Eq. (3.3) as

$$\Sigma F_x = 0 \qquad \Sigma F_y = 0 \qquad \Sigma F_z = 0 \qquad (3.5)$$

The unknowns in the problem are solved from the simultaneous scalar equations in Eqs. (3.5). This is illustrated in Example 3.10, which serves as a forerunner of the method of solution presented in Sec. 3.14.

EXAMPLE 3.10

A 13-Mg crate containing a piece of equipment is being held by three cranes with cables joined at the ring D as shown. Determine the tensions in the cables DA, DB, and DC.

Solution.   We first draw the free-body diagram of the crate and the ring combination as shown. From this, we write

$$\overrightarrow{DA} = 9\mathbf{i} + 12\mathbf{j} \qquad |\overrightarrow{DA}| = 15$$

$$\lambda_{DA} = \frac{1}{5}(3\mathbf{i} + 4\mathbf{j})$$

$$\overrightarrow{DB} = -4\mathbf{i} + 12\mathbf{j} - 6\mathbf{k} \qquad |\overrightarrow{DB}| = 14$$

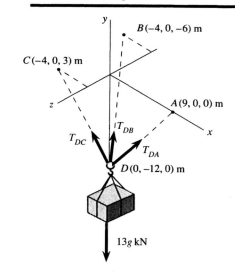

$$\lambda_{DB} = \frac{1}{7}(-2\mathbf{i} + 6\mathbf{j} - 3\mathbf{k})$$

$$\overrightarrow{DC} = -4\mathbf{i} + 12\mathbf{j} + 3\mathbf{k} \qquad |\overrightarrow{DC}| = 13$$

$$\lambda_{DC} = \frac{1}{13}(-4\mathbf{i} + 12\mathbf{j} + 3\mathbf{k})$$

$$\mathbf{T}_{DA} = T_{DA}\lambda_{DA} = \frac{T_{DA}}{5}(3\mathbf{i} + 4\mathbf{j})$$

$$\mathbf{T}_{DB} = T_{DB}\lambda_{DB} = \frac{T_{DB}}{7}(-2\mathbf{i} + 6\mathbf{j} - 3\mathbf{k})$$

$$\mathbf{T}_{DC} = T_{DC}\lambda_{DC} = \frac{T_{DC}}{13}(-4\mathbf{i} + 12\mathbf{j} + 3\mathbf{k})$$

$$\mathbf{W} = W\lambda_W = 13g(-\mathbf{j}) = -13g\mathbf{j}$$

For equilibrium, we write

$$\Sigma\mathbf{F} = \mathbf{T}_{DA} + \mathbf{T}_{DB} + \mathbf{T}_{DC} + \mathbf{W} = \mathbf{0}:$$

$$\left(\frac{3}{5}T_{DA} - \frac{2}{7}T_{DB} - \frac{4}{13}T_{DC}\right)\mathbf{i}$$

$$+ \left(\frac{4}{5}T_{DA} + \frac{6}{7}T_{DB} + \frac{12}{13}T_{DC} - 13g\right)\mathbf{j}$$

$$+ \left(-\frac{3}{7}T_{DB} + \frac{3}{13}T_{DC}\right)\mathbf{k} = \mathbf{0}$$

which entails that

$$\frac{3}{5}T_{DA} - \frac{2}{7}T_{DB} - \frac{4}{13}T_{DC} = 0 \tag{1}$$

$$\frac{4}{5}T_{DA} + \frac{6}{7}T_{DB} + \frac{12}{13}T_{DC} - 13g = 0 \tag{2}$$

$$-\frac{3}{7}T_{DB} + \frac{3}{13}T_{DC} = 0 \tag{3}$$

Solving these three equations, we get

$$T_{DA} = 49.0 \text{ kN} \qquad T_{DB} = 34.3 \text{ kN} \qquad T_{DC} = 63.8 \text{ kN} \quad \blacktriangleleft$$

Developmental Exercise

D3.28   What characterizing feature does an equilibrium problem of particles in space have?

3.14   Method of Solution: Equilibrium in Space

It is true that certain problems can best be solved by employing a special set of steps. However, the following procedure may serve as a guide in *solving equilibrium problems of particles in space*:

1. *Draw the relevant free-body diagram* of a particle where no more than three unknowns will be involved. [This is desirable because no more than three independent equilibrium equations can be obtained in step (4) below.] If this step is impossible, go to step (7) below.
2. *Write the analytical expression of each force*, which may contain known and unknown quantities.
3. *Add the analytical expressions of all of the forces* acting on the particle to obtain the analytical expression of the resultant force in terms of known and unknown quantities.
4. *Set each of the three scalar components* of the resultant force, obtained in step (3), *equal to zero* to obtain a set of three scalar equations.
5. *Solve for the unknowns* in the equations obtained in step (4).
6. *Repeat steps (1) through (5)* for other particles until the desired unknown or unknowns have been determined.
7. If the system consists of *n* particles and *the free-body diagram drawn for any of the n particles in the system involves more than three unknowns*, then we draw the free-body diagram of each particle in the system. We may write three force equilibrium equations for each particle as described in steps (2) through (4) and obtain a total of 3*n* simultaneous equations. If the total number of unknowns equals 3*n*, the unknowns may be determined.

EXAMPLE 3.11

Collars *A* and *B*, each weighing 450 lb, are connected by the wire *AB* and may slide freely on the smooth rod having the shape shown. If the collars are in equilibrium in the position shown, determine the magnitude of the applied force **P**.

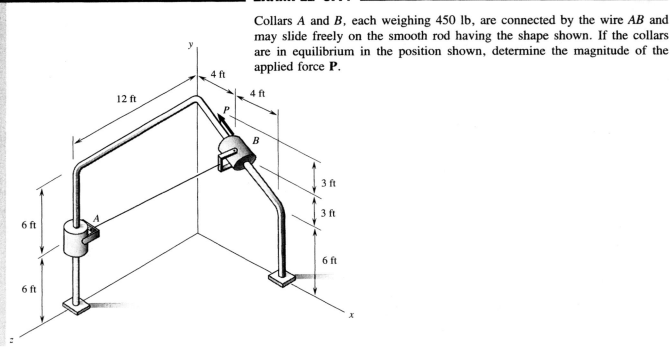

Solution. Since the collars may slide freely on the smooth rod, the reactive force from the rod to a collar must lie somewhere in the plane perpendicular to the portion of the rod at the collar. The free-body diagram of the collar A should be considered first because it contains only three unknowns: the reaction components A_x and A_z and the tension T_{AB}. Since the coordinates of the collars are $A(0, 6, 12)$ ft and $B(4, 9, 0)$ ft, we write

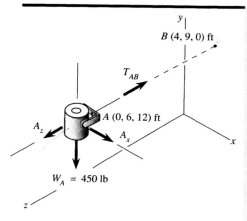

$$\overrightarrow{AB} = 4\mathbf{i} + 3\mathbf{j} - 12\mathbf{k} \qquad \lambda_{AB} = \frac{1}{13}(4\mathbf{i} + 3\mathbf{j} - 12\mathbf{k})$$

$$\mathbf{T}_{AB} = T_{AB}\lambda_{AB} = \frac{T_{AB}}{13}(4\mathbf{i} + 3\mathbf{j} - 12\mathbf{k})$$

$$\mathbf{A} = A_x\mathbf{i} + A_z\mathbf{k} \qquad \mathbf{W}_A = -450\mathbf{j}$$

For equilibrium of *the collar A,* we write

$$\mathbf{R}_A = \Sigma\mathbf{F} = \mathbf{T}_{AB} + \mathbf{A} + \mathbf{W}_A = \mathbf{0}:$$

$$\left(\frac{4}{13}T_{AB} + A_x\right)\mathbf{i} + \left(\frac{3}{13}T_{AB} - 450\right)\mathbf{j} + \left(-\frac{12}{13}T_{AB} + A_z\right)\mathbf{k} = \mathbf{0}$$

Equating the coefficient of $\mathbf{j}$ to zero, we obtain $T_{AB} = 1950$ lb. Thus,

$$\mathbf{T}_{AB} = 600\mathbf{i} + 450\mathbf{j} - 1800\mathbf{k} \text{ lb}$$

The reactive force from the rod to the collar B may be resolved into two rectangular components $\mathbf{B}_{xy}$ and $\mathbf{B}_z$, where $\mathbf{B}_{xy}$ lies in the xy plane and $\mathbf{B}_z$ is parallel to the z axis. Thus, the forces acting on the free body of the collar B are

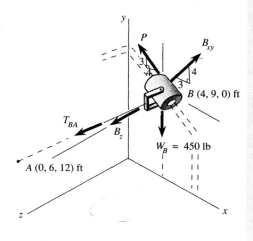

$$\mathbf{T}_{BA} = -\mathbf{T}_{AB} = -600\mathbf{i} - 450\mathbf{j} + 1800\mathbf{k}$$

$$\mathbf{B} = B_{xy}\left(\frac{3}{5}\mathbf{i} + \frac{4}{5}\mathbf{j}\right) + B_z\mathbf{k}$$

$$\mathbf{P} = P\left(-\frac{4}{5}\mathbf{i} + \frac{3}{5}\mathbf{j}\right) \qquad \mathbf{W}_B = -450\mathbf{j}$$

For equilibrium of *the collar B,* we write

$$\mathbf{R}_B = \Sigma\mathbf{F} = \mathbf{T}_{BA} + \mathbf{B} + \mathbf{P} + \mathbf{W}_B = \mathbf{0}:$$

$$\left(-600 + \frac{3}{5}B_{xy} - \frac{4}{5}P\right)\mathbf{i} + \left(-450 + \frac{4}{5}B_{xy} + \frac{3}{5}P - 450\right)\mathbf{j}$$
$$+ (1800 + B_z)\mathbf{k} = \mathbf{0}$$

Solving the three ensuing scalar equations from this vector equation, we obtain

$$B_{xy} = 1080 \text{ lb} \qquad B_z = -1800 \text{ lb} \qquad P = 60 \text{ lb} \blacktriangleleft$$

Note that the negative value for B_z above indicates that the correct sense of $\mathbf{B}_z$ is opposite to that assumed in the free-body diagram.

Developmental Exercises

D3.29 Determine the components of reaction from the rod to the collar A in Example 3.11.

D3.30 If a system of particles in equilibrium in space is to be analyzed and we cannot find a single particle whose free-body diagram involves no more than three unknowns, what should we do?

PROBLEMS

3.46 through 3.49* A body is suspended as shown. Write the analytical expression of the resultant force **R** of the force system acting on the free body, which is obtained by cutting through the supporting cables AB, AC, and AD, in terms of the unknown cable tensions.

Fig. P3.46 and P3.50

Fig. P3.47* and P3.51

Fig. P3.48 and P3.52

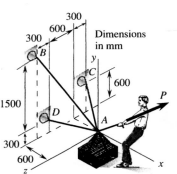

Fig. P3.49* and P3.53*

3.50 through 3.53* Determine the tension in the cable AB shown.

3.54 A 12-kg box is suspended as shown. Determine the applied force **P** which will induce cable tensions $T_{AB} = 70$ N, $T_{AC} = 45$ N, and $T_{AD} = 0$.

Fig. P3.54

Fig. P3.55

Fig. P3.56

Fig. P3.57

3.55 and 3.56   Each of the smooth collars connected as shown has a mass of 140 kg. Determine the magnitude of the horizontal force **P** which must be applied to the collar *A* to maintain equilibrium.

3.57   Three blocks are hung by a system of wires as shown. Determine the tension in each of the wires.[†]

3.15   Concluding Remarks

The central fact in the equilibrium of a particle is the equality to zero of the resultant force of all forces acting on the free body of the particle. The free-body diagram of a body is a sketch which exhibits the system of applied external forces and reactions that act on the body. From such a sketch, we can have a clear geometric view of the situation in which the relations between the known and unknown quantities in the system are to be established. Therefore, *the drawing of free-body diagrams is an important step in solving equilibrium problems*.

The equilibrium of a body acted on by a pair of collinear forces which are equal in magnitude and opposite in sense is simple and virtually trivial. Thus, the first nontrivial fundamental study of equilibrium is the study of the equilibrium of particles in a plane. Three methods of solution for equilibrium problems of particles in a plane have been presented in this chapter: the scalar method, the analytical vector method, and the semigraphical method. In dealing with forces in space, it is often awkward to apply the parallelogram law directly. Thus, we choose to use the analytical vector method in our formulation, analysis, and solution of equilibrium problems of particles in space.

It has been shown in the examples that the solution of some equilibrium problems involving several connected particles may lead to a sizable set of simultaneous equations whose solution could present a challenge. However, with the widespread availability of personal as well as mainframe computers to most engineers, such a challenge can readily be met and overcome. One

---

[†]Cf. App. D for computer solution of simultaneous equations.

may find the BASIC computer program in App. D useful. This program contains "directions" and "caution" in its beginning REMark section and automatically prints out the value of the determinant of the coefficient matrix as well as the checking of the solution by back substitution in the output. These features are particularly useful in helping beginners of statics watch out for the mistake of including "dependent" equations in formulating the solution.

Equilibrium problems of rigid bodies acted on by nonconcurrent force systems are not within the province of equilibrium of particles. The study of such problems requires other concepts (e.g., moments of forces) and is treated in later chapters.

REVIEW PROBLEMS

3.58 Determine the magnitude of the force **P** required to hold the 100-kg crate A in equilibrium as shown.

3.59 Determine the weight of the cart A if the system shown is in equilibrium.

Fig. P3.58

Fig. P3.59

Fig. P3.60

3.60 An elastic cord is held in equilibrium by a 59-lb force **P** as shown. When the cord is stretched directly between the supports A and B, the tension is 50.7 lb. Determine for the cord (a) the spring modulus k, (b) the free length L.

3.61* A 50-kg cart is held in equilibrium as shown. Determine the tension T_{AC} in the cable AC.

3.62* Two cylinders are hung by a system of wires and small rings as shown. Determine the tension in each wire.[†]

Fig. P3.61*

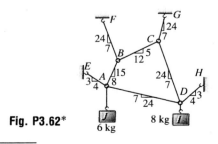

Fig. P3.62*

[†]Cf. App. D for computer solution of simultaneous equations.

3.63 Three cables DA, DB, and DC are used to tie down a balloon at D as shown. If the balloon exerts an upward force of 640 N at D, determine the tension in each cable.

Fig. P3.63

Fig. P3.64*

3.64* A 310-lb block on a smooth incline is held in equilibrium by cables AB and AC as shown. Determine the tensions T_{AB} and T_{AC} in the cables.

3.65 Two smooth collars A and B, each having a weight of 36 lb, are connected by a cable. A force **P** is applied to the collar A as shown to maintain equilibrium. Determine the magnitude P.

Fig. P3.65

Fig. P3.66

3.66 A 2.6-Mg H-shaped slab is suspended by cables and pulleys as shown. Determine the tension T_{DC} in the cable DC.

Chapter 4

Moments of Forces

KEY CONCEPTS

- Moment of a force about an axis, moment of a force about a point, moment center, moment arm, right-hand rule for moment vectors, Varignon's theorem, torque, cross product, dot product, and scalar triple product.
- Moment of a system of forces about a point.
- Moment of a couple and the addition of moments of couples.
- Moment of a system of forces about an axis.
- Shortest distance between an axis and a force.

This chapter presents the concepts and methods of computation of moments of forces as well as their applications. The moment of a force acting on a body about an axis is a measure of the effectiveness or tendency of the force to rotate the body about that axis. The turning, twisting, or bending done to a body is an effect directly related to the moments of forces acting on the body about the axis with respect to which the body is turned, twisted, or bent. The computation of moments of forces using scalar algebra is presented first so that the physical implication of the moment of a force may clearly be comprehended in the early stage of the development. Operations involving cross and dot products of vectors are later presented and utilized as an effective tool to handle moments of forces and their applications in three-dimensional space. A good understanding of moments of forces is a necessary step toward a meaningful and successful study of the conditions of rigid bodies subjected to the action of force systems.

SCALAR FORMULATION

4.1 Moment of a Force

Besides the tendency to move a body in the direction of its action, a force acting on a body has also a tendency to rotate the body about an axis, which does not intersect the line of action of the force and is not parallel to the force. The *moment of a force about an axis* is an action reflecting the effectiveness or tendency of the force to turn (or rotate) a body about that axis; such a turning action should be well understood early by all students of mechanics. Observations show that the moment of a force about an axis is *zero* if any of the following conditions is true: (a) the line of action of the force *intersects the axis*, (b) the force is *parallel to the axis*, (c) the force itself is *zero*.

Let us consider a rigid slab acted on by a force **F** in the plane of the slab as shown in Fig. 4.1. The effectiveness or tendency of **F** to rotate the slab about the axis *AB* normal to the plane of the slab is, clearly, proportional to the magnitude F of the force **F** and the shortest distance d_s between the axis *AB* and the line of action of **F**. The axis about which the moment of a force is computed is referred to as the *moment axis*. Thus, the *magnitude* of the moment $\mathbf{M}_{AB}$ of **F** about the moment axis *AB* is given by

$$M_{AB} = d_s F \tag{4.1}$$

Note that the shortest distance d_s between the line of action of the force **F** and the moment axis *AB*, as indicated in Fig. 4.1, is referred to as the *moment arm* of **F** about the axis *AB*.

To distinguish a moment vector from a force vector, we use a *hollow arrow* to represent the moment $\mathbf{M}_{AB}$ as shown in Fig. 4.1.[†] The *direction* of $\mathbf{M}_{AB}$ is defined to be the same as that of the extended thumb of the right hand when the other fingers of the right hand are curled in the direction of the tendency to rotate. This definition is elaborated upon in Sec. 4.2.

The phrase *moment of a force about a point* is frequently used in situations where the forces and the points under consideration all lie in a plane. This phrase is commonly understood to mean "moment of a force about an axis which passes through the point and is perpendicular to the plane containing the point and the line of action of the force." In Fig. 4.1, the axis *AB* is perpendicular to the plane of the slab and intersects it at the point *C*. The point about which the moment of a force is computed is referred to as the *moment center* of the force. Thus, the moment $\mathbf{M}_C$ of the force **F** about the moment center *C* is the same as the moment $\mathbf{M}_{AB}$ of **F** about the moment axis *AB*. Besides representation by a hollow arrow like $\mathbf{M}_{AB}$, the moment $\mathbf{M}_C$ is often represented graphically by a *curved arrow* pointing in the direction of the tendency of the moment to rotate about the moment center *C* as shown in Fig. 4.1. The *magnitude* of $\mathbf{M}_C$ is, of course, identical with that defined in Eq. (4.1); i.e.,

$$M_C = d_s F \tag{4.2}$$

In this case, d_s is also referred to as the *moment arm* of **F** about *C*.

The moments of coplanar forces about any given point in the plane of the forces are vectors all perpendicular to the plane of the forces and are hence parallel vectors. The senses of these parallel moment vectors point outward from the plane if they are counterclockwise moments, and inward to the plane if they are clockwise moments. Thus, the directions of these parallel moment vectors may be accounted for by letting *counterclockwise* moments be *positive* and *clockwise* moments be *negative*. Such a sign

Fig. 4.1

[†]A moment vector is also represented in other literature by a straight arrow with a curved arrow around it; e.g., $\leftarrow\!\!\psi = \Uparrow$.

convention for parallel moment vectors is generally used, although it may be defined the other way around. The addition (and hence the computation of the resultant moment) of parallel moment vectors may, therefore, be accomplished with *scalar algebra*.

From Eqs. (4.1) and (4.2), it is obvious that the moment of a force has the dimensions *length* × *force*. The primary units for the moment of a force are newton-meters (N·m) in the SI and pound-feet (lb·ft) in the U.S. customary system.

Developmental Exercises

D4.1 Define the terms: (a) moment of a force about an axis, (b) moment axis, (c) moment arm, (d) moment of a force about a point, (e) moment center.

D4.2 What are the conditions under which the moment of a force about an axis is equal to zero?

D4.3 How is the moment of a force represented in a figure so that it can readily be distinguished from a force?

D4.4 What are the primary units for the moment of a force in (a) the SI, (b) the U.S. customary system?

4.2 Moment Vectors and the Right-Hand Rule

In general, the *moment* $\mathbf{M}_Q$ *of a force* $\mathbf{F}$ *about a point* Q in space is defined as the *moment* $\mathbf{M}_{QB}$ *of* $\mathbf{F}$ *about the axis* QB which passes through Q and is perpendicular to the plane containing Q and the line of action of $\mathbf{F}$; i.e., $\mathbf{M}_Q = \mathbf{M}_{QB}$. The geometry involved in this definition is illustrated in Fig. 4.2, where the point P is the point of application of $\mathbf{F}$, the point A is any convenient point on the line of action of $\mathbf{F}$, $\mathbf{r}_{QA}$ is the position vector from Q to A, θ is the angle between the positive directions of $\mathbf{r}_{QA}$ and $\mathbf{F}$, and d_s is the shortest distance between Q and the line of action of $\mathbf{F}$. Note that $\mathbf{M}_Q$ is

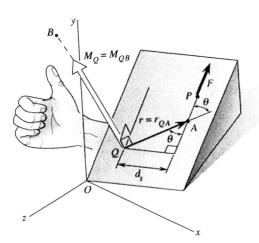

Fig. 4.2

perpendicular to both $\mathbf{r}_{QA}$ and $\mathbf{F}$. The *magnitude* of $\mathbf{M}_Q$ may be written as

$$M_Q = d_s F = (r\sin\theta)F \qquad (4.3)$$

The *sense* of $\mathbf{M}_Q$ is defined to point from the knuckle to the nail of the extended thumb of the right hand when the other fingers of the right hand are curled in the direction of the tendency of the moment of the force to rotate about the axis QB. This is called the *right-hand rule for moment of a force about a point*. Physically, the *magnitude* of $\mathbf{M}_Q$ measures the effectiveness or tendency of the force $\mathbf{F}$ to impart to a rigid body a rotational motion about an axis which passes through Q and is parallel to the direction of $\mathbf{M}_Q$.

If a shaft is *turned* by a *single force* $\mathbf{F}$, the shaft will, at the same time, be bent by $\mathbf{F}$. The turning action of $\mathbf{F}$ about the shaft is the moment of $\mathbf{F}$ about the shaft. The illustration in Fig. 4.3 shows that the moment of the force $\mathbf{F}$ at A about the shaft CD is $d_s F$ pointing to the left along the shaft, while the moment of the force $\mathbf{F}'$ at B about the shaft CD is $d_s' F'$ pointing to the right along the shaft. Since these two moments are parallel to each other, we may say that the resultant moment of $\mathbf{F}$ and $\mathbf{F}'$ about the shaft CD is either $(d_s F - d_s' F')$ pointing to the left or $(d_s' F' - d_s F)$ pointing to the right along the shaft.

Fig. 4.3

Fig. 4.4

The right-hand rule for moment vectors described above may equivalently be stated in one of the following two additional versions:

1. When the extended four fingers of the right hand point along the line of action of the force $\mathbf{F}$ and the palm of the right hand faces the point Q, the extended thumb of the right hand will point in the direction of the moment vector $\mathbf{M}_Q$ of $\mathbf{F}$ about the moment center Q. This is illustrated in Fig. 4.4.
2. If the orientation of the axis of a right-hand threaded screw at the point Q is the same as the orientation of the moment vector $\mathbf{M}_Q$ which acts on (or twists) the screw head at the point Q, the screw will advance along its axis in the same sense as that of $\mathbf{M}_Q$. This is illustrated in Fig. 4.5.

Fig. 4.5

The above equivalent versions of the right-hand rule for moment vectors are very handy because they may readily help us decide which way we should apply our moment to a right-hand threaded screw, bolt, lug nut, oil filter, etc. when we want to tighten or loosen it.

Developmental Exercises

D4.5   Describe two versions of the right-hand rule for moment vectors.

D4.6   A wrench is attached to an ordinary bolt and pushed by a horizontal force **F** as shown. (a) Which way does the bolt advance or tend to advance? (b) What is the physical implication of the magnitude of the moment M_Q of **F** about Q?

Fig. D4.6

4.3   Varignon's Theorem: Forces in a Plane

The moment of a force is often determined indirectly from the moments of its components by using the so-called *Varignon's theorem*, after the French mathematician Pierre Varignon (1654–1722).[†] *Varignon's theorem* states that *the moment of a force about any point is equal to the sum of the moments of its components about the same point*. This theorem is here proved for a force which is resolved into only two components at a point.[††]

Suppose that the force **R** acting at the point O is resolved into two components **A** and **B** as shown in Fig. 4.6. The moment arms of **A**, **B**, and **R** drawn from the moment center P are a, b, and r as shown. Letting the y axis be directed along the line OP and projecting **A**, **B**, and **R** onto the x axis, we note that $\overline{CE} = \overline{OD}$. We find from the geometry in this figure that the moment of **R** about the moment center P is

$$rR = (\overline{OP}\sin\gamma)R = (R\sin\gamma)\overline{OP} = \overline{OE}(\overline{OP}) = (\overline{OC} + \overline{CE})\overline{OP}$$

$$= (\overline{OC} + \overline{OD})\overline{OP} = (A\sin\alpha + B\sin\beta)\overline{OP}$$

$$= (\overline{OP}\sin\alpha)A + (\overline{OP}\sin\beta)B = aA + bB$$

This proves that the moment of **R** about any point P is equal to the sum of the moments of its two components **A** and **B** about the same point P. In fact, Varignon's theorem applies equally well to the case where a force has three or more components, since the number of components can always be reduced to two by direct combination. We shall note in Sec. 6.2 that Varignon's theorem is a special case of the *principle of moments*.

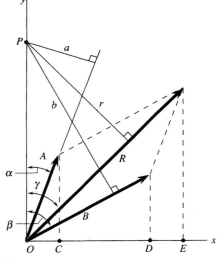

Fig. 4.6

[†]Varignon's major contribution to mechanics was his book *Projet d'uné nouvelle méchanique*, published in 1687 (the same year as Newton's *Principia*) and dedicated to the French Academy of Sciences. Cf. Sec. 4.18 for additional comments.

[††]Varignon's theorem for forces in space is presented in Sec. 4.11.

For a *moment vector* perpendicular to the *xy* plane, we often use a curved arrow in the *xy* plane to show its direction according to the right-hand rule for moment vectors; this is a *geometric notation* for moment vectors. For instance, if $\mathbf{M}_A = 200\mathbf{k}$ N·m and $\mathbf{M}_B = -300\mathbf{k}$ N·m, we may write

$$\mathbf{M}_A = 200 \text{ N}\cdot\text{m} \circlearrowright \qquad \mathbf{M}_B = 300 \text{ N}\cdot\text{m} \circlearrowleft$$

Furthermore, a vector quantity in a plane is often expressed by indicating its magnitude with a number (or a symbol) and its direction with an arrow sign at a slope or an angle. Such a notation is a *geometric* or *polar notation* for vectors in a plane. For instance, if $\mathbf{r} = -8\mathbf{i} + 15\mathbf{j}$ m, $\mathbf{F} = 3\mathbf{i} + 4\mathbf{j}$ kN, $\mathbf{P} = 5\mathbf{i} - 6\mathbf{j}$ kN, and $\mathbf{Q} = -50\mathbf{i} - 120\mathbf{j}$ lb, we may write

$$\mathbf{r} = 17 \text{ m} \quad \text{or} \quad \mathbf{r} = 17 \text{ m} \measuredangle 61.9°$$

$$\mathbf{F} = 5 \text{ kN} \quad \text{or} \quad \mathbf{F} = 5 \text{ kN}\measuredangle 53.1°$$

$$\mathbf{P} = 7.81 \text{ kN} \quad \text{or} \quad \mathbf{P} = 7.81 \text{ kN}\measuredangle 50.2°$$

$$\mathbf{Q} = 130 \text{ lb} \quad \text{or} \quad \mathbf{Q} = 130 \text{ lb}\measuredangle 67.4°$$

It is often preferable to work with the slope triangles as given if the corresponding angles are not explicitly required.

EXAMPLE 4.1

A 500-N force $\mathbf{F}$ is applied to a bracket fixed on a wall as shown. Determine the moment $\mathbf{M}_A$ of $\mathbf{F}$ about the point A.

Solution. The system is coplanar. We may take advantage of its simplicity and solve it in four different ways for illustrative purposes.

(a) *Using shortest distance.* The moment $\mathbf{M}_A$ may be determined by first calculating its *magnitude* $d_s F$ and then using the right-hand rule to ascertain its *direction*. From the sketch shown, we write

$$\theta_1 = \tan^{-1}\left(\frac{0.5}{1.2}\right) = 22.62° \qquad \theta_2 = \tan^{-1}\left(\frac{4}{3}\right) = 53.13°$$

$$\overline{AB} = [(1.2)^2 + (0.5)^2]^{1/2} = 1.3 \qquad d_s = \overline{AB}\sin(\theta_1 + \theta_2) = 1.26$$

$$M_A = -d_s F = -1.26(500) \qquad M_A = 630 \text{ N}\cdot\text{m} \circlearrowleft \blacktriangleleft$$

Note that the negative value for M_A indicates that $\mathbf{M}_A$ is clockwise.

(b) *Using Varignon's theorem at B.* We first resolve the 500-N force $\mathbf{F}$ at B into two components as shown. By Varignon's theorem and the right-hand rule, we write

$$M_A = -0.5(300) - 1.2(400) \qquad M_A = 630 \text{ N}\cdot\text{m} \circlearrowleft \blacktriangleleft$$

(b)

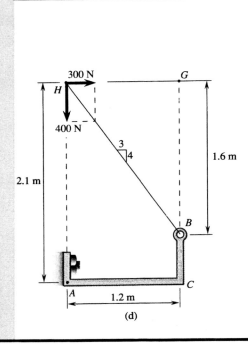

(d)

(c)

(c) *Using Varignon's theorem at E.* We apply the principle of transmissibility to transmit **F** to E and then resolve it there into two components as shown. Note that $\overline{CE} = (3/4)(0.5)$ m $= 0.375$ m. By Varignon's theorem and the right-hand rule, we write

$$M_A = 0(300) - 1.575(400) \qquad \mathbf{M}_A = 630 \text{ N} \cdot \text{m} \downarrow \blacktriangleleft$$

(d) *Using Varignon's theorem at H.* We apply the principle of transmissibility to transmit **F** to H and then resolve it there into two components as shown. Note that $\overline{BG} = (4/3)(1.2)$ m $= 1.6$ m. By Varignon's theorem and the right-hand rule, we write

$$M_A = -2.1(300) + 0(400) \qquad \mathbf{M}_A = 630 \text{ N} \cdot \text{m} \downarrow \blacktriangleleft$$

Developmental Exercises

D4.7 Describe Varignon's theorem.

D4.8 Express the following vectors in geometric notations: (a) $\mathbf{F} = 24\mathbf{i} - 7\mathbf{j}$ kN, (b) $\mathbf{r} = -200\mathbf{i} + 300\mathbf{j}$ m, (c) $\mathbf{M}_A = 4\mathbf{k}$ kN·m, (d) $\mathbf{C} = -700\mathbf{k}$ lb·ft.

EXAMPLE 4.2

The two forces acting on the bracket as shown have the same unit. Determine the resultant moment $\mathbf{M}_A^R$ of these two forces about the point A in terms of known and unknown quantities.

Solution. We first resolve the 260-lb force at C into two components as shown. By Varignon's theorem and the right-hand rule, we write

$$M_A^R = -2P + 3(240) + 4(100) = 1120 - 2P$$
$$\mathbf{M}_A^R = (1120 - 2P) \text{ lb} \cdot \text{ft} \downarrow \blacktriangleleft$$

Developmental Exercises

D4.9 Determine the moment M_A of the force shown (see right) about A.

D4.10 Determine the resultant moment M_A^R of the two forces shown (below) about A.

Fig. D4.10

Fig. D4.9

4.4 Moment of a Couple: Torque

A *couple* is a pair of parallel forces equal in magnitude, opposite in direction, and having different lines of action, such as $\mathbf{F}$ and $-\mathbf{F}$ shown in Fig. 4.7. Since $\mathbf{F} + (-\mathbf{F}) = \mathbf{0}$, a couple can have no translational (i.e., pushing or pulling) effect on a body. However, a couple does have a moment and a purely torsional (i.e., twisting) effect on a body. The *moment of a couple* about any point is defined as the sum of the moments of the two forces in the couple about that point.

Fig. 4.7

Fig. 4.8

Since only a couple can exert a purely torsional action on a shaft, the *moment of a couple* is often called the *torque of the couple*, or simply a *torque*.[†] Let a rigid slab be acted on by a couple consisting of $\mathbf{F}$ and $-\mathbf{F}$ in the plane of the slab as shown in Fig. 4.8(a). Then, the *magnitude* of the moment $\mathbf{M}_O$ of the couple about any chosen point O, as shown, is

$$M_O = \overline{OB}(F) - \overline{OA}(F) = (\overline{OB} - \overline{OA})(F) = \overline{AB}(F) = d_s F$$

where d_s is the shortest distance between the lines of action of $\mathbf{F}$ and $-\mathbf{F}$. Thus, the value of M_O contains no dimension which relates the moment center O to the couple. It follows that the moment of a couple has the same value for *all* moment centers and is, therefore, a *free vector*. It is logical that we drop the subscript O in $\mathbf{M}_O$ and write the *magnitude* of the moment of the couple as

$$M = d_s F \tag{4.4}$$

[†]The term *torque* is usually used to denote the moment of a couple and the moment about the axis of a shaft.

The moment **M** of a couple consisting of **F** and −**F** is a vector perpendicular to the plane containing **F** and −**F**. This vector points in the *direction* of the extended thumb of the right hand when the other fingers of the right hand are curled in the direction of the tendency of the couple to rotate. The moment **M** is illustrated in Fig. 4.8(b) by a *hollow arrow* and in Fig. 4.8(c) by a *curved arrow* pointing in the direction of the tendency of the couple to rotate.

In coplanar systems, torques of couples tending to rotate counterclockwise and clockwise are usually represented as illustrated in Fig. 4.9. Note that the *torque*, or *moment of a couple*, is a *vector* and is an *attribute* of the couple.[†]

Counterclockwise torque   Clockwise torque

Fig. 4.9

EXAMPLE 4.3

A force system is applied to a cantilever beam as shown. Determine the resultant moment $\mathbf{M}_A^R$ of the force system about the point A.

Solution. We note that the force system shown consists of a force and two torques, which are free vectors. Thus, referring to the given space diagram, we write

$$M_A^R = -400 - 10(200) + 300 \qquad M_A^R = -2100 \text{ lb} \cdot \text{ft}$$

$$\mathbf{M}_A^R = 2.1 \text{ kip} \cdot \text{ft} \, \curvearrowright \quad \blacktriangleleft$$

REMARK. In the above computation, only the 200-lb force is to be multiplied by its moment arm of 10 ft; *no* moment arms are used in computing the moments of the 400-lb · ft and 300-lb · ft torques about the point A.

[†]Since a couple is composed of two opposite forces which are noncollinear and have equal magnitudes, some mechanicists feel that the *moment of the couple*, rather than the *couple*, is a vector. Perhaps, a source contributing to the dispute is the fact that the terms *moment* and *couple* have been taken as interchangeable. For further discussion of this and other basic concepts in mechanics, see T. R. Kane, "Teaching of Mechanics to Undergraduates," *The International Journal of Mechanical Engineering Education (IJMEE)*, Vol. 6, No. 4, 1978, pp. 187–189.

Developmental Exercises

D4.11 Define (a) a couple, (b) the moment of a couple, (c) a torque.

D4.12 Why is the moment of a couple a free vector?

D4.13 How are couples represented in coplanar systems?

D4.14 In the system shown, A_x and A_y are in newtons and M_A is in newton-meters. Determine the resultant moment of the force system about (a) the point A, (b) the point B.

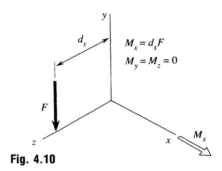

Fig. D4.14

4.5 Moments about Coordinate Axes

When the line of action of a force **F** is parallel to, or intersects, a coordinate axis, the moment of **F** about the coordinate axis is zero. On the other hand, when the line of action of **F** is perpendicular to a coordinate axis, the moment of **F** about that axis has a magnitude equal to $d_s F$ where F is the magnitude of **F** and d_s is the shortest distance between that axis and the line of action of **F**. This is illustrated in Fig. 4.10. If a moment points to the positive direction of the coordinate axis, it is a positive moment; otherwise, it is a negative moment.

If the force **F** is neither parallel nor perpendicular to a coordinate axis, we may first resolve **F** into rectangular components parallel to the coordinate axes. The moment of **F** about an axis is, by Varignon's theorem, equal to the sum of the moments of the rectangular components of **F** about that axis. Note that only those components which are perpendicular to the axis may contribute moments about that axis.

Fig. 4.10

EXAMPLE 4.4

A 350-lb force **F** is applied to a plate along the direction of the line AB as shown. Determine the moments M_x, M_y, and M_z of **F** about the coordinate axes x, y, and z, respectively.

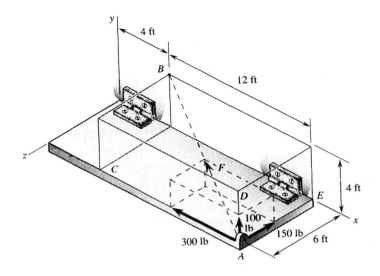

Solution.   From the plate shown above, we have

$$\overline{AB} = [(12)^2 + (4)^2 + (6)^2]^{1/2} = 14$$

The scalar components of the 350-lb force **F** are simply the projections of **F** onto the directions parallel to the *xyz* axes. By proportions of the two parallelepipeds as sketched, we write

$$F_x = -\frac{\overline{AC}}{\overline{AB}} F = -\frac{12}{14} (350) = -300$$

$$F_y = \frac{\overline{AD}}{\overline{AB}} F = \frac{4}{14} (350) = 100$$

$$F_z = -\frac{\overline{AE}}{\overline{AB}} F = -\frac{6}{14} (350) = -150$$

Thus, **F** is resolved into three rectangular components at *A* as indicated. Applying Varignon's theorem and the right-hand rule, we write

$$M_x = 0 - 6(100) + 0 \qquad\qquad M_x = -600 \text{ lb} \cdot \text{ft} \blacktriangleleft$$

$$M_y = -6(300) + 0 + 16(150) \qquad M_y = 600 \text{ lb} \cdot \text{ft} \blacktriangleleft$$

$$M_z = 0 + 16(100) + 0 \qquad\qquad M_z = 1600 \text{ lb} \cdot \text{ft} \blacktriangleleft$$

Note that M_x points to the negative direction of the *x* axis, while M_y and M_z point to the positive directions of the *y* and *z* axes, respectively.

EXAMPLE 4.5

The free body of a plate is acted on by a force system as shown, where the unknown forces are expressed in kilonewtons and the unknown moments in kilonewton-meters. Determine the resultant moments M_x^R, M_y^R, and M_z^R of the force system about the *x*, *y*, and *z* axes, respectively.

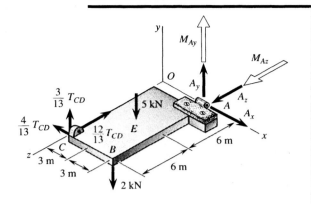

Solution. Similar to the solution in Example 4.4, we first resolve the force $\mathbf{T}_{CD}$ into three rectangular components at C as shown. By Varignon's theorem and the right-hand rule, we write

$$M_x^R = -12\left(\frac{3}{13} T_{CD}\right) + 6(5) + 12(2)$$

$$M_x^R = -\frac{36}{13} T_{CD} + 54 \quad \text{kN} \cdot \text{m} \quad \blacktriangleleft$$

$$M_y^R = M_{Ay} - 6A_z - 12\left(\frac{4}{13} T_{CD}\right)$$

$$M_y^R = M_{Ay} - 6A_z - \frac{48}{13} T_{CD} \quad \text{kN} \cdot \text{m} \quad \blacktriangleleft$$

$$M_z^R = M_{Az} + 6A_y - 3(5) - 6(2)$$

$$M_z^R = M_{Az} + 6A_y - 27 \quad \text{kN} \cdot \text{m} \quad \blacktriangleleft$$

Developmental Exercises

D4.15 Determine the moments M_x, M_y, and M_z of the 180-lb force about the x, y, and z axes as shown.

D4.16 Refer to Example 4.5. Determine the moments M_{AB} and M_{CB} of the force system about the axes along AB and CB, respectively.

Fig. D4.15

PROBLEMS

4.1 through 4.4   A force **F** acts on the body as shown. Determine the moment $\mathbf{M}_O$ of **F** about the point O.

70 mm

140 mm

60°

$F = 100$ N

IB 250 ACME

O

Fig. P4.1

3 ft

O

1 ft

$F = 260$ lb

5
12

Fig. P4.2

$F = 150$ N

200 mm

75 mm

75 mm

70°

O

560 mm

Fig. P4.3*

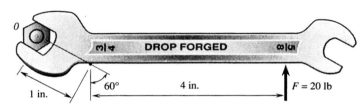

O

DROP FORGED

60°   4 in. $F = 20$ lb

1 in.

Fig. P4.4

4.5 through 4.7*   Determine the resultant moment $\mathbf{M}_A^R$ of the two forces about the point A as shown.

100 lb

60 lb

4
3

A

2 ft   3 ft

Fig. P4.5

800 N

400 N

1 m

60°

A

1 m   1 m

Fig. P4.6

2 m

250 N

1 m

A

3 m

340 N

15
8

Fig. P4.7*

3 kN   6 kN

A_x A B

A_y 2 m   2 m   2 m P

60°

Fig. P4.8 and P4.11

4.8 through 4.10   A free body is acted on by a system of five forces as shown. The unknown forces, as labeled, and the known forces have the same unit. Determine the resultant moment $\mathbf{M}_A^R$ of the force system about the point A.

1 kN   4 kN

A_x

A_y A

3 m

B

4
3
P

2 m   2 m

Fig. P4.9* and P4.12

200 lb

3 ft

A B

3
4   B_x

P 650 lb B_y

4 ft   4 ft

Fig. P4.10 and P4.13*

4.11 through 4.13* A free body is acted on by a system of five forces as shown. The unknown forces, as labeled, and the known forces have the same unit. Determine the resultant moment $\mathbf{M}_B^R$ of the force system about the point B.

4.14 For the system shown, determine (a) the angle θ for which the moment $\mathbf{M}_A$ of the 200-N force about the point A has a maximum magnitude, (b) the maximum magnitude of $\mathbf{M}_A$.

4.15* The resultant moment of the two forces about the point A as shown is zero. Determine the magnitude of $\mathbf{P}$.

Fig. P4.14

Fig. P4.15*

Fig. P4.16

4.16 The resultant moment of the two forces about the point A as shown is zero. Determine the angle θ.

4.17 through 4.19* A force system is applied to a body cantilevered at one end as shown. Determine the resultant moment $\mathbf{M}_A^R$ of the force system about the end at A.

Fig. P4.17

Fig. P4.18

Fig. P4.19*

4.20 through 4.22* A free body is acted on by a force system as shown, where the unknown and known forces have the same unit. Determine the resultant moment $\mathbf{M}_A^R$ of the force system about the point A.

Fig. P4.20 and P4.23*

Fig. P4.21 and P4.24

Fig. P4.22* and P4.25

4.23* through 4.25   A free body is acted on by a force system as shown, where the unknown and known forces have the same unit. Determine the resultant moment $\mathbf{M}_B^R$ of the force system about the point B.

Fig. P4.26

Fig. P4.27*

Fig. P4.28

4.26 through 4.28   Determine the moments M_x, M_y, and M_z of the force $\mathbf{F}$ shown about the coordinate axes x, y, and z, respectively.

4.29 through 4.32   A free body is acted on by a force system as shown, where the unknown and known forces have the same unit, and the unknown and known moments have the same units. Determine the resultant moments M_x^R, M_y^R, and M_z^R of the force system about the x, y, and z axes, respectively.

Fig. P4.29

Fig. P4.30

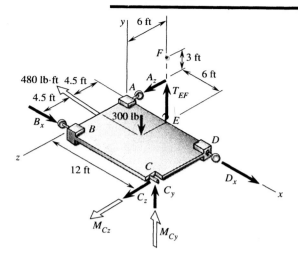

Fig. P4.31*   **Fig. P4.32**

APPLICATION OF CROSS PRODUCTS

Note. Those who are familiar with the rudiments of cross products as presented in Secs. 4.6 through 4.9 may proceed directly to Sec. 4.10.

4.6 Trihedrals

A *trihedral* is a figure formed by three noncoplanar line segments which intersect at a common point. This is illustrated by the three line segments $\overline{OA}$, $\overline{OB}$, and $\overline{OC}$ as shown in Fig. 4.11. A *directed trihedral* is a trihedral formed by three directed line segments (or three vectors). The trihedral formed by the x, y, and z axes of a rectangular coordinate system, as shown in Fig. 4.12, is a directed trihedral and is called the *coordinate trihedral*.

The three concurrent vectors **A**, **B**, **C** are said to form a *right-handed trihedral* if, when the thumb of the right hand extends from the point of concurrence along the first vector **A**, the other fingers of the right hand can curl only in the direction in which the second vector **B** could be rotated through an angle of less than 180° to coincide with the third vector **C**. This is illustrated in Fig. 4.13(a). If the vectors **A**, **B**, **C** (in the order **A**, **B**, **C**)

Fig. 4.11

Fig. 4.12

(a)   (b)   (c)

Fig. 4.13

form a right-handed trihedral, then it must be true that these vectors in the order **B**, **C**, **A** as well as in the order **C**, **A**, **B** also form a right-handed trihedral, as shown in parts (b) and (c) of Fig. 4.13. A trihedral which is not right-handed is called a left-handed trihedral. In general, coordinate trihedrals used in mechanics are right-handed trihedrals.

Developmental Exercises

D4.17 Define the terms: (a) trihedral, (b) coordinate trihedral, (c) right-handed trihedral.

D4.18 Which of the sets of vectors in the order **A**, **B**, **C** form right-handed trihedrals?

Fig. D4.18

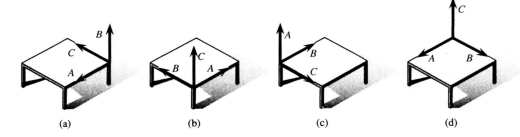

(a)   (b)   (c)   (d)

4.7   Cross Product of Two Vectors

Two basic ways for multiplying two vectors are employed in this text. The first way is presented here and the other in Sec. 4.14. The *cross product* of two vectors **P** and **Q** is defined as a vector **V** which has the following characteristics: the *magnitude* of **V** is equal to the product of the magnitudes of **P** and **Q** and the sine of the angle θ between them; the *orientation* of **V** is perpendicular to the plane of **P** and **Q**; and the *sense* of **V** is such that the three vectors **V**, **P**, **Q** (in this order) form a right-handed trihedral. Without losing generality, the cross product of **P** and **Q** may be illustrated as shown in Fig. 4.14. As in vector analysis, we write the above cross product as

$$\mathbf{V} = \mathbf{P} \times \mathbf{Q} \tag{4.5}$$

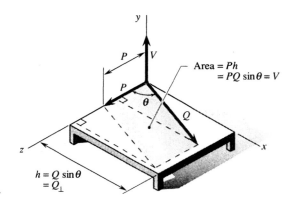

Fig. 4.14

which is read "**V** equals **P** cross **Q**." Because of the nature of the product, the cross product is also called the *vector product*.

By definition, the magnitude of $\mathbf{P} \times \mathbf{Q}$ in Eq. (4.5) is given by

$$|\mathbf{P} \times \mathbf{Q}| = PQ \sin\theta \qquad (4.6)$$

where θ is the angle between the positive directions of **P** and **Q** and $0 \le \theta \le 180°$. Writing the magnitude of $\mathbf{P} \times \mathbf{Q}$ as $P(Q \sin\theta)$, we readily obtain from Fig. 4.14 a simple *geometric interpretation* that the magnitude of the cross product of **P** and **Q** is equal to the area of the parallelogram, or twice the area of the triangle formed by the vectors **P** and **Q** as the adjacent sides. On the other hand, if **P** is a unit vector (i.e., $P = 1$), then $|\mathbf{P} \times \mathbf{Q}| = P(Q \sin\theta) = Q \sin\theta = h =$ the magnitude of the *perpendicular component* $\mathbf{Q}_\perp$ of **Q** with respect to **P**. Furthermore, when $\mathbf{P} \times \mathbf{Q} = \mathbf{0}$, it implies one or more of the following cases: (a) $\mathbf{P} = \mathbf{Q}$ (where $P = Q$ and $\theta = 0$), (b) **P** is parallel to **Q** (where $\theta = 0$ or $180°$), (c) **P** is a zero vector (i.e., $P = 0$), (d) **Q** is a zero vector (i.e., $Q = 0$).

Applying the definition of the cross product, we may readily obtain the cross products of the various possible pairs of the cartesian unit vectors **i**, **j**, and **k**, as shown in Fig. 4.15. These cross products are

$$
\begin{array}{lll}
\mathbf{i} \times \mathbf{i} = \mathbf{0} & \mathbf{i} \times \mathbf{j} = \mathbf{k} & \mathbf{i} \times \mathbf{k} = -\mathbf{j} \\
\mathbf{j} \times \mathbf{i} = -\mathbf{k} & \mathbf{j} \times \mathbf{j} = \mathbf{0} & \mathbf{j} \times \mathbf{k} = \mathbf{i} \\
\mathbf{k} \times \mathbf{i} = \mathbf{j} & \mathbf{k} \times \mathbf{j} = -\mathbf{i} & \mathbf{k} \times \mathbf{k} = \mathbf{0}
\end{array} \qquad (4.7)
$$

It may be noted from the identities in Eqs. (4.7) that the cross product of any two of the three unit vectors **i**, **j**, **k** is equal to positive or negative of the remaining third unit vector depending on whether or not they, in their order, form a right-handed trihedral. A useful mnemonic device for this rule may be constructed as shown in Fig. 4.16.

Developmental Exercises

D4.19 What is the value of $\mathbf{P} \times (s\mathbf{P})$ where s is a scalar quantity?

D4.20 A 300-lb force **Q** is shown. For $\mathbf{P} = \overrightarrow{OA} = 5\mathbf{i}$ ft and $\mathbf{V} = \mathbf{P} \times \mathbf{Q}$, determine (a) the angle θ between **P** and **Q**, (b) the magnitude V, (c) the unit vector which points in the same direction as **V**, and (d) the analytical expression of **V**.

D4.21 Discuss the cases in which $\mathbf{P} \times \mathbf{Q} = \mathbf{0}$.

D4.22 Find the cross products of the various possible pairs of the three cartesian unit vectors **i**, **j**, **k**.

4.8 Properties of Cross Products

By definition, it is obvious that the cross products $\mathbf{P} \times \mathbf{Q}$ and $\mathbf{Q} \times \mathbf{P}$ have the same magnitude and orientation but opposite senses as shown in Fig. 4.17; i.e.,

$$\mathbf{P} \times \mathbf{Q} = -\mathbf{Q} \times \mathbf{P} \qquad (4.8)$$

This shows that *cross products are not commutative*.

Fig. 4.15

Fig. 4.16

Fig. D4.20

Fig. 4.17

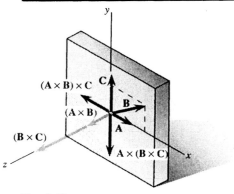

Fig. 4.18

Suppose that the vectors **A**, **B**, and **C** all lie in the *xy* plane as shown in Fig. 4.18. We readily see that both **A** × **B** and **B** × **C** are directed along the *z* axis, (**A** × **B**) × **C** is directed along the negative *x* axis, but **A** × (**B** × **C**) is directed along the negative *y* axis. Thus, we note that, in general

$$\mathbf{A} \times (\mathbf{B} \times \mathbf{C}) \neq (\mathbf{A} \times \mathbf{B}) \times \mathbf{C} \tag{4.9}$$

This shows that *cross products are not associative.*

Suppose that the vectors **Q**, **Q′**, and **P** are coplanar, the tails of **Q**, **Q′**, and **P** are coincident, and the line joining the tips of **Q** and **Q′** is parallel to **P**, as shown in Fig. 4.19. From the definition and geometric interpretation of the cross product of two vectors in Sec. 4.7, we readily see that

$$\mathbf{P} \times \mathbf{Q} = \mathbf{P} \times \mathbf{Q}' \tag{4.10}$$

This is true because *both* **P** × **Q** *and* **P** × **Q′** *have the same magnitude and the same direction and are hence equivalent.*

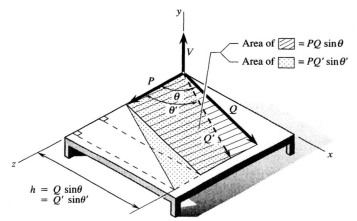

Fig. 4.19

Knowing the equivalent relation expressed in Eq. (4.10), we may proceed to prove that *cross products are distributive* over vector addition; i.e.,

$$\mathbf{A} \times (\mathbf{P} + \mathbf{Q}) = \mathbf{A} \times \mathbf{P} + \mathbf{A} \times \mathbf{Q} \tag{4.11}$$

We first, without losing generality, assume that **A** is directed along the *y* axis as shown in Fig. 4.20. Let the sum of **P** and **Q** be denoted by **R**; i.e., **R** = **P** + **Q**. Dropping perpendiculars from the tips of **P**, **Q**, and **R** onto the *zx* plane, we obtain **P′**, **Q′**, and **R′** as their projections on the *zx* plane as shown in Fig. 4.20(a). As the projection of a parallelogram onto another plane is also a parallelogram, we note that **R′** = **P′** + **Q′**. Furthermore, the quadrilateral with **A** × **P′** and **A** × **Q′** as sides and **A** × **R′** as diagonal is also a parallelogram in the *zx* plane as shown in Fig. 4.20(b); i.e.,

$$\mathbf{A} \times \mathbf{R}' = \mathbf{A} \times \mathbf{P}' + \mathbf{A} \times \mathbf{Q}'$$

Applying Eq. (4.10), we see that **A** × **R′**, **A** × **P′**, and **A** × **Q′** may, respectively, be replaced by **A** × **R**, **A** × **P**, and **A** × **Q**. Thus, we have

$$\mathbf{A} \times \mathbf{R} = \mathbf{A} \times \mathbf{P} + \mathbf{A} \times \mathbf{Q}$$

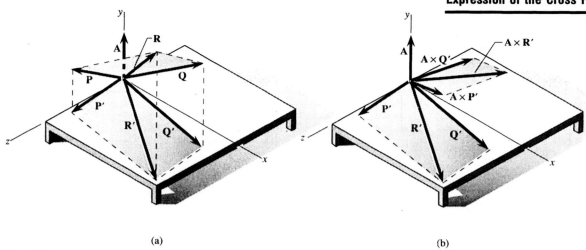

(a)   (b)

Fig. 4.20

which, upon the substitution of **R** by **P** + **Q**, leads to Eq. (4.11). This completes the proof. Moreover, applying Eqs. (4.8) and (4.11), we can readily show that

$$(\mathbf{P} + \mathbf{Q}) \times \mathbf{A} = \mathbf{P} \times \mathbf{A} + \mathbf{Q} \times \mathbf{A} \qquad (4.12)$$

Developmental Exercises

D4.23 It is known that $\mathbf{A} \times \mathbf{B} = 3\mathbf{i} - 4\mathbf{j} + 12\mathbf{k}$. What is the analytical expression of $\mathbf{B} \times \mathbf{A}$?

D4.24 Compute (a) $\mathbf{j} \times (\mathbf{j} \times \mathbf{k})$, (b) $(\mathbf{j} \times \mathbf{j}) \times \mathbf{k}$.

D4.25 What are the conditions that $\mathbf{Q}'$ has to satisfy if $\mathbf{P} \times \mathbf{Q} = \mathbf{P} \times \mathbf{Q}'$?

D4.26 Applying the distributive property of cross products, compute (a) $3\mathbf{j} \times (2\mathbf{i} + \mathbf{j} - \mathbf{k})$, (b) $(4\mathbf{i} - \mathbf{j}) \times (3\mathbf{k})$, (c) $(\mathbf{i} - \mathbf{j}) \times (\mathbf{j} - \mathbf{k})$.

4.9 Analytical Expression of the Cross Product

If the analytical expressions of **P** and **Q** are given as

$$\mathbf{P} = P_x\mathbf{i} + P_y\mathbf{j} + P_z\mathbf{k} \qquad \mathbf{Q} = Q_x\mathbf{i} + Q_y\mathbf{j} + Q_z\mathbf{k}$$

we may use the distributive property of cross products as well as the identities in Eqs. (4.7) to write

$$\mathbf{V} = \mathbf{P} \times \mathbf{Q}$$

$$= P_x\mathbf{i} \times (Q_x\mathbf{i} + Q_y\mathbf{j} + Q_z\mathbf{k}) + P_y\mathbf{j} \times (Q_x\mathbf{i} + Q_y\mathbf{j} + Q_z\mathbf{k})$$

$$+ P_z\mathbf{k} \times (Q_x\mathbf{i} + Q_y\mathbf{j} + Q_z\mathbf{k})$$

$$\mathbf{V} = (P_yQ_z - P_zQ_y)\mathbf{i} + (P_zQ_x - P_xQ_z)\mathbf{j} + (P_xQ_y - P_yQ_x)\mathbf{k} \qquad (4.13)$$

which is the analytical expression of **P** cross **Q**. The scalar components of this cross product are thus found to be

$$V_x = P_y Q_z - P_z Q_y$$
$$V_y = P_z Q_x - P_x Q_z \qquad (4.14)$$
$$V_z = P_x Q_y - P_y Q_x$$

Frequently, we make use of a determinant and write[†]

$$\mathbf{P} \times \mathbf{Q} = \begin{vmatrix} \mathbf{i} & \mathbf{j} & \mathbf{k} \\ P_x & P_y & P_z \\ Q_x & Q_y & Q_z \end{vmatrix} \qquad (4.15)$$

which, when expanded, yields the same analytical expression for **V** as that in Eq. (4.13). The use of a computer to compute the cross product of two vectors is illustrated in Fig. 4.21.

```
10 REM      CROSS PRODUCT OF TWO VECTORS, P X Q = V       (File Name: CROSSPD)
20 REM        Enter the scalar components of P and Q as data in Line 100.
30 REM      To output to the printer, change PRINT to LPRINT in the program.
40 READ PX,PY,PZ,QX,QY,QZ:VX=PY*QZ-PZ*QY:VY=PZ*QX-PX*QZ:VZ=PX*QY-PY*QX:PRINT
50 PRINT"(";PX;"i";:W=PY:GOSUB 80:PRINT"j";:W=PZ:GOSUB 80:PRINT"k ) X (";
60 PRINT QX;"i";:W=QY:GOSUB 80:PRINT"j";:W=QZ:GOSUB 80:PRINT"k )":PRINT"=";
70 PRINT VX;"i";:W=VY:GOSUB 80:PRINT"j";:W=VZ:GOSUB 80:PRINT"k":END
80 S$=" +": IF W<O THEN W=-W: S$=" -"
90 PRINT S$;W;: RETURN
100 DATA 12,-12,6,16,-12,0

( 12 i - 12 j + 6 k ) X ( 16 i - 12 j + 0 k )
= 72 i + 96 j + 48 k
```

Fig. 4.21   (Cf. the cross product $\overrightarrow{AC} \times \overrightarrow{AB}$ in the solution in Example 4.6)

EXAMPLE 4.6

A pyramid is shown. Making use of the cross product, determine (a) the area A_{ABC} of the side ABC, (b) the angle α as indicated, (c) the outward unit vector $\boldsymbol{\lambda}_{ABC}$, which is perpendicular to the side ABC.

Solution.   From the given sketch, we write

$$\overrightarrow{AC} = 12\mathbf{i} - 12\mathbf{j} + 6\mathbf{k} \qquad |\overrightarrow{AC}| = 18$$
$$\overrightarrow{AB} = 16\mathbf{i} - 12\mathbf{j} \qquad |\overrightarrow{AB}| = 20$$

$$\overrightarrow{AC} \times \overrightarrow{AB} = \begin{vmatrix} \mathbf{i} & \mathbf{j} & \mathbf{k} \\ 12 & -12 & 6 \\ 16 & -12 & 0 \end{vmatrix}$$

$$= \mathbf{i} \begin{vmatrix} -12 & 6 \\ -12 & 0 \end{vmatrix} - \mathbf{j} \begin{vmatrix} 12 & 6 \\ 16 & 0 \end{vmatrix} + \mathbf{k} \begin{vmatrix} 12 & -12 \\ 16 & -12 \end{vmatrix}$$

$$= 72\mathbf{i} + 96\mathbf{j} + 48\mathbf{k} \qquad |\overrightarrow{AC} \times \overrightarrow{AB}| = 129.24$$

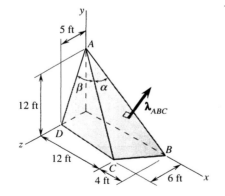

[†]Cf. Sec. B1.4 of App. B for the expansion of a determinant.

From the *geometric interpretation* and the *definition* of cross product, we write

$$A_{ABC} = \frac{1}{2} | \overrightarrow{AC} \times \overrightarrow{AB} | = \frac{1}{2}(129.24) \qquad A_{ABC} = 64.6 \text{ ft}^2 \blacktriangleleft$$

$$|\overrightarrow{AC} \times \overrightarrow{AB}| = |\overrightarrow{AC}| \, |\overrightarrow{AB}| \sin\alpha \qquad 129.24 = 18(20) \sin\alpha$$

$$\alpha = 21.0° \blacktriangleleft$$

We know that λ_{ABC} must be parallel to the cross product $\overrightarrow{AC} \times \overrightarrow{AB}$. We write

$$\lambda_{ABC} = \frac{\overrightarrow{AC} \times \overrightarrow{AB}}{|\overrightarrow{AC} \times \overrightarrow{AB}|}$$

$$\lambda_{ABC} = 0.557\mathbf{i} + 0.743\mathbf{j} + 0.371\mathbf{k} \blacktriangleleft$$

Developmental Exercise

D4.27 Refer to the pyramid in Example 4.6 and the program in Fig. 4.21. Determine (a) the area A_{ACD} of the side ACD, (b) the angle β between the edges AC and AD, (c) the outward unit vector λ_{ACD} which is perpendicular to the side ACD.

4.10 Moment of a Force about a Point by Cross Product

We recall that the moment of a force $\mathbf{F}$, acting at the point P, about a point Q in space is a vector $\mathbf{M}_Q$ as shown in Fig. 4.2, which is repeated here for convenience of reference. The *magnitude* of $\mathbf{M}_Q$ is equal to $d_s F$ where F is the magnitude of $\mathbf{F}$ and d_s is the shortest distance from the point Q to the

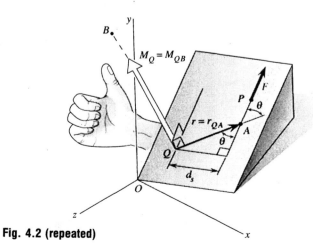

Fig. 4.2 (repeated)

line of action of the force **F** as shown. The *orientation* of $\mathbf{M}_Q$ is perpendicular to the plane containing the point Q and the line of action of **F**. Let the position vector from the point Q to *any* convenient point A on the line of action of **F** be denoted by **r** as shown. It should be noted carefully that **r** is directed *from Q to A*; i.e.,

$$\mathbf{r} = \mathbf{r}_{QA} \qquad (\mathbf{r} \neq \mathbf{r}_{AQ}) \tag{4.16}$$

The *sense* of $\mathbf{M}_Q$ is defined to be such that **r**, **F**, and $\mathbf{M}_Q$ in order form a right-handed trihedral. From Fig. 4.2 and the above definition, we have

$$d_s = r\sin\theta \qquad M_Q = d_s F \tag{4.17}$$

Making use of Eq. (4.17) and the cross product defined in Sec. 4.7, we may, therefore, write the moment vector $\mathbf{M}_Q$ as

$$\mathbf{M}_Q = \mathbf{r} \times \mathbf{F} \tag{4.18}$$

Equation (4.18) is the vector formula for computing the moment of a force **F** about a point Q. The *resultant moment of a system of forces about a point* is, of course, equal to the vector sum of the moments of the individual forces in the system about that point. If **F** is a force of magnitude one, we have, from Eq. (4.17), $M_Q = d_s$. This means that *the shortest distance between a point and a line* is numerically equal to the magnitude of the moment of a unit force, directed along that line, about that point.

EXAMPLE 4.7

A 130-N thrust **F** is generated at the point $B(0.5, 0.6, 0)$ m of a pipe and points toward the joint at $C(1.7, 1.1, 0)$ m as shown. Determine (a) the moment $\mathbf{M}_A$ of **F** about the joint at $A(0.3, -0.2, 1)$ m, (b) the shortest distance d_s from A to the line of action of **F**.

Solution. For determining the moment $\mathbf{M}_A$, we write

$$\overrightarrow{BC} = 1.2\mathbf{i} + 0.5\mathbf{j} \qquad |\overrightarrow{BC}| = 1.3 \qquad \lambda_{BC} = \frac{\overrightarrow{BC}}{|\overrightarrow{BC}|} = \frac{1}{13}(12\mathbf{i} + 5\mathbf{j})$$

$$F = 130 \qquad \mathbf{F} = F\boldsymbol{\lambda}_{BC} = 120\mathbf{i} + 50\mathbf{j} \qquad \mathbf{r} = \overrightarrow{AB} = 0.2\mathbf{i} + 0.8\mathbf{j} - \mathbf{k}$$

$$\mathbf{M}_A = \mathbf{r} \times \mathbf{F} = \begin{vmatrix} \mathbf{i} & \mathbf{j} & \mathbf{k} \\ 0.2 & 0.8 & -1 \\ 120 & 50 & 0 \end{vmatrix}$$

$$\mathbf{M}_A = 50\mathbf{i} - 120\mathbf{j} - 86\mathbf{k} \ \text{N·m} \quad \blacktriangleleft$$

Since $M_A = d_s F$, we write

$$[(50)^2 + (120)^2 + (86)^2]^{1/2} = d_s(130) \qquad d_s = 1.199 \ \text{m} \quad \blacktriangleleft$$

Developmental Exercises

D4.28 Define the magnitude, the orientation, and the sense of the moment $\mathbf{M}_Q$ of a force $\mathbf{F}$ about a point Q.

D4.29 A 150-N force $\mathbf{P}$ is shown below. (a) Write the analytical expression of $\mathbf{P}$. (b) Is the moment $\mathbf{M}_A$ of $\mathbf{P}$ about the point A given by $\mathbf{r}_{AB} \times \mathbf{P}$ or $\mathbf{r}_{BA} \times \mathbf{P}$? (c) Determine $\mathbf{M}_A$. (d) Is $\mathbf{M}_A = \mathbf{r}_{AD} \times \mathbf{P}$? (e) Determine the shortest distance d_s from A to the line of action of $\mathbf{P}$.

D4.30 Refer to D4.29. In addition to the force $\mathbf{P}$ shown, suppose that another force $\mathbf{Q} = 100\mathbf{k}$ N is acting at the point C. Determine the resultant moment of $\mathbf{P}$ and $\mathbf{Q}$ about the point A.

D4.31 Using the concept of moment of a force, describe the general procedure to find the shortest distance between a point and a line.

Fig. D4.29

4.11 Varignon's Theorem: Forces in Space

In general, *Varignon's theorem* may be stated as follows: *the moment $\mathbf{M}_P$, about any point P, of the resultant $\mathbf{R}$ of two or more concurrent forces $\mathbf{F}_1$, $\mathbf{F}_2$, . . . , $\mathbf{F}_n$ acting at the point C is equal to the sum of the moments $\mathbf{M}_1$, $\mathbf{M}_2$, . . . , $\mathbf{M}_n$ of the various forces $\mathbf{F}_1$, $\mathbf{F}_2$, . . . , $\mathbf{F}_n$ at C about the same point P.* To prove this theorem, let the mentioned system be first illustrated as shown in Fig. 4.22. Then, applying the distributive property of cross

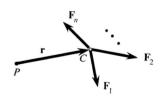

Fig. 4.22

products, we write

$$\mathbf{M}_P = \mathbf{r} \times \mathbf{R} = \mathbf{r} \times (\mathbf{F}_1 + \mathbf{F}_2 + \cdots + \mathbf{F}_n)$$
$$= \mathbf{r} \times \mathbf{F}_1 + \mathbf{r} \times \mathbf{F}_2 + \cdots + \mathbf{r} \times \mathbf{F}_n$$

Since $\mathbf{M}_1 = \mathbf{r} \times \mathbf{F}_1$, $\mathbf{M}_2 = \mathbf{r} \times \mathbf{F}_2, \ldots,$ and $\mathbf{M}_n = \mathbf{r} \times \mathbf{F}_n$, we have

$$\mathbf{M}_P = \mathbf{M}_1 + \mathbf{M}_2 + \cdots + \mathbf{M}_n \tag{4.19}$$

This completes the proof.

EXAMPLE 4.8

A guyed pole AO is shown. The tensions in the cables AB and AC are $T_{AB} = 27$ kN and $T_{AC} = 28$ kN. Determine the range of values of the tension T_{AD} in the cable AD for which the magnitude of the resultant moment developed at the base O of the pole will not exceed 200 kN·m.

Solution. Referring to the figure shown, we write

$$\mathbf{T}_{AB} = 27\lambda_{AB} = \frac{27}{3}(2\mathbf{i} - 2\mathbf{j} - \mathbf{k})$$

$$\mathbf{T}_{AC} = 28\lambda_{AC} = \frac{28}{7}(-2\mathbf{i} - 6\mathbf{j} - 3\mathbf{k})$$

$$\mathbf{T}_{AD} = T_{AD}\lambda_{AD} = \frac{T_{AD}}{5}(-4\mathbf{j} + 3\mathbf{k})$$

The resultant force $\mathbf{R}$ at A is

$$\mathbf{R} = \mathbf{T}_{AB} + \mathbf{T}_{AC} + \mathbf{T}_{AD}$$
$$= 10\mathbf{i} - \left(42 + \frac{4}{5}T_{AD}\right)\mathbf{j} + \left(-21 + \frac{3}{5}T_{AD}\right)\mathbf{k}$$

By Varignon's theorem, the resultant moment $\mathbf{M}_O$ developed at the base O is

$$\mathbf{M}_O = \mathbf{r}_{OA} \times \mathbf{R} = 12\mathbf{j} \times \mathbf{R} = \left(\frac{36}{5}T_{AD} - 252\right)\mathbf{i} - 120\mathbf{k}$$

Setting the magnitude of $\mathbf{M}_O$ equal to 200 kN·m, we write

$$\left[\left(-252 + \frac{36}{5}T_{AD}\right)^2 + (120)^2\right]^{1/2} = 200$$

Squaring both sides of this equation and solving the resulting quadratic equation, we obtain $T_{AD} = 12.78$ or 57.2. For $M_O \leq 200$ kN·m, we must have

$$12.78 \text{ kN} \leq T_{AD} \leq 57.2 \text{ kN} \quad \blacktriangleleft$$

Developmental Exercise

D4.32   Three forces act at the free end of a cantilevered bent rod as shown. Determine the resultant moment of these forces about the fixed end at O.

Fig. D4.32

4.12   Moment of a Couple by Cross Product

We know that the moment of a couple about a point in space is the sum of the moments of the two forces in the couple about that point. Thus, the moment $\mathbf{M}_O$ of the couple $\mathbf{F}$ and $-\mathbf{F}$ about the point O in space, as shown in Fig. 4.23, is

$$\mathbf{M}_O = \mathbf{r}_A \times \mathbf{F} + \mathbf{r}_B \times (-\mathbf{F}) = (\mathbf{r}_A - \mathbf{r}_B) \times \mathbf{F}$$

$$\mathbf{r} = \mathbf{r}_A - \mathbf{r}_B$$

$$\mathbf{M}_O = \mathbf{r} \times \mathbf{F} \qquad (4.20)$$

where the points A and B may be any points on the lines of action of $\mathbf{F}$ and $-\mathbf{F}$, respectively. Since $\mathbf{r}$ is actually any position vector *from* any point on the line of action of $-\mathbf{F}$ *to* any point on the line of action of $\mathbf{F}$, the moment

Fig. 4.23

$\mathbf{M}_O$ of the couple is, in fact, independent of the choice of the point O. This confirms that the moment of a couple is a *free vector*. Dropping the subscript O in $\mathbf{M}_O$, we may write this free vector as

$$\mathbf{M} = \mathbf{r} \times \mathbf{F} \qquad (4.21)$$

In using Eq. (4.21), note that *either* of the two forces in a couple can be designated as $\mathbf{F}$ and the other as $-\mathbf{F}$, even though the analytical expression of the force designated as $\mathbf{F}$ may contain negative terms. However, $\mathbf{r}$ must be drawn *from* the line of action of the force designated as $-\mathbf{F}$ *to* the line of action of the force designated as $\mathbf{F}$.

The moment $\mathbf{M}$ of a couple consisting of $\mathbf{F}$ and $-\mathbf{F}$ is a vector perpendicular to the plane containing $\mathbf{F}$, $-\mathbf{F}$, and $\mathbf{r}$. This moment points in the *direction* of the extended thumb of the right hand when the other four fingers of the right hand are curled in the direction of the tendency of the couple to rotate. The *magnitude* of $\mathbf{M}$ is

$$M = rF \sin\theta = (r \sin\theta)F \qquad M = d_s F \qquad (4.22)$$

where d_s is the shortest distance between the lines of action of $\mathbf{F}$ and $-\mathbf{F}$.

Fig. D4.34

Developmental Exercises

D4.33 The formula $\mathbf{M} = \mathbf{r} \times \mathbf{F}$ is to be used to compute the moment of a couple. (a) Which of the two forces in the couple should be designated as $\mathbf{F}$ and which as $-\mathbf{F}$? (b) Is $\mathbf{r}$ drawn from $\mathbf{F}$ to $-\mathbf{F}$ or from $-\mathbf{F}$ to $\mathbf{F}$?

D4.34 A couple consists of two 200-lb forces as shown. Compute the moments of the couple about the points (a) A, (b) B, (c) O, (d) any point P.

4.13 Equivalent Couples and Addition of Moments of Couples

We know that a couple is characterized by its moment $\mathbf{M}$. Two couples are *equivalent* if they have the *same moment* $\mathbf{M}$. The couples illustrated in Fig. 4.24 are equivalent because they have the same moment, $120\mathbf{j}$ N·m. All equivalent couples must lie in parallel planes and have the same effect of rotation on a rigid body. The plane of the couple in Fig. 4.24 may be any plane perpendicular to $\mathbf{M}$.

Fig. 4.24

Couples are added according to the parallelogram law (in the same way as forces or any other vectors are added), as illustrated in Fig. 4.25. Furthermore, the *resultant moment* $\mathbf{M}$ of any number of couples is equal to the vector sum of the moments, $\mathbf{M}_1, \mathbf{M}_2, \ldots, \mathbf{M}_n$, of these couples; i.e.,

$$\mathbf{M} = \mathbf{M}_1 + \mathbf{M}_2 + \cdots + \mathbf{M}_n \tag{4.23}$$

Fig. 4.25

EXAMPLE 4.9

A system of three couples is shown. Determine the analytical expressions of (a) the moment $\mathbf{M}_1$ produced by the pair of 100-N forces, (b) the moment $\mathbf{M}_2$ produced by the pair of 200-N forces, (c) the moment $\mathbf{M}_3$ produced by the pair of 150-N forces, (d) the resultant moment $\mathbf{M}$ of $\mathbf{M}_1$, $\mathbf{M}_2$, and $\mathbf{M}_3$.

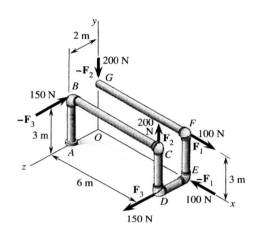

Solution. Using Eqs. (4.21) and (4.23), we write

$$\mathbf{M}_1 = \mathbf{r}_{EF} \times \mathbf{F}_1 = 3\mathbf{j} \times (100\mathbf{i})$$
$$\mathbf{M}_1 = -300\mathbf{k} \text{ N·m} \blacktriangleleft$$

$$\mathbf{M}_2 = \mathbf{r}_{GC} \times \mathbf{F}_2 = (6\mathbf{i} + 2\mathbf{k}) \times (200\mathbf{j})$$
$$\mathbf{M}_2 = -400\mathbf{i} + 1200\mathbf{k} \text{ N·m} \blacktriangleleft$$

$$\mathbf{M}_3 = \mathbf{r}_{BD} \times \mathbf{F}_3 = (6\mathbf{i} - 3\mathbf{j}) \times (150\mathbf{k})$$
$$\mathbf{M}_3 = -450\mathbf{i} - 900\mathbf{j} \text{ N·m} \blacktriangleleft$$

$$\mathbf{M} = \mathbf{M}_1 + \mathbf{M}_2 + \mathbf{M}_3$$
$$\mathbf{M} = -850\mathbf{i} - 900\mathbf{j} + 900\mathbf{k} \text{ N·m} \blacktriangleleft$$

Developmental Exercise

D4.35 For the system of forces shown, determine the analytical expressions of (a) the moment $\mathbf{M}_1$ produced by the pair of 2-kN forces, (b) the moment $\mathbf{M}_2$ produced by the pair of 4-kN forces, (c) the resultant moment $\mathbf{M}$ of $\mathbf{M}_1$ and $\mathbf{M}_2$.

Fig. D4.35

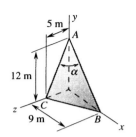

Fig. P4.35 and P4.38

PROBLEMS

4.33 If $\mathbf{P} = 2\mathbf{i} - \mathbf{j} + 3\mathbf{k}$ and $\mathbf{Q} = \mathbf{i} - 2\mathbf{j} + \mathbf{k}$, determine (a) $\mathbf{P} \times \mathbf{Q}$, (b) $\mathbf{Q} \times \mathbf{P}$, (c) $\mathbf{P} \times (\mathbf{P} - \mathbf{Q})$, (d) $|\mathbf{P} \times \mathbf{Q}|$, (e) the angle between $\mathbf{P}$ and $\mathbf{Q}$.

4.34* If $\mathbf{S} = \mathbf{i} - 3\mathbf{j} - 2\mathbf{k}$ and $\mathbf{T} = 2\mathbf{i} + \mathbf{j} + \mathbf{k}$, determine (a) $(\mathbf{T} + \mathbf{S}) \times \mathbf{T}$, (b) $(\mathbf{T} + \mathbf{S}) \times \mathbf{S}$, (c) $|\mathbf{S} \times \mathbf{T}|$, (d) the angle between $\mathbf{S}$ and $\mathbf{T}$.

4.35 through 4.37* Determine the area A_{ABC} of the side ABC of the pyramid shown.

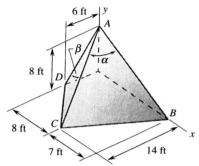

Fig. P4.36, P4.39, and P4.41

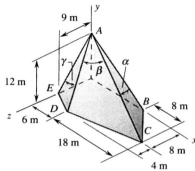

Fig. P4.37*, P4.40, P4.42, and P4.43*

4.38 through 4.40 Determine the angle α indicated on the pyramid shown.

4.41 and 4.42 For the pyramid shown, determine (a) the area A_{ACD} of the side ACD, (b) the angle β as indicated.

4.43* For the pyramid shown, determine (a) the area A_{ADE} of the side ADE, (b) the angle γ as indicated.

4.44 and 4.45 For the system shown, determine the moment $\mathbf{M}_A$ of $\mathbf{F}$ about the point A in five different ways.

Fig. P4.44

Fig. P4.45

Fig. P4.46

4.46 A car sideswipes a highway sign and breaks one of its supports, leaving the broken and twisted sign supported only by the post OA as shown. A wind produces a resultant force of $100\mathbf{i} - 400\mathbf{k}$ N acting at the center $C(-0.4, 2, -0.1)$ m of the sign. Determine the moment $\mathbf{M}_O$ of the wind force about the point O.

4.47 and 4.48* A system of four forces acts on a cantilevered bent rod as shown. Determine the moment about the fixed end A of (a) $\mathbf{P}$, (b) the entire system of four forces.

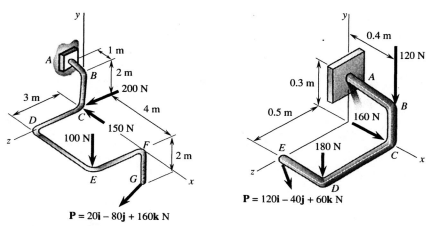

Fig. P4.47

$$P = 20i - 80j + 160k \text{ N}$$

Fig. P4.48*

$$P = 120i - 40j + 60k \text{ N}$$

4.49 and 4.50 Find the analytical expression of the resultant moment M_A^R, in terms of known and unknown quantities, of the force system acting on the free body about the point A as shown.

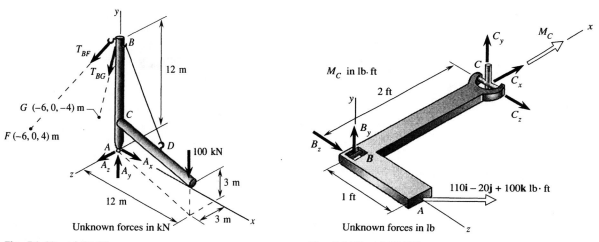

Fig. P4.49 and P4.51

Unknown forces in kN

Fig. P4.50 and P4.52*

Unknown forces in lb

$$110i - 20j + 100k \text{ lb} \cdot \text{ft}$$

M_C in lb·ft

4.51 and 4.52* Find the analytical expression of the resultant moment M_B^R of the force system acting on the free body about the point B as shown.

4.53 through 4.56* A force system consisting of couples is shown. Determine the analytical expressions of (a) the moment M_i ($i = 1, 2, \ldots$) of each individual couple, (b) the resultant moment M of all couples, (c) the resultant moment M_A of the force system about the support at A.

Fig. P4.53

Fig. P4.54

Fig. P4.55

Fig. P4.56*

APPLICATION OF DOT PRODUCTS

Note. Those who are familiar with the rudiments of dot products as presented in Secs. 4.14 and 4.15 may proceed directly to Sec. 4.16.

4.14 Dot Product and Its Properties

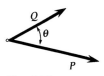

Fig. 4.26

The *dot product* of two vectors **P** and **Q**, as shown in Fig. 4.26, is defined as a scalar which is equal to the product of the magnitudes of **P** and **Q** and the cosine of the angle θ between the positive directions of **P** and **Q**. Usually, we take $0 \le \theta \le 180°$, although this is not necessary because $\cos(360° - \theta) = \cos\theta$. As in vector analysis, the dot product of **P** and **Q** is denoted by **P** · **Q**, which is read ''**P** dot **Q**.'' Thus, we write

$$\mathbf{P} \cdot \mathbf{Q} = PQ \cos\theta \tag{4.24}$$

$$PQ \cos\theta = QP \cos\theta \qquad \mathbf{P} \cdot \mathbf{Q} = \mathbf{Q} \cdot \mathbf{P} \tag{4.25}$$

Equations (4.25) show that the *dot product of two vectors is commutative*. Because of the nature of the product, the dot product is also called the *scalar product*; sometimes, it is also called the *inner product*.

From the above definition and Fig. 4.15, it can readily be shown that

$$\begin{aligned} \mathbf{i} \cdot \mathbf{i} = 1 \qquad \mathbf{j} \cdot \mathbf{j} = 1 \qquad \mathbf{k} \cdot \mathbf{k} = 1 \\ \mathbf{i} \cdot \mathbf{j} = 0 \qquad \mathbf{j} \cdot \mathbf{k} = 0 \qquad \mathbf{k} \cdot \mathbf{i} = 0 \end{aligned} \tag{4.26}$$

Furthermore, when $\mathbf{P} \cdot \mathbf{Q} = 0$, it implies one or more of the following cases: (a) **P** and **Q** are perpendicular to each other, (b) **P** is a zero vector, (c) **Q** is a zero vector.

From the definition of the dot product, we see that the projection of **Q** on **P**, as shown in Fig. 4.27, may be written as

Fig. 4.27

Projection of $\mathbf{Q}$ *on* $\mathbf{P} = Q \cos\theta = (1)Q \cos\theta = \boldsymbol{\lambda}_P \cdot \mathbf{Q}$

where λ_P is a unit vector in the direction of **P**. Similarly, the projection of **P** on **Q** may be written as

$$Projection\ of\ \mathbf{P}\ on\ \mathbf{Q} = P\cos\theta = (1)P\cos\theta = \lambda_Q \cdot \mathbf{P}$$

where λ_Q is a unit vector in the direction of **Q**.

For the vectors **A**, **B**, **C**, **B** + **C**, and the unit vector λ_A as shown in Fig. 4.28, we see that the projections on **A** have the following relationship:

$$\lambda_A \cdot (\mathbf{B} + \mathbf{C}) = \lambda_A \cdot \mathbf{B} + \lambda_A \cdot \mathbf{C} \qquad (4.27)$$

Multiplying both sides of Eq. (4.27) by A and noting that $\mathbf{A} = A\lambda_A$, we write

$$A\lambda_A \cdot (\mathbf{B} + \mathbf{C}) = A\lambda_A \cdot \mathbf{B} + A\lambda_A \cdot \mathbf{C}$$

$$\mathbf{A} \cdot (\mathbf{B} + \mathbf{C}) = \mathbf{A} \cdot \mathbf{B} + \mathbf{A} \cdot \mathbf{C} \qquad (4.28)$$

which reveals that *dot products are distributive*.

Using the above distributive property and Eqs. (4.26), we write

$$\mathbf{P} \cdot \mathbf{Q} = (P_x\mathbf{i} + P_y\mathbf{j} + P_z\mathbf{k}) \cdot (Q_x\mathbf{i} + Q_y\mathbf{j} + Q_z\mathbf{k})$$

$$\mathbf{P} \cdot \mathbf{Q} = P_xQ_x + P_yQ_y + P_zQ_z \qquad (4.29)$$

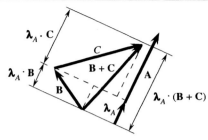

Fig. 4.28

Developmental Exercises

D4.36   Vectors **A**, **B**, **C**, and **D** as shown have magnitudes of 4, 2, 3, and 5, respectively. Compute (a) **A** · **B**, (b) **B** · **A**, (c) **A** · **C**, (d) **B** · **C**, (e) **A** · **A**, (f) **B** · **D**.

D4.37   Vectors **P** and **Q** have magnitudes of 10 and 20, respectively, and the angle θ between them is 30°. Determine (a) **P** · **Q**, (b) the projection of **P** on **Q**, (c) the projection of **Q** on **P**.

D4.38   If **P** = 2**i** − **j** + 3**k** and **Q** = **i** − 2**j** + **k**, determine **P** · **Q**.

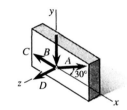

Fig. D4.36

EXAMPLE 4.10

Refer to the pyramid in Example 4.6. Making use of the cross product and the dot product, determine the interior angle $\theta_{ABC/zx}$ between the side ABC and the zx plane.

Solution.   The unit vector λ_{ABC} perpendicular to the side ABC has been found, by using the cross product, in Example 4.6 as

$$\lambda_{ABC} = 0.557\mathbf{i} + 0.743\mathbf{j} + 0.371\mathbf{k}$$

The unit vector perpendicular to the zx plane is **j**. The interior angle $\theta_{ABC/zx}$ is congruent to the angle between λ_{ABC} and **j**. Applying the dot product, we write

$$\cos\theta_{ABC/zx} = 1 \cdot 1 \cdot \cos\theta_{ABC/zx} = \lambda_{ABC} \cdot \mathbf{j} = 0.743$$

$$\theta_{ABC/zx} = 42.0° \blacktriangleleft$$

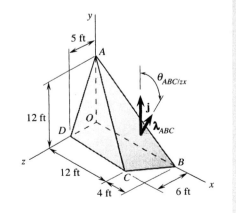

Developmental Exercises

D4.39 In the figure shown, determine, as appropriate, the analytical expression or value of (a) λ_{OA}, (b) λ_{OB}, (c) $\lambda_{OA} \cdot \lambda_{OB}$, (d) $\cos\theta$, (e) θ, (f) $\lambda_{OA} \cdot \overrightarrow{OB}$, (g) the projection of $\overrightarrow{OB}$ on $\overrightarrow{OA}$.

D4.40 Refer to Example 4.10. Determine the interior angle $\theta_{ABC/ACD}$ between the side *ABC* and the side *ACD*.

Fig. D4.39

4.15 Scalar Triple Product

The dot product of **A** with the cross product of **P** and **Q** expressed as **A** $\cdot$ (**P** $\times$ **Q**) is called the *scalar triple product* of **A**, **P**, and **Q**. In terms of scalar components, we write

$$\mathbf{A} \cdot (\mathbf{P} \times \mathbf{Q}) = A_x(P_yQ_z - P_zQ_y) + A_y(P_zQ_x - P_xQ_z)$$
$$+ A_z(P_xQ_y - P_yQ_x) \qquad (4.30)$$

In the form of a determinant, we write

$$\mathbf{A} \cdot (\mathbf{P} \times \mathbf{Q}) = \begin{vmatrix} A_x & A_y & A_z \\ P_x & P_y & P_z \\ Q_x & Q_y & Q_z \end{vmatrix} \qquad (4.31)$$

which is more easily remembered. Using the rules governing the interchanging of rows in a determinant (cf. Sec. B1.4 of App. B) and the commutation of dot products, we can readily verify that

$$\mathbf{A} \cdot (\mathbf{P} \times \mathbf{Q}) = \mathbf{P} \cdot (\mathbf{Q} \times \mathbf{A}) = \mathbf{Q} \cdot (\mathbf{A} \times \mathbf{P}) = (\mathbf{A} \times \mathbf{P}) \cdot \mathbf{Q}$$

Thus, we see that[†]

$$\mathbf{A} \cdot (\mathbf{P} \times \mathbf{Q}) = (\mathbf{A} \times \mathbf{P}) \cdot \mathbf{Q} \qquad (4.32)$$

where $\cdot$ and $\times$ can be interchanged if the cross product is computed before the dot product and the order of the vectors is maintained.

[†]Since (**A** $\cdot$ **P**) $\times$ **Q** is not defined and has no meaning, the parentheses in Eq. (4.32) may be omitted without causing ambiguity.

The scalar triple product of **A**, **P**, and **Q** is sometimes called the *mixed triple product* or *box product* and may be denoted as [**A P Q**]. Geometrically, the absolute value of the scalar triple product is equal to the volume of the parallelepiped, as illustrated in Fig. 4.29, which has the three vectors as its adjacent edges. Of course, if any two of the three vectors **A**, **P**, and **Q** are parallel to each other, or **A**, **P**, and **Q** are coplanar, then the scalar triple product [**A P Q**] must be zero. The use of a computer to compute a scalar triple product is illustrated in Fig. 4.30.

Volume = | **A** · (**P** × **Q**) |

Fig. 4.29

```
10 REM    SCALAR TRIPLE PRODUCT, A . (P X Q) = S       (File Name: STRIPPD)
20 REM    Enter the scalar components of A, P, and Q as data in Line 100.
30 REM    To output to the printer, change PRINT to LPRINT in the program.
40 READ AX,AY,AZ,PX,PY,PZ,QX,QY,QZ:S1=AX*(PY*QZ-PZ*QY)+AY*(PZ*QX-PX*QZ)
50 S=S1+AZ*(PX*QY-PY*QX): PRINT:PRINT"The components of A, P, and Q are"
60 PRINT AX,AY,AZ:PRINT PX,PY,PZ:PRINT QX,QY,QZ:PRINT"Their scalar ";
70 PRINT "triple product is ";S;"."
100 DATA -12,0,5,-4,4,0,-2,1,2

The components of A, P, and Q are
-12            0              5
-4             4              0
-2             1              2
Their scalar triple product is -76 .
```

Fig. 4.30   (Cf. the scalar triple product in the solution in Example 4.12.)

Developmental Exercise

D4.41   Let **A** = **i** + 2**j** ft, **P** = 3**i** ft, and **Q** = −**k** ft be the adjacent edges of a parallelepiped. (a) Evaluate **A** · (**P** × **Q**). (b) What is the volume of this parallelepiped? (c) Is (**A** · **P**) × **Q** defined? (d) Using the program in Fig. 4.30, verify the answer to (a).

4.16   Moment of a Force about an Axis

The *moment of a force about an axis* is equal to the projection of the moment vector of the force, computed about any point on the axis, onto that axis. Suppose that the line of action of a force **F** passes through the point *A*, and a point *D* is on the axis *BC* as shown on the next page in Fig. 4.31(a). Then, the *moment* M_{BC} of **F** about the axis *BC* is the projection of its moment $\mathbf{M}_D$ onto the axis *BC* as shown in Fig. 4.31(b). We write

$$M_{BC} = \boldsymbol{\lambda}_{BC} \cdot \mathbf{M}_D \qquad M_{BC} = \boldsymbol{\lambda}_{BC} \cdot (\mathbf{r} \times \mathbf{F}) \qquad (4.33)$$

where

$$\boldsymbol{\lambda}_{BC} = \frac{\overrightarrow{BC}}{|\overrightarrow{BC}|} = \text{unit vector pointing in the direction of } \overrightarrow{BC}.$$

  r = a position vector from any point on the axis *BC* (e.g., the point *D*) to any convenient point on the line of action of **F** (e.g., the point *A*).

The vector form of the moment M_{BC} may be written as

$$\mathbf{M}_{BC} = [\boldsymbol{\lambda}_{BC} \cdot (\mathbf{r} \times \mathbf{F})]\boldsymbol{\lambda}_{BC} \qquad (4.34)$$

(a)   (b)

Fig. 4.31

Fig. 4.32

If M_x, M_y, and M_z are the rectangular scalar components of the moment $\mathbf{M}_O$ of a force about the origin O, as shown in Fig. 4.32, then

$$\mathbf{M}_O = M_x\mathbf{i} + M_y\mathbf{j} + M_z\mathbf{k} \qquad (4.35)$$

From Eqs. (4.35) and (4.33), we note that the rectangular scalar components M_x, M_y, and M_z are in fact the moment of the force about the x, y, and z axes, respectively.

EXAMPLE 4.11

A free body is acted on by a force system as shown, where the unknown and known forces have the same unit, and $\mathbf{B}_h$ lies in the horizontal plane and is perpendicular to $\overline{AB}$. Determine (a) the resultant moment $\mathbf{M}_A^R$ of the force system about the point A, (b) the resultant moment M_{AB}^R of the force system about the axis passing through the points A and B.

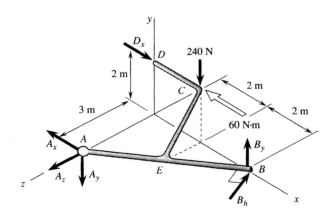

Solution.   Since B_h is perpendicular to $\overline{AB}$, we see from the top view of the free body, as shown, that

$$\mathbf{B}_h = \frac{B_h}{5}(-3\mathbf{i} - 4\mathbf{k})$$

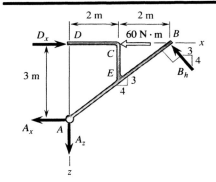

The resultant force of $\mathbf{B}_y$ and $\mathbf{B}_h$ is

$$\mathbf{B} = -\frac{3}{5}B_h\mathbf{i} + B_y\mathbf{j} - \frac{4}{5}B_h\mathbf{k}$$

The forces at the points C and D are

$$\mathbf{C} = -240\mathbf{j} \qquad \mathbf{D} = D_x\mathbf{i}$$

The relevant position vectors for computing $\mathbf{M}_A^R$ are

$$\mathbf{r}_{AB} = \overrightarrow{AB} = 4\mathbf{i} - 3\mathbf{k}$$
$$\mathbf{r}_{AC} = \overrightarrow{AC} = 2\mathbf{i} + 2\mathbf{j} - 3\mathbf{k}$$
$$\mathbf{r}_{AD} = \overrightarrow{AD} = 2\mathbf{j} - 3\mathbf{k}$$

The moment at the point C is $\mathbf{M}_C = -60\mathbf{i}$, which is a free vector. Noting that the forces at the point A have no contribution toward $\mathbf{M}_A^R$, we write

$$\mathbf{M}_A^R = \mathbf{M}_C + \mathbf{r}_{AB} \times \mathbf{B} + \mathbf{r}_{AC} \times \mathbf{C} + \mathbf{r}_{AD} \times \mathbf{D}$$

$$= -60\mathbf{i} + \begin{vmatrix} \mathbf{i} & \mathbf{j} & \mathbf{k} \\ 4 & 0 & -3 \\ -\frac{3}{5}B_h & B_y & -\frac{4}{5}B_h \end{vmatrix} + \begin{vmatrix} \mathbf{i} & \mathbf{j} & \mathbf{k} \\ 2 & 2 & -3 \\ 0 & -240 & 0 \end{vmatrix}$$

$$+ \begin{vmatrix} \mathbf{i} & \mathbf{j} & \mathbf{k} \\ 0 & 2 & -3 \\ D_x & 0 & 0 \end{vmatrix}$$

$$\mathbf{M}_A^R = (3B_y - 780)\mathbf{i} + (5B_h - 3D_x)\mathbf{j}$$
$$+ (4B_y - 2D_x - 480)\mathbf{k} \quad \text{N}\cdot\text{m} \quad \blacktriangleleft$$

The unit vector in the direction of $\overline{AB}$ is

$$\lambda_{AB} = \frac{\overrightarrow{AB}}{|\overrightarrow{AB}|} = \frac{1}{5}(4\mathbf{i} - 3\mathbf{k})$$

The resultant moment M_{AB}^R is the projection of $\mathbf{M}_A^R$ onto $\overline{AB}$. Thus, we write

$$M_{AB}^R = \lambda_{AB} \cdot \mathbf{M}_A^R \qquad M_{AB}^R = 1.2D_x - 336 \quad \text{N}\cdot\text{m} \quad \blacktriangleleft$$

Developmental Exercise

D4.42   Describe the procedure for finding the moment of a force about a given axis.

★4.17   Shortest Distance between an Axis and a Force[†]

Physically, the *moment* M_{BC} about the axis BC of a force $\mathbf{F}$ acting on a body measures the effectiveness or tendency of $\mathbf{F}$ to rotate the body about that axis. If M_{BC} is *positive*, it means that the sense of the effectiveness or tendency to rotate the body about the axis BC is, according to the right-hand rule, the same as that of $\boldsymbol{\lambda}_{BC}$; *otherwise*, it is opposite to the sense of $\boldsymbol{\lambda}_{BC}$.

By resolving each of the vectors $\mathbf{r}$ and $\mathbf{F}$ into two rectangular components, one parallel to the axis BC and the other perpendicular to the axis BC as shown in Fig. 4.31, we have

$$\mathbf{r} = \mathbf{r}_\parallel + \mathbf{r}_\perp \qquad \mathbf{F} = \mathbf{F}_\parallel + \mathbf{F}_\perp \tag{4.36}$$

where the subscripts $\parallel$ and $\perp$ denote parallel and perpendicular components, respectively. Noting that both $\mathbf{r}_\parallel$ and $\mathbf{F}_\parallel$ are parallel to $\boldsymbol{\lambda}_{BC}$ in Fig. 4.31 and applying Eqs. (4.36) and the conditions for a scalar triple product to have zero value in Sec. 4.15, we may write the moment M_{BC} as

$$\begin{aligned} M_{BC} &= \boldsymbol{\lambda}_{BC} \cdot (\mathbf{r} \times \mathbf{F}) = \boldsymbol{\lambda}_{BC} \cdot [(\mathbf{r}_\parallel + \mathbf{r}_\perp) \times (\mathbf{F}_\parallel + \mathbf{F}_\perp)] \\ &= \boldsymbol{\lambda}_{BC} \cdot (\mathbf{r}_\parallel \times \mathbf{F}_\parallel) + \boldsymbol{\lambda}_{BC} \cdot (\mathbf{r}_\parallel \times \mathbf{F}_\perp) + \boldsymbol{\lambda}_{BC} \cdot (\mathbf{r}_\perp \times \mathbf{F}_\parallel) \\ &\quad + \boldsymbol{\lambda}_{BC} \cdot (\mathbf{r}_\perp \times \mathbf{F}_\perp) \\ &= 0 + 0 + 0 + \boldsymbol{\lambda}_{BC} \cdot (\mathbf{r}_\perp \times \mathbf{F}_\perp) \end{aligned}$$

$$M_{BC} = \boldsymbol{\lambda}_{BC} \cdot (\mathbf{r}_\perp \times \mathbf{F}_\perp) \tag{4.37}$$

Since both $\mathbf{r}_\perp$ and $\mathbf{F}_\perp$ are perpendicular to the axis BC, the direction of $\mathbf{r}_\perp \times \mathbf{F}_\perp$ must be parallel to $\boldsymbol{\lambda}_{BC}$. Thus, from Fig. 4.31 and the definition of the cross product, we write

$$\mathbf{r}_\perp \times \mathbf{F}_\perp = (r_\perp F_\perp \sin\theta)(\pm\boldsymbol{\lambda}_{BC})$$

$$\begin{aligned} |M_{BC}| &= |\boldsymbol{\lambda}_{BC} \cdot (r_\perp F_\perp \sin\theta)(\pm\boldsymbol{\lambda}_{BC})| = r_\perp F_\perp \sin\theta |\pm\boldsymbol{\lambda}_{BC} \cdot \boldsymbol{\lambda}_{BC}| \\ &= (r_\perp \sin\theta)F_\perp = d_s F_\perp \end{aligned}$$

$$\boxed{|M_{BC}| = d_s F_\perp} \tag{4.38}$$

where d_s is the *shortest distance* between the axis BC and the line of action of $\mathbf{F}$.

EXAMPLE 4.12

A T-shaped lever is supported by bearings at B and C and leans against a wall at A. The lever is found to have an impending rotation about the axis BC when a 50.7-lb force $\mathbf{F}$ is applied to it as shown. Determine (a) the moment M_{BC} of $\mathbf{F}$ about the axis BC when rotation impends, (b) the analytical expression of $\mathbf{M}_{BC}$, (c) the shortest distance d_s between the axis BC and the line AD.

[†]A five-point star preceding a section number indicates that the section is an optional section as described in the Preface.

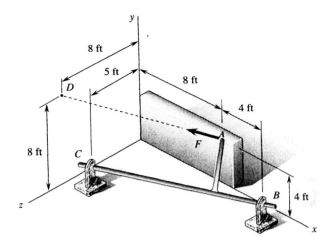

Solution. For determining M_{BC}, we may choose $\mathbf{r}$ to be $\overrightarrow{BA}$ and write

$$\mathbf{r} = \overrightarrow{BA} = -4\mathbf{i} + 4\mathbf{j}$$

$$\overrightarrow{BC} = -12\mathbf{i} + 5\mathbf{k} \qquad \boldsymbol{\lambda}_{BC} = \frac{1}{13}(-12\mathbf{i} + 5\mathbf{k})$$

$$\overrightarrow{AD} = -8\mathbf{i} + 4\mathbf{j} + 8\mathbf{k} \qquad \boldsymbol{\lambda}_{AD} = \frac{1}{3}(-2\mathbf{i} + \mathbf{j} + 2\mathbf{k})$$

$$\mathbf{F} = 50.7\boldsymbol{\lambda}_{AD} = 16.9(-2\mathbf{i} + \mathbf{j} + 2\mathbf{k})$$

$$M_{BC} = \boldsymbol{\lambda}_{BC} \cdot (\mathbf{r} \times \mathbf{F}) = \frac{16.9}{13} \begin{vmatrix} -12 & 0 & 5 \\ -4 & 4 & 0 \\ -2 & 1 & 2 \end{vmatrix} = 1.3\,(-76)$$

$$M_{BC} = -98.8 \ \text{lb}\cdot\text{ft} \quad \blacktriangleleft$$

The negative sign in the value of M_{BC} indicates that the sense of the moment of $\mathbf{F}$ about the axis BC is opposite to the sense of $\boldsymbol{\lambda}_{BC}$. This may be confirmed by studying the figure and applying the right-hand rule. For the vector form of the moment, we write

$$\mathbf{M}_{BC} = M_{BC}\boldsymbol{\lambda}_{BC} = \frac{-98.8}{13}(-12\mathbf{i} + 5\mathbf{k})$$

$$\mathbf{M}_{BC} = 91.2\mathbf{i} - 38\mathbf{k} \ \ \text{lb}\cdot\text{ft} \quad \blacktriangleleft$$

For finding the shortest distance d_s, we need to compute $F_\perp$. Placing $\boldsymbol{\lambda}_{BC}$ at A and letting the angle between $\mathbf{F}$ and $\boldsymbol{\lambda}_{BC}$ be θ, we write

$$F_\perp = F\sin\theta = 1 \cdot F \cdot \sin\theta = |\boldsymbol{\lambda}_{BC} \times \mathbf{F}|$$
$$= |1.3(-5\mathbf{i} + 14\mathbf{j} - 12\mathbf{k})| = 24.84$$

By Eq. (4.38), we have

$$|-98.8| = d_s\,(24.84) \qquad d_s = 3.98 \ \text{ft} \quad \blacktriangleleft$$

Fig. D4.45

Developmental Exercises

D4.43   The moment of a force $\mathbf{F}$ about an axis BC is 100 lb·ft. If the component of $\mathbf{F}$ perpendicular to BC is $F_\perp = 20$ lb, determine the shortest distance d_s between the axis BC and the line of action of $\mathbf{F}$.

D4.44   The moments of a force $\mathbf{F}$ about the x, y, and z axes are 450 lb·ft, −700 lb·ft, and 500 lb·ft, respectively. What is the analytical expression of the moment $\mathbf{M}_O$ of $\mathbf{F}$ about the origin O?

D4.45   A force $\mathbf{F}$ and an axis BC are shown. Determine (a) the unit vector $\boldsymbol{\lambda}_{BC}$, (b) the moment M_{BC} of $\mathbf{F}$ about the axis BC, (c) $\mathbf{M}_{BC}$.

D4.46   Refer to D4.45. Determine (a) $\boldsymbol{\lambda}_{BC} \times \mathbf{F}$, (b) $|\boldsymbol{\lambda}_{BC} \times \mathbf{F}|$, (c) the perpendicular component $F_\perp$ of $\mathbf{F}$ with respect to the axis BC, (d) the shortest distance d_s between the axis BC and the line of action of $\mathbf{F}$.

D4.47   Which of the following statements are true?
(a) The moment of a couple is a free vector.
(b) The moment of a couple about a point in space is the sum of the moments of the two forces in the couple about that point.
(c) The moment of a couple about any point has the same value.
(d) The moment of a couple about any axis has the same value.
(e) The moment of a couple is a vector perpendicular to the plane containing the two forces in the couple.

PROBLEMS

4.57   Refer to the pyramid in Prob. 4.36. Determine (a) the interior angle $\theta_{ABC/xy}$ between the side ABC and the xy plane, (b) the interior angle $\theta_{ABC/ACD}$ between the side ABC and the side ACD.

4.58*   Refer to the pyramid in Prob. 4.37*. Determine (a) the interior angle $\theta_{ABC/ACD}$ between the side ABC and the side ACD, (b) the interior angle $\theta_{ACD/ADE}$ between the side ACD and the side ADE.

4.59* and 4.60   For the system shown, determine (a) the moment $\mathbf{M}_O$ of $\mathbf{F}$ about the point O, (b) the projection of $\mathbf{M}_O$ onto the axis OB, (c) the moment M_{OB} of $\mathbf{F}$ about the axis OB, (d) $\mathbf{M}_{OB}$.

Fig. P4.59*

Fig. P4.60 and P4.61

4.61   Using the scalar triple product, determine the moment M_{OB} of $\mathbf{F}$ about the axis OB as shown.

4.62 The moment of the force **P** acting at the point $A(5, 10, 20)$ mm about the z axis, as shown, shall not exceed the limit of 0.24 N·m. Determine the allowable magnitude of **P** which acts in the direction of

$$\lambda_P = -\frac{5}{13}\mathbf{i} + \frac{12}{13}\mathbf{k}$$

4.63* Solve Prob. 4.62 if **P** acts in the direction of

$$\lambda_P = -\frac{3}{5}\mathbf{i} + \frac{4}{5}\mathbf{k}$$

4.64 The faucet at the end of the pipe DF is to be replaced. In order to loosen the joint at the faucet and at the same time to prevent the pipe DF from rotating, a plumber uses two pipe wrenches EG and FH and exerts two 300-N forces parallel to the yz plane as shown. All threads at the joints are right-handed. (a) Compute the moment M_{BC} of the two 300-N forces. (b) Determine whether the two 300-N forces exerted by the plumber tend to tighten or loosen the joint between the pipe BC and the elbow B.

Fig. P4.62

Fig. P4.64

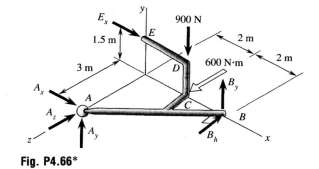

Fig. P4.66*

4.65 Refer to Prob. 4.64. Employing the concept of moments, determine the shortest distance d_s between the x axis and the line of action of the 300-N force acting at H.

4.66* The unknown and known forces in the force system shown have the same unit. Determine (a) the resultant moment $\mathbf{M}_A^R$ of the force system about the point A, (b) the resultant moment M_{AB}^R of the force system about the axis AB.

4.67 A laser beam emanates from the point P through the point Q, while another laser beam emanates from the point R through the point S. Employing the concept of moments, determine the shortest distance d_s between these two laser beams.

Fig. P4.67

4.18   Concluding Remarks

Three basic cases of moments of forces have been treated in this chapter. They are (1) the moment of a force about an axis, (2) the moment of a force about a point, and (3) the moment of a couple, which is a free vector having a purely torsional effect on a body. Note that the *moment* of a couple is an *attribute* of the couple and is often called the *torque of the couple*, or simply a *torque*. The primary objective of this chapter is to instill in engineering students a good understanding of moments of forces and their applications. A basic knowledge of moments of forces is a prerequisite to the study of the conditions of rigid bodies under the action of forces. The several equivalent versions of the right-hand rule for moment vectors, as presented in Sec. 4.2, should be very helpful in gaining a good understanding of the physical implications of the moments of forces.

Varignon's theorem as presented in Secs. 4.3 and 4.11 is, indeed, a direct consequence of the *parallelogram law* and the *distributive property of the cross product*. Historically, we note that the technique of composing forces by the *parallelogram law* had undergone more than a century of development before it was simultaneously published in 1687 by Newton in his *Philosophiæ Naturalis Principia Mathematica* and by Varignon in his *Projet d'uné nouvelle méchanique*. Although Varignon's *Projet* is not to be compared with Newton's *Principia*, Varignon did contribute his share to mechanics.

In general, any moment acting on a right cross section of a longitudinal member (e.g., a shaft, beam, column, or post) may be resolved into two rectangular components: one perpendicular and the other parallel to the right cross section. We recognize that the moment acting perpendicular to the right cross section of a longitudinal member is called a *torque,* which is a moment parallel to the axis of the member. The torque is a measure of the effectiveness or tendency of an action to rotate or twist a longitudinal member about its axis. However, the moment acting parallel to a right cross section of a longitudinal member is a measure of the effectiveness or tendency of an action to bend or flex the member at that cross section and is called the *bending moment*. An elementary study of bending moments in beams is presented in Chap. 8. A more in-depth study of bending moments is usually found in texts on mechanics of materials.

REVIEW PROBLEMS

4.68   A free body is acted on by a force system as shown, where unknown and known forces have the same unit. Determine the resultant moment $\mathbf{M}_A^R$ of the force system about the point A.

4.69   The end D of a pipeline is being adjusted with a 270-N force $\mathbf{F}$ as shown. Determine (a) the moment $\mathbf{M}_A$ of $\mathbf{F}$ about the joint A, (b) the moment M_{AB} of $\mathbf{F}$ about the axis of the pipe AB, (c) whether the action of $\mathbf{F}$ tends to tighten or loosen the joint A where the threads are right-handed, (d) the shortest distance between the point A and the line of action of $\mathbf{F}$, (e) the shortest distance between the line containing the axis AB and the line of action of $\mathbf{F}$.

Fig. P4.68

A (0,4,0)
E (0,2,4)
D (4,0,4)
B (2,2,4)

600 mm

400 mm

400 mm

200 mm

200 mm

y

A

E

z

B

C

F D

200 mm
200 mm
200 mm

x

Fig. P4.69

500 N 2 m

200 N

3 m

y

C

500 N

B

D

6 m

O

A

P

z

E

F

200 N

x

Fig. P4.70

4.70 A system of forces consists of a 210-N force **P** and two couples as shown. Determine (a) the resultant moment **M** of the two couples, (b) the moment M_z of **P** and the two couples about the z axis.

4.71 A free body is acted on by a force system as shown, where the unknown and known forces have the same unit. Determine (a) the resultant moment $\mathbf{M}_A^R$ of the force system about the point A, (b) the resultant moment M_{AB}^R of the force system about the axis passing through the points A and B.

4.72* A laser beam emanates from the point P through the point Q, while another laser beam emanates from the point R through the point S. Employing the concept of moments, determine the shortest distance d_s between these two laser beams.

y

E

B_y

4 m

B_z

A_y

T_{DE}

B

B_x

A_x

A

D

2 m

z

2 m

C

2 m

4 m

x

300 N

600 N·m

Fig. P4.71

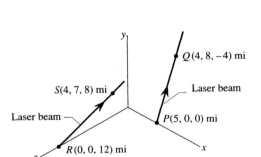

y

$Q(4, 8, -4)$ mi

$S(4, 7, 8)$ mi

Laser beam

Laser beam

$P(5, 0, 0)$ mi

z

$R(0, 0, 12)$ mi

x

Fig. P4.72*

Chapter 5

Equilibrium of Rigid Bodies

KEY CONCEPTS

- Wrench, equivalent systems of forces, equipollent systems of forces, equilibrium of a rigid body, two-force body, three-force body, statically indeterminate, partially constrained, and improperly constrained.
- Reduction of a system of forces to a force-moment system at a chosen point.
- Reactions from supports and connections for rigid bodies in a plane and in space, where the force systems are nonconcurrent.
- Maximum number of statically determinate unknowns.
- Methods of solution for equilibrium problems of simple rigid bodies in a plane and in space.

When a body has no translational acceleration, the resultant force of the force system shown in the free-body diagram of the body must be equal to zero. Similarly, if a body has no rotational acceleration, the resultant moment, about a point, of that force system must be equal to zero. In general, a rigid body may be caused to have either translational or rotational acceleration, or both. Thus, the equilibrium of a rigid body requires that the aforementioned resultant force and resultant moment be both equal to zero. This chapter is focused on the basic study of equilibrium problems of *simple* rigid bodies acted on by forces in a plane and by forces in space, where the solution of each problem generally entails the drawing of only *one* free-body diagram. Equilibrium problems which require more advanced considerations are treated in later chapters.

5.1 Resolution of a Given
Force into a Force-Moment
System at Another Point

125

EQUIVALENT SYSTEMS OF FORCES

5.1 Resolution of a Given Force into a Force-Moment System at Another Point[†]

The action of a force **F** applied at a given point A of a rigid body is *equivalent to the combined action* of the force **F** applied at another point B of the rigid body and an applied moment $\mathbf{M}_B$ which is equal to the moment of **F** about B when **F** was applied at A. This may be verified step by step as shown in Fig. 5.1, where we first apply the *rigid-body principle*[††] and then employ the fact that the action of a couple can be replaced by the action of the moment of the couple.

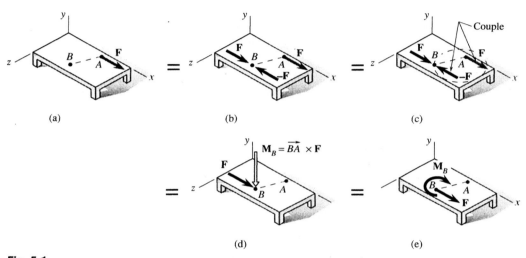

Fig. 5.1

Note that the moment $\mathbf{M}_B$ may be represented graphically by a hollow arrow or a curved arrow, as shown in Fig. 5.1. The moment $\mathbf{M}_B$ is given by

$$\mathbf{M}_B = \overrightarrow{BA} \times \mathbf{F} \tag{5.1}$$

We note from the definition of the cross product that the moment $\mathbf{M}_B$ is perpendicular to $\overrightarrow{BA}$ and **F**. The moment $\mathbf{M}_B$ is actually a free vector, but it is, for convenience, shown to act at the point B. Furthermore, it can be shown that the new force system in Fig. 5.1(e) and the original force in Fig. 5.1(a) do have the *same resultant moment* $\mathbf{M}_P$ about any point P as well as the *same resultant force* **F**.

[†]A *force-moment system* is also called a *force-couple system*. Cf. the footnote on p. 90.
[††]Cf. Sec. 1.12.

Fig. D5.2

Developmental Exercises

D5.1 The moment of the force $\mathbf{F}$ at the point A in Fig. 5.1(a) about any point P is $\mathbf{M}_P = \overrightarrow{PA} \times \mathbf{F}$, while the moment of the force-moment system in Fig. 5.1(e) about the point P is $\mathbf{M}_P' = \overrightarrow{PB} \times \mathbf{F} + \mathbf{M}_B$, where $\mathbf{M}_B = \overrightarrow{BA} \times \mathbf{F}$. Show that $\mathbf{M}_P = \mathbf{M}_P'$.

D5.2 A 100-N vertical force $\mathbf{P}$ acts at the point A as shown. (a) Resolve $\mathbf{P}$ into a force-moment system $\mathbf{F}_O$ and $\mathbf{M}_O$ acting at the origin O. (b) Resolve $\mathbf{P}$ into a force-moment system $\mathbf{F}_B$ and $\mathbf{M}_B$ acting at the point B. (c) Does each of the force-moment systems obtained in (a) and (b) produce a moment about the point C equal to $\overrightarrow{CA} \times \mathbf{P}$?

5.2 Reduction of a System of Forces to a Force-Moment System

It follows from Sec. 5.1 that a system of forces $\mathbf{F}_1, \mathbf{F}_2, \ldots, \mathbf{F}_n$ acting on a rigid body may be moved from their respective points of application $A_1, A_2, \ldots, A_n$ to a chosen point O on the rigid body, as shown in Fig. 5.2(a), if the moments $\mathbf{M}_1, \mathbf{M}_2, \ldots, \mathbf{M}_n$ equal to the moments of the forces $\mathbf{F}_1, \mathbf{F}_2, \ldots, \mathbf{F}_n$ about O are also added to act on the rigid body, as shown in Fig. 5.2(b), where

$$\mathbf{M}_i = \overrightarrow{OA_i} \times \mathbf{F}_i \qquad (i = 1, 2, \ldots n) \qquad (5.2)$$

The forces $\mathbf{F}_1, \mathbf{F}_2, \ldots, \mathbf{F}_n$ and the moments $\mathbf{M}_1, \mathbf{M}_2, \ldots, \mathbf{M}_n$ acting at O in Fig. 5.2(b) may, in turn, be combined to yield a single resultant force $\mathbf{R}$ and a single resultant moment $\mathbf{M}_O^R$, as shown in Fig. 5.2(c), where

$$\mathbf{R} = \mathbf{F}_1 + \mathbf{F}_2 + \cdots + \mathbf{F}_n$$

$$\mathbf{M}_O^R = \mathbf{M}_1 + \mathbf{M}_2 + \cdots + \mathbf{M}_n \qquad (5.3)$$

Thus, any system of forces may be reduced to a force-moment system as defined by Eqs. (5.2) and (5.3) and illustrated in Fig. 5.2.

A force-moment system consisting of a force $\mathbf{F}$ and a moment $\mathbf{M}$ which are parallel to each other is, by tradition, called a *wrench* (perhaps, a *screw-*

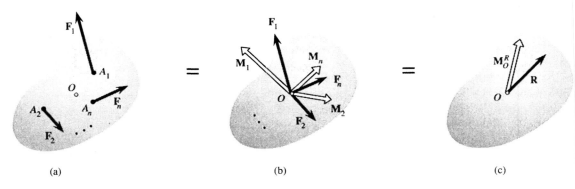

(a)   (b)   (c)

Fig. 5.2

**5.2 Reduction of a
System of Forces to a
Force-Moment System**

127

driver is a better name).[†] The action of a wrench is physically equivalent to a push or pull along an axis plus a twist about that axis. The line of action of **F** is called the *axis of the wrench*, and the ratio M/F is called the *pitch of the wrench*. If **F** and **M** of the wrench are of the same sense, the wrench is a *positive wrench*; otherwise, the wrench is a *negative wrench*. This is illustrated in Fig. 5.3.

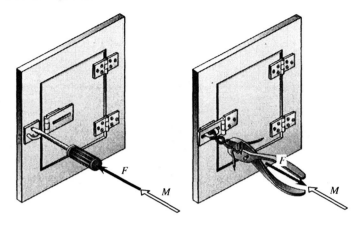

Positive Wrench   Negative Wrench

Fig. 5.3

When the moment **M** is zero, the wrench degenerates into a force **F**; when the force **F** is zero, the wrench degenerates into a moment **M**. A *zero* (or *null*) *wrench* is a condition in which both the force **F** and the moment **M** are zero. It will be shown in Sec. 5.4 that any force system acting on a rigid body can be reduced to a wrench, or one of its degenerate cases, at an appropriate location of the rigid body (or its massless extension).

[†]The combined action of a force **F** and a moment **M** parallel to **F** is more closely simulated by the action transmitted by a *screwdriver* or a pair of *pliers*, as illustrated in Fig. 5.3, than the action transmitted by a common *wrench*, which usually turns, but does not push or pull.

EXAMPLE 5.1

Reduce the system of forces shown to a force-moment system consisting of a force **R** and a moment $\mathbf{M}_O^R$ at the origin O.

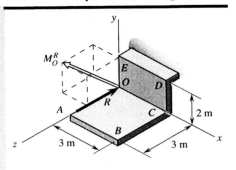

Solution. For the forces at the points A, B, C, D, and E, as shown, we write

$$\mathbf{R} = \Sigma \mathbf{F} = -\mathbf{j} + 2\mathbf{i} + \mathbf{j} - 2\mathbf{i} - 4\mathbf{k}$$

$$\mathbf{R} = -4\mathbf{k} \text{ kN} \quad \blacktriangleleft$$

$$\mathbf{M}_O^R = \Sigma(\mathbf{r} \times \mathbf{F})$$
$$= 3\mathbf{k} \times (-\mathbf{j}) + (3\mathbf{i} + 3\mathbf{k}) \times (2\mathbf{i}) + 3\mathbf{i} \times \mathbf{j}$$
$$+ (3\mathbf{i} + 2\mathbf{j}) \times (-2\mathbf{i}) + 2\mathbf{j} \times (-4\mathbf{k})$$

$$\mathbf{M}_O^R = -5\mathbf{i} + 6\mathbf{j} + 7\mathbf{k} \text{ kN} \cdot \text{m} \quad \blacktriangleleft$$

Developmental Exercises

D5.3 The forces $\mathbf{F}_1 = 3\mathbf{j}$ kN and $\mathbf{F}_2 = 4\mathbf{i}$ kN act at the points $A_1(3, 0, 0)$ m and $A_2(0, 0, 2)$ m, respectively. They are to be reduced to a force $\mathbf{R}$ acting at the point $B(1, 0, 1)$ m and a moment $\mathbf{M}_B^R$. Determine (a) $\mathbf{R}$, (b) the moment $\mathbf{M}_1$ equal to the moment of $\mathbf{F}_1$ about B, (c) the moment $\mathbf{M}_2$ equal to the moment of $\mathbf{F}_2$ about B, (d) $\mathbf{M}_B^R$.

D5.4 Define the terms: (a) wrench, (b) axis of the wrench, (c) pitch of the wrench, (d) positive wrench, (e) negative wrench, (f) zero wrench.

D5.5 A negative wrench consists of a force $\mathbf{F} = 2\mathbf{i} - \mathbf{j} - 2\mathbf{k}$ lb and a moment $\mathbf{M}$ of 15 lb·ft. What is the analytical expression of $\mathbf{M}$?

5.3   Equivalent and Equipollent Systems of Forces

We have learned in Sec. 5.2 that any system of forces acting on a rigid body may be reduced to a force-moment system at a given point. The action of the original system of forces is shown to be equivalent to the action of such a force-moment system. Therefore, *equivalent systems of forces* are force systems which act on a rigid body and are reducible to the same force-moment system at a given point, and either system may be referred to as a *replacement* of the other.

If the unprimed system of forces $\mathbf{F}_1$, $\mathbf{F}_2$, . . . , $\mathbf{F}_n$ is equivalent to the primed system of forces $\mathbf{F}_1'$, $\mathbf{F}_2'$, . . . , $\mathbf{F}_m'$ then their resultant forces $\mathbf{R}$ and $\mathbf{R}'$ and their resultant moments $\mathbf{M}_P^R$ and $(\mathbf{M}_P^R)'$ about a given point P, as shown in Fig. 5.4, must respectively be equal; i.e.,

$$\mathbf{R} = \mathbf{R}' \qquad \mathbf{M}_P^R = (\mathbf{M}_P^R)' \tag{5.4}$$

$$\mathbf{R} = \mathbf{F}_1 + \mathbf{F}_2 + \cdots + \mathbf{F}_n \tag{5.5a}$$

$$\mathbf{R}' = \mathbf{F}_1' + \mathbf{F}_2' + \cdots + \mathbf{F}_m' \tag{5.5b}$$

$$\mathbf{M}_P^R = \mathbf{M}_1 + \mathbf{M}_2 + \cdots + \mathbf{M}_n \tag{5.6a}$$

$$(\mathbf{M}_P^R)' = \mathbf{M}_1' + \mathbf{M}_2' + \cdots + \mathbf{M}_m' \tag{5.6b}$$

where $\mathbf{M}_1$, $\mathbf{M}_2$, . . . , $\mathbf{M}_n$ are the moments of $\mathbf{F}_1$, $\mathbf{F}_2$, . . . , $\mathbf{F}_n$ about the given point P, respectively; and $\mathbf{M}_1'$, $\mathbf{M}_2'$, . . . , $\mathbf{M}_m'$ are the moments of $\mathbf{F}_1'$, $\mathbf{F}_2'$, . . . , $\mathbf{F}_m'$ about P, respectively. Equations (5.4) represent the

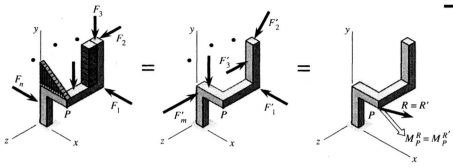

Fig. 5.4

necessary and sufficient conditions for equivalent systems of forces acting on a rigid body.

If two systems of forces satisfy Eqs. (5.4) but act on a set of nonrigidly connected bodies, where different forces may act on the various individual bodies, then they are called *equipollent systems of forces*, which may have *different effects* on the individual nonrigidly connected bodies. This is illustrated for systems containing only one force in Fig. 5.5, where = indicates *equivalence* and ⇒ indicates *equipollence*. The scalar version of Eqs. (5.4) may be written as

$$\Sigma F_x = \Sigma F'_x \qquad \Sigma F_y = \Sigma F'_y \qquad \Sigma F_z = \Sigma F'_z$$
$$\Sigma (M_P)_x = \Sigma (M_P)'_x \quad \Sigma (M_P)_y = \Sigma (M_P)'_y \quad \Sigma (M_P)_z = \Sigma (M_P)'_z \tag{5.7}$$

If the forces in the system all lie in the xy plane, Eqs. (5.7) reduce to

$$\Sigma F_x = \Sigma F'_x \qquad \Sigma F_y = \Sigma F'_y \qquad \Sigma M_P = \Sigma M'_P \tag{5.8}$$

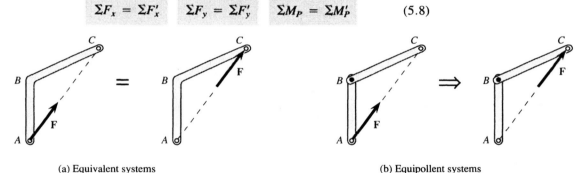

(a) Equivalent systems   (b) Equipollent systems   Fig. 5.5

EXAMPLE 5.2

Suppose that there are two equivalent systems of forces as shown. Determine the magnitudes of the forces **F** and **Q** and the moment **M**.

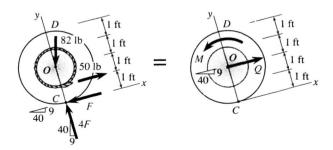

Solution.   Since the hypotenuse of the slope triangle is 41, we write

$$+\nwarrow \Sigma F_y = +\nwarrow \Sigma F_y': \qquad 4F - 82\left(\frac{40}{41}\right) = 0$$

$$+\nearrow \Sigma F_x = +\nearrow \Sigma F_x': \qquad -F + 50 - 82\left(\frac{9}{41}\right) = Q$$

$$+\,\circlearrowleft \Sigma M_C = +\,\circlearrowleft \Sigma M_C': \qquad -1(50) + 2(82)\left(\frac{9}{41}\right) = M - 2Q$$

Solving these three equations, we obtain

$$F = 20 \text{ lb} \qquad Q = 12 \text{ lb} \qquad M = 10 \text{ lb·ft} \quad \blacktriangleleft$$

For a check on these answers, we write

$$+\,\circlearrowleft \ \Sigma M_D = +\,\circlearrowleft \ \Sigma M_D':$$

$$-2(82)\left(\frac{9}{41}\right) + 3(50) - 4(20) = 10 + 2(12)$$

Developmental Exercises

D5.6   Define the terms: (a) equivalent systems of forces, (b) equipollent systems of forces.

D5.7   Which of the six force systems shown are equivalent?

(a)   (b)   (c)   (d)   (e)   (f)

Fig. D5.7

★5.4   Reduction of a System of Forces to a Wrench

We saw in Sec. 5.2 that a system of forces may be reduced to a force-moment system consisting of a resultant force **R** acting at a chosen point O and a resultant moment $\mathbf{M}_O^R$ about O as illustrated in Fig. 5.6(a). If $\mathbf{M}_O^R$ is not parallel to **R** and neither of them is zero, we may resolve $\mathbf{M}_O^R$ into two rectangular components: $\mathbf{M}_\parallel$ and $\mathbf{M}_\perp$, where $\mathbf{M}_\parallel \parallel \mathbf{R}$ and $\mathbf{M}_\perp \perp \mathbf{R}$, as shown in Fig. 5.6(b). Letting $\lambda_R = \mathbf{R}/R$, we write

$$\mathbf{M}_O^R = \mathbf{M}_\parallel + \mathbf{M}_\perp \qquad \mathbf{M}_\parallel = (\lambda_R \cdot \mathbf{M}_O^R)\lambda_R \qquad \mathbf{M}_\perp = \mathbf{M}_O^R - \mathbf{M}_\parallel \qquad (5.9)$$

By Sec. 5.1, we may equivalently replace **R** acting at the point O with **R** acting at a point P plus a moment $\mathbf{M}_P$ such that

$$\mathbf{M}_P = \overrightarrow{PO} \times \mathbf{R} = -\mathbf{M}_\perp \qquad (5.10)$$

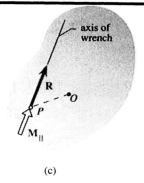

(a)   (b)   (c)

Fig. 5.6

Then, $\mathbf{M}_P$ cancels $\mathbf{M}_\perp$ and the original system of forces is reduced to a *wrench* consisting of $\mathbf{M}_\parallel$ and $\mathbf{R}$ acting at the point P of the rigid body (or its massless extension), as shown in Fig. 5.6(c). If $\mathbf{M}_\parallel = \mathbf{0}$, then the system of forces is equivalent to a single force $\mathbf{R}$ at P.

EXAMPLE 5.3

Reduce the system of forces in Example 5.1 to a wrench.

Solution. From the solution in Example 5.1, we know that the given system of forces is equivalent to a force-moment system $\mathbf{R}$ and $\mathbf{M}_O^R$ at the origin O, where

$$\mathbf{R} = -4\mathbf{k} \text{ kN} \qquad \mathbf{M}_O^R = -5\mathbf{i} + 6\mathbf{j} + 7\mathbf{k} \text{ kN·m}$$

As illustrated, we first resolve $\mathbf{M}_O^R$ into $\mathbf{M}_\parallel$ and $\mathbf{M}_\perp$ as follows:

$$\boldsymbol{\lambda}_R = \frac{\mathbf{R}}{R} = -\mathbf{k} \qquad \mathbf{M}_\parallel = (\boldsymbol{\lambda}_R \cdot \mathbf{M}_O^R)\boldsymbol{\lambda}_R = 7\mathbf{k}$$

$$\mathbf{M}_\perp = \mathbf{M}_O^R - \mathbf{M}_\parallel = -5\mathbf{i} + 6\mathbf{j}$$

Next, we move the resultant force $\mathbf{R}$ to the point $P(x, y, z)$ and add a moment equal to $\overrightarrow{PO} \times \mathbf{R}$ to keep the effect of the force system on the body unchanged. If the point P is chosen so that the added moment is $\mathbf{M}'_\perp = -\mathbf{M}_\perp$ as shown, then a wrench is obtained. Thus, we write

$$\overrightarrow{PO} \times \mathbf{R} = -\mathbf{M}_\perp: \quad (-x\mathbf{i} - y\mathbf{j} - z\mathbf{k}) \times (-4\mathbf{k}) = -(-5\mathbf{i} + 6\mathbf{j})$$

$$4y\mathbf{i} - 4x\mathbf{j} = 5\mathbf{i} - 6\mathbf{j}$$

The ensuing scalar equations are

$$4y = 5 \qquad \text{and} \qquad -4x = -6$$

which yield $x = 1.5$, $y = 1.25$, and z can have any value (including zero). Thus, we may say that the wrench acts through the point $P(1.5, 1.25, 0)$ m and consists of

$$\mathbf{R} = -4\mathbf{k} \text{ kN} \qquad \text{and} \qquad \mathbf{M}_\| = 7\mathbf{k} \text{ kN} \cdot \text{m} \blacktriangleleft$$

Fig. D5.8

Developmental Exercise

D5.8 The force system shown is to be reduced first to a force-moment system at the origin O and then to a wrench at a point P. Determine (a) the resultant force $\mathbf{R}$, (b) the resultant moment $\mathbf{M}_O^R$ about O, (c) λ_R, (d)$\mathbf{M}_\|$, (e) $\mathbf{M}_\perp$, (f) $\vec{PO}$ from $\vec{PO} \times \mathbf{R} = -\mathbf{M}_\perp$, (g) the coordinates of P.

PROBLEMS

5.1 and 5.2 Resolve the force shown into a force-moment system at the origin O.

Fig. P5.1 and P5.3

Fig. P5.2 and P5.4*

Fig. P5.5, P5.9, and P5.21

5.3 and 5.4* Resolve the force shown into a force-moment system at the point C.

5.5 through 5.8* For the system of forces shown, determine its equivalent force-moment system at the origin O.

Fig. P5.6, P5.10, and P5.22

Fig. P5.7, P5.11, and P5.23

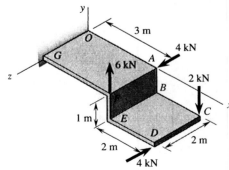

Fig. P5.8*, P5.12*, and P5.24*

5.9 through 5.12* For the system of forces shown, determine its equivalent force-moment system at the point A.

5.13 and 5.14* Four equivalent force systems are shown. Determine the magnitudes of **P**, **Q**, and **M** in the force systems (b), (c), and (d).

Fig. P5.13

Fig. P5.14*

5.15 Suppose that there are two equivalent systems of forces as shown. Determine the magnitudes of the forces **P** and **Q** and the moment **M**.

Fig. P5.15

Fig. P5.16

5.16 The two force systems shown are known to be equivalent. Determine the magnitudes of the forces **P** and **Q** and the distance d.

5.17 Three men have attempted twice to free their boat from a sand bar as shown in the left and right figures. If the effects of their efforts to free the boat during these two attempts are equivalent and $A_y = 100$ N, $C = 150$ N, determine the values of A_x, B, P_y, P_z, Q, and S.

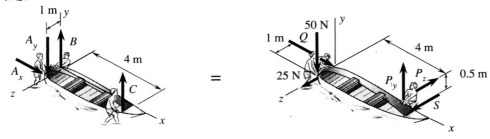

Fig. P5.17

5.18* Solve Prob. 5.17 if $A_y = 80$ N, $C = 125$ N.

5.19 and 5.20* Two equivalent force systems are shown. Determine (a) the magnitudes of **P**, **Q**, and **S**, (b) the wrench equivalent to either of the force systems.

Fig. P5.19

Fig. P5.20*

5.21 through 5.24* *(The figures for these problems are shown and specified along with those for Probs. 5.5 through 5.8* on p. 132.)* Reduce the system of forces shown to either a wrench or a single force.

EQUILIBRIUM IN A PLANE

5.5   Reactions from Supports and Connections in a Plane

Reactions from the supports or connections of a rigid body are the *constraining forces and moments* acting on the body at the points of support or connection. Each *reaction* must reflect the *restriction of movement* imposed on the body at the point of support or connection. Therefore, the number of components of reaction must always correspond to the number of ways in which the support or connection is *capable of restricting* the movement of the body at the point of support or connection. If the sense of a component of reaction is initially not known from the conditions of constraint and loading, it may first be assumed in the analysis and checked later in the subsequent solution.[†]

Note that, unless otherwise indicated, the weights of rigid bodies are usually assumed to be negligibly small compared with the applied loads.

[†]Cf. the footnote on p. 59.

Furthermore, the forces acting on a rigid body in *equilibrium in a plane* are *coplanar* but generally *nonconcurrent*. A rigid body acted on by coplanar but nonconcurrent forces is referred to as a *rigid body in a plane*. Likewise, supports or connections of a rigid body in a plane are referred to as *supports* or *connections in a plane*. The reactions from supports and connections in a plane may be classified into the following three types:

TYPE I: REACTIONS WITH ONE UNKNOWN. This type of reaction is equivalent to a force with either unknown magnitude but known line of action or known magnitude but unknown line of action in the plane of the system. Supports and connections producing this type of reaction include *rollers, rockers, smooth surfaces, collars on smooth rods, pins in smooth slots, sliders in smooth slots, cordlike elements*, and *short links*. Each of these supports and connections *allows rotation about the point of support but restricts translation in only one direction*. The unknown magnitude of the reaction is usually represented by an appropriate letter as shown in Fig. 5.7.

Fig. 5.7 Reactions with one unknown.

TYPE II: REACTIONS WITH TWO UNKNOWNS. This type of reaction is equivalent to a force with two unknown rectangular components, or equivalently, a force with an unknown magnitude and an unknown direction in the plane of the system. Supports and connections producing this type of reaction include *knife edges, rough surfaces, smooth corners, hinges, pins*, and *pivots*. Each of these supports and connections *allows rotation about the point of support but restricts translation in two directions*. Therefore, this type of reaction is usually represented by *two rectangular force components* as shown in Fig. 5.8. Its resultant form is a force at an angle.

Fig. 5.8 Reactions with two unknowns.

Fig. 5.9 Reactions with three unknowns.

TYPE III: REACTIONS WITH THREE UNKNOWNS. This type of reaction is equivalent to a force with two unknown rectangular components (or equivalently, a force with an unknown magnitude and an unknown direction) plus a moment with an unknown magnitude but a known orientation which is perpendicular to the plane of the system. This type of reaction is caused by a *fixed support* which is also called a *clamped* or *built-in support*, such as the support of a cantilever beam. A fixed support *restricts both translation and rotation* at the location of the support. In this case, the rotation is resisted by the moment developed at the fixed support. Thus, we usually represent this type of reaction by *two rectangular force components and a reaction moment* as shown in Fig. 5.9.

Developmental Exercises

D5.9 (a) Describe reactions from supports and connections for a rigid body. (b) What is a rigid body in a plane? (c) What are supports or connections in a plane?

D5.10 Give two examples of supports or connections that produce reactions with (a) one unknown, (b) two unknowns.

EXAMPLE 5.4

Three rigid bodies in equilibrium in a plane are shown. Draw the free-body diagram of each body.

(a)   (b)   (c)

Solution. Recalling the guides for drawing a good free-body diagram as stated in Sec. 3.8 and recognizing the types of reactions from the supports, we may draw the free-body diagram of each rigid body as shown. Note that

5.6 Equilibrium of a
Rigid Body in a Plane
137

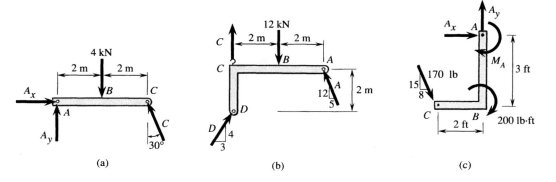

(a)　　　　　(b)　　　　　(c)

we use the letters naming the points of supports, with appropriate subscripts
as needed, to label the magnitudes of the reaction forces. A reaction moment
is labeled with the letter M, and the letter naming the point of support is
added to M as a subscript. Moreover, note that the senses of the unknown
reactions are assumed at this stage of the study.[†]

Developmental Exercises

D5.11   What are the possible ways in which we may represent the reaction from
a fixed support for a rigid body in a plane?

D5.12   Draw the free-body diagram of each of the rigid bodies shown.

(a)　　　　　(b)　　　　　(c)　　　　　(d)

Fig. D5.12

5.6 ˙Equilibrium of a Rigid Body in a Plane

The *equilibrium of a rigid body* is a state of the rigid body in which *all
particles* of the body are at rest or moving uniformly along straight lines in
an inertial reference frame (cf. Sec. 3.1). The equilibrium of a rigid body is
maintained if it is subjected to a balanced system of external forces. In this
balanced system, the resultant force $\mathbf{R}$ and resultant moment $\mathbf{M}_P^R$ of the force
system $\mathbf{F}_1, \mathbf{F}_2, \ldots, \mathbf{F}_n$ about an arbitrary point P are both equal to zero

---

[†]Continued in Example 5.5.

as illustrated in Fig. 5.10,[†] where $\mathbf{R} = \Sigma\mathbf{F}$ and $\mathbf{M}_P^R = \Sigma\mathbf{M}_P$. Therefore, the *necessary conditions* for equilibrium of a rigid body are

$$\Sigma\mathbf{F} = \mathbf{0} \qquad \Sigma\mathbf{M}_P = \mathbf{0} \qquad (5.11)$$

As an example, note that the force system in the free-body diagram of a circular disk rotating frictionlessly and uniformly about its axis of symmetry will satisfy Eqs. (5.11). However, *all particles* away from the axis of this disk are moving along circular paths with accelerations directed inward toward that axis. By the preceding definition, this rotating disk is *not* in equilibrium. Thus, Eqs. (5.11) do *not* constitute the *sufficient conditions* for equilibrium of a rigid body in the kinematic sense.[††]

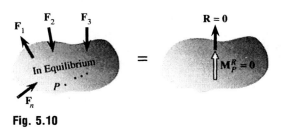

Fig. 5.10

An *equilibrium problem of a rigid body in a plane* is characterized by the feature that a set of *coplanar but nonconcurrent* unknown forces will be encountered in the solution; however, no noncoplanar unknown forces will be encountered. For a rigid body in equilibrium in the *xy* plane, the *scalar equations of equilibrium* may be written from Eqs. (5.11) as

$$\Sigma F_x = 0 \qquad \Sigma F_y = 0 \qquad \Sigma M_P = 0 \qquad (5.12)$$

Although equations of equilibrium are satisfied by the force systems acting on all bodies in equilibrium, the quality or degree of being in equilibrium may be different for different situations. This fact is illustrated in Fig. 5.11 for three pin-supported uniform rods in a vertical plane. Suppose that

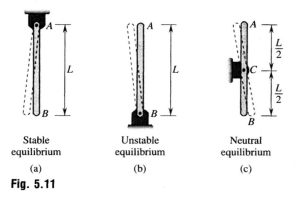

Fig. 5.11

[†]This means that the system of forces in the free-body diagram of a rigid body in equilibrium is equivalent to a *zero wrench*.

[††]In some texts, *equilibrium* was defined in kinetic (or force) terms, rather than in kinematic (or motion) terms. Consequently, Eqs. (5.11) were taken as the necessary and sufficient conditions for rigid-body equilibrium.

each rod is slightly displaced from its upright position of equilibrium and then released. It is obvious that the rod in Fig. 5.11(a) will move back toward its original position, the rod in Fig. 5.11(b) will move farther away from its original position, and the rod in Fig. 5.11(c) will stay in its new position. These three rods as shown are said to be in *stable*, *unstable*, and *neutral equilibrium*, respectively.

The *instability* exhibited by the rod in Fig. 5.11(b) is triggered by a small external disturbance given to the equilibrium configuration of the rod. This type of instability is different from the nonequilibrium of a rigid body due to insufficient or improper constraints on the rigid body. Recognitions of partially and improperly constrained rigid bodies are discussed in Secs. 5.8 and 5.12.

Developmental Exercises

D5.13 Define the equilibrium of a rigid body.

D5.14 What is the characterizing feature in an equilibrium problem of a rigid body in a plane?

EXAMPLE 5.5

Applying the equations of equilibrium, determine the reactions from the supports for each rigid body shown in Example 5.4. (The example here serves only as a preliminary study. Cf. Sec. 5.10.)

Solution. The free-body diagram of each rigid body is drawn as shown.

(a)

(b)

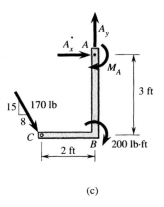
(c)

(1) Referring to the free-body diagram in Fig. (a), we write

$$+\, \circlearrowleft \, \Sigma M_A = 0: \quad 4(C \cos 30°) - 2(4) = 0 \qquad C = 2.3094$$

$$\xrightarrow{+}\, \Sigma F_x = 0: \quad A_x - C \sin 30° = 0 \qquad A_x = 1.155$$

$$+\uparrow \Sigma F_y = 0: \quad A_y + C \cos 30° - 4 = 0 \qquad A_y = 2$$

$$\mathbf{A}_x = 1.555 \text{ kN} \rightarrow \qquad \mathbf{A}_y = 2 \text{ kN} \uparrow \qquad \mathbf{C} = 2.31 \text{ kN } \measuredangle \ 60° \, \blacktriangleleft^\dagger$$

---

†We may alternatively combine the components $\mathbf{A}_x$ and $\mathbf{A}_y$ and report the reaction in its resultant form as $\mathbf{A} = 2.31$ kN $\measuredangle$ 60° or in its analytical expression as $\mathbf{A} = 1.155\mathbf{i} + 2\mathbf{j}$ kN.

(2) Referring to the free-body diagram in Fig. (b), we write

$$+\,\circlearrowright\ \Sigma M_D = 0: \quad 2\!\left(\frac{5}{13}A\right) + 4\!\left(\frac{12}{13}A\right) - 2(12) = 0$$

$$\xrightarrow{+}\ \Sigma F_x = 0: \quad \frac{3}{5}D - \frac{5}{13}A = 0$$

$$+\uparrow\ \Sigma F_y = 0: \quad \frac{4}{5}D + C + \frac{12}{13}A - 12 = 0$$

These three equations yield $A = 5.379$, $D = 3.448$, and $C = 4.276$. We write

$$\mathbf{A} = 5.38 \text{ kN} \searrow 67.4° \quad \mathbf{C} = 4.28 \text{ kN} \uparrow \ \blacktriangleleft$$

$$\mathbf{D} = 3.45 \text{ kN} \measuredangle 53.1° \ \blacktriangleleft$$

(3) Referring to the free-body diagram in Fig. (c), we write

$$+\,\circlearrowright\ \Sigma M_A = 0: \quad -M_A - 200 + 3\!\left(\frac{8}{17}\right)(170) + 2\!\left(\frac{15}{17}\right)(170) = 0$$

$$\xrightarrow{+}\ \Sigma F_x = 0: \quad A_x + \frac{8}{17}(170) = 0$$

$$+\uparrow\ \Sigma F_y = 0: \quad A_y - \frac{15}{17}(170) = 0$$

These three equations yield $M_A = 340$, $A_x = -80$, and $A_y = 150$. Since A_x is found to be negative, we write

$$\mathbf{A}_x = 80 \text{ lb} \leftarrow \quad \mathbf{A}_y = 150 \text{ lb} \uparrow \quad \mathbf{M}_A = 340 \text{ lb} \cdot \text{ft} \ \circlearrowright \ \blacktriangleleft$$

Developmental Exercises

D5.15　The frames shown were at rest. Upon the action of the applied loads, which will still be in equilibrium?

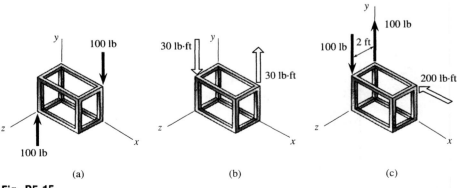

(a)　　　　　(b)　　　　　(c)

Fig. D5.15

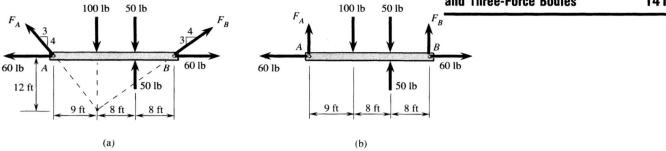

Fig. D5.16 and D5.17

D5.16 If equilibrium exists and F_A and F_B are to be determined, identify the type of equilibrium problem for each case as shown.

D5.17 The body AB is in equilibrium. Determine the unknown F_A and F_B in each case as shown.

D5.18 Three cylindrical pipes are in equilibrium on the supports shown. Identify the type of equilibrium in each case.

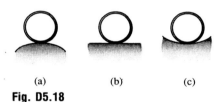

Fig. D5.18

5.7 Equilibrium of Two- and Three-Force Bodies

We know that a set of concurrent forces can always be combined to form a single resultant force at the point of concurrence. A *two-force body* is a rigid body acted on by a force system composed of, or reducible to, two forces at two different points. For instance, the body AB in Fig. 5.12(a) is a two-force body, whose free-body diagram is shown in Fig. 5.12(b) and (c). *When a two-force body is in equilibrium, the forces acting on it must be (i) directed along the line joining their points of application, (ii) opposite in direction, and (iii) equal in magnitude.* These three properties are illustrated in Fig. 5.12(c). The first property must be true; otherwise, a summation of moments of those two forces about either one of their points of application will not be zero. The second and third properties must be true; otherwise, a summation of forces in the direction of either one of those two forces will not be zero.

Fig. 5.12

Fig. 5.13

(a)

(b)

(c)

Fig. D5.20

A *three-force body* is a rigid body acted on by a force system composed of, or reducible to, three forces at three different points. *If a three-force body is in equilibrium, the lines of action of the three forces must be (1) coplanar*, and *(2) either concurrent or parallel to each other*. These properties are illustrated in Fig. 5.13. Their truth can be seen from the following reasoning: (1') if those three forces were concurrent but noncoplanar, the summation of forces in a direction perpendicular to a plane containing two of the three forces would not be zero; (2') if those three forces were coplanar but neither concurrent nor parallel, the summation of moments about the point of intersection of the lines of action of two of those three forces would not be zero.

Developmental Exercises

D5.19 When equilibrium exists, describe the properties of the forces acting on (a) a two-force body, (b) a three-force body.

D5.20 Ascertain whether each of the bodies shown is a two-force body or a three-force body.

D5.21 Name the two-force body as well as the three-force body, if any, in each of the systems shown.

(a)

(b)

(c)

Fig. D5.21

5.8 Statical Indeterminacy and Constraint Conditions

An unknown, or a problem, associated with a body in equilibrium is said to be *statically determinate* if it can analytically be determined, or solved, solely from the equations of equilibrium and appropriate geometric relations in the system, without using any material properties of the system. Otherwise, it is said to be *statically indeterminate*. Meantime, a rigid body in equilibrium is said to be *statically indeterminate* if its free-body diagram contains one or more statically indeterminate unknowns. In general, the number of unknowns (in the free-body diagram of the body) in excess of the number of independent scalar equilibrium equations available is called the *degree of statical indeterminacy* of the body. For illustrations, see Fig. 5.14.

(a) Statically
determinate

(b) Statically indeterminate
to the first degree

(c) Statically indeterminate
to the second degree

Fig. 5.14

A rigid body is said to be *partially constrained* for the loading if (i) there is at least one prospective equilibrium equation which cannot be satisfied by all of the forces and moments in the free-body diagram of the body, and (ii) the total number of unknowns is *smaller* than the total number of prospective independent equilibrium equations to be applied. A partially constrained rigid body is not in equilibrium because the loading is unbalanced and will either accelerate the body into *motion* or cause its supports and connections to *deform* or *rupture*.

A rigid body is said to be *improperly constrained* for the loading if (i) there is at least one prospective equilibrium equation which cannot be satisfied by all of the forces and moments in the free-body diagram of the body, and (ii) the total number of unknowns is *equal to* or *greater than* the total number of prospective independent equilibrium equations to be applied. The unbalanced loading will either accelerate the improperly constrained body into *motion* or cause its supports and connections to *deform* or *rupture*.

Developmental Exercise

D5.22   State the degree of statical indeterminacy of each of the beams shown.

Fig. D5.22

EXAMPLE 5.6

Verify that the beam AB is partially constrained for the loading as shown.

Solution. The free-body diagram of the beam reveals that there are *three* prospective independent equilibrium equations, $\Sigma F_x = 0$, $\Sigma F_y = 0$, and $\Sigma M_A = 0$, to be applied. We note that (i) the prospective equilibrium equation $\Sigma F_x = 0$, cannot be satisfied, and (ii) there are *two* unknown reactions, A and B. The beam AB is, therefore, *partially constrained* for the loading. This beam will be accelerated to *move* to the right.

EXAMPLE 5.7

Determine whether or not the beam ABC is improperly constrained for the loading as shown if (a) $\theta = 60°$, (b) $\theta = 90°$.

Solution. This problem may be solved as follows:

Case (a) $\theta = 60°$. From the free-body diagram drawn below for this case, we see that there are *three* prospective independent equilibrium equations, $\Sigma F_x = 0$, $\Sigma F_y = 0$, and $\Sigma M_A = 0$, to be applied. We find (i) $\Sigma F_x \neq 0$; and (ii) there are *three* unknowns, A, B, and C. Thus, if $\theta = 60°$, the beam ABC is *improperly constrained* for the loading. This beam will be accelerated to *move* to the right.

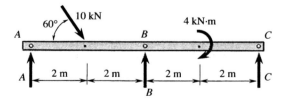

Case (b) $\theta = 90°$. From the free-body diagram drawn for this case, we see that there are only *two* prospective independent equilibrium equations, $\Sigma F_y = 0$ and $\Sigma M_A = 0$, to be applied. We find the following: (i) *no* prospective equilibrium equation which cannot be satisfied; and (ii) there are *three*

unknowns, A, B, and C. Thus, if $\theta = 90°$, the beam ABC is *not* improperly constrained for the loading. Rather, this beam is *statically indeterminate to the first degree* and may remain *in equilibrium*.

Developmental Exercises

D5.23 A rigid body is shown. (a) Draw the free-body diagram of the body. (b) Is $\Sigma M_B = 0$? (c) Is the body in equilibrium? (d) Is the body *partially* or *improperly* constrained for the loading?

D5.24 Ascertain whether the beam AB is *partially* or *improperly* constrained for the loading as shown. Give reasons for your answer.

D5.25 A plate $ABCD$ is shown. (a) Draw the free-body diagram of the plate. (b) Is the plate *partially* or *improperly* constrained for the loading? (c) Give reasons for your answer to part (b).

Fig. D5.23

Fig. D5.24

Fig. D5.25

5.9 Possible Sets of Independent Scalar Equilibrium Equations

When a rigid body is in equilibrium, we know that Eqs. (5.11) must be satisfied. Consequently, there exist *two necessary conditions for equilibrium of any body*, which are very useful concepts for solving equilibrium problems and may be stated as follows:

1. *The sum of the components of forces* (in the free-body diagram) *in any desirable direction must be zero.*
2. *The sum of the moments of forces and couples*, if any (in the free-body diagram), *about any desirable point or axis must be zero.*

By repeated applications of one or both of the above *two necessary conditions for equilibrium of any body*, it is possible to solve a statically deter-

Space diagram   Free-body diagram

Fig. 5.15

minate problem with *different sets of independent scalar equilibrium equa-tions.* For instance, let us consider the equilibrium of the simple beam in Fig. 5.15. We see that there are three possible sets of independent scalar equilibrium equations for solving the three unknown reaction components: A_x, A_y, and B. They are listed in Table 5.1.

Table 5.1   Possible Sets of Independent Scalar Equilibrium Equations for the Beam in Fig. 5.15

Set	1	2	3	Unknowns Determined
Equations of equilibrium	$\Sigma M_A = 0$	$\Sigma M_A = 0$	$\Sigma M_A = 0$	$B_y = 24$ lb
	$\Sigma F_y = 0$	$\Sigma M_B = 0$	$\Sigma M_B = 0$	$A_y = 36$ lb
	$\Sigma F_x = 0$	$\Sigma F_x = 0$	$\Sigma M_C = 0$	$A_x = 80$ lb

Note that *in each set of independent scalar equilibrium equations, the number of equations cannot exceed that obtained by applying Eqs. (5.11).*[†] Furthermore, *each such set should include a moment equilibrium equation if the unknown force components are nonconcurrent.* This is necessary be-cause only moment equilibrium equations can account for the effects of mo-ments and the positions of the forces. In this respect, a moment equilibrium equation is *more efficacious* than a force equilibrium equation.

Developmental Exercise

Fig. D5.26

D5.26 A beam is supported and loaded as shown. (a) How many unknown reaction components are there in the free-body diagram of the beam? (b) How many independent scalar equilibrium equations are there for the solution? (c) What will $\Sigma M_A = 0$ yield? (d) What will $\Sigma M_B = 0$ yield? (e) Does $\Sigma F_y = 0$ yield any additional nontrivial information? (f) Does $\Sigma M_C = 0$ yield any addi-tional nontrivial information? (g) Can we ever solve for A_x and B_x by repeated applications of the two necessary conditions for equilibrium stated in Sec. 5.9? (h) Is the beam statically determinate?

[†]For instance, the *maximum number* of independent scalar equilibrium equations for *a* rigid body is equal to *three* if the body is in equilibrium in a plane; it is equal to *six* if the body is in equilibrium in space.

5.10   Method of Solution: Equilibrium in a Plane

Equilibrium problems of rigid bodies which are statically determinate and relatively simple are considered in this chapter. Those that are more complex are treated in Chaps. 8, 9, and 10. We know that certain problems can best be solved by employing a special set of steps. However, the following procedure may serve as a guide in *solving equilibrium problems of simple rigid bodies in a plane*:

1. *Identify two-force bodies*, if any, in the system.
2. *Indicate the chosen xy rectangular coordinate axes* if they are not in the horizontal and vertical directions.
3. *Draw a good free-body diagram* of the rigid body. (Cf. Sec. 3.8.)
4. *Sum to zero the moments of all forces and couples about a chosen point* (preferably, a point where the lines of action of two unknown forces intersect) to obtain a scalar moment equilibrium equation. If this equation contains only one unknown, it is often solved right away.
5. *Sum to zero the scalar components of all forces in the x and y directions*, *respectively*, to obtain two scalar force equilibrium equations. However, only one independent equation will be obtained here if the unknown forces are parallel forces.
6. *Solve for the unknowns* contained in all equations obtained.

Note that the total number of unknowns and the total number of independent scalar equilibrium equations must match if *all* unknowns are to be determined. If only a partial solution of a rigid-body equilibrium problem is desired, we may appropriately apply the *two necessary conditions for equilibrium of any body* stated in Sec. 5.9.

EXAMPLE 5.8

A 600-lb crate is suspended as shown. Determine the reaction from the support at A.

Solution. We first note that the member $BCDE$ is a two-force body. Thus, the reaction from the support at B must be directed along the line containing the points E and B. Since the system is in equilibrium, the tension in the cable fastened at J and passing around the pulley at K must be

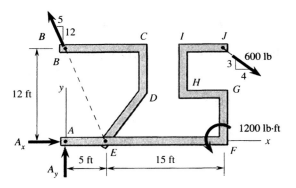

equal to the weight of the block.[†] The free-body diagram of the entire frame may be drawn as shown. In this diagram, there are three unknowns: A_x, A_y, and B. Since we need to determine only A_x and A_y, we write

$$+ \circlearrowleft \Sigma M_B = 0: \quad 12A_x - 20\left(\frac{3}{5}\right)(600) + 1200 = 0$$

$$+ \circlearrowleft \Sigma M_E = 0: \quad -5A_y - 12\left(\frac{4}{5}\right)(600) - 15\left(\frac{3}{5}\right)(600) + 1200 = 0$$

These two equations yield $A_x = 500$ and $A_y = -1992$. Since A_y is negative, the true sense of $\mathbf{A}_y$ is opposite to that assumed in the free-body diagram. Thus, we write

$$\mathbf{A}_x = 500 \text{ lb} \rightarrow \qquad \mathbf{A}_y = 1992 \text{ lb} \downarrow \quad \blacktriangleleft$$

Developmental Exercise

D5.27   Determine the reaction from the support at B in Example 5.8.

EXAMPLE 5.9

A frame $ABCD$ supports a 330-lb crate E with the cable and pulley system as shown. Determine the reaction from the support at D.

Solution.   The cable and pulley system acts on the frame at B. To determine this action, we draw the free-body diagram of the pulley at B and readily note that $B_x = T$ from $\Sigma F_x = 0$ and $B_y = T$ from $\Sigma F_y = 0$, where T is the tension in the cable. Referring to the free-body diagram of the crate E together with the lowest pulley as shown, we write

$$+ \uparrow \Sigma F_y = 0: \quad 3T - 330 = 0 \qquad B_x = B_y = T = 110 \text{ lb}$$

[†]In dynamics, the tension in a cable is generally *not* equal to the weight of the body to which the cable is connected.

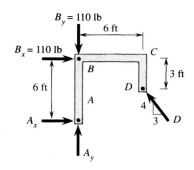

We next draw the free-body diagram of the frame $ABCD$ as shown, where the forces at B are consistent with Newton's third law. We write

$$+\,\circlearrowright\,\Sigma M_A = 0: \quad -6(110) + 6\left(\frac{4}{5}D\right) + 3\left(\frac{3}{5}D\right) = 0$$

$$\mathbf{D} = 100 \text{ lb} \measuredangle 53.1° \quad \blacktriangleleft$$

Developmental Exercises

D5.28 If a rigid-body equilibrium problem requires only a partial solution, what should we do?

D5.29 Refer to Example 5.9. Determine $\mathbf{A}_x$ and $\mathbf{A}_y$.

EXAMPLE 5.10

Using the concept of rigid-body equilibrium, determine the tensions in the wires AD, BE, and CF of the system in Example 3.3.

Solution. For ease of reference, the space diagram for Example 3.3 is repeated here. We first draw the free-body diagram of the block at G. Referring to the free-body diagram of the block, we write

$$+\uparrow \Sigma F_y = 0: \quad T_{CG} - 10g = 0 \qquad T_{CG} = 10g \text{ N}$$

Next, we draw the free-body diagram of the body ABC obtained by cutting the wires AD, BE, CF, and CG. To facilitate the solution, the horizontal and vertical projections of the wire AC are assumed to be $24L$ and $7L$ units long, respectively, where L is a constant. The length of $\overline{BH}$ may be determined from the fact that

$$\frac{3}{4}\overline{BH} + \frac{5}{12}(\overline{BH} + 7L) = 24L$$

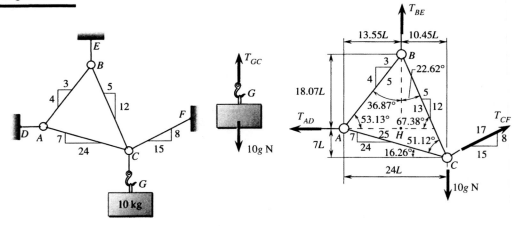

which yields $\overline{BH} = 18.07L$. In the free-body diagram of the body ABC, there are three unknowns: T_{AD}, T_{BE}, and T_{CF}. Treating ABC as a rigid body in equilibrium, we write

$$+\circlearrowleft \Sigma M_H = 0: \quad 7L\left(\frac{15}{17}\right)T_{CF} + 10.45L\left(\frac{8}{17}\right)T_{CF} - 10.45L(10g) = 0$$

$$T_{CF} = 9.419g \qquad\qquad T_{CF} = 92.4 \text{ N} \quad \blacktriangleleft$$

$$\xrightarrow{+} \Sigma F_x = 0: \quad -T_{AD} + \frac{15}{17}T_{CF} = 0 \qquad\qquad T_{AD} = 81.5 \text{ N} \quad \blacktriangleleft$$

$$+\uparrow \Sigma F_y = 0: \quad T_{BE} + \frac{8}{17}T_{CF} - 10g = 0 \qquad\qquad T_{BE} = 54.6 \text{ N} \quad \blacktriangleleft$$

Note that these answers agree with those obtained in Example 3.8.

Developmental Exercise

D5.30   Refer to Example 5.10. Verify the dimensions indicated for ABC.

PROBLEMS

5.25 through 5.27* Draw the free-body diagram of the body AB as shown. (*Postpone the determination of reactions until Probs. 5.34 through 5.36.*)

Fig. P5.25 and P5.34   (a)   (b)   (c)

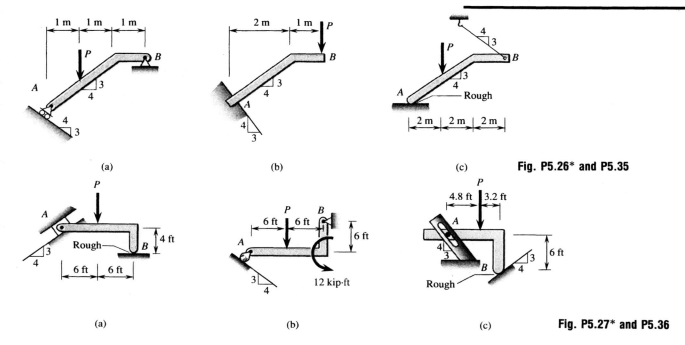

(a)   (b)   (c)   **Fig. P5.26* and P5.35**

(a)   (b)   (c)   **Fig. P5.27* and P5.36**

5.28 through 5.30* For $P = 120$ lb, determine the reactions from the supports of the body AB as shown.

Fig. P5.28   **Fig. P5.29**   **Fig. P5.30***

5.31 through 5.33* For $P = 8$ kN, determine the reactions from the supports at A and C as shown.

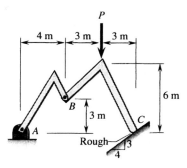

Fig. P5.31   **Fig. P5.32**   **Fig. P5.33***

Fig. P5.37*

5.34 and 5.35 *(The figures for these problems are shown and specified along with those for Probs. 5.25 and 5.26*.)* For $P = 6$ kN, determine the reactions from the supports of the body AB as shown.

5.36 *(The figures for this problem are shown and specified along with those for Prob. 5.27*.)* For $P = 600$ lb, determine the reactions from the supports of the body AB as shown.

5.37* A car is parked on a driveway. It has a wheelbase b and a total weight W which acts at a distance a forward of the rear axle as shown. Derive the expression of the reaction $\mathbf{R}$ to the pair of rear wheels in terms of a, b, and W.

5.38 and 5.39 For the frame shown, determine the reactions from the supports at A and D.

Fig. P5.38

Fig. P5.39

5.40 A truck carries a load of coal at a constant speed up a grade as shown. The resultant weight force of the load is 61 kN acting through the point G_1 and the resultant weight force of the empty truck is 30.5 kN acting through the point G_2. Determine the reaction from the ground to the pair of rear driving wheels.

Fig. P5.40

Fig. P5.41

5.41 A 4.1-Mg rubber-tired tractor is pulling with constant speed a load $\mathbf{P}$ up a grade as shown. The resultant weight force of the tractor acts through the point G, and the friction force exerted by the ground on each of the four wheels is three-fourths of the normal reaction to the wheel. Determine (a) the magnitude of $\mathbf{P}$, (b) the reaction from the ground to the pair of front wheels.

5.42 When no counterweight is provided, the resultant weight force of the 10-Mg hoisting crane acts at the point G as shown. If the crane is to pick up a load of 5 Mg in the position shown, what is the minimum amount of mass m_C of the counterweight that must be provided at the point C?

5.43 and 5.44* A plane truss, which may here be considered as a rigid body, is loaded and supported as shown. Determine the reaction from the support at A.

Fig. P5.42

Fig. P5.43

Fig. P5.44*

5.45 A model airplane is being displayed as shown. (a) Using a single moment equilibrium equation, determine the tension T_{CF} in the wire CF. (b) Determine the tensions in the other wires.

5.46 The 100-lb resultant weight force of the rod AB acts through the point G and the rod is in equilibrium as shown. Determine (a) the value of θ which the rod makes with the horizontal, (b) the tension T_{AD} in the wire AD.

Fig. P5.45

Fig. P5.46

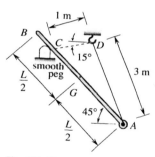

Fig. P5.47

5.47 The resultant weight force of the 10-kg rod AB acts through the point G and the rod is in equilibrium as shown. Determine (a) the length L of the rod, (b) the tension T_{AD} in the wire AD.

EQUILIBRIUM IN SPACE

5.11 Reactions from Supports and Connections in Space

When a rigid body is in *equilibrium in space*, the force system acting on it is generally *noncoplanar* and *nonconcurrent*. A rigid body acted on by noncoplanar and nonconcurrent forces is referred to as a *rigid body in space*.

Likewise, supports or connections of a rigid body in space are referred to as *supports* or *connections in space*. We recall from Sec. 5.5 that *reactions* from supports or connections of a rigid body are the *constraining forces and moments* acting on the body at the points of support or connection. Each *reaction* must reflect the *restriction of movement* imposed on the body at the point of support or connection. Consequently, the number of unknowns associated with the reaction from a support or connection in space may range from *one* to *six*, which are illustrated and labeled in Figs. 5.16 through 5.21.

TYPE I: REACTIONS WITH ONE UNKNOWN. See illustrations in Fig. 5.16.

| Ball | Smooth surface | | Cable | Short link | |

Fig. 5.16

TYPE II: REACTIONS WITH TWO UNKNOWNS. See illustrations in Fig. 5.17.

Roller on rough surface   Shallow flanged wheel on rail   Ball and socket on a slider   Narrow bearing   Eyebolt

Fig. 5.17

TYPE III: REACTIONS WITH THREE UNKNOWNS. See illustrations in Fig. 5.18.

Rough Surface   Smooth Surface   Ball and socket   Narrow bearing with stop

Fig. 5.18

TYPE IV: REACTIONS WITH FOUR UNKNOWNS. See illustrations in Fig. 5.19.

Universal Joint   Wide hinge   Wide bearing

Fig. 5.19 Note that the reaction moment developed in a universal joint is perpendicular to the *crosspiece* (or "spider") in the joint.

TYPE V: REACTIONS WITH FIVE UNKNOWNS. See illustrations in Fig. 5.20.

Pin and
bracket

Wide hinge
with stop

Wide bearing
with stop

Fig. 5.20

TYPE VI: REACTIONS WITH SIX UNKNOWNS. See the illustration in Fig. 5.21. Of course, the supports or connections in Figs. 5.16 through 5.21 may be modified to give a certain type of reaction as illustrated in Fig. 5.22.

Fixed support

Fig. 5.21   **Fig. 5.22**

EXAMPLE 5.11

A 70-kg plate is supported and loaded as shown (below left). Assume that the resultant weight force of the plate acts through the point G. Draw the free-body diagram of the plate.

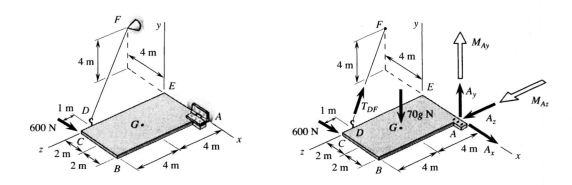

Solution. We note that the support at A is a wide hinge with stop and the support at D is a cable. The reactions from these two supports are, respectively, type V and type I reactions. Thus, the free-body diagram of the plate is drawn as shown (above right).[†]

Fig. D5.32

Fig. D5.33

Developmental Exercises

D5.31 (a) What is a rigid body in space? (b) What are supports or connections in space?

D5.32 A horizontal shaft is supported by two *narrow* bearings as shown. If one of the bearings is removed, will the shaft remain horizontal?

D5.33 A horizontal shaft supported by a single bearing remains horizontal before and after a force is applied as shown. Should the bearing be considered as *narrow* or *wide*?

5.12 Maximum Number of Statically Determinate Unknowns

When a rigid body is in equilibrium in space, the force system acting on it must satisfy Eqs. (5.11), which may also be written as

$$(\Sigma F_x)\mathbf{i} + (\Sigma F_y)\mathbf{j} + (\Sigma F_z)\mathbf{k} = \mathbf{0}$$
$$(\Sigma M_P)_x\mathbf{i} + (\Sigma M_P)_y\mathbf{j} + (\Sigma M_P)_z\mathbf{k} = \mathbf{0}$$

(5.13)

[†]Continued in Example 5.18.

These two vector equations of equilibrium are equivalent to the following six *scalar equations of equilibrium*:

$$\Sigma F_x = 0 \qquad \Sigma F_y = 0 \qquad \Sigma F_z = 0$$

$$(\Sigma M_P)_x = 0 \qquad (\Sigma M_P)_y = 0 \qquad (\Sigma M_P)_z = 0$$

(5.14)

The first three of Eqs. (5.14) are referred to as the *force equilibrium equations,* while the other three are referred to as the *moment equilibrium equations.* We recall that an independent equilibrium equation cannot be obtained through a linear combination of other equilibrium equations. If the total number of unknowns associated with the equilibrium of a rigid body in space is greater than the total number of independent scalar equilibrium equations available, the rigid body is *statically indeterminate* and *some* or *all* of the unknowns *cannot* be determined solely from those equations.

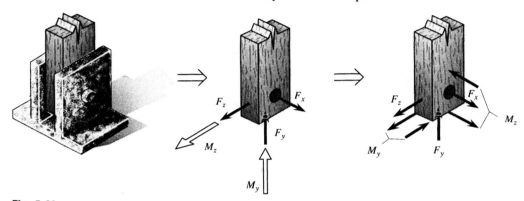

Fig. 5.23

For ascertaining the total number of statically determinate unknowns, each unknown component of *reaction moment* may be replaced by an equivalent unknown *couple* where the parallel forces are set at a unit distance from each other as illustrated in Fig. 5.23. The number of statically determinate unknowns in an equilibrium problem of a *simple* rigid body in space is dependent on the configuration of the noncoplanar and nonconcurrent unknown forces in the free-body diagram of the rigid body. It is important to recognize that the *maximum number of statically determinate unknowns* allowed by that free-body diagram is equal to *six or less* as described below:

1. The maximum number of statically determinate unknowns allowed is equal to *three* if the unknown forces are all parallel to each other. In this case, two force equilibrium equations and one moment equilibrium equation are trivial. (Cf. Example 5.14.)
2. It is equal to *four* if the unknown forces are all parallel to a plane and all intersect a common straight line. In this case, one force equilibrium equation and one moment equilibrium equation are trivial. (Cf. Example 5.15.)
3. It is equal to *five* if the unknown forces are all parallel to a plane but are not all parallel to a line and do not all intersect a common straight line. In this case, one force equilibrium equation is trivial. (Cf. Example 5.16.)

4. It is also equal to *five* if the unknown forces either are parallel to or intersect a common straight line but are not all parallel to a plane. In this case, one moment equilibrium equation is trivial. (Cf. Example 5.17.)
5. It is equal to *six* if the unknown forces are not all parallel to a plane nor all intersect a common straight line. (Cf. Example 5.18.)

If the total number of unknowns in the free-body diagram exceeds that described in the list above, the equilibrium problem is *statically indeterminate*. However, it may still be possible to determine *some*, but not all, of the unknowns by using the *two necessary conditions for equilibrium of any body* as stated in Sec. 5.9. (Cf. Example 5.19.) Furthermore, note that the definitions for the *degree of statical indeterminacy*, the *partially constrained rigid body*, and the *improperly constrained rigid body* in Sec. 5.8 continue to be applicable to rigid bodies in space.

EXAMPLE 5.12

Verify that the bent bar *ABC* is partially constrained for the loading as shown at left.

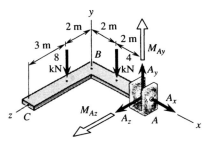

Solution. From the free-body diagram shown above, we see that there are *six* prospective independent equilibrium equations, $\Sigma F_x = 0$, $\Sigma F_y = 0$, $\Sigma F_z = 0$, $\Sigma M_x = 0$, $\Sigma M_y = 0$, and $\Sigma M_z = 0$, to be applied. We find that (i) $\Sigma M_x \neq 0$; and (ii) there are *five* unknowns, A_x, A_y, A_z, M_{Ay}, and M_{Az}. Thus, the bent bar *ABC* is *partially constrained* for the loading, which will cause it to rotate about the *x* axis.

EXAMPLE 5.13

Determine whether the bent rod *ABC* is partially or improperly constrained for the loading as shown.

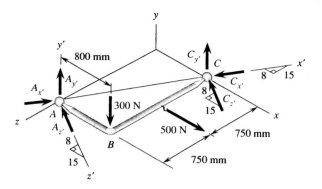

Solution. Introducing the $x'y'z'$ coordinate system in the free-body diagram as shown, we see that there are *six* prospective independent equilibrium equations, $\Sigma F_{x'} = 0$, $\Sigma F_{y'} = 0$, $\Sigma F_{z'} = 0$, $\Sigma M_{x'} = 0$, $\Sigma M_{y'} = 0$, and $\Sigma M_{z'} = 0$, to be applied. We find the following: (i) $\Sigma M_{x'} \neq 0$; and (ii) there are *six* unknowns, $A_{x'}$, $A_{y'}$, $A_{z'}$, $C_{x'}$, $C_{y'}$, and $C_{z'}$. Thus, the bent rod *ABC* is *improperly constrained* for the loading, which will cause the bent rod to rotate about the x' axis.

Developmental Exercises

D5.34 For each of the bent bars shown, indicate (a) the total number of unknowns in the free-body diagram of the bent bar, (b) the total number of independent scalar equilibrium equations available, (c) the statical indeterminacy.

D5.35 For ascertaining the total number of statically determinate unknowns, how is each unknown component of reaction moment replaced?

(a)   (b)

Fig. D5.34

5.13 Method of Solution: Equilibrium in Space

The characterizing feature of an *equilibrium problem of a rigid body in space* is that a set of *noncoplanar* and *nonconcurrent* unknown forces acting on the rigid body will be encountered in the solution. Although certain problems can best be solved by a special set of steps, the following procedure may serve as a guide in *solving equilibrium problems of simple rigid bodies in space*:

1. *Identify two-force bodies*, if any, in the system.
2. *Draw correctly the free-body diagram* of the rigid body.
3. *Write the analytical expression* of each external force and moment acting on the free body.
4. *Add the analytical expressions* of all forces acting on the free body to obtain the resultant force in terms of known and unknown quantities.
5. *Set each of the scalar components of the resultant force*, obtained in step (4), *equal to zero* to obtain the force equilibrium equations.
6. *Compute the resultant moment* of the force system about a chosen point (preferably, a point where several of the lines of action of the unknown forces intersect). This resultant moment is now expressed in terms of known and unknown quantities. Sometimes, a partial solution is facilitated if the resultant moment is taken about an axis.
7. *Set each of the scalar components of the resultant moment*, obtained in step (6), *equal to zero* to obtain the moment equilibrium equations.
8. *Solve for the unknowns* in the equilibrium equations obtained. Note that the total number of unknowns and the total number of independent scalar equilibrium equations must match if all unknowns are statically determinate.

If only a *partial solution* of a rigid-body equilibrium problem is desired, we may appropriately apply the *two necessary conditions for equilibrium of any body* stated in Sec. 5.9. Of course, a *complete solution* can also be achieved by obtaining several *partial solutions*.

EXAMPLE 5.14

A horizontal rectangular plate is supported by three vertical cables as shown. The weight force of the plate and the loads on the plate are equivalent to a 1000-lb force acting at the point D as shown. Are the three tensions in the cables statically determinate? If so, determine their values.

Solution. The free-body diagram of the plate may be drawn as shown. We note that the unknown reaction forces acting on the free body are all parallel to each other as well as to the y axis (or the xy and yz planes). Therefore, $(\Sigma M_D)_y = 0$, $\Sigma F_x = 0$, and $\Sigma F_z = 0$ are all trivial equations.

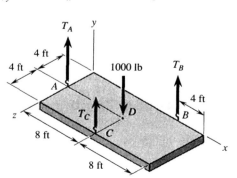

We expect to have *three* independent equilibrium equations available for the solution of the *three* unknowns: T_A, T_B, and T_C. [Cf. case (1) in Sec. 5.12.] Thus, the three tensions are *statically determinate*. Referring to the free-body diagram, we write

$$\mathbf{T}_A = T_A\mathbf{j} \quad \mathbf{T}_B = T_B\mathbf{j} \quad \mathbf{T}_C = T_C\mathbf{j} \quad \mathbf{D} = -1000\mathbf{j}$$
$$\overrightarrow{DA} = -8\mathbf{i} \quad \overrightarrow{DB} = 4\mathbf{i} - 4\mathbf{k} \quad \overrightarrow{DC} = 4\mathbf{k}$$

For equilibrium, we write

$$\Sigma\mathbf{F} = \mathbf{T}_A + \mathbf{T}_B + \mathbf{T}_C + \mathbf{D} = \mathbf{0}:$$
$$(T_A + T_B + T_C - 1000)\mathbf{j} = \mathbf{0} \qquad (1)$$
$$\Sigma\mathbf{M}_D = \overrightarrow{DA} \times \mathbf{T}_A + \overrightarrow{DB} \times \mathbf{T}_B + \overrightarrow{DC} \times \mathbf{T}_C = \mathbf{0}:$$
$$(4T_B - 4T_C)\mathbf{i} + (-8T_A + 4T_B)\mathbf{k} = \mathbf{0} \qquad (2)$$

Equating the coefficients of $\mathbf{i}$, $\mathbf{j}$, and $\mathbf{k}$ to zero in the above vector equations (1) and (2), we get a set of three simultaneous scalar equations (as expected), which yield

$$T_A = 200 \text{ lb} \qquad T_B = T_C = 400 \text{ lb} \blacktriangleleft$$

Developmental Exercise

D5.36   Describe the situation as manifested in Example 5.14 in which the maximum number of statically determinate unknowns is equal to *three*.

EXAMPLE 5.15

A beam is supported by two eyebolts at A and C and is loaded as shown. Determine the reactions from the supports.

Solution.   The free-body diagram of the beam may be drawn as shown. We note that the unknown forces acting on the free body are all parallel to the yz plane and all intersect the x axis. Therefore, $(\Sigma M_A)_x = 0$ and $\Sigma F_x = 0$ are trivial. We expect to have *four* independent equilibrium equations available for the solution of the *four* unknowns: A_y, A_z, C_y, and C_z. [Cf. case (2) in Sec. 5.12.] We have

$$\mathbf{A} = A_y\mathbf{j} + A_z\mathbf{k} \quad \mathbf{B} = 50\mathbf{k} \quad \mathbf{C} = C_y\mathbf{j} + C_z\mathbf{k} \quad \mathbf{D} = -40\mathbf{j}$$
$$\overrightarrow{AB} = 2\mathbf{i} \quad \overrightarrow{AC} = 5\mathbf{i} \quad \overrightarrow{AD} = 7\mathbf{i}$$

For equilibrium, we write

$$\Sigma\mathbf{F} = \mathbf{A} + \mathbf{B} + \mathbf{C} + \mathbf{D} = \mathbf{0}:$$
$$(A_y + C_y - 40)\mathbf{j} + (A_z + C_z + 50)\mathbf{k} = \mathbf{0} \qquad (1)$$
$$\Sigma\mathbf{M}_A = \overrightarrow{AB} \times \mathbf{B} + \overrightarrow{AC} \times \mathbf{C} + \overrightarrow{AD} \times \mathbf{D} = \mathbf{0}:$$
$$(-5C_z - 100)\mathbf{j} + (5C_y - 280)\mathbf{k} = \mathbf{0} \qquad (2)$$

The above vector equations (1) and (2) together give a set of four simultaneous scalar equations (as expected), which yield

$$A_y = -16 \qquad A_z = -30 \qquad C_y = 56 \qquad C_z = -20$$

$$\mathbf{A} = -16\mathbf{j} - 30\mathbf{k} \text{ lb} \qquad \mathbf{C} = 56\mathbf{j} - 20\mathbf{k} \text{ lb} \blacktriangleleft$$

Developmental Exercise

D5.37   Describe the situation as manifested in Example 5.15 in which the maximum number of statically determinate unknowns is equal to *four*.

EXAMPLE 5.16

A ladder in a warehouse is supported by an unflanged wheel A resting against a smooth *shallow* track fixed to the wall and by two *shallow* flanged wheels B and C on a rail fixed to the floor. A man stands on the ladder and leans to one side to pick up a machine part from a high shelf. The combined mass of the ladder and the man holding the machine part is 96 kg, and the weight force acts as shown. Determine the reactions from the supports at A, B, and C.

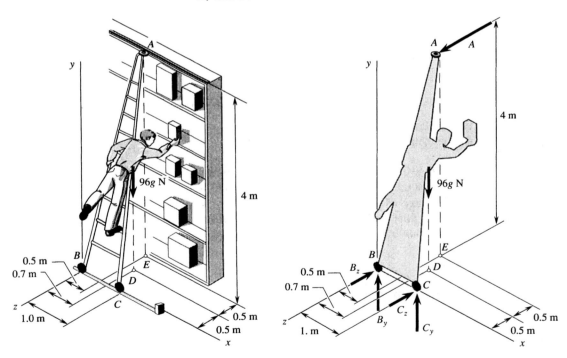

Solution.   Since the ladder can move freely in the direction parallel to the x axis, there are no reaction components in the x direction. The free-body diagram of the ladder, the man, and the machine part may be drawn as shown. We note that the noncoplanar and nonconcurrent unknown reaction forces are all parallel to the yz plane but are not all parallel to a line and do not all intersect a common straight line. Therefore, only $\Sigma F_x = 0$ is trivial. We expect to have *five* independent equilibrium equations available for the solution of the *five* unknowns: A, B_y, B_z, C_y, and C_z. [Cf. case (3) in Sec. 5.12.] We have

$$\mathbf{A} = A\mathbf{k} \qquad \vec{BA} = 0.5\mathbf{i} + 4\mathbf{j} - \mathbf{k}$$

$$\mathbf{B} = B_y\mathbf{j} - B_z\mathbf{k} \qquad \vec{BC} = \mathbf{i}$$

$$\mathbf{C} = C_y\mathbf{j} - C_z\mathbf{k} \qquad \vec{BD} = 0.7\mathbf{i} - 0.5\mathbf{k} \qquad \mathbf{W} = -96g\mathbf{j}$$

For equilibrium, we write

$$\Sigma \mathbf{F} = \mathbf{A} + \mathbf{B} + \mathbf{C} + \mathbf{W} = \mathbf{0}:$$

$$(B_y + C_y - 96g)\mathbf{j} + (A - B_z - C_z)\mathbf{k} = \mathbf{0} \qquad (1)$$

$$\Sigma \mathbf{M}_B = \vec{BA} \times \mathbf{A} + \vec{BC} \times \mathbf{C} + \vec{BD} \times \mathbf{W} = \mathbf{0}:$$

$$(4A - 48g)\mathbf{i} + (-0.5A + C_z)\mathbf{j} + (C_y - 67.2g)\mathbf{k} = \mathbf{0} \qquad (2)$$

The above vector equations (1) and (2) together give a set of five simultaneous scalar equations (as expected), which yield

$$A = 12g \qquad B_y = 28.8g \qquad B_z = 6g \qquad C_y = 67.2g \qquad C_z = 6g$$

$$A = 117.7\mathbf{k} \text{ N} \qquad \mathbf{B} = 283\mathbf{j} - 58.9\mathbf{k} \text{ N} \quad \blacktriangleleft$$

$$\mathbf{C} = 659\mathbf{j} - 58.9\mathbf{k} \text{ N} \quad \blacktriangleleft$$

Developmental Exercise

D5.38 Describe the situation as manifested in Example 5.16 in which the maximum number of statically determinate unknowns is equal to *five*.

EXAMPLE 5.17

The system shown carries a 110-lb crate at B and is supported by a ball-and-socket joint at A and two cables at C and D. Determine the tensions in cables CE and DF and the reaction from the support at A.

110 lb

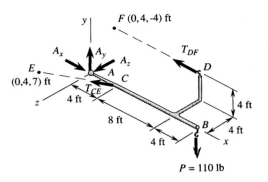

Solution. The free-body diagram may be drawn as shown. We note that the noncoplanar and nonconcurrent unknown reaction forces either intersect or are parallel to the x axis but are not all parallel to a plane. Therefore, $(\Sigma M_A)_x = 0$ is trivial. We expect to have *five* independent equilibrium equations available for the solution of the *five* unknowns: T_{CE}, T_{DF}, A_x, A_y, and A_z. [Cf. case (4) in Sec. 5.12.] We have

$$\overrightarrow{CE} = -4\mathbf{i} + 4\mathbf{j} + 7\mathbf{k} \qquad \lambda_{CE} = \frac{1}{9}(-4\mathbf{i} + 4\mathbf{j} + 7\mathbf{k})$$

$$\mathbf{T}_{CE} = \frac{T_{CE}}{9}(-4\mathbf{i} + 4\mathbf{j} + 7\mathbf{k}) \qquad \mathbf{T}_{DF} = -T_{DF}\mathbf{i}$$

$$\mathbf{A} = A_x\mathbf{i} + A_y\mathbf{j} + A_z\mathbf{k} \qquad \mathbf{P} = -110\mathbf{j}$$

$$\overrightarrow{AC} = 4\mathbf{i} \qquad \overrightarrow{AB} = 16\mathbf{i} \qquad \overrightarrow{AF} = 4\mathbf{j} - 4\mathbf{k}$$

For equilibrium, we write

$$\Sigma\mathbf{F} = \mathbf{T}_{CE} + \mathbf{T}_{DF} + \mathbf{A} + \mathbf{P} = 0:$$

$$\left(-\frac{4}{9}T_{CE} - T_{DF} + A_x\right)\mathbf{i} + \left(\frac{4}{9}T_{CE} + A_y - 110\right)\mathbf{j}$$

$$+ \left(\frac{7}{9}T_{CE} + A_z\right)\mathbf{k} = 0 \tag{1}$$

$$\Sigma\mathbf{M}_A = \overrightarrow{AC} \times \mathbf{T}_{CE} + \overrightarrow{AB} \times \mathbf{P} + \overrightarrow{AF} \times \mathbf{T}_{DF} = 0:$$

$$\left(-\frac{28}{9}T_{CE} + 4T_{DF}\right)\mathbf{j} + \left(\frac{16}{9}T_{CE} + 4T_{DF} - 1760\right)\mathbf{k} = 0 \tag{2}$$

The above vector equations (1) and (2) together give a set of five simultaneous scalar equations (as expected), which yield

$$T_{CE} = 360 \qquad T_{DF} = 280 \qquad A_x = 440 \qquad A_y = -50 \qquad A_z = -280$$

$$T_{CE} = 360 \text{ lb} \qquad T_{DF} = 280 \text{ lb} \qquad \mathbf{A} = 440\mathbf{i} - 50\mathbf{j} - 280\mathbf{k} \text{ lb} \blacktriangleleft$$

Developmental Exercise

D5.39 Describe the situation as manifested in Example 5.17 in which the maximum number of statically determinate unknowns is equal to *five*.

EXAMPLE 5.18

Refer to the plate in Example 5.9. Determine the tension in the cable DF and the reaction from the support at A.

Solution. The free-body diagram of the plate is first drawn as shown. Following Sec. 5.12, we may replace each of the two unknown moment reaction components $\mathbf{M}_{Ay}$ and $\mathbf{M}_{Az}$ by a pair of forces which are at a small

distance from each other and form a couple as shown. Thus, we see that the noncoplanar and nonconcurrent unknown reaction forces are not all parallel to a plane and their lines of action do not all intersect a common straight line. From case (5) in Sec. 5.12, we expect to have *six* independent scalar equilibrium equations for the solution of the *six* unknowns: T_{DF}, A_x, A_y, A_z, M_{Ay}, and M_{Az}. We have

$$\overrightarrow{DF} = -4\mathbf{i} + 4\mathbf{j} - 7\mathbf{k} \qquad \boldsymbol{\lambda}_{DF} = \frac{1}{9}(-4\mathbf{i} + 4\mathbf{j} - 7\mathbf{k})$$

$$\mathbf{T}_{DF} = \frac{T_{DF}}{9}(-4\mathbf{i} + 4\mathbf{j} - 7\mathbf{k}) \qquad \mathbf{A} = A_x\mathbf{i} + A_y\mathbf{j} + A_z\mathbf{k}$$

$$\mathbf{C} = 600\mathbf{i} \qquad \mathbf{G} = -70g\mathbf{j} \qquad \mathbf{M}_A = M_{Ay}\mathbf{j} + M_{Az}\mathbf{k}$$

$$\overrightarrow{AD} = -4\mathbf{i} + 7\mathbf{k} \qquad \overrightarrow{AB} = 8\mathbf{k} \qquad \overrightarrow{AG} = -2\mathbf{i} + 4\mathbf{k}$$

For equilibrium, we write

$$\Sigma\mathbf{F} = \mathbf{T}_{DF} + \mathbf{A} + \mathbf{C} + \mathbf{G} = \mathbf{0}:$$

$$\left(-\frac{4}{9}T_{DF} + A_x + 600\right)\mathbf{i} + \left(\frac{4}{9}T_{DF} + A_y - 70g\right)\mathbf{j}$$

$$+ \left(-\frac{7}{9}T_{DF} + A_z\right)\mathbf{k} = \mathbf{0} \qquad\qquad (1)$$

$$\Sigma \mathbf{M}_A = \overrightarrow{AD} \times \mathbf{T}_{DF} + \mathbf{M}_A + \overrightarrow{AB} \times \mathbf{C} + \overrightarrow{AG} \times \mathbf{G} = 0:$$

$$\left(-\frac{28}{9} T_{DF} + 280g \right)\mathbf{i} + \left(-\frac{56}{9} T_{DF} + M_{Ay} + 4800 \right)\mathbf{j}$$

$$+ \left(-\frac{16}{9} T_{DF} + M_{Az} + 140g \right)\mathbf{k} = 0 \qquad (2)$$

The above vector equations (1) and (2) together give a set of six simultaneous scalar equations (as expected), which yield

$$T_{DF} = 90g \qquad M_{Ay} = 693.6 \qquad M_{Az} = 20g$$

$$A_x = -207.6 \qquad A_y = 30g \qquad A_z = 70g$$

$$T_{DF} = 883 \text{ N} \qquad \mathbf{M}_A = 694\mathbf{j} + 196.2\mathbf{k} \text{ N} \cdot \text{m} \blacktriangleleft$$

$$\mathbf{A} = -208\mathbf{i} + 294\mathbf{j} + 687\mathbf{k} \text{ N} \blacktriangleleft$$

Developmental Exercise

D5.40 Describe the situation as manifested in Example 5.18 in which the maximum number of statically determinate unknowns is equal to *six*.

EXAMPLE 5.19

The rigid body *ABCDEF* is supported by a cable *EG* and two ball-and-socket supports at *A* and *F*. The allowable tension in the cable is 6 kN. Determine the maximum mass *m* that can be supported at *D* as shown.

Solution. When the maximum mass *m* is supported at *D*, the tension in the cable *EG* will be equal to 6000 N as indicated in the free-body diagram shown. We note that there are *seven* unknowns: A_x, A_y, A_z, F_x, F_y, F_z, and *m*; however, the maximum number of statically determinate unknowns for a rigid body in space is *six*. Therefore, the rigid body is statically indeterminate and we *may* be able to determine some, but not all, of the unknowns.

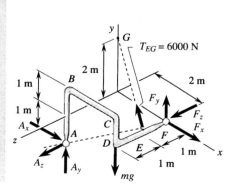

By inspection, we find that all unknowns except m can be prevented from entering an equilibrium equation when we equate to zero the sum of the moments of all forces about the axis AF. From the sketch, we have

$$\overrightarrow{EG} = -2\mathbf{i} + 2\mathbf{j} - \mathbf{k} \qquad \lambda_{EG} = \frac{1}{3}(-2\mathbf{i} + 2\mathbf{j} - \mathbf{k}) \qquad T_{EG} = 6000$$

$$\mathbf{T}_{EG} = T_{EG}\lambda_{EG} = 2000(-2\mathbf{i} + 2\mathbf{j} - \mathbf{k}) \qquad \mathbf{W} = -mg\mathbf{j} \qquad \overrightarrow{FE} = \mathbf{k}$$

$$\overrightarrow{FD} = 2\mathbf{k} \qquad \overrightarrow{AF} = 2\mathbf{i} + \mathbf{j} - 2\mathbf{k} \qquad \lambda_{AF} = \frac{1}{3}(2\mathbf{i} + \mathbf{j} - 2\mathbf{k})$$

For equilibrium, we write

$$\Sigma M_{AF} = \lambda_{AF} \cdot (\overrightarrow{FE} \times \mathbf{T}_{EG}) + \lambda_{AF} \cdot (\overrightarrow{FD} \times \mathbf{W}) = -4000 + \frac{4}{3}mg = 0$$

$$m = 3000/g \qquad m = 306 \text{ kg} \blacktriangleleft$$

REMARK. Without using vector algebra to take moment about the axis AF, it would be rather difficult to solve the problem in this example.

Developmental Exercise

D5.41 If a rigid-body equilibrium problem requires only a partial solution, what should we do?

PROBLEMS

5.48 and 5.49* Is the bent bar ABC improperly or partially constrained for a vertical force $\mathbf{P}$ applied as shown?

Fig. P5.48   **Fig. P5.49***

5.50 Do Prob. 5.48 if the flanged wheel on rail at the end C is replaced with a ball-and-socket joint.

5.51* Do Prob. 5.49* if the wide bearing with stop at the end C is replaced with a wide bearing without stop.

Fig. P5.52   **Fig. P5.53***

5.52 and 5.53* A polygonal table has a total weight W and three vertical legs as shown. The table has equal sides and identical corners. Determine the maximum magnitude of a vertical force **P** which may act anywhere on the table without tipping it over.

5.54 It is known that the resultant weight force of the 200-kg ventilating duct shown passes through the point G. If $d = 3$ m, determine the tensions in the vertical wires supporting the duct at the points A, B, and C.

Fig. P5.56

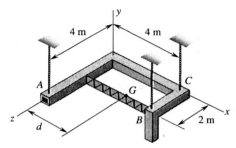

Fig. P5.54

5.55* Solve Prob. 5.54 if $d = 2$ m.

5.56 A 5-kg pipe is slipped over a wrench to loosen a stubborn bolt as shown. The friction between the pipe and the wrench handle is negligible. For $P = 400$ N and $L = 800$ mm, determine the reactions from the wrench handle to the pipe at the points B and C.

5.57 Solve Prob. 5.56 if $P = 300$ N and $L = 1$ m.

5.58 A uniform conic shell having a weight W and a base of radius r is suspended with three vertical wires of equal lengths L as shown. Determine the magnitude of the vertical moment **M** which is needed to rotate the shell through a small angle θ.

5.59* A 16-ft steel boom AB which weighs 100 lb is acted on by a force **P** of 340 lb as shown. Assuming the weight force acts through the midpoint of the boom, determine the tension in each cable and the reaction from the support at A.

5.60 A derrick carries a 12-Mg crate as shown. Determine the tensions in the cables BF and BG and the reaction from the support at A.

Fig. P5.58

Fig. P5.59*

Fig. P5.60

5.61   Determine the reactions from the supporting rails at A, B, and C as shown.

5.62 through 5.67* *Using an appropriate moment equilibrium equation*, determine the tension in the cable supporting the rigid body shown.

Fig. P5.61

Fig. P5.62 and P5.68

Fig. P5.63 and P5.69

Fig. P5.64 and P5.70

Fig. P5.65 and P5.71

Fig. P5.66* and P5.72*

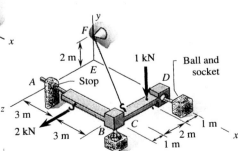

Fig. P5.67* and P5.73*

5.68 through 5.73* For the rigid body shown, determine the reaction from the support at A.

Fig. P5.74

Fig. P5.75

Fig. P5.76*

Fig. P5.77*

5.74 through 5.77* For the rigid body shown, determine the tension in the cable and the reactions from the supports at A and B.

5.78 Determine the reactions from the supports at A and B as shown if $\mathbf{M} = \mathbf{0}$.

5.79* Solve Prob. 5.78 if $\mathbf{M} = 450\mathbf{i} - 400\mathbf{k}$ N·m.

5.80 A rigid body in the shape of a triad is supported by three eyebolts as shown. Determine the reactions from the supports if $\mathbf{F} = -600\mathbf{j}$ lb and $\mathbf{M} = \mathbf{0}$.

Fig. P5.78

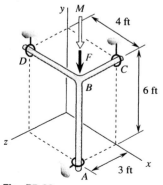

Fig. P5.80

5.81* Solve Prob. 5.80 if $\mathbf{F} = -600\mathbf{j}$ lb and $\mathbf{M} = -120\mathbf{j}$ lb·ft.

5.82 Having disconnected a clogged drainpipe, a plumber introduced a pipe auger (or power snake) through the opening of the end at A as shown. The cutting head of the auger is connected by a heavy-duty flexible cable to an electric motor which uniformly rotates the cable the plumber forces into the pipe. The supports at B, C, and D may be taken as narrow bearings. The cleaning efforts at A are equivalent to a wrench of $\mathbf{F} = 60\mathbf{i}$ N and $\mathbf{M} = 200\mathbf{i}$ N·m. Determine the reactions from the supports at B, C, and D due to the cleaning operation.

Fig. P5.82

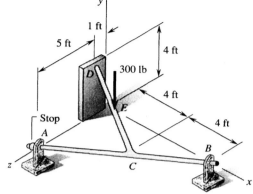

Fig. P5.83

5.83 A lever leans against a smooth vertical wall at D and is supported at A and B by narrow bearings as shown. A vertical force of 300 lb is applied at the midpoint E of the rod CD. Determine the reactions from the supports at A, B, and D.

5.84 A 30-kg rod AB is supported by a ball and socket at A and leans against the top of a vertical wall at E and the surface of a taller vertical wall at B as shown. Assume that friction forces between the rod and the walls are negligible and the weight force acts through the midpoint of the rod. Determine (a) the coordinates (x_E, y_E, z_E) of the point E, (b) the unit vector $\boldsymbol{\lambda}_E$ in the direction of the reaction from the support at E, (c) the reactions from the supports of the rod.

5.85 through 5.88* A rigid body is shown. (a) How many unknowns are there in the reactions from the supports? (b) How many independent equilibrium equations are there available for the solution of these unknowns? (c) Are there any statically indeterminate unknowns? (d) If any unknowns are statically determinate, calculate their values.

Fig. P5.84

Fig. P5.85

Fig. P5.86

Fig. P5.87

Fig. P5.88*

5.89 Three decorations are suspended by a system of wires as shown. Using the concept of equilibrium of a rigid body in space, determine the tensions in wires AD, AE, BF, BG, CH, and CI.

Fig. P5.89

5.14   Concluding Remarks

The fact surrounding the condition of equilibrium of a rigid body is that the *resultant force* and the *resultant moment,* about a point, of the force system acting on its free body are *equal to zero.* This fact provides us with the basis upon which the relations among the applied forces and the reactions from the supports and connections of the rigid body are established. The free-body diagram of a rigid body exhibits to us all the external forces that act together on the free body to bring about the condition or state of the rigid body as it existed in its space diagram. Thus, the establishment of those relations is efficaciously accomplished through the use of the free-body diagram of the rigid body.

The drawing of a good free-body diagram of a rigid body usually involves the exercise of good judgment on the way the supports or connections impose restrictions on the movements of the points of support of the body which is subjected to an applied force system. The recognition of the types of reactions, as discussed and illustrated in Secs. 5.5 and 5.11, is of fundamental importance in the drawing of correct free-body diagrams of rigid bodies in a plane and in space. Furthermore, it is well to *regard the drawing of appropriate free-body diagrams as an integral part of the effective study of the conditions of rigid bodies under the action of force systems whether they are in equilibrium or not*. Bodies not in equilibrium are studied in dynamics.

REVIEW PROBLEMS

5.90 and 5.91* For each rigid body shown, state (a) the unknowns in its free-body diagram; (b) the prospective equilibrium equations to be applied; (c) the prospective equilibrium equations, if any, which cannot be satisfied; (d) the one term (statically determinate, statically indeterminate, partially constrained, or improperly constrained) which best describes the body.

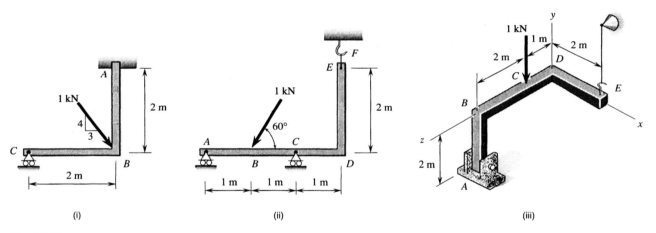

(i)   (ii)   (iii)

Fig. P5.90

(i)   (ii)   (iii)

Fig. P5.91*

5.92* In each of the following rigid bodies, (a) identify (by the number of unknown components) the type of reaction from the support at A, (b) draw the free-body diagram of the body AB.

(i)

(ii)   (iii)

Fig. P5.92* and P5.93

5.93 For each of the rigid bodies shown, determine the reaction from the support at A.

5.94* The resultant weight force $\mathbf{W}$ of the rod AB acts through the point G as shown. Verify that the transcendental equation governing the geometric parameters a, b, L, α and θ of the rod in equilibrium is

$$\left[\left(\frac{b}{a}\right)^2 - \sin^2(\theta + \alpha)\right]^{1/2}\left[\frac{L}{2a} + \sin(\theta + \alpha)\tan\theta - \cos(\theta + \alpha)\right]$$
$$+ \frac{1}{2}\tan\theta\,\sin 2(\theta + \alpha) + \sin^2(\theta + \alpha) - \left(\frac{b}{a}\right)^2 = 0$$

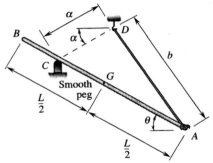

Fig. P5.94* and P5.95*

5.95* The resultant weight force $\mathbf{W}$ of the rod AB acts through the point G as shown. If $b = 2a$, $L = 3a$, $\alpha = 45°$, and the rod is in equilibrium, determine the angle θ. (*Hint.* $20° < \theta < 30°$.)[†]

5.96* Use the one term (statically determinate, statically indeterminate, partially constrained, or improperly constrained) that best describes the body shown if (a) the cable DE is removed, (b) the cable DF is removed, (c) both cables are removed, (d) nothing is removed.

Fig. P5.96*

[†]Cf. App. E or Sec. B5.8 of App. B for the solution of a transcendental equation.

Fig. P5.97

5.97 For the bent rod shown, determine the tension in the cable *EG*.

5.98 Determine the reactions from the supports of the bent bar shown.

5.99 A plate *ABCD* is shown. (a) Determine whether the plate is statically indeterminate. (b) If any unknowns are statically determinate, calculate their values.

Fig. P5.98

Fig. P5.99

Chapter 6

First Moments: Centroids and Centers of Gravity

KEY CONCEPTS

- Resultant of parallel forces, first moment of a quantity, principle of moments, geographic center, and population center.
- Centroids of lines and centers of gravity of slender members.
- Centroids of areas and centers of gravity of laminas.
- Distributed loads on beams.
- First and second theorems of Pappus-Guldinus.
- Centroids of volumes, centers of gravity of bodies, and centers of mass of bodies.
- Composite volumes and center of pressure.

This chapter deals with the first moments of various quantities and is focused on the *principle of moments*, which is synthesized from the concept of *equivalent systems of forces*. This principle is identified as the key to locating the resultant of parallel forces or any other force system, the centroids of geometric figures, and the centers of gravity as well as centers of mass of physical bodies. Common applications of this principle include those in determining (a) the resultant of distributed forces on a body; (b) the geographic center and the population center of a country; (c) the centroids of cross sections of beams and columns; and (d) the centers of gravity as well as the centers of mass of wires, plates, shells, machine parts, and various other physical bodies. Naturally, this chapter provides a prerequisite for the study of structures withstanding distributed loads or weight forces of given bodies.

6.1 Dimensional Homogeneity in Equations

The equation of a curve or surface is a mathematical statement of the general relationship to be satisfied by the values of the coordinates of any point on the curve or surface, as shown in Fig. 6.1. The equation of a curve or

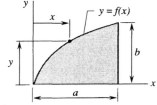

Fig. 6.1

surface is frequently used in engineering to describe the intensity of a distributed load, the shape of a figure or body, and other relations. However, the resulting *dimensions* (and hence units) *of each term* in such an equation must be the same. This requirement is usually met if the equation is mathematically homogeneous.[†] In any case, the constants including the implicit 1's (such as those in $y = x^2$) in an equation should be understood to carry whatever dimensions necessary to ensure that *each term has the same resulting dimensions*.

Developmental Exercises

D6.1   A cantilever beam carries a load whose intensity w, in kN/m, varies continuously along its axis as shown. (a) What are the value and units of the constant c? (b) What is the loading intensity w at $x = 1$ m?

D6.2   A paraboloid is shown. The equation of its curved surface is given by $ay = x^2 + z^2$. What are the value and units of the constant a?

Fig. D6.1

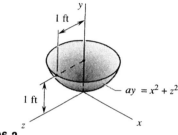

Fig. D6.2

6.2   Resultant of Parallel Forces: Principle of Moments

We saw in Sec. 2.2 that, if the lines of action of two force vectors intersect as shown in Fig. 2.8, the line of action of the resultant of these two forces must pass through the point of intersection. However, the *resultant of a system of parallel forces* cannot be located in that manner; it is located by using Eqs. (5.4), which define equivalent systems of forces. As a generalization of the concept of the moment of a force, the *first moment*, or more briefly the *moment*, *of a quantity* is defined as the product of an appropriate distance and the quantity. This generalized definition of *moment* and the concepts of *equivalence* and *equipollence* between two systems of forces in Sec. 5.3 may here be synthesized into a general rule as follows: *The moment of the "resultant" with respect to a point (or an axis, or a plane) is equal to the sum of the moments of the "components" with respect to the same point (or axis, or plane), where the "resultant" is equal to the sum of the "components."* This *synthesized rule* is referred to as the *principle of moments*; i.e.,

> *Resultant* = Sum of the *Components*
>
> Moment of the *Resultant*
> = Sum of the Moments of the *Components*   (6.1)

where the moment may be taken with respect to any point, axis, or plane, as appropriate. If the *components* are scalar quantities (e.g., lengths, areas, volumes, or masses), the *resultant* is a total scalar quantity (e.g., total length, total area, total volume, or total mass) equal to the algebraic sum of the component scalars. If the *components* are vector quantities (e.g., forces),

---

[†]An equation is mathematically homogeneous if each term of the equation will contain t to the same power as a factor when each variable in the equation is replaced by t times the variable, where $t \neq 0$.

the *resultant* is a resultant vector quantity (e.g., resultant force) equal to the vector sum of the component vectors. When the components are *concurrent forces*, the principle of moments degenerates into *Varignon's theorem* presented in Secs. 4.3 and 4.11.

The principle of moments owes its origin to the concept of *equivalent systems of forces* and is primarily applied to *locate the position of the resultant relative to the components* in many situations. The centroids of lines, areas, and volumes, as well as the centers of mass and centers of gravity of bodies, are their respective "centers," which are defined and located according to the *principle of moments* in the sequel. Note that the *geographic center* of an area is simply the centroid of the area, whereas the *population center* of an area is a point determined according to the *principle of moments*, where the components are the entities of inhabitants and the resultant is the total number of inhabitants in the area.

EXAMPLE 6.1

If the resultant force **R** of the three vertical forces shown acts through the point $P(\bar{x}, 0, \bar{z})$, determine $\bar{x}$ and $\bar{z}$.

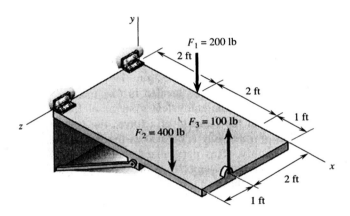

Solution. The component forces are

$$\mathbf{F}_1 = -200\mathbf{j} \text{ lb} \qquad \mathbf{F}_2 = -400\mathbf{j} \text{ lb} \qquad \mathbf{F}_3 = 100\mathbf{j} \text{ lb}$$

The resultant force **R** is

$$\mathbf{R} = \mathbf{F}_1 + \mathbf{F}_2 + \mathbf{F}_3 \qquad \mathbf{R} = -500\mathbf{j} \text{ lb}$$

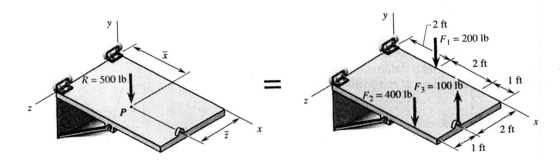

Applying the *principle of moments* and using the right-hand rule, we write

$$\Sigma M_z: \quad -\bar{x}(500) = -2(200) - 4(400) + 5(100) \qquad \bar{x} = 3 \text{ ft} \blacktriangleleft$$

$$\Sigma M_x: \quad \bar{z}(500) = 0(200) + 3(400) - 2(100) \qquad \bar{z} = 2 \text{ ft} \blacktriangleleft$$

Developmental Exercises

D6.3 Determine the distance $\bar{x}$ between the y axis and the line of action of the resultant force of the forces shown.

D6.4 The major towns in the county *PQRST*, as shown, are Oshkosh, Akin, and Butler, which have populations of 5000, 3000, and 2000, and coordinates of locations $O(0, 0)$ km, $A(-10, 20)$ km, and $B(10, -2)$ km, respectively. For this county, locate (a) its geographic center, (b) its approximate population center.

D6.5 (a) Define the *first moment of a quantity*. (b) Describe the *principle of moments*.

D6.6 Define the terms: (a) geographic center, (b) population center.

Fig. D6.3

Fig. D6.4

LINES AND SLENDER MEMBERS

6.3   Centroids of Lines[†]

A line may be either a *straight line* or a *curved line*, which is often simply called a *curve*. The *centroid of a line* is the "center" of the line and is defined in accordance with the *principle of moments*. Let the length of the infinitesimal differential line element between the points $P(x, y, z)$ and P' $(x + dx, y + dy, z + dz)$ be dL as shown in Fig. 6.2. By the Pythagorean theorem, we can show that

$$dL = [(dx)^2 + (dy)^2 + (dz)^2]^{1/2}$$

The centroid $\bar{P}(\bar{x}_{el}, \bar{y}_{el}, \bar{z}_{el})$ of this line element may be taken as being at the midpoint of the line element between the points P and P', where the sub-

[†] A *centroid* indicates the "center" of a geometric figure, not of a physical body. Note that the suffix *-oid* usually means "resembling." Cf. Secs. 6.5, 6.9, 6.14, and 6.15.

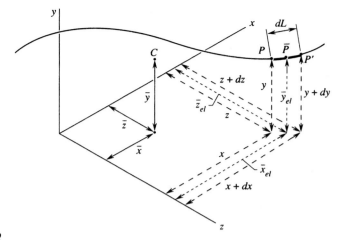

Fig. 6.2

script "*el*" stands for "element" and overbars are used with symbols which are associated with the centroid. Since the line element dL is an infinitesimal quantity, the coordinates of the centroid $\overline{P}$ may be written as

$$\overline{x}_{el} = x + \frac{dx}{2} \approx x \qquad \overline{y}_{el} = y + \frac{dy}{2} \approx y \qquad \overline{z}_{el} = z + \frac{dz}{2} \approx z$$

The *centroid* $C(\overline{x}, \overline{y}, \overline{z})$ of the line of total length L is, according to the *principle of moments*, defined by the following equations:

$$L = \int dL$$

$$\Sigma M_{yz}: \quad \overline{x}L = \int x \, dL$$

$$\Sigma M_{zx}: \quad \overline{y}L = \int y \, dL \tag{6.2}$$

$$\Sigma M_{xy}: \quad \overline{z}L = \int z \, dL$$

The right-hand side integrals in the last three equations of Eqs. (6.2) give the so-called *first moments* of the line with respect to the yz, zx, and xy planes, respectively. Thus, the coordinates $(\overline{x}, \overline{y}, \overline{z})$ of the centroid C of the line can be obtained by dividing the total length L into the first moments of the line with respect to the coordinate planes. For a line lying in the xy plane, we have $\overline{z} = \overline{z}_{el} = z = 0$ and the last equation in Eqs. (6.2) becomes trivial. The first moments $\overline{x}L$ and $\overline{y}L$ may then be considered as being taken with respect to the y and x axes, respectively. For a line lying in a plane, dL is given by one of the following forms:

$$dL = \left[1 + \left(\frac{dy}{dx} \right)^2 \right]^{1/2} dx \tag{6.3}$$

$$dL = \left[\left(\frac{dx}{dy} \right)^2 + 1 \right]^{1/2} dy \tag{6.4}$$

$$dL = \left[r^2 + \left(\frac{dr}{d\theta} \right)^2 \right]^{1/2} d\theta \tag{6.5}$$

where (r, θ) are the polar coordinates.

The centroid C of a line is not necessarily located on the line. However, if the line is straight, the centroid is at the midpoint of the line. It is to be noted that the dimensions of the first moment of a line are *(length)*^{2}. Furthermore, for the purpose of locating the centroid of a line, the *differential line element dL* and the *total length L* of a line segment are always made *positive*, even though the coordinates of the centroids of *dL* and *L* may be positive, negative, or zero.

EXAMPLE 6.2

Locate the centroid $C(\bar{x}, \bar{y})$ of the semicircular *arc* of radius R as shown.

Solution.   The semicircular arc is symmetrical about the y axis. Therefore, its centroid $C(\bar{x}, \bar{y})$ must lie on the y axis; i.e.,

$$\bar{x} = 0 \quad \blacktriangleleft$$

The length L of the semicircular arc is well-known; i.e.,

$$L = \pi R$$

For illustrative purposes, we will determine $\bar{y}$ in two different ways:[†]

(a) *Using rectangular coordinates*. The equation of the semicircular arc is

$$x^2 + y^2 = R^2 \qquad \text{(same as that of a circle)}$$

By differentiation, we have

$$2x + 2y\frac{dy}{dx} = 0 \qquad \frac{dy}{dx} = -\frac{x}{y}$$

$$dL = \left[1 + \left(\frac{dy}{dx}\right)^2\right]^{1/2} dx = \frac{(y^2 + x^2)^{1/2}}{y}\, dx = \frac{R}{y}\, dx$$

Applying the *principle of moments*, we write

$$\bar{y}L = \int y\, dL: \quad \bar{y}(\pi R) = \int_{-R}^{R} y\left(\frac{R}{y}\right) dx = 2R^2 \qquad \bar{y} = \frac{2R}{\pi} \quad \blacktriangleleft$$

[†]A third way to determine this $\bar{y}$ is given in Example 6.14.

(b) *Using polar coordinates.* In this case, we have

$$dL = R\,d\theta \qquad y = R\sin\theta$$

Applying the *principle of moments*, we write

$$\bar{y}L = \int y\,dL: \quad \bar{y}(\pi R) = R^2 \int_0^\pi \sin\theta\,d\theta = 2R^2 \qquad \bar{y} = \frac{2R}{\pi} \blacktriangleleft$$

EXAMPLE 6.3

Let the point $C(\bar{x}, \bar{y})$ be the centroid and L be the length of the curve OA as shown. Making use of integral formulas (cf. Sec. B5.2 of App. B), determine the values of L, $\bar{x}$, and $\bar{y}$.

Solution. Differentiating the equation of the curve OA and using Eq. (6.3) as well as formula (t) in Sec. B5.2 of App. B, we write

$$\frac{dy}{dx} = 2x \qquad dL = (1 + 4x^2)^{1/2}\,dx$$

$$L = \int_0^1 (1 + 4x^2)^{1/2}\,dx = \frac{1}{4}\left[2\sqrt{5} + \ln(2 + \sqrt{5})\right] = 1.4789$$

$$L = 1.479 \text{ m} \blacktriangleleft$$

By the *principle of moments* and formulas (a) and (u) in Sec. B5.2 of App. B, we write

$$\bar{x}L = \int x\,dL: \quad \bar{x}(1.4789) = 2\int_0^1 x\left(x^2 + \frac{1}{4}\right)^{1/2}dx = \frac{1}{12}(5\sqrt{5} - 1)$$

$$\bar{y}L = \int y\,dL: \quad \bar{y}(1.4789) = 2\int_0^1 x^2\left(x^2 + \frac{1}{4}\right)^{1/2}dx$$

$$= \frac{1}{64}\left[18\sqrt{5} - \ln(2 + \sqrt{5})\right]$$

$$\bar{x} = 0.574 \text{ m} \qquad \bar{y} = 0.410 \text{ m} \blacktriangleleft$$

Developmental Exercise

D6.7 Let the point $C(\bar{x}, \bar{y})$ be the centroid and L be the length of the curve OA as shown. Determine (a) dL in terms of $d\theta$, (b) L, (c) $\bar{x}L$, (d) $\bar{x}$, (e) $\bar{y}L$, (f) $\bar{y}$.

Fig. D6.7

6.4   Composite Lines

A *composite line* of total length L is one which is composed of line segments of lengths $L_1, L_2, \ldots, L_n$, where

$$L = L_1 + L_2 + \cdots + L_n \tag{6.6}$$

If the centroids of the line segments of lengths $L_1, L_2, \ldots, L_n$, as shown in Fig. 6.3, are located at the points $C_1(\bar{x}_1, \bar{y}_1, \bar{z}_1)$, $C_2(\bar{x}_2, \bar{y}_2, \bar{z}_2)$, ..., $C_n(\bar{x}_n, \bar{y}_n, \bar{z}_n)$, respectively, then the *centroid* $C(\bar{x}, \bar{y}, \bar{z})$ of the composite line is, according to the *principle of moments*, defined by the following equations:

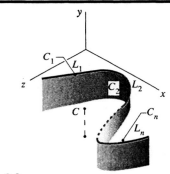

Fig. 6.3

$$\Sigma M_{yz}: \quad \bar{x}L = \bar{x}_1 L_1 + \bar{x}_2 L_2 + \cdots + \bar{x}_n L_n$$

$$\Sigma M_{zx}: \quad \bar{y}L = \bar{y}_1 L_1 + \bar{y}_2 L_2 + \cdots + \bar{y}_n L_n \qquad (6.7)$$

$$\Sigma M_{xy}: \quad \bar{z}L = \bar{z}_1 L_1 + \bar{z}_2 L_2 + \cdots + \bar{z}_n L_n$$

For a line lying in the xy plane, we have $\bar{z} = \bar{z}_1 = \bar{z}_2 = \cdots = \bar{z}_n = 0$ and the last equation in Eqs. (6.7) becomes trivial. Note that the centroid of a composite line, as well as the centroid of a curved line, is generally not located on the line. Moreover, a composite line may have an empty (or negative) line as a component. In this case, the length of the empty line is treated as negative. For use in determining the centroids of composite lines, the centroids of some circular arcs are shown in Fig. 6.4. If desired, a short computer program may be used to locate the centroid of a composite line as shown in Fig. 6.5.

Quarter-circular arc

$$\bar{x} = \bar{y} = \frac{2R}{\pi}$$

$$L = \frac{\pi R}{2}$$

Fig. 6.4

Semicircular arc

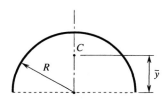

$$\bar{y} = \frac{2R}{\pi}$$

$$L = \pi R$$

Segment of circular arc

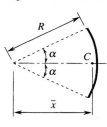

$$\bar{x} = \frac{R \sin \alpha}{\alpha}$$

$$L = 2\alpha R$$

```
10 REM   A COMPOSITE LINE WITH N COMPONENT LINES: Enter N, L1, x1, y1, z1, L2,
15 REM   x2, y2, z2, L3, x3, y3, z3, L4, x4, y4, z4, ..., as data in Line 100.
20 REM   To output to the printer, change PRINT to LPRINT in the program.
25 READ N:DIM L(N),X(N),Y(N),Z(N):FOR I=1 TO N:READ L(I),X(I),Y(I),Z(I):L=L+L(I)
30 XL=XL+X(I)*L(I):YL=YL+Y(I)*L(I):ZL=ZL+Z(I)*L(I):NEXT I:X=XL/L:Y=YL/L:Z=ZL/L
35 PRINT:PRINT"The component lines and the coordinates of their centroids are"
40 FOR I = 1 TO N: PRINT "L ("; I; ") =";: PRINT USING " ##.######^^^^"; L(I),
45 PRINT SPC(8); "x, y, z = "; X(I); Y(I); Z(I): NEXT I: S$ = "          "
50 PRINT"The centroid of the composite line is located at":PRINT "x = ";X;S$;
55 PRINT "y = ";Y;: PRINT S$;"z = ";Z:                  REM (File Name: CENTL)
100 DATA 3,1.479,0.574,0.410,0,1,-0.4,0,0.3,3.141593,-1.43662,-1,0.6
```

```
The component lines and the coordinates of their centroids are
L ( 1 ) =    1.479000E+00        x, y, z =  .574   .41   0
L ( 2 ) =    1.000000E+00        x, y, z = -.4    0   .3
L ( 3 ) =    3.141593E+00        x, y, z = -1.43662 -1   .6
The centroid of the composite line is located at
x = -.723114       y = -.4510561       z =  .3887412
```

Fig. 6.5 (Cf. the solution in Example 6.4.)

EXAMPLE 6.4

Making use of the solutions in Examples 6.2 and 6.3, locate the centroid $C(\bar{x}, \bar{y}, \bar{z})$ of the composite line $PQRST$ as shown.

 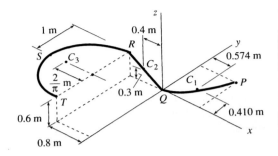

Solution. Let L_1, L_2, and L_3 be the lengths of the component lines PQ, QR, and RST. From the solutions in Examples 6.2 and 6.3 and for the coordinate system shown, we have $L_1 = 1.4789$ m, $L_2 = 1$ m, $L_3 = \pi$ m, and the centroids of these component lines are respectively located at $C_1(0.574, 0.410, 0)$ m, $C_2(-0.4, 0, 0.3)$ m, and $C_3(-0.8 - 2/\pi, -1, 0.6)$ m as shown. The total length is

$$L = L_1 + L_2 + L_3 \qquad L = 5.621 \text{ m}$$

Applying the *principle of moments* in the form of Eqs. (6.7), we write

$$\Sigma M_{yz}: \quad \bar{x}(5.621) = 0.574(1.479) + (-0.4)(1) + (-0.8 - 2/\pi)\pi$$

$$\Sigma M_{zx}: \quad \bar{y}(5.621) = 0.410(1.479) + 0(1) + (-1)(\pi)$$

$$\Sigma M_{xy}: \quad \bar{z}(5.621) = 0(1.479) + 0.3(1) + 0.6(\pi)$$

These three equations yield

$$\bar{x} = -0.723 \text{ m} \qquad \bar{y} = -0.451 \text{ m} \qquad \bar{z} = 0.389 \text{ m} \quad \blacktriangleleft$$

Fig. D6.8

Developmental Exercise

D6.8 Locate the centroid $C(\bar{x}, \bar{y})$ of the composite line $OPQR$ as shown.

6.5 Centers of Gravity of Slender Members

The *center of gravity* of a body is the center of the gravitational attraction experienced by the body. In other words, the resultant weight force of the body acts at the center of gravity of the body. A slender member is a body where any linear measurement of its transverse cross section (e.g., diameter) is negligibly small compared with its length. Slender members may include wires, bars, rods, tubes, and pipes.

The *mass density* ρ_L of a slender member is expressed as mass per unit length, while its *weight density* γ_L is expressed as weight per unit length.

By Newton's second law, we note that

$$\gamma_L = \rho_L g \qquad (6.8)$$

where g is the gravitational acceleration at the location of the slender member. Unless otherwise stated, each slender member will be assumed to have a constant value of γ_L and be situated near the surface of the earth where $g = 9.81 \text{ m/s}^2$. Thus, the *centers of gravity* of slender members are taken to be coincident with the centroids of their center lines or axes, which are located as explained in Secs. 6.3 and 6.4. If the slender members were not uniform as assumed, they should be treated as bodies which are studied in Sec. 6.14.

EXAMPLE 6.5

A slender tube *ABCDE* is bent into the shape as shown. The tube has a mass density of 1 kg/m and is held in equilibrium by three vertical wires at *A*, *B*, and *D*. Determine the tension in each wire.

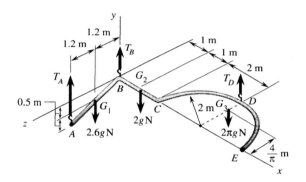

Solution. The center of gravity of the slender tube may be taken as coincident with the centroid of its center line. By the *principle of moments*, the resultant weight force of the tube is equivalent to the sum of the weight forces of the segments *AB*, *BC*, and *CDE*, which are located at G_1, G_2, and G_3 as indicated. Since the tube has a mass density of 1 kg/m, the weight forces of these segments are $2.6g$ N, $2g$ N, and $2\pi g$ N, respectively, as indicated, where the distance between G_3 and the x axis is $2R/\pi = 4/\pi$ m (see Fig. 6.4). For equilibrium of the free body drawn, we write

$$\Sigma M_z = 0: \quad 4T_D - 4(2\pi g) - 1(2g) = 0 \qquad T_D = 66.543$$

$$\Sigma M_x = 0: \quad 2T_D - \frac{4}{\pi}(2\pi g) + 1.2(2.6g) - 2.4T_A = 0 \qquad T_A = 35.506$$

$$\Sigma F_y = 0: \quad T_A + T_B + T_D - 2.6g - 2g - 2\pi g = 0 \qquad T_B = 4.715$$

$$T_A = 35.5 \text{ N} \qquad T_B = 4.72 \text{ N} \qquad T_D = 66.5 \text{ N} \blacktriangleleft$$

REMARK. The above results may similarly be obtained by replacing the component weight forces with the resultant weight force of the entire tube.

Fig. D6.10

Developmental Exercises

D6.9   Locate the center of gravity $G(\bar{x}, \bar{y}, \bar{z})$ of the slender tube in Example 6.5. Use it to check the answers obtained in that example.

D6.10   A bent rod $ABCDE$ weighing 10 lb/ft is suspended as shown. Determine the tensions in the wires at A, B, and C.

PROBLEMS

6.1   For the curves shown, what are the values and units of the constants c and k in each case?

Fig. P6.1

Fig. P6.2

6.2   For the beams shown, what is the loading intensity w at $x = 0.5$ m in each case?

6.3 through 6.5   Using the *principle of moments*, locate the resultant force **R** of the parallel forces acting as shown.

Fig. P6.3

Fig. P6.4

Fig. P6.5

6.6 through 6.8   Let the point $C(\bar{x}, \bar{y})$ be the centroid of the curved line shown. Determine by *integration* the value of $\bar{y}$.

Fig. P6.6

Fig. P6.7

Fig. P6.8

6.9 through 6.17* Locate the centroid of the composite line shown.

Fig. P6.9

Fig. P6.10

Fig. P6.11*

Fig. P6.12

Fig. P6.13

Fig. P6.14*

Fig. P6.15

Fig. P6.16

Fig. P6.17*

6.18 A bent rod ABC is suspended by a wire at B as shown. Determine the length L for which the portion AB of the bent rod is horizontal.

6.19 A bent rod ABC is suspended by a wire at B as shown. Determine the length L for which the portion BC of the bent rod is horizontal.

6.20* A wire is bent as shown and is suspended by a string at O. Determine the angle ϕ for which the plane of the wire is horizontal.

Fig. P6.18 and P6.19

Fig. P6.20*

Fig. P6.21

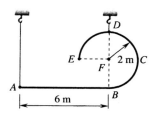

Fig. P6.22

6.21 through 6.23* The rigid body shown has a mass density of 1 kg/m. Determine the reactions from the supports.

Fig. P6.23*

6.24 through 6.26* The rigid body shown has a weight density of 20 lb/ft. Determine the reactions from the supports.

Fig. P6.24

Fig. P6.25

Fig. P6.26*

AREAS AND LAMINAS

6.6   Orders of Differential Area Elements

A differential area element dA may be of the first or second order. A *first-order differential area element* is an *infinitesimal areal strip* which is expressed in terms of a differential of a coordinate variable. It may be illustrated as shown in Fig. 6.6.

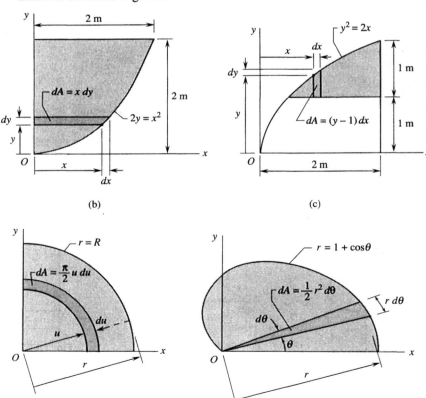

Fig. 6.6

A *second-order differential area element* is an *infinitesimal areal patch* which is expressed in terms of the product of two differentials of coordinate variables. Second-order differential area elements are generally of the shapes of a quadrilateral area and may be illustrated as shown in Fig. 6.7. In Fig. 6.7(a), note that the variables $(\xi, \eta)^\dagger$ denote the rectangular coordinates of the point E where the *second-order differential area element dA* is located, while the variables (x, y) denote the rectangular coordinates of *a point on the boundary curve* of the area A. In Fig. 6.7(b), note similarly that the variables $(u, \theta)^\dagger$ denote the polar coordinates of the point E where the *second-order differential area element dA* is located, while the variables (r, θ) denote the polar coordinates of *a point on the boundary curve* of the area A.

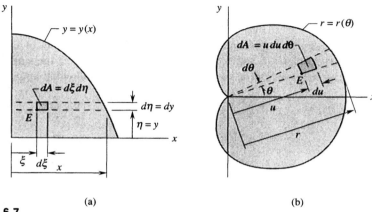

(a)   (b)

Fig. 6.7

Developmental Exercises

D6.11   Define the terms: (a) first-order differential area element, (b) second-order differential area element.

D6.12   Write, in terms of y and dy, the first-order differential area element dA parallel to the x axis for each of the areas shown.

D6.13   Write, in terms of x and dx, the first-order differential area element dA parallel to the y axis for each of the areas shown.

(a)   (b)

Fig. D6.12 and D6.13

6.7   Centroids of Areas

An *area* may be a plane area or a curved area. A curved area is generally called a *surface area*, or simply a *surface*. The *centroid of an area* is the "center" of the area and is defined in accordance with the principle of moments. Let the point $P(\overline{x}_{el}, \overline{y}_{el}, \overline{z}_{el})$ be the centroid of the differential area *element dA* of a given surface of total area A as shown in Fig. 6.8, where dA may be of the *second* or the *first* order. It is important to make sure that

---

†These variables can be used as the running (or dummy) variables for the integration (cf. Example 6.8).

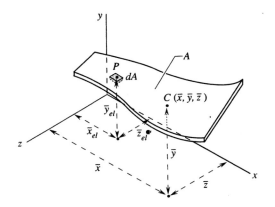

Fig. 6.8

$(\bar{x}_{el}, \bar{y}_{el}, \bar{z}_{el})$ are the *coordinates of the centroid* of dA if dA is of the *first order* (cf. Examples 6.6 and 6.8). The centroid $C(\bar{x}, \bar{y}, \bar{z})$ of the surface is, according to the *principle of moments*, defined by the following equations:

$$
\begin{aligned}
A &= \int dA \\
\Sigma M_{yz}: \quad \bar{x}A &= \int \bar{x}_{el}\, dA \\
\Sigma M_{zx}: \quad \bar{y}A &= \int \bar{y}_{el}\, dA \\
\Sigma M_{xy}: \quad \bar{z}A &= \int \bar{z}_{el}\, dA
\end{aligned}
\tag{6.9}
$$

The integrals in the last three equations in Eqs. (6.9) give the *first moments* of the surface with respect to the yz, zx, and xy planes, respectively.[†] Thus, by dividing the total area A into the first moments of the surface, we can obtain the coordinates $(\bar{x}, \bar{y}, \bar{z})$ of the centroid C of the surface. For an area lying in the xy plane, we have $z = \bar{z} = \bar{z}_{el} = 0$ and the last equation in Eqs. (6.9) becomes trivial. It should be noted that, for the purpose of locating the centroid of an area with no holes, the *values* of dA and A are always made *positive*. However, the coordinates of the centroids of dA and A may be positive, negative, or zero.

EXAMPLE 6.6

Determine the ordinate $\bar{y}$ of the centroid $C(\bar{x}, \bar{y})$ of the triangular area PQR as shown.

Solution. From geometry, we know that the area of the triangle is

$$A = \frac{1}{2} bh$$

By similar triangles in the sketch shown, we write

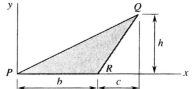

[†]*Second moments* of areas (i.e., moments of first moments) are presented in Chap. 7.

$$\frac{\xi}{b} = \frac{h - y}{h} \qquad \xi = b - \frac{b}{h} y$$

$$dA = \xi \, dy = \left(b - \frac{b}{h} y \right) dy \qquad \bar{y}_{el} = y + \frac{dy}{2} \approx y$$

Applying the *principle of moments*, we write

$$\bar{y} A = \int \bar{y}_{el} \, dA: \qquad \bar{y} \left(\frac{1}{2} bh \right) = \int_0^h y \left(b - \frac{b}{h} y \right) dy = \frac{1}{6} bh^2$$

$$\bar{y} = \frac{h}{3} \quad \blacktriangleleft^\dagger$$

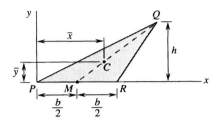

REMARK. The solution shows that, for any triangular area, the distance between its centroid C and a base is equal to one third of the corresponding height. Notice that any side of a triangle may be regarded as a *base*, and the length of the altitude to that base is the *corresponding height*. Furthermore, the centroid C must lie on any median drawn. The answer in this example is widely used as a *formula*.

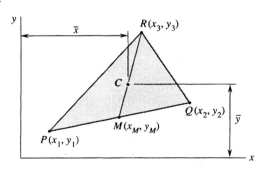

EXAMPLE 6.7

The vertices of an arbitrary triangle are located at $P(x_1, y_1)$, $Q(x_2, y_2)$, and $R(x_3, y_3)$. For the triangular area PQR, determine the formulas for (a) its area A, (b) the coordinates $\bar{x}$ and $\bar{y}$ of its centroid C.

Solution. Since $A = \frac{1}{2} |\overrightarrow{PQ} \times \overrightarrow{PR}|$ as shown (cf. Sec. 4.7), we write

$$\overrightarrow{PQ} = (x_2 - x_1)\mathbf{i} + (y_2 - y_1)\mathbf{j}$$

$$\overrightarrow{PR} = (x_3 - x_1)\mathbf{i} + (y_3 - y_1)\mathbf{j}$$

$$\overrightarrow{PQ} \times \overrightarrow{PR} = \begin{vmatrix} \mathbf{i} & \mathbf{j} & \mathbf{k} \\ x_2 - x_1 & y_2 - y_1 & 0 \\ x_3 - x_1 & y_3 - y_1 & 0 \end{vmatrix} = \begin{vmatrix} x_2 - x_1 & y_2 - y_1 \\ x_3 - x_1 & y_3 - y_1 \end{vmatrix} \mathbf{k}$$

$$= \begin{vmatrix} 1 & x_1 & y_1 \\ 0 & x_2 - x_1 & y_2 - y_1 \\ 0 & x_3 - x_1 & y_3 - y_1 \end{vmatrix} \mathbf{k} = \begin{vmatrix} 1 & x_1 & y_1 \\ 1 & x_2 & y_2 \\ 1 & x_3 & y_3 \end{vmatrix} \mathbf{k}$$

$$A = \frac{1}{2} |x_1(y_2 - y_3) + x_2(y_3 - y_1) + x_3(y_1 - y_2)| \quad \blacktriangleleft$$

[†]The centroid of a triangular *area* and the centroid of the triangular *line* coinciding with the boundaries of that triangular area are generally *different*.

The coordinates of the midpoint M between P and Q are given by

$$x_M = \frac{1}{2}(x_1 + x_2) \qquad y_M = \frac{1}{2}(y_1 + y_2)$$

By the solution in Example 6.6, we know that the centroid C lies on the median $\overline{MR}$, and $\overline{MC} = \frac{1}{3}\overline{MR}$. Thus, we write

$$\overline{x} = x_M + \frac{1}{3}(x_3 - x_M) = \frac{1}{2}(x_1 + x_2) + \frac{1}{3}\left[x_3 - \frac{1}{2}(x_1 + x_2)\right]$$

$$\overline{y} = y_M + \frac{1}{3}(y_3 - y_M) = \frac{1}{2}(y_1 + y_2) + \frac{1}{3}\left[y_3 - \frac{1}{2}(y_1 + y_2)\right]$$

$$\overline{x} = \frac{1}{3}(x_1 + x_2 + x_3) \qquad \overline{y} = \frac{1}{3}(y_1 + y_2 + y_3) \quad \blacktriangleleft$$

REMARK. The above formulas for A, $\overline{x}$, and $\overline{y}$ are useful in locating the centroid of *any* plane area using *triangular finite elements* (e.g., Probs. 6.40 through 6.42).

EXAMPLE 6.8

Locate the centroid $C(\overline{x}, \overline{y})$ of the semicircular *area* of radius R as shown.

Solution. By symmetry, the centroid $C(\overline{x}, \overline{y})$ must lie on the y axis; i.e.,

$$\overline{x} = 0 \quad \blacktriangleleft$$

The area A of the semicircle is well-known; i.e., $A = \frac{1}{2}\pi R^2$. For illustrative purposes, $\overline{y}$ will here be determined in four different ways:[†]

(a) *Using rectangular coordinates and single integration.* We first choose a first-order differential area element dA as shown. We see that

$$dA = y\,dx \qquad \overline{y}_{el} \approx \frac{1}{2}y$$

Applying the *principle of moments*, we write

$$\overline{y}A = \int \overline{y}_{el}\,dA:$$

$$\overline{y}\left(\frac{1}{2}\pi R^2\right) = \frac{1}{2}\int_{-R}^{R} y^2\,dx$$

$$= \frac{1}{2}\int_{-R}^{R}(R^2 - x^2)\,dx = \frac{2}{3}R^3$$

$$\overline{y} = \frac{4R}{3\pi} \quad \blacktriangleleft$$

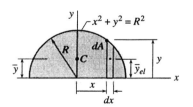

[†]A fifth way to determine the ordinate $\overline{y}$ is given in Example 6.15.

(b) *Using polar coordinates and single integration.* In polar coordinates, the equation of the curved boundary is $r = R$. The first-order differential area element dA as shown may be taken as a triangular area. Thus,

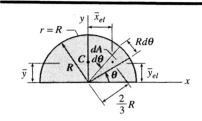

$$dA = \frac{1}{2}(R\,d\theta)R = \frac{1}{2}R^2\,d\theta$$

Making use of the solution in Example 6.6, we write

$$\bar{y}_{el} \approx \left(\frac{2}{3}R\right)(\sin\theta) = \frac{2}{3}R\sin\theta$$

Applying the *principle of moments*, we write

$$\bar{y}A = \int \bar{y}_{el}\,dA: \quad \bar{y}\left(\frac{1}{2}\pi R^2\right) = \frac{1}{3}R^3\int_0^\pi \sin\theta\,d\theta = \frac{2}{3}R^3$$

$$\bar{y} = \frac{4R}{3\pi} \blacktriangleleft$$

(c) *Using rectangular coordinates and double integration.* We here choose a second-order differential area element dA as shown. We see that

$$dA = d\xi\,d\eta = dx\,d\eta = d\eta\,dx \qquad \bar{y}_{el} = \eta + \frac{1}{2}d\eta \approx \eta$$

Applying the *principle of moments*, we write

$$\bar{y}A = \int \bar{y}_{el}\,dA: \quad \bar{y}\left(\frac{1}{2}\pi R^2\right) = \int_{x=-R}^{R}\int_{\eta=0}^{y}\eta\,d\eta\,dx = \int_{x=-R}^{R}\frac{1}{2}y^2\,dx$$

$$= \frac{1}{2}\int_{-R}^{R}(R^2 - x^2)\,dx = \frac{2}{3}R^3$$

$$\bar{y} = \frac{4R}{3\pi} \blacktriangleleft$$

(d) *Using polar coordinates and double integration.* The second-order differential area element dA as shown may be taken as a rectangle with sides du and $u\,d\theta$. Thus,

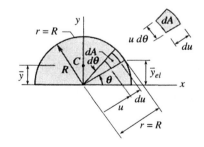

$$dA = du(u\,d\theta) = u\,du\,d\theta \qquad \bar{y}_{el} \approx u\sin\theta$$

Applying the *principle of moments*, we write

$$\bar{y}A = \int \bar{y}_{el}\,dA: \quad \bar{y}\left(\frac{1}{2}\pi R^2\right) = \int_{\theta=0}^{\pi}\int_{u=0}^{R}u^2\sin\theta\,du\,d\theta$$

$$= \int_{\theta=0}^{\pi}\frac{1}{3}R^3\sin\theta\,d\theta = \frac{2}{3}R^3$$

$$\bar{y} = \frac{4R}{3\pi} \blacktriangleleft$$

REMARK. From the answers in Examples 6.2 and 6.8, it is to be noted that the centroid of a *semicircular arc* does not coincide with the centroid of a *semicircular area* with the same radius.

Fig. D6.14

Developmental Exercise

D6.14   The point C is the centroid of the area of the circular sector shown. Using integration in polar coordinates, (a) verify that $\bar{x} = (2R \sin\alpha)/(3\alpha)$, (b) check the value of $\bar{x}$ for $\alpha = \pi/2$.

EXAMPLE 6.9

Locate the centroid $C(\bar{x}, \bar{y}, \bar{z})$ of the curved surface of a hemispherical thin shell of radius R as shown.

Solution.   By symmetry, we know that the centroid C must be situated on the x axis. Thus, we have

$$\bar{y} = \bar{z} = 0 \blacktriangleleft$$

The surface of this shell can be generated by revolving the quarter circular arc PQ about the x axis one revolution as shown. The arc PQ is given by

$$x^2 + y^2 = R^2$$

The area element dA may be taken as the surface of a circular band obtained by revolving the element dL about the x axis one revolution. Thus, we write

$$2x + 2y\frac{dy}{dx} = 0 \qquad \frac{dy}{dx} = -\frac{x}{y} \qquad \bar{x}_{el} \approx x$$

$$dL = \left[1 + \left(\frac{dy}{dx}\right)^2\right]^{1/2} dx = \frac{1}{y}(y^2 + x^2)^{1/2} dx = \frac{R}{y} dx$$

$$dA = 2\pi y\, dL = 2\pi R\, dx \qquad A = \int dA = 2\pi R \int_0^R dx = 2\pi R^2$$

Applying the *principle of moments*, we write

$$\bar{x}A = \int \bar{x}_{el}\, dA: \quad \bar{x}(2\pi R^2) = \int x\, dA = 2\pi R \int_0^R x\, dx = \pi R^3$$

$$\bar{x} = \frac{R}{2} \blacktriangleleft$$

Fig. D6.15

Developmental Exercise

D6.15   The point C is the centroid of the surface area A of the right circular conic thin shell with an open base as shown. Determine (a) dA in terms of x and dx, (b) A, (c) $\bar{x}A$, (d) $\bar{x}$.

6.8   Composite Areas

A *composite area* of total surface area A is one which is composed of component surface areas $A_1, A_2, \ldots, A_n$, where

$$A = A_1 + A_2 + \cdots + A_n \tag{6.10}$$

If the centroids of the component surface areas $A_1, A_2, \ldots, A_n$, as shown in Fig. 6.9, are located at the points $C_1(\bar{x}_1, \bar{y}_1, \bar{z}_1), C_2(\bar{x}_2, \bar{y}_2, \bar{z}_2), \ldots, C_n(\bar{x}_n, \bar{y}_n, \bar{z}_n)$, respectively, then the centroid $C(\bar{x}, \bar{y}, \bar{z})$ of the composite surface area is, according to the *principle of moments*, defined by the following equations:

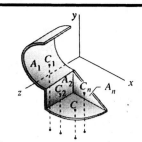

$$\Sigma M_{yz}: \quad \bar{x}A = \bar{x}_1 A_1 + \bar{x}_2 A_2 + \cdots + \bar{x}_n A_n$$

$$\Sigma M_{zx}: \quad \bar{y}A = \bar{y}_1 A_1 + \bar{y}_2 A_2 + \cdots + \bar{y}_n A_n \qquad (6.11)$$

$$\Sigma M_{xy}: \quad \bar{z}A = \bar{z}_1 A_1 + \bar{z}_2 A_2 + \cdots + \bar{z}_n A_n$$

Fig. 6.9

For a composite area lying in the *xy* plane, we have $\bar{z} = \bar{z}_1 = \bar{z}_2 = \cdots = \bar{z}_n = 0$ and the last equation in Eqs. (6.11) becomes trivial. Furthermore, note that one or more of the component areas of a composite area may be negative. For use in determining the centroids of composite areas, the centroids of areas of some common shapes are shown in Fig. 6.10.

Fig. 6.10 Centroids of areas of common shapes.

If desired, a short computer program may be used to locate the centroid of a composite area as shown in Fig. 6.11, where the routine is the same as that shown in Fig. 6.5. Note in Line 100 of the program in Fig. 6.11 that the data are entered in the following order: N, A_1, $\bar{x}_1$, $\bar{y}_1$, $\bar{z}_1$, A_2, $\bar{x}_2$, $\bar{y}_2$, $\bar{z}_2$, A_3, $\bar{x}_3$, $\bar{y}_3$, $\bar{z}_3$, A_4, $\bar{x}_4$, $\bar{y}_4$, $\bar{z}_4$, A_5, $\bar{x}_5$, $\bar{y}_5$, and $\bar{z}_5$.

```
10 REM  A COMPOSITE AREA WITH N COMPONENT AREAS: Enter N, A1, x1, y1, z1, A2,
15 REM  x2, y2, z2, A3, x3, y3, z3, A4, x4, y4, z4, ..., as data in Line 100.
20 REM  To output to the printer, change PRINT to LPRINT in the program.
25 READ N:DIM A(N),X(N),Y(N),Z(N):FOR I=1 TO N:READ A(I),X(I),Y(I),Z(I):A=A+A(I)
30 XA=XA+X(I)*A(I):YA=YA+Y(I)*A(I):ZA=ZA+Z(I)*A(I):NEXT I:X=XA/A:Y=YA/A:Z=ZA/A
35 PRINT:PRINT"The component areas and the coordinates of their centroids are"
40 FOR I = 1 TO N: PRINT "A ("; I; ") =";: PRINT USING " ##.######^^^^"; A(I),
45 PRINT SPC(8); "x, y, z = "; X(I); Y(I); Z(I): NEXT I: S$ = "          "
50 PRINT"The centroid of the composite area is located at":PRINT "x = ";X;S$;
55 PRINT "y = ";Y;: PRINT S$;"z = ";Z:                    REM  (File Name: CENTA)
100 DATA 5,9,2,1,0,36,3,0,3,-14.1372,3,0,4.72676,48,6,-4,3,-3.141593,6,-2,2
```

```
The component areas and the coordinates of their centroids are
A ( 1 ) =   9.000000E+00        x, y, z =  2   1   0
A ( 2 ) =   3.600000E+01        x, y, z =  3   0   3
A ( 3 ) =  -1.413720E+01        x, y, z =  3   0   4.72676
A ( 4 ) =   4.800000E+01        x, y, z =  6  -4   3
A ( 5 ) =  -3.141593E+00        x, y, z =  6  -2   2
The centroid of the composite area is located at
x =   4.658389        y = -2.333782        z =   2.362531
```

Fig. 6.11   (Cf. the solution in Example 6.11.)

EXAMPLE 6.10

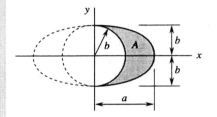

Making use of the formulas in Fig. 6.10, locate the centroid $C(\bar{x}, \bar{y})$ of the area A between the semicircle and the semiellipse shown.

Solution.   By symmetry, the centroid C must be on the x axis; i.e.,

$$\bar{y} = 0 \quad \blacktriangleleft$$

The area A may be treated as a composite area equal to a semielliptic

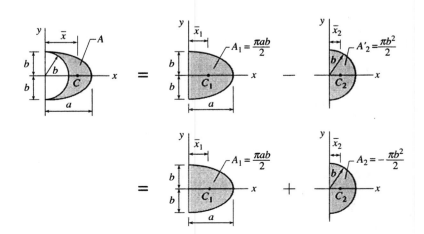

area A_1 minus a semicircular area A_2' or, equivalently, a semielliptic area A_1 plus a *negative* semicircular area A_2 as shown. Using the formulas in Fig. 6.10 and treating A_2 as negative, we write

$$\bar{x}_1 = \frac{4a}{3\pi} \qquad A_1 = \frac{\pi ab}{2} \qquad \bar{x}_2 = \frac{4b}{3\pi} \qquad A_2 = -\frac{\pi b^2}{2}$$

$$A = A_1 + A_2 = \frac{\pi ab}{2} - \frac{\pi b^2}{2} = \frac{\pi b}{2}(a - b)$$

Applying the *principle of moments*, we write

$$\bar{x}A = \bar{x}_1A_1 + \bar{x}_2A_2: \quad \bar{x}\left[\frac{\pi b}{2}(a - b)\right] = \frac{4a}{3\pi}\left(\frac{\pi ab}{2}\right) + \frac{4b}{3\pi}\left(-\frac{\pi b^2}{2}\right)$$

$$\bar{x} = \frac{4}{3\pi}(a + b) \quad \blacktriangleleft$$

REMARK. As $a \to b$, the above answer leads to $\bar{x} = 8b/(3\pi) = 0.849b$, instead of $2b/\pi = 0.637b$, which is the distance between the centroid of a semicircular arc of radius b and its center. This is to be expected because as $a \to b$ the above area approaches a thin strip of the shape of a *crescent*, instead of the shape of a uniform semicircular arc.

Developmental Exercise

D6.16 Making use of the formulas in Fig. 6.10, locate the centroid $C(\bar{x}, \bar{y})$ of the area as shown.

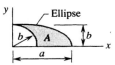

Fig. D6.16

6.9 Centers of Gravity of Laminas

We know that a *lamina* (such as a thin plate or a thin shell) is characterized by the feature that its thickness is negligibly small compared with its width or height. The *mass density* ρ_A of a lamina is expressed as mass per unit area, while the *weight density* γ_A of the lamina is expressed as weight per unit area. By Newton's second law, we note that

$$\gamma_A = \rho_A g \qquad (6.12)$$

where g is the gravitational acceleration.

Unless otherwise stated, each lamina will be assumed to have a constant value of γ_A and be situated near the surface of the earth where $g = 9.81$ m/s^2. Thus, the *centers of gravity of laminas* may be taken to be coincident with the centroids of the surface areas of the corresponding laminas. If the laminas were not uniform as assumed, they should be treated as bodies which are studied in Sec. 6.14.

EXAMPLE 6.11

Locate the center of gravity of the thin sheet of aluminum $OPQRSTUO$ which is cut and bent into the shape as shown.

Solution. The center of gravity $G(\bar{x}, \bar{y}, \bar{z})$ of the thin sheet of aluminum $OPQRSTUO$ is coincident with the centroid of the composite area $OPQRSTUO$. We may treat this composite area as the algebraic sum of five component areas as shown. The centroid of each component area is first determined and entered in the accompanying table, where A_3 and A_5 are negative because they are to be subtracted from the other areas. The coordinates of the centroid of the composite area may be determined from the *principle of moments* as follows:

Component	A, ft^2	$\bar{x}$, ft	$\bar{y}$, ft	$\bar{z}$, ft	$\bar{x}A$, ft^3	$\bar{y}A$, ft^3	$\bar{z}A$, ft^3
A_1	9	2	1	0	18	9	0
A_2	36	3	0	3	108	0	108
A_3	$-9\pi/2$	3	0	$6 - (4/\pi)$	-42.41	0	-66.82
A_4	48	6	-4	3	288	-192	144
A_5	$-\pi$	6	-2	2	-18.85	6.28	-6.28
Sum	$\Sigma A = 75.72$	$\cdots$	$\cdots$	$\cdots$	$\Sigma\bar{x}A = 352.74$	$\Sigma\bar{y}A = -176.72$	$\Sigma\bar{z}A = 178.90$

$$\bar{x} = \frac{\Sigma\bar{x}A}{\Sigma A} = \frac{352.74}{75.72} \qquad \bar{y} = \frac{\Sigma\bar{y}A}{\Sigma A} = \frac{-176.72}{75.72} \qquad \bar{z} = \frac{\Sigma\bar{z}A}{\Sigma A} = \frac{178.90}{75.72}$$

$$\bar{x} = 4.66 \text{ ft} \qquad \bar{y} = -2.33 \text{ ft} \qquad \bar{z} = 2.36 \text{ ft} \quad \blacktriangleleft^\dagger$$

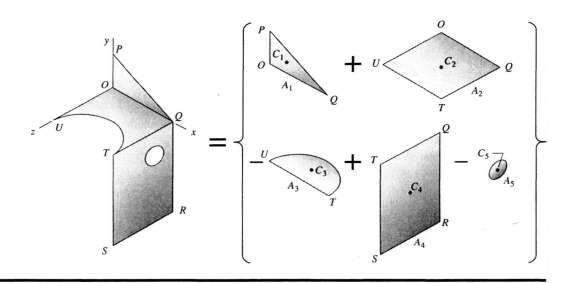

---

†Cf. the answers obtained via a computer program in Fig. 6.11.

EXAMPLE 6.12

A plate having a mass density of 10 kg/m² is cut into the shape *ABCDEFA* and is suspended as shown. Determine the tension in each cable.

Solution. The center of gravity of the plate may be taken as coincident with the centroid of its surface area. By treating the plate as comprising a semicircular part, a rectangular part, and a *negative* triangular part, we write the areas and weight forces of these three parts as follows:

$$A_1 = \frac{1}{2}\pi(2)^2 = 2\pi \qquad A_2 = 6(4) = 24 \qquad A_3 = -\frac{1}{2}(3)(2) = -3$$

$$W_1 = 20\pi g \text{ N} \downarrow \qquad W_2 = 240g \text{ N} \downarrow \qquad W_3 = 30g \text{ N} \uparrow$$

The direction of W_3 is upward because it is contributed from the negative triangular part. Recalling the formulas specifying the locations of the centroids of semicircular and triangular areas, we draw the free-body diagram of the plate as shown. For equilibrium, we write

$$+\circlearrowleft \Sigma M_A = 0: \quad 6T_B + \frac{8}{3\pi}(20\pi g) - 3(240g) + 4(30g) = 0$$

$$T_A = 1783 \text{ N} \blacktriangleleft$$

$$+\circlearrowleft \Sigma M_B = 0: \quad -6T_A + (6 + \frac{8}{3\pi})(20\pi g) + 3(240g) - 2(30g) = 0$$

$$T_B = 894 \text{ N} \blacktriangleleft$$

Check $\quad +\uparrow \Sigma F_y = 0: \quad 1783 + 894 - 20\pi g - 240g + 30g \approx 0$

REMARK. For finding reactions from the supports, it is *not* necessary to locate the center of gravity of the *entire* composite body.

Developmental Exercise

D6.17 A thin plate *ABCDEA* having a mass density of 5 kg/m² is suspended as shown. Determine the tension in each wire.

Fig. D6.17

PROBLEMS

6.27 through 6.35* Determine by integration the centroid of the area shown.

Fig. P6.27

Fig. P6.28

Fig. P6.29*

Fig. P6.30

Fig. P6.31

Fig. P6.32*

Fig. P6.33

Fig. P6.34

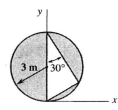

Fig. P6.35*

6.36 through 6.44* Locate the centroid of the area shown.

Fig. P6.36

Fig. P6.37

Fig. P6.38*

Fig. P6.39*

Fig. P6.40

Fig. P6.41*

Fig. P6.42†

Fig. P6.43†

Fig. P6.44*

---

†*Suggestion.* Use *triangular finite elements* and apply the formulas in Example 6.7. (The result may be checked by using the program AREA described in App. F.)

6.45 through 6.47* The body shown is cut from a thin plate having a mass density of 5 kg/m². Determine the tension in each wire.

Fig. P6.45

Fig. P6.46

Fig. P6.47*

6.48 and 6.49 Determine by *integration* the centroid of the *surface* of revolution generated by revolving the line *AB* as shown about the *x* axis one revolution.

$y = \frac{1}{2}x$

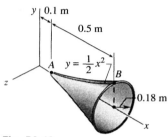

$y = \frac{1}{2}x^2$

Fig. P6.48

Fig. P6.49

6.50 through 6.53 Locate the center of gravity of the sheet metal form shown.

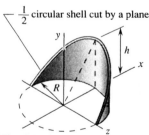

$\frac{1}{2}$ circular shell cut by a plane

$\frac{1}{4}$ spherical shell

Fig. P6.50

Fig. P6.51*

Fig. P6.52

Fig. P6.53

6.10   Distributed Loads on Beams[†]

We know that a *load* is an applied force or moment carried by a body. A load whose intensity of action either remains constant or varies along the axis of a beam is called a *distributed load* on the beam. Generally, a distributed load on a beam is represented by a *loading diagram* indicating the load intensity $w = w(x)$ as a function of the position x on the beam as shown in Fig. 6.12, where the direction of the load is indicated by the arrow signs in the diagram. The load intensity is expressed as the load per unit length of the beam (e.g., N/m or lb/ft).

Fig. 6.12

As illustrated in Fig. 6.12, the *resultant weight W* supported by the beam is given by the *total area of the loading diagram*; i.e.,

$$W = \int_0^L w \, dx \qquad (6.13)$$

The resultant weight force **W** *acts through the centroid C of the loading diagram*, where $\bar{x}$ is given by the *principle of moments*; i.e.,

$$\Sigma M_O: \quad \bar{x}W = \int_0^L xw \, dx \qquad (6.14)$$

Clearly, centroids are useful in the study of distributed loads.

EXAMPLE 6.13

Sacks of cement are stacked on a plank which is supported at each end so that it acts like a beam. The distributed load exerted by the sacks of cement

---

[†]Note that distributed loads do not act on beams only, but may also act on plates, shells, gates, and others. Cf. Example 6.17.

is essentially constant in the transverse direction of the plank and may be represented by the loading diagram shown. Determine the reactions from the supports at A and B.

Solution. By treating the area of the loading diagram as a composite area, we may replace the resultant force $\mathbf{W}$ at C with three concentrated forces $\mathbf{P}_1$, $\mathbf{P}_2$, and $\mathbf{P}_3$ at C_1, C_2, and C_3 as shown in the free-body diagram.

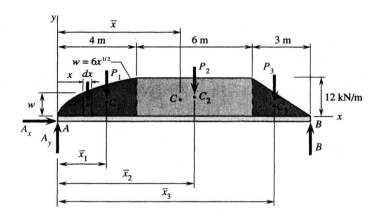

$$P_1 = \int dP_1 = \int_0^4 w \, dx = \int_0^4 6x^{1/2} \, dx = 32$$

$$P_2 = 12(6) = 72 \qquad P_3 = \frac{1}{2}(12)(3) = 18$$

Applying the *principle of moments*, we write

$$+ \smallint \Sigma M_A: \quad \bar{x}_1 P_1 = \int \bar{x}_{el} \, dP_1 = \int_0^4 xw \, dx = 6 \int_0^4 x^{1.5} \, dx = 76.8$$

$$\bar{x}_2 = 4 + \frac{6}{2} = 7 \qquad \bar{x}_3 = 4 + 6 + \frac{3}{3} = 11$$

For equilibrium of the beam, we write

$$+ \smallint \Sigma M_A = 0: \quad 13B - \bar{x}_1 P_1 - \bar{x}_2 P_2 - \bar{x}_3 P_3 = 0$$

$$\mathbf{B} = 59.9 \text{ kN } \uparrow \quad \blacktriangleleft$$

$$\xrightarrow{+} \Sigma F_x = 0: \quad A_x = 0$$

$$\mathbf{A}_x = \mathbf{0} \quad \blacktriangleleft$$

$$+ \uparrow \Sigma F_y = 0: \quad A_y + B - P_1 - P_2 - P_3 = 0$$

$$\mathbf{A}_y = 62.1 \text{ kN } \uparrow \quad \blacktriangleleft$$

REMARK. The abscissas $\bar{x}_1$ and $\bar{x}$ of $\mathbf{P}_1$ and $\mathbf{W}$ do not need to be explicitly found in this particular problem because our aim is to find the reactions from A and B, not the location of $\mathbf{W}$. Since the areas and centroids of rectangles and triangles are well-known, no integration is needed in finding P_2, P_3, $\bar{x}_2$, and $\bar{x}_3$.

Developmental Exercises

D6.18   Refer to Example 6.13. Determine the value of $\bar{x}$.

D6.19 and D6.20   A beam *AB* supports a distributed load as shown. Determine the reaction from the support at *B*.

Fig. D6.19   **Fig. D6.20**

6.11   Theorems of Pappus-Guldinus[†]

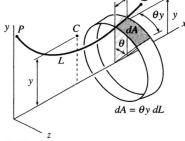

Fig. 6.13

A *surface of revolution* is a curved area which can be generated by revolving a line segment in a plane, called the *generating curve*, about a nonintersecting axis in that plane. This axis is called the *axis of revolution*. Let the line *PQ* of length *L* in the *xy* plane, as shown in Fig. 6.13, be the generating curve and the *x* axis be the axis of revolution. When the line *PQ* is revolved through an angle θ (in radians) about the *x* axis, the line element *dL* on the line *PQ* will generate an areal strip whose area *dA* is

$$dA = \theta y\, dL \qquad (6.15)$$

By the *principle of moments*, the total area generated may be written as

$$A = \theta \int y\, dL \qquad A = \theta \bar{y} L \qquad (6.16)$$

where $\bar{y}$ is the ordinate of the centroid of the generating curve *PQ*, which *does not cross* the *x* axis. If the generating curve *PQ* is revolved one revolution, then $\theta = 2\pi$. Since the distance traveled by the centroid *C* of the generating curve is equal to $\theta \bar{y}$, we have the following *first theorem of Pappus-Guldinus*: *The area of any surface of revolution is equal to the length of the generating curve times the distance traveled by the centroid of the generating curve in generating the area.*[††]

A *body of revolution* is a body whose volume can be generated by revolving a plane area, called the *generating area*, about a nonintersecting axis in the plane of the area. This axis is called the *axis of revolution*. Let *A* be the total area of the generating area in the *xy* plane, as shown in Fig. 6.14, and the *x* axis be the axis of revolution. When the generating area is revolved through an angle θ (in radians) about the *x* axis, the area element

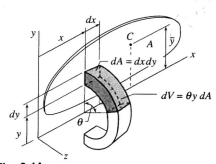

Fig. 6.14

[†]These two theorems are named after the geometer Pappus (born in Alexandria, Egypt; flourished A.D. 300–350 in the silver age of Greek mathematics) and the mathematician Habakkuk Guldinus (or Paul Guldin, born 1577 in Saint Gall, Switzerland; died 1643 in Graz, Austria). Cf. Sec. 6.15 for additional comments.

[††]The projection of the centroid of the generating curve on the axis of revolution is, in general, not coincident with the centroid of the surface of revolution, even when $\theta = 2\pi$.

dA on the generating area will generate a volumetric strip whose length is θy and whose volume is

$$dV = \theta y \, dA \qquad (6.17)$$

By the *principle of moments*, the total volume generated may be written as

$$V = \theta \int y \, dA \qquad V = \theta \overline{y} A \qquad (6.18)$$

where $\overline{y}$ is the ordinate of the centroid of the generating area, which *does not cross* the x axis. If the generating area is revolved one revolution, then $\theta = 2\pi$. Recognizing that $\theta \overline{y}$ is the distance traveled by the centroid C of the generating area, we have the following *second theorem of Pappus-Guldinus*: *The volume of any body of revolution is equal to the generating area times the distance traveled by the centroid of the generating area in generating the volume.*[†]

EXAMPLE 6.14

Using the *first theorem of Pappus-Guldinus*, determine the ordinate $\overline{y}$ of the centroid $C(0, \overline{y})$ of the semicircular *arc* of radius R as shown in Example 6.2.

Solution. We know that the surface area of a sphere of radius R is $4\pi R^2$ and the length of the semicircular arc of radius R as shown is πR. By considering the spherical area as a surface of revolution obtained by revolving the semicircular arc as the generating curve about the x axis through $\theta = 2\pi$ rad, we write, according to the *first theorem of Pappus-Guldinus*,

$$A = \theta \overline{y} L: \quad 4\pi R^2 = 2\pi(\overline{y})(\pi R) \qquad \overline{y} = \frac{2R}{\pi} \quad \blacktriangleleft$$

Developmental Exercises

D6.21 Describe the *first theorem of Pappus-Guldinus*.

D6.22 The point C as shown is the centroid of $\overline{OP}$, which is to be revolved one revolution about the x axis to generate a conic surface of area A. Determine (a) $\overline{y}$ in terms of R, (b) the distance traveled by C, (c) the area A.

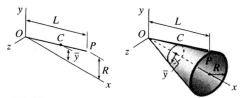

Fig. D6.22

[†]The projection of the centroid of the generating area on the axis of revolution is, in general, not coincident with the centroid of the volume of the body of revolution, even when $\theta = 2\pi$.

EXAMPLE 6.15

Using the *second theorem of Pappus-Guldinus*, determine the ordinate $\bar{y}$ of the centroid $C(0, \bar{y})$ of the semicircular area of radius R as shown in Example 6.8.

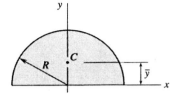

Solution. We know that the volume of a sphere of radius R is $\frac{4}{3}\pi R^3$ and the area of a semicircular area of radius R as shown is $\frac{1}{2}\pi R^2$. By considering the spherical volume as the volume of a body of revolution obtained by revolving the semicircular area as the generating area about the x axis through $\theta = 2\pi$ rad, we write, according to the *second theorem of Pappus-Guldinus*,

$$V = \theta\bar{y}A: \quad \frac{4}{3}\pi R^3 = 2\pi(\bar{y})\left(\frac{1}{2}\pi R^2\right) \qquad \bar{y} = \frac{4R}{3\pi} \blacktriangleleft$$

Developmental Exercises

D6.23 Describe the *second theorem of Pappus-Guldinus*.

D6.24 The point C as shown is the centroid of the area A of the triangle OPQ, which is revolved one revolution about the x axis to generate a conic volume V. Determine (a) $\bar{y}$ in terms of R, (b) the distance traveled by C, (c) the volume V.

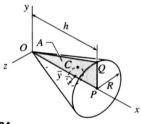

Fig. D6.24

PROBLEMS

6.54 through 6.59* Determine the reactions from the supports of the beam loaded as shown.

Fig. P6.54

Fig. P6.55

Fig. P6.56

Fig. P6.57

Fig. P6.58

Fig. P6.59*

6.60 through 6.62* Making use of a theorem of Pappus-Guldinus, determine the area of the *entire* surface of the body of revolution which has the hatched area shown as its generating area.

Fig. P6.60 and P6.63

Fig. P6.61 and P6.64

Fig. P6.62* and P6.65*

6.63 through 6.65* Making use of a theorem of Pappus-Guldinus, determine the volume of the body of revolution which has the hatched area shown as its generating area.

6.66 A machine part is of the same shape as that obtained by revolving the area shown about the *y* axis one revolution. It is made of an alloy whose mass density is 7 Mg/m³. Determine the total mass of the part.

Fig. P6.66

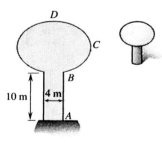

Fig. P6.67

6.67 The water storage tank shown is a shell of revolution with the composite line *ABCD* as the generating curve. The curved line segment *BCD* has a length of 12 m and its centroid is 3.5 m from the centerline of the tank. If the tank is to be sprayed with a coat of paint at a coverage of 10 m² per liter, how many liters of paint will be used?

VOLUMES AND BODIES

6.12   Centroids of Volumes

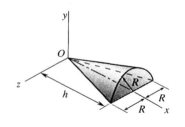

Fig. 6.15

We know that a *volume* is an amount of space and is measured in cubic lengths (e.g., m^3). Let the point $P(\bar{x}_{el}, \bar{y}_{el}, \bar{z}_{el})$ be the centroid of the differential volume element dV of a given volume V as shown in Fig. 6.15, where dV is usually of the *third* or the *first* order. It is important to make sure that $(\bar{x}_{el}, \bar{y}_{el}, \bar{z}_{el})$ are the *coordinates of the centroid* of dV if dV is of the *first order* (cf. Example 6.16). The centroid $C(\bar{x}, \bar{y}, \bar{z})$ of the volume is, according to the *principle of moments*, defined by the following equations:

$$V = \int dV$$

$$\Sigma M_{yz}: \quad \bar{x}V = \int \bar{x}_{el}\, dV$$

$$\Sigma M_{zx}: \quad \bar{y}V = \int \bar{y}_{el}\, dV$$

$$\Sigma M_{xy}: \quad \bar{z}V = \int \bar{z}_{el}\, dV$$

(6.19)

where the last three integrals give the *first moments* of the volume about the coordinate planes. The *values* of dV and V are, for the purpose of locating the centroid of a volume, always made *positive*. However, the coordinates of the centroids of dV and V may be positive, negative, or zero.

EXAMPLE 6.16

Locate the centroid $C(\bar{x}, \bar{y}, \bar{z})$ of the volume of the half right circular cone with a height h and a base of radius R as shown.

Solution.   The volume shown is symmetrical about the xy plane. So, we have

$$\bar{z} = 0 \quad \blacktriangleleft$$

From the similar triangles *OEF* and *OIJ* in the sketch, we write

$$\frac{r}{R} = \frac{x}{h} \qquad r = \frac{R}{h}x$$

The volume element dV is in the form of a semicircular disk. Thus, we write

$$dV = \frac{1}{2}\pi r^2\, dx = \frac{\pi}{2}\left(\frac{R}{h}\right)^2 x^2\, dx$$

$$V = \int dV = \frac{\pi}{2}\left(\frac{R}{h}\right)^2 \int_0^h x^2\, dx = \frac{\pi R^2 h}{6}$$

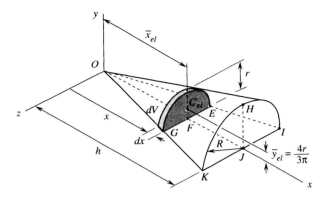

The coordinates of the centroid $C_{el}(\bar{x}_{el}, \bar{y}_{el}, \bar{z}_{el})$ of dV are

$$\bar{x}_{el} \approx x \qquad \bar{y}_{el} \approx \frac{4r}{3\pi} = \frac{4R}{3\pi h} \quad \text{(Cf. Example 6.8.)} \qquad \bar{z}_{el} = 0$$

Applying the *principle of moments*, we write

$$\bar{x}V = \int \bar{x}_{el} \, dV: \quad \bar{x}\left(\frac{\pi R^2 h}{6}\right) = \frac{\pi}{2}\left(\frac{R}{h}\right)^2 \int_0^h x^3 \, dx = \frac{\pi R^2 h^2}{8}$$

$$\bar{y}V = \int \bar{y}_{el} \, dV: \quad \bar{y}\left(\frac{\pi R^2 h}{6}\right) = \frac{2R^3}{3h^3} \int_0^h x^3 \, dx = \frac{R^3 h}{6}$$

$$\bar{x} = \frac{3}{4} h \qquad \bar{y} = \frac{R}{\pi} \quad \blacktriangleleft$$

REMARK. The volume of the half right circular cone *OHIJK* may be obtained by revolving the right triangular area *OIJ* as the generating area about the x axis through π rad. The centroid of this generating area is at a point which is $h/3$ from its base *IJ*; however, the centroid of the half right circular cone is at a point which is $h/4$ from its base *HIJK*.

Developmental Exercises

D6.25 The point C as shown is the centroid of the right circular solid cone of volume V. Determine (a) dV in terms of x and dx, (b) V, (c) $\bar{x}V$, (d) $\bar{x}$.

D6.26 Refer to the conic shell in D6.15 and the solid cone in D6.25. Does the centroid of the curved surface area of a conic shell with an open base coincide with the centroid of the volume of a solid cone with identical height and base?

Fig. D6.25

6.13   Composite Volumes and Center of Pressure

A *composite volume* of total volume V is one which has component volumes $V_1, V_2, \ldots, V_n$, where

$$V = V_1 + V_2 + \cdots + V_n \tag{6.20}$$

Let the centroids of the component volumes V_1, V_2, . . . , V_n be located at the points $C_1(\bar{x}_1, \bar{y}_1, \bar{z}_1)$, $C_2(\bar{x}_2, \bar{y}_2, \bar{z}_2)$, . . . , $C_n(\bar{x}_n, \bar{y}_n, \bar{z}_n)$, respectively. Then, the centroid $C(\bar{x}, \bar{y}, \bar{z})$ of the composite volume is, according to the *principle of moments*, defined by the following equations:

$$\Sigma M_{yz}: \quad \bar{x}V = \bar{x}_1 V_1 + \bar{x}_2 V_2 + \cdots + \bar{x}_n V_n$$

$$\Sigma M_{zx}: \quad \bar{y}V = \bar{y}_1 V_1 + \bar{y}_2 V_2 + \cdots + \bar{y}_n V_n \qquad (6.21)$$

$$\Sigma M_{xy}: \quad \bar{z}V = \bar{z}_1 V_1 + \bar{z}_2 V_2 + \cdots + \bar{z}_n V_n$$

Similar to the case of a composite area, one or more of the component volumes of a composite volume may be negative in some situations. For use in determining the centroids of composite volumes, the centroids of volumes of some common shapes are listed as shown in Fig. 6.16.

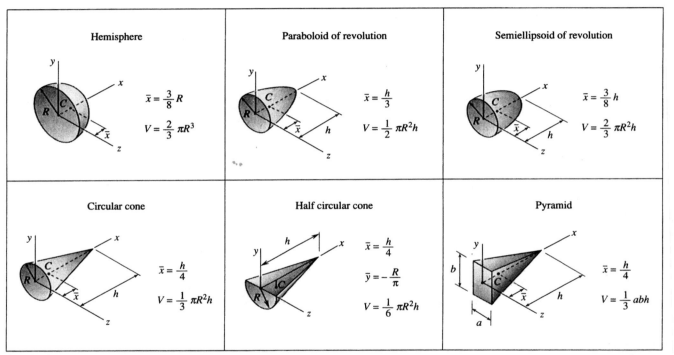

| Hemisphere | Paraboloid of revolution | Semiellipsoid of revolution |

$$\bar{x} = \frac{3}{8}R \qquad \bar{x} = \frac{h}{3} \qquad \bar{x} = \frac{3}{8}h$$

$$V = \frac{2}{3}\pi R^3 \qquad V = \frac{1}{2}\pi R^2 h \qquad V = \frac{2}{3}\pi R^2 h$$

| Circular cone | Half circular cone | Pyramid |

$$\bar{x} = \frac{h}{4} \qquad \bar{x} = \frac{h}{4} \qquad \bar{x} = \frac{h}{4}$$

$$V = \frac{1}{3}\pi R^2 h \qquad \bar{y} = -\frac{R}{\pi} \qquad V = \frac{1}{3}abh$$

$$V = \frac{1}{6}\pi R^2 h$$

Fig. 6.16   Centroids of volumes of common shapes.

Fig. 6.17

Fluids that are essentially incompressible are called *liquids*; otherwise, *gases*. The statics of liquids is generally referred to as *hydrostatics*,[†] which is usually treated in detail in *fluid mechanics* and *hydraulics*. An important property in hydrostatics is that the *pressure p* (in force per unit area) exerted by the liquid (e.g., water) on a submerged surface is *perpendicular* to the submerged surface as shown in Fig. 6.17 and is given by

$$p = \gamma h \qquad (6.22)^{††}$$

[†]The history of the science of *hydrostatics* may trace back to the oft-quoted anecdotal bathtub discovery of the solution by Archimedes (287–212 B.C.) of Syracuse, Sicily, for King Hieron's problem concerning the purity of his gold crown.

[††]The pressure p is usually expressed in N/m^2 or kN/m^2 in the SI, and in lb/ft^2 or lb/in^2 in the U.S. customary system of units. Note that 1 N/m^2 = 1 pascal (Pa) and 1 lb/in^2 = 1 psi.

where γ is the specific weight (in force per unit volume) of the liquid and h is the vertical distance (or depth) measured from the free surface of the liquid to the point on the submerged surface under consideration. By Newton's second law, we note that

$$\gamma = \rho g \qquad (6.23)$$

where ρ is the mass density of the liquid and g is the gravitational acceleration. The *total volume of the pressure diagram* constructed according to this property gives the *magnitude of the resultant force of the pressure*. Furthermore, *the resultant force of the pressure acts through the centroid of volume of the pressure diagram*. In hydrostatics, the intersection of the line of action of the resultant force of the pressure and the submerged surface is known as the *center of pressure*.

EXAMPLE 6.17

A dam is holding back impounded water of mass density $\rho = 1 \text{ Mg/m}^3$. At the bottom of the dam as shown, there is a vertical gate which is 0.6 m high and 1 m wide. The gate is hinged along its top A and stopped along the bottom B. Determine (a) the magnitude of the reaction from the bottom B of the gate, (b) the distance $\bar{y}$ between the center of pressure and the bottom edge of the gate.

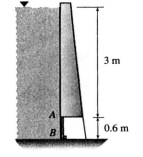

Solution. The specific weight of the water is $\gamma = \rho g = g \text{ kN/m}^3$. Note that the depths at A and B are $h_A = 3$ m and $h_B = 3.6$ m, respectively. Thus, the pressures at A and B are

$$p_A = \gamma h_A = g(3) = 3g \text{ (kN/m}^2) \qquad p_B = \gamma h_B = g(3.6) = 3.6g \text{ (kN/m}^2)$$

The resultant force $\mathbf{R}$ exerted by the water on the gate is a force which acts through the centroid C of the volume of the pressure diagram shown and which has a magnitude equal to the volume of the pressure diagram.

By treating the volume of the pressure diagram as a composite volume, we may replace the resultant force $\mathbf{R}$ at C with two concentrated forces $\mathbf{F}_1$ and $\mathbf{F}_2$ at C_1 and C_2 as shown in the free-body diagrams, where

$$F_1 = p_A(0.6)(1) = 1.8g \text{ (kN)} \qquad F_2 = \tfrac{1}{2}(p_B - p_A)(0.6)(1) = 0.18g \text{ (kN)}$$

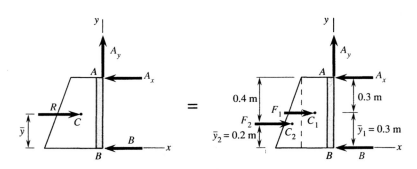

Pressure diagram on the gate   Free-body diagrams of the gate

For equilibrium of the gate, we write

$$+ \circlearrowleft \; \Sigma M_A = 0: \qquad 0.3F_1 + 0.4F_2 - 0.6B = 0$$

$$B = 1.02g \qquad \blacktriangleleft \; B = 10.01 \text{ kN} \; \blacktriangleleft$$

To determine $\bar{y}$, we apply the *principle of moments* as follows:

$$R = F_1 + F_2 = 1.8g + 0.18g = 1.98g$$

$$\bar{y}R = \bar{y}_1 F_1 + \bar{y}_2 F_2 = 0.3(1.8g) + 0.2(0.18g)$$

$$\bar{y} = \frac{0.576g}{1.98g} \qquad \bar{y} = 0.291 \text{ m} \; \blacktriangleleft$$

Developmental Exercises

D6.27 Refer to Example 6.17. Determine the reaction component $\mathbf{A}_x$.

D6.28 Making use of the formulas in Fig. 6.16, locate the centroid of the composite volume consisting of a cone and a cylinder as shown.

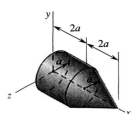

Fig. D6.28

6.14 Centers of Mass and Centers of Gravity of Bodies

A *body* can be viewed as a mass occupying a volume and having a shape. Let ρ (expressed as mass per unit volume) be the *mass density* of a body of total mass m as shown in Fig. 6.18. The *center of mass* $G(\bar{x}, \bar{y}, \bar{z})$ of the body is, according to the *principle of moments*, defined by the following equations:

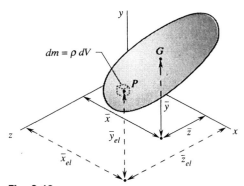

Fig. 6.18

$$m = \int dm = \int \rho \, dV$$

$$\Sigma M_{yz}: \quad \bar{x}m = \int \bar{x}_{el} \, dm = \int \bar{x}_{el} \, \rho \, dV$$

$$\Sigma M_{zx}: \quad \bar{y}m = \int \bar{y}_{el} \, dm = \int \bar{y}_{el} \, \rho \, dV \qquad (6.24)$$

$$\Sigma M_{xy}: \quad \bar{z}m = \int \bar{z}_{el} \, dm = \int \bar{z}_{el} \, \rho \, dV$$

If the mass density of the body is constant throughout, the *center of mass* of the body is clearly coincident with the centroid of the volume occupied by the body. Applying the *principle of moments*, we can determine the center of mass of a composite body in a similar manner as explained in Secs. 6.4, 6.8, and 6.13.

The weight force **W** of a body and hence the location of its center of gravity G is dependent on the gravitational acceleration vector **g**. In situations where **g** varies significantly within the space occupied by a body of total weight W as shown in Fig. 6.19, the center of gravity $G(\bar{x}, \bar{y}, \bar{z})$ of the body is, according to the *principle of moments*, defined by the following equations:

$$W = \int dW = \int \rho g \, dV$$

$$\Sigma M_{yz}: \quad \bar{x}W = \int \bar{x}_{el} \, dW = \int \bar{x}_{el}\rho g \, dV$$

$$\Sigma M_{zx}: \quad \bar{y}W = \int \bar{y}_{el} \, dW = \int \bar{y}_{el}\rho g \, dV \qquad (6.25)$$

$$\Sigma M_{xy}: \quad \bar{z}W = \int \bar{z}_{el} \, dW = \int \bar{z}_{el}\rho g \, dV$$

If **g** has negligibly small variations throughout the space occupied by the body, the *center of gravity G* of the body is clearly *coincident* with the *center of mass* of the same body. In most engineering applications, the center of gravity and the center of mass may be taken as the same point without introducing appreciable errors. The determination of the *center of gravity of a composite body* can similarly be carried out according to the *principle of moments* as in Secs. 6.4, 6.8, and 6.13.

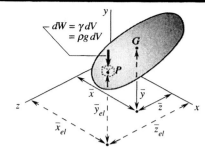

Fig. 6.19

EXAMPLE 6.18

A tetrahedral body has a mass density $\rho = 15z$ kg/m³ and is bounded by the inclined plane $x + 2y + 4z = 8$ and the three coordinate planes, where the unit of length is meter. Locate the center of mass of the body.

Solution.   Since $\rho = 15z$ kg/m³, the center of mass of the body will not coincide with the centroid of its volume. We first let dm be situated at (x, y, z) such that

$$\bar{x}_{el} \approx x \qquad \bar{y}_{el} \approx y \qquad \bar{z}_{el} \approx z$$

$$dm = \rho \, dV = 15z \, dx \, dy \, dz$$

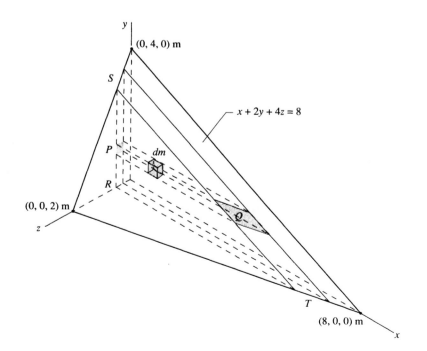

Then, we use the following procedure to obtain the total mass m:

1. Keeping y and z constant, integrate with respect to x from P (where $x = 0$) to Q (where $x = 8 - 2y - 4z$) to add the third-order mass elements and obtain the mass of the filament PQ.
2. Keeping z constant, integrate with respect to y from RT (where $y = 0$) to S (where $x = 0$, $2y + 4z = 8$ or $y = 4 - 2z$) to add the masses of the filaments and obtain the mass of the slab RST.
3. Integrate with respect to z from the back (where $z = 0$) to the front (where $z = 2$) of the tetrahedron to add the masses of the slabs and obtain the total mass m of the tetrahedral body. Thus, we write

$$m = \int_{z=0}^{2} \int_{y=0}^{4-2z} \int_{x=0}^{8-2y-4z} 15z \, dx \, dy \, dz$$

$$= \int_{z=0}^{2} \int_{y=0}^{4-2z} 15z(8 - 2y - 4z) \, dy \, dz$$

$$= \int_{z=0}^{2} 15z(16 - 16z + 4z^2) \, dz = 80 \quad \text{(kg)}$$

By the *principle of moments*, the mass center $G(\bar{x}, \bar{y}, \bar{z})$ of the tetrahedral body may similarly be determined by triple integration as follows:

$$\Sigma M_{yz}: \quad \bar{x}m = \int_{z=0}^{2} \int_{y=0}^{4-2z} \int_{x=0}^{8-2y-4z} 15zx \, dx \, dy \, dz$$

$$= 128 \quad \text{(kg} \cdot \text{m)} \qquad\qquad \bar{x} = 1.6 \text{ m} \quad \blacktriangleleft$$

$$\Sigma M_{zx}: \quad \bar{y}m = \int_{z=0}^{2} \int_{y=0}^{4-2z} \int_{x=0}^{8-2y-4z} 15zy \, dx \, dy \, dz$$

$$= 64 \quad \text{(kg} \cdot \text{m)} \qquad\qquad \bar{y} = 0.8 \text{ m} \quad \blacktriangleleft$$

$$\Sigma M_{xy}: \quad \bar{z}m = \int_{z=0}^{2} \int_{y=0}^{4-2z} \int_{x=0}^{8-2y-4z} 15z^2 \, dx \, dy \, dz$$

$$= 64 \quad \text{(kg} \cdot \text{m)} \qquad\qquad \bar{z} = 0.8 \text{ m} \quad \blacktriangleleft$$

Developmental Exercise

D6.29   Work Example 6.18 if the mass density of the body is (a) $\rho = 3$ kg/m^3, (b) $\rho = y$ kg/m^3.

EXAMPLE 6.19

The body of a machine part is of the same shape as that obtained by revolving the area shown about the x axis. The machine part has a mass density $\rho = 7.7$ Mg/m^3. Determine the tension in each cable.

Solution. We first note that ρ is constant and **g** is essentially constant throughout the space occupied by the body. Thus, the center of mass, the center of gravity, and the centroid of volume of the body may be taken as coincident. Let V_1, V_2, and V_3 be the component volumes of the portions of the machine part lying in the ranges $0 \leq x \leq 0.3$ m, 0.3 m $\leq x \leq 0.5$ m, and 0.5 m $\leq x \leq 0.8$ m, respectively. From the sketch shown, we have

$$V_2 = \pi(0.1)^2(0.2) = 0.002\pi$$

$$dV = \pi y^2 dx = 0.2\pi x\, dx \qquad \bar{x}_{el} \approx x$$

$$V_1 = \int dV = 0.2\pi \int_0^{0.3} x\, dx = 0.009\pi$$

$$V_3 = \int dV = 0.2\pi \int_{0.5}^{0.8} x\, dx = 0.039\pi$$

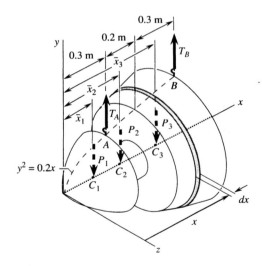

Applying the *principle of moments*, we write

$$\bar{x}_1 V_1 = \int \bar{x}_{el}\, dV:$$

$$\bar{x}_1(0.009\pi) = 0.2\pi \int_0^{0.3} x^2\, dx = 0.0018\pi \qquad \bar{x}_1 = 0.2 \text{ m}$$

$$\bar{x}_3 V_3 = \int \bar{x}_{el}\, dV:$$

$$\bar{x}_3(0.039\pi) = 0.2\pi \int_{0.5}^{0.8} x^2\, dx = 0.0258\pi \qquad \bar{x}_3 = \frac{43}{65} \text{ m}$$

By inspection, we have

$$\bar{x}_2 = \left(0.3 + \frac{0.2}{2}\right) \text{ m} = 0.4 \text{ m}$$

Since $\rho = 7.7 \text{ Mg/m}^3$, the three equivalent concentrated weights at C_1, C_2, and C_3 are

$$P_1 = \rho V_1 g = 0.0693\pi g \quad \text{(kN)} \qquad P_2 = \rho V_2 g = 0.0154\pi g \quad \text{(kN)}$$

$$P_3 = \rho V_3 g = 0.3003\pi g \quad \text{(kN)}$$

For equilibrium, we write

$$+\,\circlearrowright\, \Sigma M_A = 0: \quad (0.3 - 0.2)(0.0693\pi g) - (0.4 - 0.3)(0.0154\pi g)$$

$$-\left(\frac{43}{65} - 0.3\right)(0.3003\pi g) + 0.5T_B = 0$$

$$+\,\circlearrowright\, \Sigma M_B = 0: \quad (0.8 - 0.2)(0.0693\pi g) - 0.5T_A$$

$$+ (0.8 - 0.4)(0.0154\pi g) + \left(0.8 - \frac{43}{65}\right)(0.3003\pi g) = 0$$

$$T_A = 5.51 \text{ kN} \qquad T_B = 6.36 \text{ kN} \quad \blacktriangleleft$$

Check $+\uparrow \Sigma F_y = T_A + T_B - P_1 - P_2 - P_3 \approx 0$

Developmental Exercise

D6.30 Locate the center of gravity of the machine part described in Example 6.19.

PROBLEMS

6.68 and **6.69** Determine by integration the centroid of volume of the body whose curved surface is defined by the equation as indicated. (*Hint.* Let dV be a thin slab perpendicular to the z axis.)

Fig. P6.68

Fig. P6.69

Fig. P6.70

6.70 A body is obtained by cutting a circular cylinder of radius 4 ft along an oblique plane as shown. Locate the centroid of volume of the body. (*Hint.* Let dV be a thin slab parallel to the xy plane.)

6.71 A body of revolution is of the shape of that obtained by revolving a quarter circular area about the y axis through 180° as shown. Locate the centroid of volume of the body. (*Hint.* Let dV be a half cylindrical shell parallel to the y axis.)

Fig. P6.71

Fig. P6.72

Fig. P6.73*

6.72 Locate the centroid of volume of the spherical wedge shown. (*Hint.* Use spherical coordinates and triple integration.)

6.73* Determine by integration the centroid of volume generated by revolving the area shown in the xy plane about the x axis one revolution.

6.74 The homogeneous body shown is composed of a solid circular cone of mass m_C and a solid hemispherical base of mass m_B. For $r = 80$ mm and $h = 480$ mm, determine the minimum ratio of m_B/m_C if the body is to return to an upright position after its longitudinal axis has been pushed to a horizontal position.

Fig. P6.74

6.75 and 6.76 Locate the center of gravity of the homogeneous body shown.

Fig. P6.75

Fig. P6.76

NOTE. In Probs. 6.77 through 6.85*, take the specific weight and mass density for water to be $\gamma = 62.4$ lb/ft^3 and $\rho = 1$ Mg/m^3, and those for concrete to be $\gamma_c = 150$ lb/ft^3 and $\rho_c = 2.4$ Mg/m^3, respectively.

6.77 through 6.79* The cross section of a concrete dam is shown where a seal exists at A and no water pressure is present under the dam. Determine the maximum value of the depth d of water for which the dam will not overturn about B.

Fig. P6.77

Fig. P6.78

Fig. P6.79*

6.80 through 6.82   The gate AB for holding water has a uniform width w as indicated and is hinged at A. Determine the minimum compressive force to be exerted by the hydraulic cylinder CD to keep the gate shut.

Fig. P6.80

Fig. P6.81*

Fig. P6.82

6.83   The automatic valve ABC consists of a square plate pivoted at B as shown. Determine the depth d of water which will cause the valve to open.

Fig. P6.83

Fig. P6.84

Fig. P6.85*

6.84 and 6.85*   The tank shown is filled with water. Determine the magnitude of the horizontal force $\mathbf{P}$ required to keep the gate closed.

6.86   An industrial oil of mass density ρ is stored in a cylindrical tank to capacity as shown. Determine (a) the magnitude of the resultant force $\mathbf{P}$ of the oil pressure acting on one of the circular ends of the tank, (b) the distance between the center of pressure and the bottom of the tank. (Cf. Prob. 6.70.)

Fig. P6.86

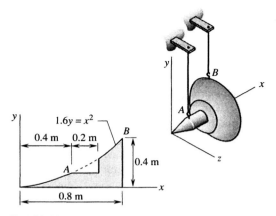

$1.6y = x^2$

0.4 m   0.2 m

B

0.4 m

A

0.8 m

Fig. P6.87

6.87 and 6.88* A body of revolution is of the same shape as that obtained by revolving the area shown about the *x* axis one revolution. The body has a mass density of 5 Mg/m³ and is hung by two wires as shown. Determine the tension in each wire.

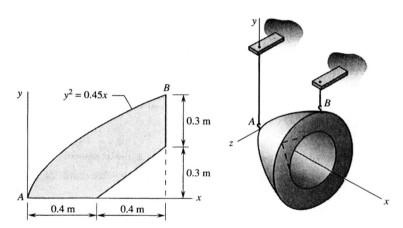

$y^2 = 0.45x$

B

0.3 m

A

0.3 m

0.4 m   0.4 m

Fig. P6.88*

6.89 and 6.90 The body shown has a variable mass density ρ as indicated. Using triple integration, locate the center of mass of the body.

$\rho = x^3 y^2 z$

b

a   *c*

Fig. P6.89

(0, 6, 0) ft

$\rho = 2z$

(0, 0, 2) ft

(12, 0, 0) ft

Fig. P6.90

6.15   Concluding Remarks

The focus throughout this chapter has been the *principle of moments*. It may roughly be stated as follows: *The resultant is equal to the sum of the components, and the moment of the resultant is equal to the sum of the moments of the components*. This principle is synthesized from the concept of *equivalent systems* of forces and degenerates into Varignon's theorem as a special case in which the components are concurrent forces. The resultants of any force systems (including distributed loads), the geographic center, and the population center are defined and located according to the *principle of moments*. Moreover, the centroids of lines, areas, and volumes as well as the centers of gravity of bodies are their respective "centers" which are defined and located according to the *principle of moments*.

The distances traveled by the centroids of generating curves and generating areas are related to the corresponding surface areas of revolution and volumes of revolution by the first and the second theorems of Pappus-Guldinus, respectively. Historical studies indicate that Pappus lived in Alexandria, Egypt around the end of the third century. The essence of those two theorems was first enunciated by him in Book VII of his work *Synagoge* (Συναγωγή) or *Collection*. The *Collection* contains his mathematical papers arranged in eight books of which Book I and the first 13 propositions (out of 26) of Book II have been lost. However, the said two theorems received little attention until Guldinus published Volume II of his four-volume work *Centrobaryca seu de centro gravitatis trium specierum quantitatis continuae* (in Vienna, Austria, 1635–1641) without reference to, and possibly without knowledge of, the work by Pappus. In some books, those two theorems are named after Pappus only.

REVIEW PROBLEMS

6.91   Locate the resultant force **R** of the parallel forces acting as shown.

6.92   The body shown is made of a thin rod weighing 4 lb/ft. Determine the reactions from the hinge support at A and the smooth support at D.

Fig. P6.91

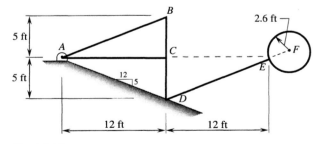

Fig. P6.92

6.93   A bent rod having a mass density of 1 kg/m and a sheet of metal having a mass density of 5 kg/m^2 are welded together and suspended from a rope in such a manner that the metal sheet is horizontal as shown. If $h = 100$ mm and $\bar{x} = 350$ mm, determine (a) L, (b) $\bar{z}$.

400 mm

100 mm

200 mm

200 mm

$\bar{x}$ $\bar{z}$

h

L

Fig. P6.93

$y^2 = 18x$

6 kN/m

A B

2 m 3 m

Fig. P6.95

6.94* Solve Prob. 6.93 if $h = 200$ mm and $\bar{x} = 360$ mm.

6.95 A beam is under the action of a distributed load as shown. Determine the reactions from the supports at A and B.

6.96* Locate the centroid of the area shown. (*Hint*. Cf. the footnote on p. 200.)

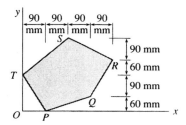

90 mm | 90 mm | 90 mm | 90 mm

S

90 mm

R 60 mm

T

90 mm

Q

60 mm

O P

Fig. P6.96*

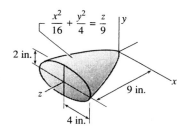

$\frac{x^2}{16} + \frac{y^2}{4} = \frac{z}{9}$

2 in.

9 in.

4 in.

Fig. P6.97

6.97 Determine by integration the centroid of volume of the body whose curved surface is defined by the equation as indicated.

6.98* Locate the centroid of volume of the half torus shown.

6.99 Locate the centroid of volume of the conic wedge shown.

40 mm

120 mm

120 mm

40 mm

Fig. P6.98*

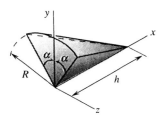

α α

R h

Fig. P6.99

6.100   A 4-m long tank is used to store a pool of oil with a specific weight of 9 kN/m^3 and a pool of salt water with a specific weight of 10 kN/m^3 by using a partition whose cross section is shown. The supporting cables are not stressed when the tank is empty. Neglecting the weight of the partition, determine the tension in each cable when $a = b = 1$ m.

Fig. P6.100

6.101*   Refer to Prob. 6.100. Determine the ratio of a/b for which the cables are unstressed.

6.102   A homogeneous body of revolution is of the same shape as that obtained by revolving the area shown about the x axis one revolution. Locate the center of gravity of the body.

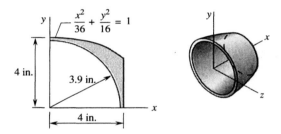

Fig. P6.102

Second Moments: Moments of Inertia[†]

Chapter 7

KEY CONCEPTS

- Terms related to a plane area with respect to certain axes: moment of inertia, polar moment of inertia, radius of gyration, and product of inertia.
- Parallel-axis theorems for area moments and products of inertia.
- Moments and products of inertia of composite areas.
- Rotation of axes for area moments and products of inertia.
- Mohr's circle for area moments and products of inertia.
- Moment of inertia and radius of gyration of a mass.
- Parallel-axis theorem for mass moments of inertia.

The study of many situations in engineering involves the use of quantities referred to as second moments, or moments of inertia, of areas and masses. For instance, moments of inertia of areas are used in expressing the stresses and strains experienced by beams in bending and shafts in torsion, the critical loads for the buckling of columns, and the locations of centers of pressure on nonhorizontal submerged surfaces. In kinetics, moments of inertia of masses are used in expressing the effective moments, the kinetic energies, and the momenta associated with bodies having rotational motion. One can see that moments of inertia of areas and masses are useful concepts which are employed in the study of mechanics of materials, fluid mechanics, and dynamics.

[†]Although it is logical for second moments to follow first moments in the sequence of study, this chapter may be covered later, such as after Chap. 9, without affecting continuity. If circumstances warrant, one may skip the mass moments of inertia in the last part of this chapter and take up that topic, as provided in the first part of Chap. 16, in *Dynamics*.

Moment

Bending stress

Fig. 7.1

Shearing stress

Fig. 7.2

Critical load

Fig. 7.5

AREA MOMENTS OF INERTIA[†]

7.1 Situations Involving Area Moments of Inertia

From time to time, engineers encounter situations in which the intensity of a distributed force (e.g., pressure or stress) over an area is proportional to the distance between the force and a certain axis. Consequently, the resultant moment of the forces distributed over the area about that axis involves an integral of the form $\int (\text{distance})^2\, d(\text{area})$. This integral is known as the *second moment*, or the *moment of inertia*, of the area about that axis.[††] An area has *no mass* and, hence, *no inertia*. However, because of mathematical similarity in the definitions (cf. Secs. 7.2 and 7.11), the term *moment of inertia* continues to be used by most engineers for both *areas* and *masses*. Moreover, note that an area moment of inertia is usually defined for a *plane area* with respect to an axis either in or perpendicular to the plane of the area.

In mechanics, there are many situations which involve the use of moments of inertia of an area in expressing certain physical quantities. We shall here mention some of those situations without dwelling on their details. In mechanics of materials, (a) the moment of inertia of the cross-sectional area of a beam about a centroidal axis is a quantity used in expressing the stresses and strains developed in the beam due to bending and shearing as shown in Figs. 7.1 and 7.2, (b) the polar moment of inertia of the cross-sectional area of a circular shaft about its center as the pole is a quantity used in expressing torsional shearing stresses and strains and the angle of twist in the shaft as shown in Figs. 7.3 and 7.4, and (c) the smallest moment of inertia of the

Torque

Torsional shearing stress

Fig. 7.3

Torque

Angle of twist

Fig. 7.4

cross-sectional area of a column about a centroidal axis is a quantity used in expressing the critical load for buckling of a column as shown in Fig. 7.5. In fluid mechanics, the moment of inertia of the area of a nonhorizontal

---

[†]Also called *second moments of areas*. Cf. Sec. 7.1.

[††]Note that $\int (\text{distance})\, d(\text{area})$ gives the *first moment* of the area; $\int (\text{distance})^2\, d(\text{area})$ is, therefore, termed the *second moment* of the area.

submerged plane surface about an axis in the free surface of the liquid is a
quantity used in locating the center of pressure on the submerged surface as
shown in Fig. 7.6.

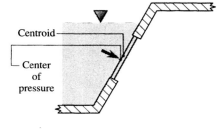

Fig. 7.6

Developmental Exercise

D7.1 Name two situations in which area moments of inertia are used.

7.2 Moments of Inertia of an Area

The *moment of inertia of an area* with respect to (or about) an axis is defined as the integral of the second-order differential area element times the
square of the distance between that axis and the area element, as shown in
Fig. 7.7. When it is feasible to subdivide a first-order differential area element of an area into an infinite number of second-order differential area

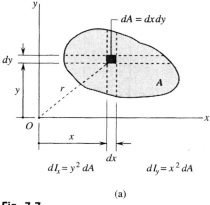

$$dI_x = y^2\,dA \qquad dI_y = x^2\,dA$$

(a)

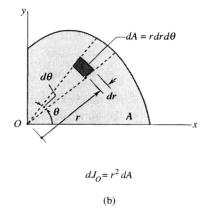

$$dJ_O = r^2\,dA$$

(b)

Fig. 7.7

elements which are *equidistant* from an axis, the *moment of inertia* of the
area about that axis may be defined as the integral of the first-order differential area element times the square of that *equidistance*, as shown in Fig.
7.8. The moments of inertia of an area with respect to the x and y axes,
which lie in the plane of the area, are customarily denoted by I_x and I_y,
respectively.[†] Referring to Figs. 7.7 and 7.8, we write

[†]Sometimes, I_x and I_y are written with double subscripts as I_{xx} and I_{yy}, respectively, to signify
that they are components of a second-order tensor (cf. the footnotes on p. 237).

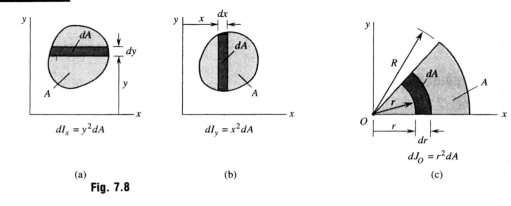

$$dI_x = y^2 dA \qquad\qquad dI_y = x^2 dA \qquad\qquad dJ_O = r^2 dA$$

(a)   (b)   (c)

Fig. 7.8

$$I_x = \int y^2 \, dA \qquad\qquad (7.1)$$

$$I_y = \int x^2 \, dA \qquad\qquad (7.2)$$

The moments of inertia I_x and I_y defined in Eqs. (7.1) and (7.2) constitute a set of *rectangular moments of inertia* of the area.

The moment of inertia of an area about an axis perpendicular to the plane of the area is called the *polar moment of inertia* of the area about that axis. The polar moment of inertia of the area A about the z axis (or the pole at the origin O) is usually denoted by J_O. Referring to Figs. 7.7 and 7.8, we write

$$J_O = \int r^2 \, dA \qquad\qquad (7.3)$$

where r denotes a polar coordinate.[†] It is clear from Eqs. (7.1) through (7.3) that the dimensions of moments of inertia of an area are $(length)^4$.

Although the differential area element dA in Eqs. (7.1) through (7.3) may be of the *first* or the *second* order, as illustrated in Figs. 7.7 and 7.8, *it is important to make sure that dA is an areal strip parallel to the axis of interest if dA is of the first order for single integration*. Of course, double integration is required when dA is of the second order. The equation relating J_O, I_x, and I_y may be derived as follows:

$$J_O = \int r^2 \, dA = \int (x^2 + y^2) \, dA = \int x^2 \, dA + \int y^2 \, dA = I_y + I_x$$

$$J_O = I_x + I_y \qquad\qquad (7.4)$$

Note that the area element dA in Eqs. (7.1) through (7.3) must always be made positive regardless of the quadrant in which it is situated. Since the square of the distance is always positive, the rectangular as well as polar moments of inertia of areas must always be *positive*. Furthermore, the moment of inertia of an area about an axis which passes through its centroid is called a *centroidal moment of inertia* of the area, which is frequently marked with an overbar; e.g., $\bar{I}_x$, $\bar{I}_y$, and $\bar{J}_O$ signify that x and y are centroidal axes and O is the centroid of the given area.

[†]This is why J_O was given the name *polar moment of inertia*.

Developmental Exercises

EXAMPLE 7.1

Determine the centroidal moments of inertia $\bar{I}_x$ and $\bar{I}_y$ of a rectangular area with base b and height h parallel to the x and y centroidal axes, respectively.

Solution.   For illustrative purposes, $\bar{I}_x$ and $\bar{I}_y$ will here be determined in two different ways:

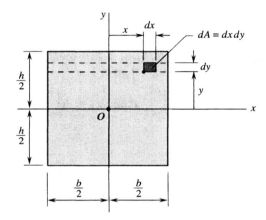

(a) *Using double integration.* We here choose a second-order differential area element dA as shown. By definition, we write

$$\bar{I}_x = \int y^2 \, dA = \int_{-h/2}^{h/2} \int_{-b/2}^{b/2} y^2 \, dx \, dy = \int_{-h/2}^{h/2} by^2 \, dy$$

$$\bar{I}_x = \frac{1}{12} bh^3 \blacktriangleleft$$

$$\bar{I}_y = \int x^2 \, dA = \int_{-h/2}^{h/2} \int_{-b/2}^{b/2} x^2 \, dx \, dy = \int_{-h/2}^{h/2} \frac{1}{12} b^3 \, dy$$

$$\bar{I}_y = \frac{1}{12} b^3 h \blacktriangleleft$$

(b) *Using single integration.* For determining $\bar{I}_x$, we choose dA to be a first-order areal strip parallel to the x axis as shown. By definition, we write

$$\bar{I}_x = \int y^2 \, dA = \int_{-h/2}^{h/2} y^2 \, (b \, dy) = b \int_{-h/2}^{h/2} y^2 \, dy$$

$$\bar{I}_x = \frac{1}{12} bh^3 \blacktriangleleft$$

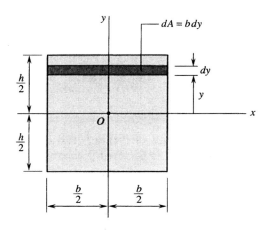

For determining $\bar{I}_y$, we choose dA to be a first-order areal strip parallel to the y axis as shown. By definition, we write

$$\bar{I}_y = \int x^2 \, dA = \int_{-b/2}^{b/2} x^2 \, (h \, dx) = h \int_{-b/2}^{b/2} x^2 \, dx$$

$$\bar{I}_y = \frac{1}{12} b^3 h \qquad \blacktriangleleft$$

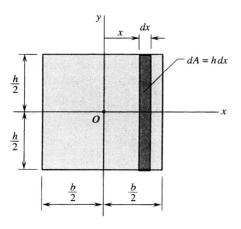

REMARK. The above results are well-known and widely used as *formulas*.

Developmental Exercise

D7.4 Refer to Example 7.1 and use Eq. (7.4). What is the centroidal polar moment of inertia $\bar{J}_O$ of a rectangle of base b and height h?

EXAMPLE 7.2

Determine the centroidal polar moment of inertia $\bar{J}_O$ of a circular area of radius R.

Solution. For illustrative purposes, $\bar{J}_O$ will here be determined in two different ways:

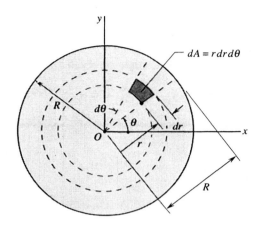

(a) *Using double integration*. We here choose a second-order dA in polar coordinates as shown. By definition, we write

$$\bar{J}_O = \int r^2 \, dA = \int_0^{2\pi} \int_0^R r^2 \, (r \, dr \, d\theta) = \int_0^{2\pi} \frac{1}{4} R^4 \, d\theta$$

$$\bar{J}_O = \frac{1}{2} \pi R^4 \quad \blacktriangleleft$$

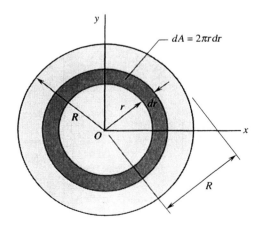

(b) *Using single integration*. We first choose a *ring-shaped* first-order differential area element dA as shown. By definition, we write

$$\bar{J}_O = \int r^2 \, dA = \int_0^R r^2 (2\pi r \, dr)$$

$$= 2\pi \int_0^R r^3 \, dr \qquad\qquad \bar{J}_O = \frac{1}{2}\pi R^4 \blacktriangleleft$$

REMARK. If we choose to integrate with respect to θ first, we write

$$\bar{J}_O = \int r^2 \, dA = \int_0^R \int_0^{2\pi} r^3 \, d\theta \, dr = \int_0^R r^3 (2\pi) \, dr = \frac{1}{2}\pi R^4$$

Since the circular area is symmetrical about the x and y axes, we know that $\bar{I}_x = \bar{I}_y$. Thus, based on the above result for $\bar{J}_O$ and Eq. (7.4), we obtain for a circular area with its centroid at the origin that

$$\bar{I}_x = \bar{I}_y = \frac{1}{2}\bar{J}_O = \frac{1}{4}\pi R^4$$

Fig. D7.5

Developmental Exercise

D7.5   Determine the polar moment of inertia J_O of the semicircular area shown.

EXAMPLE 7.3

The cross-sectional area of a slender rod is shown. Determine (a) the moments of inertia I_x and I_y of the area about the x and y axes, respectively; (b) the polar moment of inertia J_O of the area about the origin O.

Solution.   To determine I_x, we choose a first-order dA which is *parallel* to the x axis as shown. We write

$$dA = x \, dy = \left(\frac{64}{5}y\right)^{1/2} dy$$

$$I_x = \int y^2 \, dA = \int_0^5 y^2 \left(\frac{64}{5}y\right)^{1/2} dy = \frac{2000}{7}$$

$$I_x = 286 \text{ mm}^4 \blacktriangleleft$$

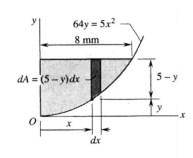

To determine I_y, we choose a first-order dA which is *parallel* to the y axis as shown. We write

$$dA = (5 - y)\,dx = \left[5 - \left(\frac{5}{64}x^2\right)\right] dx$$

$$I_y = \int x^2\,dA = \int_0^8 x^2\left[5 - \left(\frac{5}{64}x^2\right)\right] dx = \frac{1024}{3}$$

$$I_y = 341 \text{ mm}^4 \quad \blacktriangleleft$$

By Eq. (7.4), we write

$$J_O = I_x + I_y = \frac{2000}{7} + \frac{1024}{3} = \frac{13168}{21}$$

$$J_O = 627 \text{ mm}^4 \quad \blacktriangleleft$$

Developmental Exercises

D7.6 Determine the moments of inertia I_x and I_y of the triangular area shown.

D7.7 In Fig. 7.8(a), if $\overline{x}_{el}$ is the abscissa of the centroid of the first-order area element dA, is it correct to write $I_y = \int (\overline{x}_{el})^2\,dA$?

D7.8 In Fig. 7.8(b), if $\overline{y}_{el}$ is the ordinate of the centroid of the first-order area element dA, is it correct to write $I_x = \int (\overline{y}_{el})^2\,dA$?

D7.9 Refer to Fig. 6.6(a). Is $I_x = \int y^2\,(y\,dx)$?

D7.10 Refer to Fig. 6.6(b). Is $I_y = \int x^2\,(x\,dy)$?

D7.11 Refer to Fig. 6.6(f). Is $J_O = \int (2r/3)^2\,(\frac{1}{2}r^2\,d\theta)$?

Fig. D7.6

7.3 Radius of Gyration of an Area

The *radius of gyration* of an area about an axis is a distance whose square multiplied by the area gives the moment of inertia of the area about that axis. For an area A in the xy plane as shown in Fig. 7.9, we write

$$I_x = k_x^2 A \tag{7.5}$$

$$I_y = k_y^2 A \tag{7.6}$$

$$J_O = k_O^2 A \tag{7.7}$$

where k_x, k_y, and k_O are the radii of gyration of A about the x axis, the y axis, and the pole O, respectively. Recalling from Eq. (7.4) that J_O is equal to the sum of I_x and I_y, we readily observe that

$$k_O^2 = k_x^2 + k_y^2 \tag{7.8}$$

For computing the moment of inertia about an axis, an area may be imagined to be a *thin areal strip*, with the same total area, whose elements are

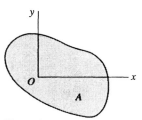

Fig. 7.9

equidistant from that axis and the *equidistance* is equal to the radius of gyration of the area about that axis as shown in Fig. 7.10. The radius of gyration of an area about an axis is a measure of the spread of the area from that axis.[†]

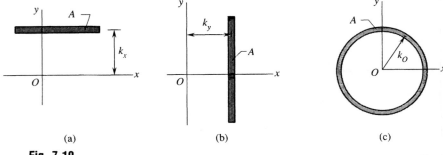

(a)   (b)   (c)

Fig. 7.10

EXAMPLE 7.4

Determine the radii of gyration k_x, k_y, and k_O of the cross-sectional area in Example 7.3.

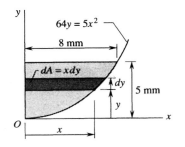

Solution. We first determine the area A of that cross section as follows:

$$dA = x \, dy = \left(\frac{64}{5}\right)^{1/2} y^{1/2} \, dy$$

$$A = \int dA = \left(\frac{64}{5}\right)^{1/2} \int_0^5 y^{1/2} \, dy \qquad A = \frac{80}{3} \text{ mm}^2$$

From the solution in Example 7.3, we write

$$I_x = \frac{2000}{7} \text{ mm}^4 \qquad I_y = \frac{1024}{3} \text{ mm}^4 \qquad J_O = \frac{13168}{21} \text{ mm}^4$$

$$\frac{2000}{7} = k_x^2 \left(\frac{80}{3}\right) \qquad \frac{1024}{3} = k_y^2 \left(\frac{80}{3}\right) \qquad \frac{13168}{21} = k_O^2 \left(\frac{80}{3}\right)$$

$$k_x = 3.27 \text{ mm} \qquad k_y = 3.58 \text{ mm} \qquad k_O = 4.85 \text{ mm} \blacktriangleleft$$

Developmental Exercises

D7.12 Define the radius of gyration of an area about an axis.

D7.13 Determine the centroidal radii of gyration $\bar{k}_x$ and $\bar{k}_y$ of the rectangular area in Example 7.1.

D7.14 Determine the centroidal polar radius of gyration $\bar{k}_O$ of the circular area in Example 7.2.

[†]A centroidal radius of gyration of an area represents the *standard deviation* of the distances of the area elements from the centroidal axis.

PROBLEMS

7.1 through 7.6 Using *double integration*, determine the moment of inertia I_x of the area shown.

**Fig. P7.1, P7.7*,
P7.13, and P7.19***

**Fig. P7.2, P7.8,
P7.14, and P7.20***

**Fig. P7.3, P7.9,
P7.15*, and P7.21***

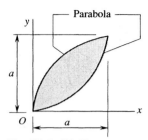

**Fig. P7.4, P7.10,
P7.16, and P7.22***

7.7* through 7.12* Using *double integration*, determine the moment of inertia I_y of the area shown.

7.13 through 7.18* Using *double integration*, determine the polar moment of inertia J_O of the area shown.

7.19* through 7.24* Using *single integration*, determine the moments of inertia I_x and I_y of the area shown.

Fig. P7.5, P7.11, P7.17, and P7.23*

**Fig. P7.6, P7.12*,
P7.18*, and P7.24***

Fig. P7.25

Fig. P7.26

Fig. P7.27*

7.25 through 7.27* Determine the moment of inertia I_x and the radius of gyration k_x of the area shown.

7.28 through 7.30* Determine the moment of inertia I_y and the radius of gyration k_y of the area shown.

Fig. P7.28

Fig. P7.29

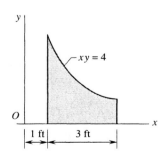

Fig. P7.30*

7.31 through 7.33* Determine the moment of inertia I_x and the radius of gyration k_x of the area shown.

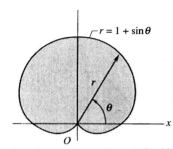

Fig. P7.31, P7.34, and P7.37   **Fig. P7.32, P7.35, and P7.38**   **Fig. P7.33*, P7.36*, and P7.39***

7.34 through 7.36* Determine the moment of inertia I_y and the radius of gyration k_y of the area shown.

7.37 through 7.39* Determine the polar moment of inertia J_O and the polar radius of gyration k_O of the area shown.

7.4 Parallel-Axis Theorem: Area Moments of Inertia

Let the point $C(\bar{x}, \bar{y})$ be the centroid of an area A in the xy plane as shown in Fig. 7.11, where the centroidal axes x' and y' are parallel to the coordinate axes x and y, respectively. From this figure, we write

$$y = y' + \bar{y} \qquad x = x' + \bar{x}$$

By definitions in Sec. 7.2 and the *principle of moments*, we write

$$I_x = \int y^2 \, dA = \int (y' + \bar{y})^2 \, dA$$
$$= \int y'^2 \, dA + 2\bar{y} \int y' \, dA + \bar{y}^2 \int dA$$
$$= \bar{I}_{x'} + 2\bar{y}(\bar{y}'A) + \bar{y}^2 A$$
$$I_y = \int x^2 \, dA = \int (x' + \bar{x})^2 \, dA$$
$$= \int x'^2 \, dA + 2\bar{x} \int x' \, dA + \bar{x}^2 \int dA$$
$$= \bar{I}_{y'} + 2\bar{x}(\bar{x}'A) + \bar{x}^2 A$$
$$J_O = I_x + I_y = (\bar{I}_{x'} + \bar{I}_{y'}) + 2(\bar{x}\bar{x}' + \bar{y}\bar{y}')A + (\bar{x}^2 + \bar{y}^2)A$$

where $\bar{x}'$ and $\bar{y}'$ are the coordinates of the centroid of the area in the $Cx'y'$ coordinate system. Since the point C is the centroid of the area A, we have $\bar{x}' = \bar{y}' = 0$. We note from Fig. 7.11 that $\bar{J}_C = \bar{I}_{x'} + \bar{I}_{y'}$ and $\bar{r}^2 = \bar{x}^2 + \bar{y}^2$. Thus, the above equations reduce to

$$\boxed{I_x = \bar{I}_{x'} + A\bar{y}^2} \qquad \boxed{I_y = \bar{I}_{y'} + A\bar{x}^2} \qquad \boxed{J_O = \bar{J}_C + A\bar{r}^2} \qquad (7.9)$$

These results are referred to as the *parallel-axis theorem for area moments of inertia*, which may be stated as follows: *The moment of inertia I of an*

Fig. 7.11

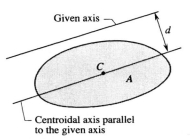

Fig. 7.12

area A about a given axis is equal to the moment of inertia $\bar{I}$ of the area about a centroidal axis parallel to the given axis plus the product Ad^2 of the area A and the square of the distance d between the two axes, as shown in Fig. 7.12; i.e.,

$$I = \bar{I} + Ad^2 \tag{7.10}$$

If a, b, and c are three parallel axes and the axis c is a centroidal axis of the area A as shown in Fig. 7.13, then the moments of inertia I_a, I_b, and $\bar{I}_c$ of the area A have the following relations:

$$I_a = \bar{I}_c + Ad_{ca}^2 \qquad I_b = \bar{I}_c + Ad_{cb}^2$$

$$I_a = I_b + A(d_{ca}^2 - d_{cb}^2) \tag{7.11}$$

where d_{ca} and d_{cb} are the distances from the centroidal axis c to the parallel axes a and b, respectively. Equations (7.10) and (7.11) are useful *transfer formulas* for moments of inertia.

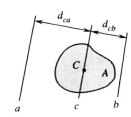

Fig. 7.13

EXAMPLE 7.5

The area considered in Example 7.3 is shown here for further study. Determine (a) I_x by choosing dA to be an areal strip perpendicular to the x axis, (b) I_y by choosing dA to be an areal strip perpendicular to the y axis. (Note in Example 7.3 that the areal strip dA in each case was chosen to be *parallel* to the axis about which the moment of inertia is computed.)

Solution.   The areal strip dA perpendicular to the x axis as shown may be treated as a rectangular area of width $b = dx$ and height $h = (5 - y)$. The distance between the horizontal centroidal x'' axis of dA and the x axis is $d = \frac{1}{2}(5 + y)$. Thus, applying the formula $I_x = \frac{1}{12} bh^3$ obtained in Example 7.1 and the parallel-axis theorem as expressed in Eq. (7.10), we write

$$dI_x = d\bar{I}_{x''} + (dA)(d^2)$$

$$= \frac{1}{12}(dx)(5 - y)^3 + [(5 - y) dx]\left[\frac{1}{2}(5 + y)\right]^2$$

$$= \frac{1}{3}(125 - y^3) dx = \frac{1}{3}\left[125 - \left(\frac{5}{64}x^2\right)^3\right] dx$$

$$= \frac{125}{3}\left(1 - \frac{x^6}{262144}\right) dx$$

$$I_x = \int dI_x = \frac{125}{3}\int_0^8 \left(1 - \frac{x^6}{262144}\right) dx = \frac{2000}{7}$$

$$I_x = 286 \text{ mm}^4 \blacktriangleleft$$

Next, we similarly let the areal strip dA be perpendicular to the y axis as shown. Treating dA as a rectangular area, we write

$$dI_y = d\bar{I}_{y''} + (dA)(d^2)$$

$$= \frac{1}{12}(dy)(x^3) + (x \, dy)\left(\frac{x}{2}\right)^2 = \frac{1}{3}x^3 \, dy$$

$$= \frac{1}{3}\left(\frac{64}{5}y\right)^{3/2} dy = \frac{512}{15\sqrt{5}}y^{3/2} \, dy$$

$$I_y = \int dI_y = \frac{512}{15\sqrt{5}}\int_0^5 y^{3/2} \, dy = \frac{1024}{3}$$

$$I_y = 341 \text{ mm}^4 \blacktriangleleft$$

Fig. D7.17

Fig. D7.18

Developmental Exercises

D7.15 Describe the parallel-axis theorem for area moments of inertia.

D7.16 Using Eqs. (7.9), show that the coordinates $\bar{x}$ and $\bar{y}$ of the centroid of an area are generally less than its radii of gyration k_y and k_x, respectively (i.e., $\bar{y} < k_x, \bar{x} < k_y$).

D7.17 A rectangular area is shown. Using the results in Example 7.1 and the parallel-axis theorem, determine (a) I_x, (b) I_y, and (c) J_O of this rectangular area.

D7.18 For the area A with its centroid at C and the parallel axes a and b as shown, it is known that $I_a = 25$ ft⁴. What is the value of I_b?

7.5　Product of Inertia of an Area

The product of inertia of an area is defined with respect to a specified pair of perpendicular axes which lie in the same plane as that of the area. The *product of inertia of an area A* with respect to the x and y axes, as shown in Fig. 7.14, is denoted by I_{xy} and is defined as the integral of the product

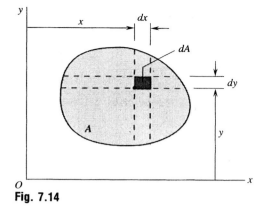

Fig. 7.14

of the *second-order* differential area element dA and the coordinates x and y defining the location of dA; i.e.,

$$I_{xy} = \int xy \, dA \qquad\qquad (7.12)^{\dagger}$$

Since dA in Eq. (7.12) is *not of the first order*, the basic definition for I_{xy} would call for *double integration*. However, I_{xy} may be determined with *single integration* via a theorem in Sec. 7.6. Note that dA is inherently positive, but x and y may be positive, negative, or zero. Consequently, the product of inertia of an area may be *positive*, *negative*, or *zero*. The dimensions of the product of inertia of an area are clearly $(length)^4$.

When an area A is symmetrical with respect to one or both of the x and y axes, the product of inertia I_{xy} of A is zero. This is true because the contributions to the product of inertia from each pair of area elements dA and dA', as shown in Fig. 7.15, which are on the opposite sides of, and located at equal distances from, the (x or y) axis of symmetry cancel out. Thus, the integral for this product of inertia reduces to zero.[††]

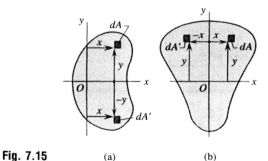

Fig. 7.15   (a)   (b)

[†]In tensor analysis, the product of inertia is mathematically defined as $I_{xy} = -\int xy \, dA$ so that the laws of transformation of second-order tensors may be followed by the moments and products of inertia.

[††]The product of inertia I_{xy} may be taken as a measure of *dissymmetry* or *imbalance* of the area with respect to the x and y axes.

EXAMPLE 7.6

Determine the product of inertia I_{xy} of the cross-sectional area in Example 7.3.

Solution. The area with a *second-order dA* is sketched as shown. By definition, we write

$$I_{xy} = \int xy \, dA = \int_{y=0}^{5} \int_{x=0}^{(64y/5)^{1/2}} xy \, dx \, dy$$

$$= \int_0^5 \left(\frac{1}{2} x^2 y \right) \Bigg|_{x=0}^{x=(64y/5)^{1/2}} dy$$

$$= \int_0^5 \frac{1}{2} \left(\frac{64}{5} y \right) y \, dy = \frac{800}{3} \qquad I_{xy} = 267 \text{ mm}^4 \quad \blacktriangleleft$$

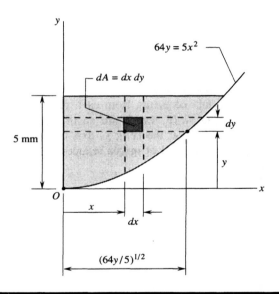

Developmental Exercises

D7.19 Define the product of inertia I_{xy} of an area A.

D7.20 Determine the product of inertia I_{xy} of each of the areas shown.

D7.21 Determine the product of inertia I_{xy} of the area shown.

Fig. D7.20

Fig. D7.21

7.6 Parallel-Axis Theorem: Area Products of Inertia

An area A in the xy plane with its centroid at the point $C(\overline{x}, \overline{y})$ is shown in Fig. 7.16. Noting the dA in this figure and Eq. (7.12), we write

$$
\begin{aligned}
I_{xy} &= \int xy \, dA = \int (x' + \overline{x})(y' + \overline{y}) \, dA \\
&= \int x'y' \, dA + \overline{y} \int x' \, dA + \overline{x} \int y' \, dA + \overline{x}\,\overline{y} \int dA \\
&= \overline{I}_{x'y'} + \overline{y}(\overline{x}'A) + \overline{x}(\overline{y}'A) + \overline{x}\,\overline{y}A \\
&= \overline{I}_{x'y'} + 0 + 0 + \overline{x}\,\overline{y}A
\end{aligned}
$$

$$
\boxed{I_{xy} = \overline{I}_{x'y'} + A\overline{x}\,\overline{y}} \tag{7.13}
$$

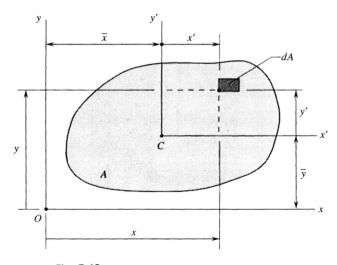

Fig. 7.16

This result is referred to as the *parallel-axis theorem* for area products of inertia. This theorem may be stated as follows: *The product of inertia I_{xy} of an area A with respect to any two perpendicular axes x and y in the plane of the area is equal to the product of inertia $\bar{I}_{x'y'}$ of the area with respect to two perpendicular centroidal axes x' and y', which are parallel to the x and y axes, respectively, plus the product $A\bar{x}\bar{y}$ of the area A and the two coordinates $\bar{x}$ and $\bar{y}$ of the centroid of the area in the xy coordinate system.* Note that $\bar{x}$ and $\bar{y}$ may be positive, negative, or zero depending on the location of the centroid C; however, the area A is always positive unless it is an empty area of a composite area.

EXAMPLE 7.7

Using *single integration*, determine the product of inertia I_{xy} of the cross-sectional area in Example 7.3. (Cf. Example 7.6.)

Solution. The area with a first-order dA is shown. The centroid of dA is at $(x/2, y)$, and the differential centroidal product of inertia $d\bar{I}_{x'y'}$ of dA is, by symmetry, equal to zero. Applying the *parallel-axis theorem* for area products of inertia, we write

$$dI_{xy} = d\bar{I}_{x'y'} + (dA)\left(\frac{x}{2}\right)(y) = 0 + \frac{1}{2}x^2 y\, dy = \frac{32}{5}y^2\, dy$$

$$I_{xy} = \int dI_{xy} = \frac{32}{5}\int_0^5 y^2\, dy = \frac{800}{3} \qquad I_{xy} = 267\ \text{mm}^4 \blacktriangleleft$$

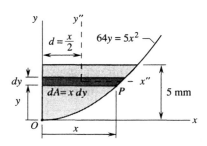

EXAMPLE 7.8

For the area shown, it is known from the solutions in Examples 7.3, 7.4, and 7.6 that $I_x = 2000/7$ mm^4, $I_y = 1024/3$ mm^4, $J_O = 13168/21$ mm^4, $A = 80/3$ mm^2, and $I_{xy} = 800/3$ mm^4. Determine for the area and axes shown the following quantities: (a) $\bar{I}_{x'}$, (b) $\bar{I}_{y'}$, (c) $\bar{J}_C$, (d) $\bar{I}_{x'y'}$, (e) $\bar{k}_{x'}$, (f) $\bar{k}_{y'}$, (g) $\bar{k}_C$.

Solution. Applying the parallel-axis theorems, we write

$$\frac{2000}{7} = \bar{I}_{x'} + \frac{80}{3}(3)^2 \qquad \bar{I}_{x'} = \frac{320}{7} \qquad \bar{I}_{x'} = 45.7 \text{ mm}^4 \quad \blacktriangleleft$$

$$\frac{1024}{3} = \bar{I}_{y'} + \frac{80}{3}(3)^2 \qquad \bar{I}_{y'} = \frac{304}{3} \qquad \bar{I}_{y'} = 101.3 \text{ mm}^4 \quad \blacktriangleleft$$

$$\frac{13168}{21} = \bar{J}_C + \frac{80}{3}(3^2 + 3^2) \qquad \bar{J}_C = \frac{3088}{21} \qquad \bar{J}_C = 147.0 \text{ mm}^4 \quad \blacktriangleleft$$

$$\frac{800}{3} = \bar{I}_{x'y'} + \frac{80}{3}(3)(3) \qquad \bar{I}_{x'y'} = \frac{80}{3} \qquad \bar{I}_{x'y'} = 26.7 \text{ mm}^4 \quad \blacktriangleleft$$

Note that $\bar{J}_C = \bar{I}_{x'} + \bar{I}_{y'}$ is satisfied. The centroidal radii of gyration $\bar{k}_{x'}$, $\bar{k}_{y'}$, and $\bar{k}_C$ may be determined as follows:

$$\bar{I}_{x'} = \bar{k}_{x'}^2 A: \qquad \frac{320}{7} = \bar{k}_{x'}^2\left(\frac{80}{3}\right) \qquad \bar{k}_{x'} = 1.309 \text{ mm} \quad \blacktriangleleft$$

$$\bar{I}_{y'} = \bar{k}_{y'}^2 A: \qquad \frac{304}{3} = \bar{k}_{y'}^2\left(\frac{80}{3}\right) \qquad \bar{k}_{y'} = 1.949 \text{ mm} \quad \blacktriangleleft$$

$$\bar{J}_C = \bar{k}_C^2 A: \qquad \frac{3088}{21} = \bar{k}_C^2\left(\frac{80}{3}\right) \qquad \bar{k}_C = 2.35 \text{ mm} \quad \blacktriangleleft$$

Developmental Exercises

D7.22 Describe the parallel-axis theorem for area products of inertia.

D7.23 For the area and the area element dA shown, determine (a) dA, (b) dI_{xy}, (c) I_{xy}.

Fig. D7.23

7.7 Composite Areas

We know that a composite area has two or more component areas. The integrals representing the moments and the product of inertia of a composite area A are to be evaluated over its component areas $A_1, A_2, \ldots, A_n$. Thus, the moment of inertia I of the composite area A about a given axis is equal to the sum of the moments of inertia $I_1, I_2, \ldots, I_n$ of its component areas about the same axis; i.e.,

$$I = I_1 + I_2 + \cdots + I_n \tag{7.14}$$

Table 7.1 Properties of Some Rolled-Steel Structural Sections* (Refer to Fig. 7.17.)

Designation	Area A, in^2	Depth d, in.	Width b, in.	I_x in^4	I_y in^4	$\bar{x}$ in.	y in.
W 21 × 44	13.0	20.66	6.500	843	20.7		
W 16 × 31	9.12	15.88	5.525	375	12.4		
W 12 × 26	7.65	12.22	6.490	204	17.3		
W 10 × 30	8.84	10.47	5.810	170	16.7		
W 8 × 21	6.16	8.28	5.270	75.3	9.77		
S 18 × 70	20.6	18.00	6.251	926	24.1		
S 12 × 50	14.7	12.00	5.477	305	15.7		
S 12 × 35	10.3	12.00	5.078	229	9.87		
S 8 × 23	6.77	8.00	4.171	64.9	4.31		
S 7 × 20	5.88	7.00	3.860	42.4	3.17		
C 15 × 40	11.8	15.00	3.520	349	9.23	0.777	
C 15 × 33.9	9.96	15.00	3.400	315	8.13	0.787	
C 10 × 20	5.88	10.00	2.739	78.9	2.81	0.606	
C 9 × 15	4.41	9.00	2.485	51.0	1.93	0.586	
C 6 × 13	3.83	6.00	2.157	17.4	1.05	0.514	
		Weight lb/ft					
L 8 × 8 × 1	15.0	51.0		89.0	89.0	2.37	2.37
L 6 × 6 × 1	11.0	37.4		35.5	35.5	1.86	1.86
L 5 × 5 × $\frac{1}{2}$	4.75	16.2		11.3	11.3	1.43	1.43
L 8 × 6 × $\frac{3}{4}$	9.94	33.8		63.4	30.7	1.56	2.56
L 6 × 4 × $\frac{3}{4}$	6.94	23.6		24.5	8.68	1.08	2.08

*Courtesy of the American Institute of Steel Construction, Chicago, Ill.

If I is the product of inertia of the composite area A with respect to two given perpendicular axes, it may similarly be determined by adding algebraically the products of inertia I_1, I_2, . . . , I_n of its component areas with respect to those two axes; i.e., Eq. (7.14) holds true for both the moments and the product of inertia of a composite area.

Sometimes, it is desirable to let a composite area have *empty areas* (i.e., negative areas) as some of its component areas. In such a case, the values of the moments and products of inertia of the *empty areas* should change their signs before they are algebraically added to the values of those of the other component areas to obtain the moments and the product of inertia of the composite area.

The cross sections of many structural members are of the form of composite areas. The moment of inertia of the cross-sectional area of a structural member is an important property of the member. Some typical rolled-steel structural sections are shown in Fig. 7.17, and their properties for designing are shown in Table 7.1. Furthermore, the formulas for centroidal moments and products of inertia in Fig. 7.18 along with the parallel-axis theorems may be applied to determine the moments and products of inertia of composite areas in many problems.

Wide-Flange Shapes (W Shapes)

A Wide-Flange Shape is designated by the letter W followed first by its *nominal* depth in inches and then its weight in lb/ft; e.g., W 16 x 31 designates a Wide-Flange Shape which has a *nominal* depth of 16 in. and weighs 31 lb/ft. Its actual depth may be greater or less than its nominal depth. (Cf. Table 7.1.)

American Standard Shapes (S Shapes)

An American Standard Shape is designated by the letter S followed first by its depth in inches and then its weight in lb/ft; e.g., S 18 x 70 designates an American Standard Shape which is 18 in. deep and weighs 70 lb/ft.

American Standard Channels (C Shapes)

An American Standard Channel is designated by the letter C followed first by its depth in inches and then its weight in lb/ft; e.g., C 15 x 33.9 designates an American Standard Channel which is 15 in. deep and weighs 33.9 lb/ft.

Angles (L Shapes)

An Angle is designated by the letter L followed first by the lengths d and b of its legs and then the thickness of its legs, all in inches; e.g., L 8 x 6 x $\frac{3}{4}$ designates an Angle which has one leg equal to 8 in. long, the other leg equal to 6 in. long, and the thickness of its legs equal to $\frac{3}{4}$ in. Some angles may have equal legs.

Fig. 7.17 Some rolled-steel structural sections.

Parallelogram and Rectangular area

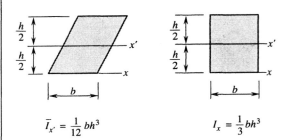

$$\overline{I}_{x'} = \frac{1}{12} bh^3 \qquad I_x = \frac{1}{3} bh^3$$

Elliptic area

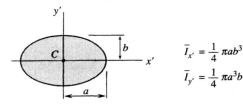

$$\overline{I}_{x'} = \frac{1}{4} \pi ab^3$$

$$\overline{I}_{y'} = \frac{1}{4} \pi a^3 b$$

Triangular area

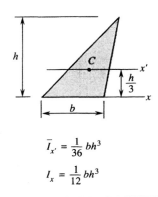

$$\overline{I}_{x'} = \frac{1}{36} bh^3$$

$$I_x = \frac{1}{12} bh^3$$

Semicircular area

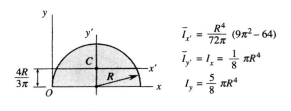

$$\overline{I}_{x'} = \frac{R^4}{72\pi} (9\pi^2 - 64)$$

$$\overline{I}_{y'} = I_x = \frac{1}{8} \pi R^4$$

$$I_y = \frac{5}{8} \pi R^4$$

Circular area

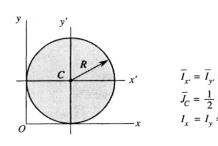

$$\overline{I}_{x'} = \overline{I}_{y'} = \frac{1}{4} \pi R^4$$

$$\overline{J}_C = \frac{1}{2} \pi R^4$$

$$I_x = I_y = \frac{5}{4} \pi R^4$$

Right triangular areas

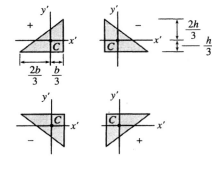

$$\overline{I}_{x'y'} = (-1)^n \left(\frac{1}{72} b^2 h^2 \right)$$

where n is the quadrant in which the right angle is situated in the $x'y'$ plane.

Quarter circular areas

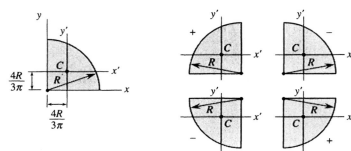

$$\overline{I}_{x'} = \overline{I}_{y'} = \frac{R^4}{144\pi} (9\pi^2 - 64)$$

$$I_x = I_y = \frac{1}{16} \pi R^4$$

$$\overline{I}_{x'y'} = (-1)^n \frac{R^4}{72\pi} (32 - 9\pi)$$

where n is the quadrant in which the right angle is situated in the $x'y'$ plane.

Fig. 7.18 Moments and products of inertia of areas of common shapes.

EXAMPLE 7.9

To increase the strength of the S 7×20 rolled-steel beam, a rectangular plate is fastened to its upper flange as shown. Determine $\bar{I}_{x'}$ and $\bar{k}_{x'}$ of the composite section of the beam with respect to its centroidal axis x' as indicated.

Solution. The area of the rectangular plate is $A_1 = 1.6$ in². From Table 7.1, the area of S 7×20 is $A_2 = 5.88$ in². The total area is $A = A_1 + A_2 = 7.48$ in². By inspection, $\bar{y}_1 = 3.7$ in. and $\bar{y}_2 = 0$. The ordinate $\bar{y}$ is determined from the *principle of moments*: $\bar{y}A = \bar{y}_1 A_1 + \bar{y}_2 A_2$. We write

$$\bar{y}(7.48) = 3.7(1.6) + 0(5.88) \qquad \bar{y} = 0.791 \text{ in.}$$

As shown in the figure, we have

$$d_1 = \bar{y}_1 - \bar{y} = 2.909 \text{ in.} \qquad d_2 = \bar{y} = 0.791 \text{ in.}$$

From Table 7.1, the centroidal moment of inertia of S 7×20 is $(\bar{I}_x)_2 = 42.4$ in⁴. Applying the *parallel-axis theorem*, we write

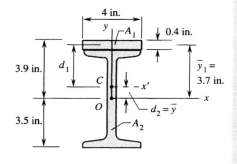

$$(\bar{I}_{x'})_1 = \frac{1}{12} bh^3 + A_1 d_1^2 = \frac{1}{12}(4)(0.4)^3 + 1.6(2.909)^2 = 13.56$$

$$(\bar{I}_{x'})_2 = (\bar{I}_x)_2 + A_2 d_2^2 = 42.4 + 5.88(0.791)^2 = 46.08$$

For the composite section, we write

$$\bar{I}_{x'} = (\bar{I}_{x'})_1 + (\bar{I}_{x'})_2 \qquad \bar{I}_{x'} = 59.6 \text{ in}^4 \blacktriangleleft$$

$$\bar{I}_{x'} = \bar{k}_{x'}^2 A \qquad \bar{k}_{x'} = 2.82 \text{ in.} \blacktriangleleft$$

Developmental Exercises

D7.24 Determine I_x and k_x of the composite area shown by treating the areas *ABCD* and *GDEF* as the component areas.

D7.25 Determine I_y and k_y of the composite area shown.

Fig. D7.24 and D7.25

EXAMPLE 7.10

A composite area is shown. Making use of the solutions in Examples 7.1, 7.3, 7.4, and 7.6, determine for the composite area and axes shown the following quantities: (a) I_x, (b) I_{xy}, (c) k_x.

Solution. We may treat this composite area as the algebraic sum of three component areas A_1, A_2, and A_3 as shown, where A_3 is an empty area. From the solutions in Examples 7.1, 7.3, 7.4, and 7.6, we write

$$A_1 = \frac{80}{3} \text{ mm}^2 \qquad (I_x)_1 = \frac{2000}{7} \text{ mm}^4 \qquad (I_{xy})_1 = \frac{800}{3} \text{ mm}^4$$

$$A_2 = 24 \text{ mm}^2 \qquad A_3 = 8 \text{ mm}^2$$

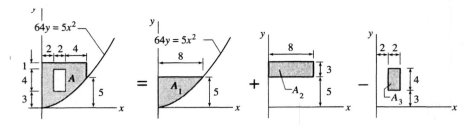

Dimensions in mm

$$(I_x)_2 = \frac{1}{12}(8)(3)^3 + A_2\left(5 + \frac{3}{2}\right)^2 \qquad (I_x)_2 = 1032 \text{ mm}^4$$

$$(I_{xy})_2 = 0 + A_2\left(\frac{8}{2}\right)\left(5 + \frac{3}{2}\right) \qquad (I_{xy})_2 = 624 \text{ mm}^4$$

$$(I_x)_3 = \frac{1}{12}(2)(4)^3 + A_3\left(3 + \frac{4}{2}\right)^2 \qquad (I_x)_3 = \frac{632}{3} \text{ mm}^4$$

$$(I_{xy})_3 = 0 + A_3\left(2 + \frac{2}{2}\right)\left(3 + \frac{4}{2}\right) \qquad (I_{xy})_3 = 120 \text{ mm}^4$$

Thus, for the composite area, we write

$$A = A_1 + A_2 - A_3 = 42.67 \text{ mm}^2$$

$$I_x = (I_x)_1 + (I_x)_2 - (I_x)_3 \qquad\qquad I_x = 1107 \text{ mm}^4 \;\blacktriangleleft$$

$$I_{xy} = (I_{xy})_1 + (I_{xy})_2 - (I_{xy})_3 \qquad I_{xy} = 771 \text{ mm}^4 \;\blacktriangleleft$$

$$I_x = k_x^2 A \qquad\qquad\qquad\qquad k_x = 5.09 \text{ mm} \;\blacktriangleleft$$

Fig. D7.27

Developmental Exercises

D7.26 Refer to Example 7.10. Determine (a) I_y, (b) k_y.

D7.27 Determine the values of I_x, I_y, and I_{xy} of the composite area shown.

EXAMPLE 7.11[†]

The vertices of an arbitrary triangle are located at $P(x_1, y_1)$, $Q(x_2, y_2)$, and $R(x_3, y_3)$. Determine I_x of the triangular area PQR.

Solution. Treating $\triangle PQR$ as a composite area as shown, we write

$$\triangle PQR = \square PSTU - \triangle PSQ - \triangle QTR - \triangle RUP$$

[†]The formulas for I_x, I_y, and I_{xy} in Example 7.11 and D7.28 and D7.29 are useful in determining the moments and the product of inertia of *any* plane area using *triangular finite elements* (e.g., Probs. 7.67 through 7.78*).

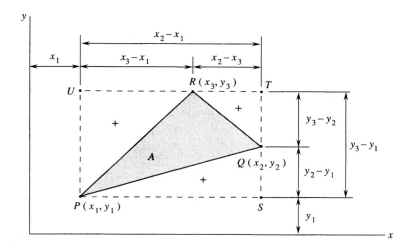

Applying appropriate formulas in Fig. 7.18 and the parallel-axis theorems, we write

$$I_x = \left[\frac{1}{12}(x_2 - x_1)(y_3 - y_1)^3 + (x_2 - x_1)(y_3 - y_1)\left(\frac{y_3 + y_1}{2}\right)^2\right]$$

$$- \left[\frac{1}{36}(x_2 - x_1)(y_2 - y_1)^3 + \frac{1}{2}(x_2 - x_1)(y_2 - y_1)\left(y_1 + \frac{y_2 - y_1}{3}\right)^2\right]$$

$$- \left[\frac{1}{36}(x_2 - x_3)(y_3 - y_2)^3 + \frac{1}{2}(x_2 - x_3)(y_3 - y_2)\left(y_3 - \frac{y_3 - y_2}{3}\right)^2\right]$$

$$- \left[\frac{1}{36}(x_3 - x_1)(y_3 - y_1)^3 + \frac{1}{2}(x_3 - x_1)(y_3 - y_1)\left(y_3 - \frac{y_3 - y_1}{3}\right)^2\right]$$

which can be simplified and expressed as

$$I_x = \frac{A}{12}[(y_1 + y_2)^2 + (y_2 + y_3)^2 + (y_3 + y_1)^2] \quad \blacktriangleleft$$

where A is the area of $\triangle PQR$ and is (cf. Example 6.7) given by

$$A = \frac{1}{2}|x_1(y_2 - y_3) + x_2(y_3 - y_1) + x_3(y_1 - y_2)|$$

Developmental Exercises

D7.28 Refer to Example 7.11. Show that

$$I_y = \frac{A}{12}[(x_1 + x_2)^2 + (x_2 + x_3)^2 + (x_3 + x_1)^2]$$

D7.29 Refer to Example 7.11. Show that

$$I_{xy} = \frac{A}{12}[(x_1 + x_2 + x_3)(y_1 + y_2 + y_3) + x_1y_1 + x_2y_2 + x_3y_3]$$

PROBLEMS

7.40 through 7.42* Determine the moment of inertia I_y of the area shown. (*Hint.* Choose dA to be an areal strip perpendicular to the y axis.)

7.43 through 7.45* Determine the moment of inertia I_x of the area shown. (*Hint.* Choose dA to be an areal strip perpendicular to the x axis.)

Fig. P7.40 and P7.46   Fig. P7.41 and P7.47   Fig. P7.42* and P7.48*   Fig. P7.43 and P7.49

Fig. P7.44* and P7.50   Fig. P7.45* and P7.51*   Fig. P7.52 and P7.58   Fig. P7.53 and P7.59

7.46 through 7.51* Determine the product of inertia I_{xy} of the area shown. (*Hint.* One way is easier than the other in the integration involved.)

7.52 through 7.57* Determine the centroidal moment of inertia $\bar{I}_x$ and the centroidal radius of gyration $\bar{k}_x$ of the area shown.

Fig. P7.54* and P7.60*   Fig. P7.55, P7.61, and P7.64   Fig. P7.56, P7.62, and P7.65   Fig. P7.57*, P7.63* and P7.66*

7.58 through 7.63*   Determine the centroidal moment of inertia $\bar{I}_y$ and the centroidal radius of gyration $\bar{k}_y$ of the area shown.

7.64 through 7.66*   Determine the centroidal product of inertia $\bar{I}_{xy}$ of the area shown.

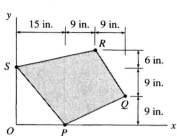

Fig. P7.67, P7.70*,
P7.73*, and P7.76

Fig. P7.68*, P7.71,
P7.74, and P7.77*

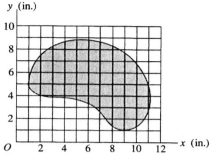

Fig. P7.69, P7.72*
P7.75, and P7.78*

7.67 through 7.69   Determine the moment of inertia I_x of the composite area shown.[†]

7.70* through 7.72*   Determine the moment of inertia I_y of the composite area shown.[†]

7.73* through 7.75* Determine the product of inertia I_{xy} of the composite area shown.[†]

7.76 through 7.78*   Determine the polar moment of inertia J_O of the composite area shown.[†]

7.79 through 7.81   Determine the centroidal moments of inertia $\bar{I}_x$ and $\bar{I}_y$ and polar moment of inertia $\bar{J}_C$ of the section shown.

Fig. P7.79

Fig. P7.80

Fig. P7.81 and P7.82

7.82   For the section shown, determine the distance a for which $\bar{I}_x = \bar{I}_y$.

[†]*Suggestion.* Use *triangular finite elements* and apply the formulas in Example 7.11, D7.28, and D7.29. (The result may be checked by using the program AREA described in App. F.)

★7.8 Rotation of Axes

An area A in the xy plane is shown in Fig. 7.19, where the set of axes u and v is obtained by rotating the set of axes x and y about the origin O through an angle θ as indicated. We shall here derive the moments and the

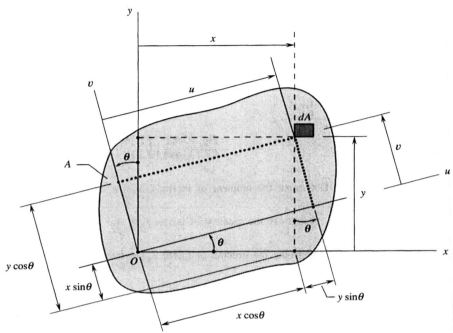

Fig. 7.19

product of inertia I_u, I_v, and I_{uv} in terms of I_x, I_y, I_{xy}, and θ. From Fig. 7.19, we have

$$u = x \cos \theta + y \sin \theta \qquad v = -x \sin \theta + y \cos \theta \qquad (7.15)$$

By the definitions of moments and products of inertia, we write

$$I_u = \int v^2 \, dA = \int (-x \sin \theta + y \cos \theta)^2 \, dA$$

$$I_u = I_x \cos^2 \theta - 2I_{xy} \sin \theta \cos \theta + I_y \sin^2 \theta \qquad (7.16)$$

$$I_v = \int u^2 \, dA = \int (x \cos \theta + y \sin \theta)^2 \, dA$$

$$I_v = I_y \cos^2 \theta + 2I_{xy} \sin \theta \cos \theta + I_x \sin^2 \qquad (7.17)$$

$$\theta I_{uv} = \int uv \, dA = \int (x \cos \theta + y \sin \theta)(-x \sin \theta + y \cos \theta) \, dA$$

$$I_{uv} = (I_x - I_y) \sin \theta \cos \theta + (\cos^2 \theta - \sin^2 \theta) I_{xy} \qquad (7.18)$$

Applying the trigonometric relations $2 \sin \theta \cos \theta = \sin 2\theta$, $\cos^2 \theta - \sin^2 \theta = \cos 2\theta$, $\cos^2 \theta = \frac{1}{2}(1 + \cos 2\theta)$, and $\sin^2 \theta = \frac{1}{2}(1 - \cos 2\theta)$ to Eqs. (7.16) through (7.18), we get the following relations:

$$I_u = \frac{1}{2}(I_x + I_y) + \frac{1}{2}(I_x - I_y)\cos 2\theta - I_{xy}\sin 2\theta$$

$$I_v = \frac{1}{2}(I_x + I_y) - \frac{1}{2}(I_x - I_y)\cos 2\theta + I_{xy}\sin 2\theta$$

(7.19)

$$I_{uv} = \frac{1}{2}(I_x - I_y)\sin 2\theta + I_{xy}\cos 2\theta$$

$$I_u + I_v = I_x + I_y$$

Developmental Exercises

D7.30 Is it true that the formula for I_v in Eqs. (7.19) may be obtained by replacing θ with $\theta + \pi/2$ in the formula for I_u?

D7.31 Refer to the area A in Fig. 7.19. Suppose that $I_x = 260$ mm^4, $I_y = 240$ mm^4, and $I_u = 200$ mm^4. What is the value of I_v?

★7.9 Representation by Mohr's Circle

The formulas in Eqs. (7.19) may graphically be represented by the so-called *Mohr's circle*,[†] after the German engineer Otto Mohr (or Christian Otto Mohr, 1835–1918). Mohr's circle for an area A in the xy plane, as shown in Fig. 7.20, is a circle which is drawn with the line segment $\overline{XY}$ as a diameter as shown in Fig. 7.21, where *the coordinates of X and Y are*

Fig. 7.20

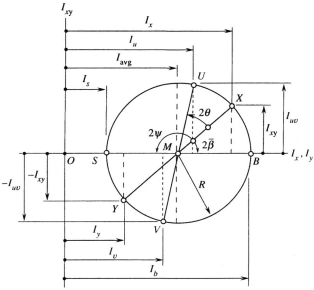

Fig. 7.21

[†]The graphical scheme of Mohr's circle is also used in studying stresses and strains in mechanics of materials, elasticity, etc.

(I_x, I_{xy}) and $(I_y, -I_{xy})$, respectively. The midpoint M of the diameter $\overline{XY}$ is on the horizontal axis and is the center of Mohr's circle. Thus, the points X and Y on Mohr's circle are related to the x and y axes on the area A, respectively. The points U and V in Fig. 7.21 are reached by moving the points X and Y counterclockwise along Mohr's circle until the angles XMU and YMV are both equal to 2θ. The points U and V on Mohr's circle are, therefore, related to the u and v axes on the area A, respectively. For the area A in Fig. 7.20, Mohr's circle in Fig. 7.21 indicates the following: *The abscissas of the points U and V are equal to I_u and I_v, respectively, and the ordinate of the point U (or the negative of the ordinate of the point V) is equal to I_{uv}.*

I. SIGN CONVENTION.

On the area A and on Mohr's circle, the *positive directions* for corresponding angles are the *same*, although the *magnitudes* of the angles on Mohr's circle are *twice* as large as the corresponding angles on the area A. Note that the points B and S on Mohr's circle in Fig. 7.21 are related to the b and s axes on the area A in Fig. 7.20, respectively.[†] Furthermore, each set of coordinate axes in Fig. 7.20 and the z axis will form a right-handed system. We observe that (1) θ is always measured *counterclockwise* from the x axis to the u axis, (2) β is always measured *clockwise* from the x axis to the b axis, and (3) positive ψ is measured *counterclockwise* from the x axis to the s axis. In Figure 7.21, we note *correspondingly* that $(1')$ 2θ is always measured *counterclockwise* from the radius $\overline{MX}$ to the radius $\overline{MU}$, $(2')$ 2β is always measured *clockwise* from the radius $\overline{MX}$ to the radius $\overline{MB}$, and $(3')$ positive 2ψ is measured *counterclockwise* from the radius $\overline{MX}$ to the radius $\overline{MS}$. From Fig. 7.21, we have

$$2\beta + 2\psi = 180° \tag{7.20}$$

Since 2θ and 2β are always positive, we see that 2ψ may be negative.[††]

II. PROOF.

Referring to Fig. 7.21, we can verify the validity of representing the formulas in Eqs. (7.19) by Mohr's circle as follows:

$$I_{\text{avg}} = \tfrac{1}{2}(I_x + I_y) \tag{7.21}$$

$$R = [(I_x - I_{\text{avg}})^2 + I_{xy}^2]^{1/2} = \tfrac{1}{2}[(I_x - I_y)^2 + 4I_{xy}^2]^{1/2} \tag{7.22}$$

$$
\begin{aligned}
I_u &= I_{\text{avg}} + R\cos(2\theta + 2\beta) \\
&= I_{\text{avg}} + R(\cos 2\theta \cos 2\beta - \sin 2\theta \sin 2\beta) \\
&= I_{\text{avg}} + [(R\cos 2\beta)\cos 2\theta - (R\sin 2\beta)\sin 2\theta] \\
&= I_{\text{avg}} + [(I_x - I_{\text{avg}})\cos 2\theta - I_{xy}\sin 2\theta] \\
&= \tfrac{1}{2}(I_x + I_y) + \tfrac{1}{2}(I_x - I_y)\cos 2\theta - I_{xy}\sin 2\theta \qquad \text{Q.E.D.}^{†††}
\end{aligned}
$$

[†]The names B and S are chosen to indicate that the moment of inertia I_b about the b axis is the *biggest*, while the moment of inertia I_s about the s axis is the *smallest*.

[††]Since 2β is always positive, we have $0 \le 2\beta < 360°$. Thus, we find from Eq. (7.20) that $2\psi < 0$ if $2\beta > 180°$. When $2\psi < 0$, the actual rotation from $\overline{MX}$ to $\overline{MS}$ through -2ψ would be *clockwise* (cf. Example 7.12).

[†††]The abbreviation "Q.E.D." stands for the Latin words "quod erat demonstrandum," meaning "which was to be proved."

$$I_v = I_{avg} - R\cos(2\theta + 2\beta)$$
$$= I_{avg} - [(I_x - I_{avg})\cos 2\theta - I_{xy}\sin 2\theta]$$
$$= \tfrac{1}{2}(I_x + I_y) - \tfrac{1}{2}(I_x - I_y)\cos 2\theta + I_{xy}\sin 2\theta \qquad \text{Q.E.D.}$$

$$I_{uv} = R\sin(2\theta + 2\beta) = R(\sin 2\theta \cos 2\beta + \cos 2\theta \sin 2\beta)$$
$$= (R\cos 2\beta)\sin 2\theta + (R\sin 2\beta)\cos 2\theta$$
$$= (I_x - I_{avg})\sin 2\theta + I_{xy}\cos 2\theta$$
$$= \tfrac{1}{2}(I_x - I_y)\sin 2\theta + I_{xy}\cos 2\theta \qquad \text{Q.E.D.}$$

$$I_u + I_v = [\tfrac{1}{2}(I_x + I_y) + \tfrac{1}{2}(I_x - I_y)\cos 2\theta - I_{xy}\sin 2\theta]$$
$$+ \tfrac{1}{2}(I_x + I_y) - \tfrac{1}{2}(I_x - I_y)\cos 2\theta + I_{xy}\sin 2\theta]$$
$$= I_x + I_y \qquad \text{Q.E.D.}$$

EXAMPLE 7.12

For the Z section shown, it is known that $I_x = 592$ in^4, $I_y = 92$ in^4, and $I_{xy} = -150$ in^4. Using Mohr's circle, determine the moments and product of inertia I_u, I_v, and I_{uv} with respect to the axes indicated.

Solution. Using $X(I_x, I_{xy})$ and $Y(I_y, -I_{xy})$ as the end points of a diameter, we may draw Mohr's circle for the Z section about O as shown.[†] We see that

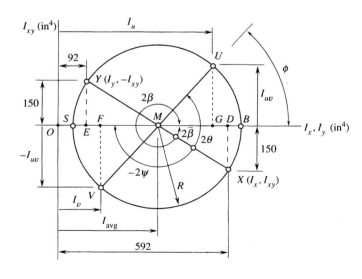

[†]The angles 2β and 2ψ discussed in Sec. 7.9 are also indicated on this circle for reference, where $2\beta = 360° - 2\overline{\beta} = 329.04°$, $2\beta + 2\psi = 180°$, and $2\psi = -149.04°$. These angles may be used as examples in studying Sec. 7.10, particularly Table 7.2.

$$I_{avg} = \tfrac{1}{2}(592 + 92) = 342$$

$$\overline{MD} = 592 - I_{avg} = 250 \qquad \overline{DX} = 150$$

$$R = (\overline{MD}^2 + \overline{DX}^2)^{1/2} = 291.5$$

$$\tan 2\overline{\beta} = \frac{\overline{DX}}{\overline{MD}} = \frac{150}{250} \qquad 2\overline{\beta} = 30.96°$$

$$\theta = 40° \qquad \phi = 2\theta - 2\overline{\beta} = 49.04°$$

Thus, we have

$$I_u = \overline{OG} = \overline{OM} + \overline{MG} = I_{avg} + R\cos\phi = 533.1$$

$$I_v = \overline{OF} = \overline{OM} - \overline{FM} = I_{avg} - R\cos\phi = 150.9$$

$$I_{uv} = \overline{GU} = R\sin\phi = 220.1$$

$$I_u = 533 \text{ in}^4 \qquad I_v = 150.9 \text{ in}^4 \qquad I_{uv} = 220 \text{ in}^4 \quad \blacktriangleleft^\dagger$$

Developmental Exercises

D7.32 The points X and Y are the end points of a diameter of Mohr's circle for an area A and are related to the x and y axes on the area A. What are the coordinates of X and Y?

D7.33 Using the formulas in Eqs. (7.19), show that (a) $(I_u - I_{avg})^2 + I_{uv}^2 = R^2$, (b) $(I_v - I_{avg})^2 + I_{uv}^2 = R^2$, where I_{avg} and R are as defined in Eqs. (7.21) and (7.22).

D7.34 The equations established in D7.33 indicate that the locus of the points (I_u, I_{uv}) and $(I_v, -I_{uv})$ is a circle, called Mohr's circle, with its radius equal to R. What are the coordinates of the center of this circle?

★7.10 Principal Moments and Axes of Inertia

Referring to Fig. 7.21, we note that the maximum and minimum values of the moment of inertia I are given by the abscissas of the extreme right point B and the extreme left point S on Mohr's circle. They are

$$I_{max} = I_b = I_{avg} + R \qquad I_{min} = I_s = I_{avg} - R$$

$$I_{max} = \tfrac{1}{2}(I_x + I_y) + \left\{ [\tfrac{1}{2}(I_x - I_y)]^2 + I_{xy}^2 \right\}^{1/2} \qquad (7.23)$$

$$I_{min} = \tfrac{1}{2}(I_x + I_y) - \left\{ [\tfrac{1}{2}(I_x - I_y)]^2 + I_{xy}^2 \right\}^{1/2} \qquad (7.24)$$

These two extremum values of I are called the *principal moments of inertia* of the area A about the point O in Fig. 7.20.

---

†Cf. the answers obtained via a computer program in Fig. 7.22.

The axes b and s with respect to which the moments of inertia are maximum and minimum are called the *principal axes of inertia* of the area A about the point O. The directions of the principal axes are called the *principal directions*. Letting θ_b and θ_s be the *directional angles* of the principal axes b and s with respect to the x axis in Fig. 7.20, we write

$$\theta_b = 360° - \beta \qquad \theta_s = \psi$$

From Fig. 7.21, we see that

$$\tan 2\beta = \frac{I_{xy}}{I_x - I_{avg}} \qquad \tan 2\beta = \frac{2I_{xy}}{I_x - I_y} \qquad (7.25)$$

where the ranges of values of 2β are summarized in Table 7.2.

Table 7.2   Ranges of Values of 2β

$I_{xy} \geq 0$	$I_x \geq I_y$	$0 \leq 2\beta \leq 90°$
	$I_x < I_y$	$90° < 2\beta \leq 180°$
$I_{xy} < 0$	$I_x \leq I_y$	$180° < 2\beta \leq 270°$
	$I_x > I_y$	$270° < 2\beta < 360°$

Using Eqs. (7.19) and setting $\frac{d}{d\theta}(I_u) = 0$ or $I_{uv} = 0$, we readily find that

$$\tan 2\theta = -\frac{2I_{xy}}{I_x - I_y} \qquad (7.26)$$

From Eqs. (7.25) and (7.26), we note that $\tan 2\theta = -\tan 2\beta = \tan(-2\beta)$ or $\theta = -\beta$. This means that a principal axis (i.e., the b axis) is reached when the x axis in Fig. 7.20 is rotated *clockwise* about the origin O through an angle of β. Thus, Eqs. (7.20) and (7.25) conform to the result in Eq. (7.26) as derived with calculus. As the ordinates of the points B and S on Mohr's circle are zero, it is important to note that *the product of inertia is zero with respect to the principal axes*. If desired, a computer program may be used to compute I_{max}, I_{min}, β, and ψ, as well as I_u, I_v, and I_{uv} from given values of I_x, I_y, I_{xy}, and θ as shown in Fig. 7.22.

```
10 REM            AREA MOMENTS OF INERTIA         (File Name: AMOI)
15 REM   Enter Ix, Iy, Ixy, and Theta (in degrees) as data in Line 100
20 REM   To output to the printer, change PRINT to LPRINT in the program.
25 READ IX,IY,XY,TH: I1=(IX+IY)/2: I2=(IX-IY)/2: T2=TH*3.141593/90
30 I3=I2*COS(T2)-XY*SIN(T2):IU=I1+I3:IV=I1-I3:UV=I2*SIN(T2)+XY*COS(T2)
35 R = SQR(I2^2+XY^2): MX = I1 + R: MN = I1 - R: S$ = "     ": PRINT
40 PRINT "Ix =";IX;S$; "Iy =";IY;S$; "Ixy = ";XY;S$; "Theta = ";TH
45 PRINT "Iu =";IU;S$;"Iv =";IV;S$;"Iuv = ";UV:B=ATN(XY/I2)*90/3.141593
50 IF XY >= 0 AND IX >= IY THEN P = 90 - B: GOTO 70
55 IF XY >= 0 AND IX < IY THEN B = B + 90: P = 90 - B: GOTO 70
60 IF XY < 0 AND IX <= IY THEN B = B + 90: P = 90 - B: GOTO 70
65 IF XY < 0 AND IX > IY THEN B = B + 180: P = 90 - B
70 PRINT "Imax =";MX;S$; "Imin =";MN;S$; "Beta =";B;S$; "Psi = ";P
100 DATA 592,92,-150,40

Ix = 592     Iy = 92     Ixy = -150     Theta =   40
Iu = 533.1332     Iv = 150.8669     Iuv =   220.1548
Imax = 633.5476     Imin = 50.4524     Beta = 164.5181     Psi = -74.51813
```

Fig. 7.22 (Cf. the solution in Example 7.12.)

EXAMPLE 7.13

A slender aluminum rod with *round ends* has a cross-sectional area A as shown. It is known from mechanics of materials that the magnitude of the

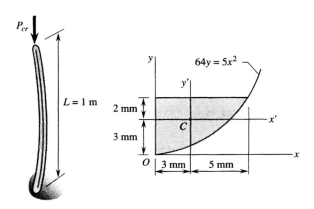

critical force $\mathbf{P}_{cr}$ for the buckling of the rod is given by $P_{cr} = (\pi^2 E \bar{I}_{min})/L^2$ where L is the length of the rod, E is the elastic modulus of the rod and is equal to 69 GPa, and $\bar{I}_{min}$ is the minimum moment of inertia of A about an axis through its centroid C as shown. This value of P_{cr} is known as Euler's buckling load. It is also known from the solution in Example 7.8 that $\bar{I}_{x'} = 45.71$ mm^4, $\bar{I}_{y'} = 101.33$ mm^4, and $\bar{I}_{x'y'} = 26.67$ mm^4. Determine (a) the value of P_{cr}, (b) the values of β and ψ specifying the directions of the principal axes of inertia b' and s' about the centroid C. Show the principal axes b' and s' on the area A in a sketch.

Solution. The data show that $L = 1$ m $= 10^3$ mm and $E = 69$ GPa $= 69 \times 10^9$ N/m^2 $= 69 \times 10^3$ N/mm^2. Thus, the determination of P_{cr} now requires a prior determination of $\bar{I}_{min}$. Using the points $(\bar{I}_{x'}, \bar{I}_{x'y'})$ and $(\bar{I}_{y'}, -\bar{I}_{x'y'})$ as the end points of the diameter $\overline{X'Y'}$, we may draw Mohr's

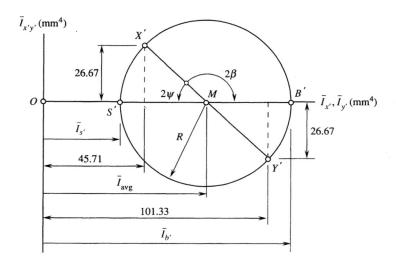

circle for the area A about C as shown. From this circle, we write

$$\bar{I}_{avg} = \frac{1}{2}(45.71 + 101.33) \qquad \bar{I}_{avg} = 73.52 \text{ mm}^4$$

$$R^2 = (\bar{I}_{avg} - 45.71)^2 + (26.67)^2 \qquad R = 38.53 \text{ mm}^4$$

$$\bar{I}_{min} = \bar{I}_{s'} = \bar{I}_{avg} - R \qquad \bar{I}_{min} = 34.99 \text{ mm}^4$$

Thus, we have

$$P_{cr} = \frac{\pi^2(69 \times 10^3 \text{ N/mm}^2)(34.99 \text{ mm}^4)}{(10^3 \text{ mm})^2}$$

$$\tan 2\psi = \frac{26.67}{73.52 - 45.71}$$

$$2\beta = 180° - 2\psi$$

$$P_{cr} = 23.8 \text{ N} \qquad \psi = 21.9° \qquad \beta = 68.1° \quad \blacktriangleleft$$

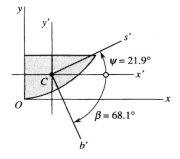

Any force greater than $\mathbf{P}_{cr}$ plus a small lateral disturbance will cause the rod to bow out (i.e., buckle) and bend about the s' axis which makes an angle of $\psi = 21.9°$ with the x' axis as shown in the sketch.

Developmental Exercises

D7.35 Refer to Example 7.12. Using Mohr's circle as drawn in the solution of that example, (a) determine for the Z section the principal moments of inertia I_b and I_s about the origin O, (b) show the principal axes b and s on the Z section in a sketch.

D7.36 Letting 45.71, 101.33, 26.67, and 0 be the data for Ix, Iy, Ixy, and θ, run the program in Fig. 7.22 to verify the values obtained for $\bar{I}_{min}$, ψ, and β in Example 7.13.

PROBLEMS

7.83* through 7.91*   Using Eqs. (7.19), determine the centroidal moments of inertia $\bar{I}_u$ and $\bar{I}_v$ and the centroidal product of inertia $\bar{I}_{uv}$ for the area and axes shown.

Fig. P7.83* and P7.92*

Fig. P7.84 and P7.93*

Fig. P7.85 and P7.94*

Fig. P7.86, P7.95*, and P7.101

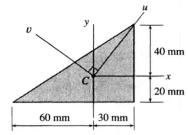

Fig. P7.87*, P7.96*, and P7.102*

7.92* through 7.100* Using Mohr's circle, determine the centroidal moments of inertia $\bar{I}_u$ and $\bar{I}_v$ and the centroidal product of inertia $\bar{I}_{uv}$ for the area and axes shown.

7.101 through 7.106 For the area shown, determine (a) the centroidal principal moments of inertia $\bar{I}_b$ and $\bar{I}_s$, (b) the angles β and ψ locating the centroidal principal axes of inertia b and s on the area, respectively. Indicate the b and s axes on the area in a sketch.

7.107 For a given area in the xy plane, it is known that $I_x = 14 \times 10^6$ mm^4, $I_y = 8 \times 10^6$ mm^4, $I_{max} = 16 \times 10^6$ mm^4, and $I_{xy} > 0$. Using Mohr's circle, determine (a) I_{xy}, (b) I_{min}, (c) the angles β and ψ locating the principal axes of inertia b and s, respectively. Indicate the b and s axes on the xy plane.

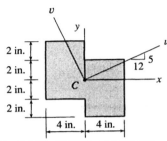

Fig. P7.88, P7.97*, and P7.103

Fig. P7.89*, P7.98, and P7.104

Dimensions in millimeters

Fig. P7.90*, P7.99, and P7.105*

Fig. P7.91*, P7.100*, and P7.106

MASS MOMENTS OF INERTIA[†]

7.11 Moments of Inertia of a Mass

In addition to having a mass, a rigid body has a definite shape and size. Thus, under the action of nonconcurrent forces $\mathbf{F}_1$, $\mathbf{F}_2$, . . . , $\mathbf{F}_n$ as shown in Fig. 7.23, a rigid body will generally be accelerated to rotate as well as

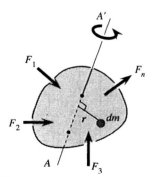

Fig. 7.23

[†]Cf. the footnote on p. 223.

translate. The tendency of a body to resist a translational acceleration is, of course, measured by the *mass* of the body. On the other hand, the tendency of a body to resist a rotational acceleration about an axis is measured by a quantity called the *moment of inertia* of the body about that axis. This quantity, encountered in kinetics of rigid bodies having rotational motion, is an integral of the form $\int$ (distance)^2 d(mass) or

$$I = \int r^2 \, dm \qquad (7.27)$$

where *dm* is a differential mass element of the body, *r* is the shortest distance between *dm* and the axis of rotation, say *AA'*, and the integration extends over the entire body. The moment of inertia of a body defined in Eq. (7.27) is often referred to as the *mass moment of inertia*.

The differential mass element *dm* illustrated in Fig. 7.23 for Eq. (7.27) is of the *third order*. However, *dm* in Eq. (7.27) often takes the form of a *first-order* slender strip, thin ring, or thin shell if the particles in *dm* can be taken as equidistant from the axis of rotation, with *r* being the *equidistance* within the framework of first-order accuracy.

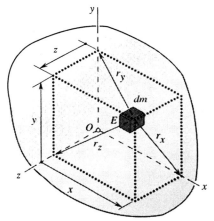

Fig. 7.24

The moments of inertia of a body about the *x*, *y*, and *z* axes are customarily denoted by I_x, I_y, and I_z, respectively.[†] As shown in Fig. 7.24, the rectangular coordinates of the differential mass element *dm* at the point *E* are (*x*, *y*, *z*), and the shortest distances between *dm* and the *x*, *y*, and *z* axes are r_x, r_y, and r_z, respectively. We write

$$\begin{aligned}
I_x &= \int r_x^2 \, dm = \int (y^2 + z^2) \, dm \\
I_y &= \int r_y^2 \, dm = \int (z^2 + x^2) \, dm \qquad (7.28) \\
I_z &= \int r_z^2 \, dm = \int (x^2 + y^2) \, dm
\end{aligned}$$

[†]Sometimes they are written with double subscripts as I_{xx}, I_{yy}, and I_{zz} to signify that they are components of a second-order tensor.

Note from Eqs. (7.27) and (7.28) that the moment of inertia of a body about any axis is always a *positive scalar quantity*, whose dimensions are *mass* × *(length)²*, because dm is always taken as positive regardless of the position in which it is situated. The primary units for the moment of inertia of a body are $kg \cdot m^2$ in the SI and $slug \cdot ft^2$ (or $lb \cdot ft \cdot s^2$) in the U.S. customary system. Unless otherwise stated, it will be assumed that each body considered is homogeneous and has a constant mass density.

Developmental Exercise

D7.37   Show that $1 \ slug \cdot ft^2 = 1 \ lb \cdot ft \cdot s^2 = 1.356 \ kg \cdot m^2$.

EXAMPLE 7.14

A slender rod of length L and mass m is shown. Determine the moments of inertia I_x, I_y, and I_z of the rod about the axes as indicated.

Solution.   The mass density of the rod may be expressed as $\rho_L = m/L$. We note that the y and z coordinates of the mass element dm may be taken as zero throughout the rod as shown. Thus, we write

$$I_x = \int r_x^2 \, dm = \int (y^2 + z^2) \, dm = 0$$

$$I_y = \int r_y^2 \, dm = \int_{-L/2}^{L/2} x^2 \, \rho_L \, dx = \frac{1}{12} \rho_L L^3 = \frac{1}{12}\left(\frac{m}{L}\right)L^3$$

By symmetry, we note that $I_z = I_y$. Thus, we have

$$I_x = 0 \qquad I_y = I_z = \frac{1}{12} mL^2 \quad \blacktriangleleft$$

Developmental Exercise

D7.38   Define the moment of inertia of a body about an axis.

EXAMPLE 7.15

A thin rectangular plate of mass m lies in the zx plane as shown. Determine I_z of the plate.

Solution. The mass density of the plate may be expressed as $\rho_A = m/(ca)$. Choosing dm to be a slender rod as shown, we write

$$I_z = \int r_z^2 \, dm = \int_{-a/2}^{a/2} x^2 \rho_A c \, dx = \frac{1}{12} \rho_A c a^3$$

$$I_z = \frac{1}{12} ma^2 \quad \blacktriangleleft$$

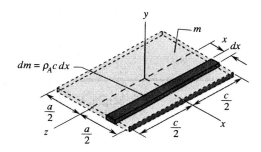

REMARK. For the thin plate in this example, we note that $y \approx 0$ throughout the plate. By Eq. (7.28), we write

$$I_x = \int z^2 \, dm \qquad I_y = \int (z^2 + x^2) \, dm \qquad I_z = \int x^2 \, dm$$

Thus, for such a plate, we have $I_y = I_z + I_x$.

Developmental Exercise

D7.39 Determine I_x of the plate in Example 7.15.

EXAMPLE 7.16

A thin circular disk of radius r and mass m lies in the xy plane as shown. Determine I_x, I_y, and I_z of the disk.

Solution. Since the thin circular disk lies in the xy plane with $z \approx 0$, we readily deduce from Eqs. (7.28) that

$$I_z = I_x + I_y$$

By symmetry, we note that $I_x = I_y$. Thus, we have

$$I_z = 2I_x \qquad I_x = I_y = \frac{1}{2} I_z$$

The mass density of the disk may be expressed as $\rho_A = m/(\pi r^2)$. Referring to the sketch, we write

$$I_z = \int r_z^2 \, dm = \int_0^r u^2 \rho_A(2\pi u) \, du = \frac{1}{2} \rho_A \pi r^4 = \frac{1}{2}\left(\frac{m}{\pi r^2}\right)\pi r^4$$

$$I_z = \frac{1}{2} mr^2 \qquad I_x = I_y = \frac{1}{4} mr^2 \blacktriangleleft$$

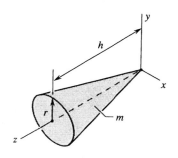

Fig. D7.40

Developmental Exercise

D7.40 A circular cylinder of total mass m and radius r is shown. Applying the results obtained in Example 7.16, determine I_z of the cylinder.

EXAMPLE 7.17

A circular cone having a mass m, a height h, and a base of radius r is shown. Using the solution for I_z in Example 7.16 as a formula, determine I_z of the cone.

Solution. Choosing dm to be a thin disk as shown in a different perspective for better clarity, we first determine the mass density ρ and dm in terms

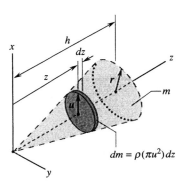

of z as follows:

$$u = \frac{r}{h} z \qquad dm = \rho(\pi u^2) \, dz = \frac{\rho \pi r^2}{h^2} z^2 \, dz$$

$$m = \int dm = \frac{\rho \pi r^2}{h^2} \int_0^h z^2 \, dz = \frac{\rho \pi r^2 h}{3} \qquad \rho = \frac{3m}{\pi r^2 h} \qquad dm = \frac{3m}{h^3} z^2 \, dz$$

Using the solution for I_z in Example 7.16 as a formula, we write

$$dI_z = \frac{1}{2} (dm) \, u^2 = \frac{3mr^2}{2h^5} z^4 \, dz$$

$$I_z = \int dI_z = \frac{3mr^2}{2h^5} \int_0^h z^4 \, dz = \frac{3mr^2}{2h^5}(\frac{1}{5} h^5) \qquad I_z = \frac{3}{10} mr^2 \quad \blacktriangleleft$$

Developmental Exercise

D7.41 For the circular cone in Example 7.17, is $I_z = I_x + I_y$?

EXAMPLE 7.18

A tetrahedron of mass density $\rho = 15$ slugs/ft^3 is bounded by the planes
$x + 2y + 4z = 8$, $x = 0$, $y = 0$, and $z = 0$, where the lengths are in
feet. Determine I_x of the tetrahedron.

Solution. Let the differential mass element dm of the tetrahedron be chosen as sketched. We have

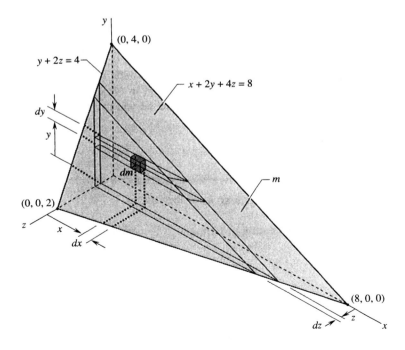

$$dm = \rho \, dV = \rho \, dx \, dy \, dz = 15 \, dx \, dy \, dz$$

$$I_x = \int (y^2 + z^2) \, dm$$

$$= 15 \int_{z=0}^{2} \int_{y=0}^{4-2z} \int_{x=0}^{8-2y-4z} (y^2 + z^2) \, dx \, dy \, dz$$

$$= 15 \int_{z=0}^{2} \int_{y=0}^{4-2z} (y^2 + z^2)(8 - 2y - 4z) \, dy \, dz$$

$$= 20 \int_{0}^{2} (5z^4 - 28z^3 + 60z^2 - 64z + 32) \, dz$$

$$I_x = 320 \text{ slug} \cdot \text{ft}^2 \quad \blacktriangleleft$$

Developmental Exercise

D7.42 Determine I_y of the tetrahedron in Example 7.18.

7.12 Radius of Gyration of a Mass

The *radius of gyration* of a body about an axis is a distance whose square multiplied by the mass of the body gives the moment of inertia of the body about that axis. For a body of mass m as shown in Fig. 7.25, we write

$$\boxed{I_x = k_x^2 m} \quad \boxed{I_y = k_y^2 m} \quad \boxed{I_z = k_z^2 m} \qquad (7.29)$$

where k_x, k_y, and k_z are the radii of gyration of the body about the x, y, and z axes, respectively.

The radius of gyration $k_{AA'}$ of a body of mass m about an axis AA' is illustrated in Fig. 7.26, where $I_{AA'}$ is the moment of inertia of the body about the axis AA'. Note that the moment of inertia of a body about an axis will remain unchanged if the entire mass of the body is concentrated at a point which is away from that axis at a *distance* equal to the *radius of gyration* of the body about that axis. Clearly, the dimension of the radius of gyration is *length*; and it is measured in meters (m) in the SI and in feet (ft) in the U.S. customary system.

Fig. 7.25

$$k_{AA'} = \sqrt{\frac{I_{AA'}}{m}}$$

Fig. 7.26

EXAMPLE 7.19

Refer to Example 7.14. Determine k_y of the rod.

Solution. From the solution in Example 7.14, we know that $I_y = \frac{1}{12} mL^2$. Thus, we write

$$\frac{1}{12} mL^2 = k_y^2 m$$

$$k_y = \frac{\sqrt{3}}{6} L = 0.289L \quad \blacktriangleleft$$

EXAMPLE 7.20

Determine k_z of the disk in Example 7.16.

Solution. From the solution in Example 7.16, we know that $I_z = \frac{1}{2} mr^2$. Thus, we write

$$\frac{1}{2} mr^2 = k_z^2 m$$

$$k_z = \frac{\sqrt{2}}{2} r = 0.707r \quad \blacktriangleleft$$

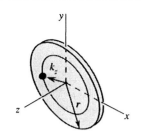

Developmental Exercises

D7.43 Define the radius of gyration of a body about an axis.

D7.44 Determine k_z of the cone in Example 7.17.

D7.45 The radius of gyration of a 50-kg flywheel about its axis of rotation A is $k_A = 0.4$ m. Determine I_A of the flywheel.

D7.46 Does it make sense for the radius of gyration of any wheel about its axle to be greater than its radius? (*Hint.* Cf. Example 7.20.)

PROBLEMS

NOTE. The following problems are to be solved *by integration*.

7.108* and **7.109*** Determine I_x and I_y of the slender rod of mass m as shown.

7.110 Determine I_x, I_y, and I_z of the thin plate of mass m as shown.

7.111* through **7.116*** Determine I_z and k_z of the body of mass m as shown.

Fig. P7.108*

Fig. P7.109*

Fig. P7.110

Fig. P7.111*

Fig. P7.112*

Hemispherical shell
Fig. P7.113*

Spherical shell
Fig. P7.114*

Hemisphere
Fig. P7.115*

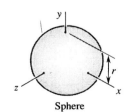

Sphere
Fig. P7.116*

7.117 A body of revolution and its generating area are shown. If the specific weight is $\gamma = 322$ lb/ft^3, determine I_x and k_x of the body.

7.118 An ellipsoid and its generating area are shown. If the mass density is $\rho = 6$ Mg/m^3, determine I_x and k_x of the ellipsoid.

Fig. P7.117

Fig. P7.118

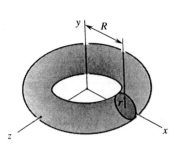

Fig. P7.119*

7.119* Determine I_y of the torus shown if its mass is m. (*Hint.* Choose dm to be a thin shell around the y axis and refer to Sec. B5.2 of App. B. if necessary.)

7.120 and 7.121* Determine I_x of the body shown if its mass is m.

Fig. P7.120 and P7.122

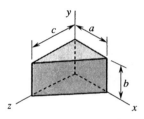

Fig. P7.121* and Fig. P7.123

7.122 and 7.123 Determine I_y and k_y of the body shown if its mass is m.

7.124 The mass density of the prism shown is $\rho = 4$ Mg/m^3. If $a = 0.4$ m, $b = 0.3$ m, and $c = 0.6$ m, determine I_x and k_x of the prism.

7.125* The mass density of the prism shown is $\rho = 5$ Mg/m^3. If $a = 80$ mm, $b = 60$ mm, and $c = 120$ mm, determine I_y and k_y of the prism.

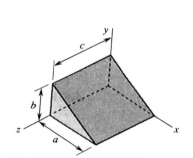

Fig. P7.124 and P7.125*

7.13 Parallel-Axis Theorem:
Mass Moments of Inertia

Let the point $G(\bar{x}, \bar{y}, \bar{z})$ be the center of mass of a body of mass m as shown in Fig. 7.27, where the *central axes* x', y', and z' of the body are parallel to the fixed coordinate axes x, y, and z, respectively.[†] The coordinates of

---

[†] The word *central* indicates something associated with a center (e.g., center of mass or center of gravity), while the word *centroidal* indicates something associated with the *centroid* of a geometric figure (e.g., line, area, or volume). Thus, *central axes* of a body are axes passing through the center of mass of the body, while *centroidal axes* of a body are axes passing through the centroid of the volume of the body. Note that the *center of mass* of a body does not necessarily coincide with the *centroid* of the volume of the body. If these two points coincide in a body, then the *centroidal axes* are also the *central axes* of the body.

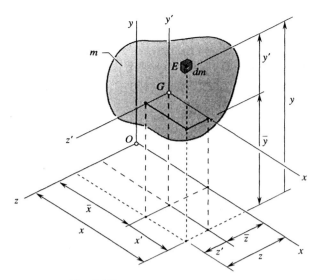

Fig. 7.27

the differential mass element dm at the point E are (x, y, z) in the $Oxyz$ fixed coordinate system but are (x', y', z') in the $Gx'y'z'$ central coordinate system as indicated. Thus, we write

$$x = \bar{x} + x' \qquad y = \bar{y} + y' \qquad z = \bar{z} + z' \qquad (7.30)$$

$$I_x = \int (y^2 + z^2)\, dm = \int [(\bar{y} + y')^2 + (\bar{z} + z')^2]\, dm$$

$$= \int (y'^2 + z'^2)\, dm + 2\bar{y} \int y'\, dm + 2\bar{z} \int z'\, dm + (\bar{y}^2 + \bar{z}^2) \int dm$$

$$= \bar{I}_{x'} + 2\bar{y}\bar{y}'m + 2\bar{z}\bar{z}'m + (\bar{y}^2 + \bar{z}^2)m$$

where $\bar{I}_{x'}$ represents the moment of inertia of the body about the central axis x', $\bar{y}'$ represents the distance between G and the $z'x'$ plane, and $\bar{z}'$ represents the distance between G and the $x'y'$ plane. Since G lies in the $z'x'$ and $x'y'$ planes, we see that $\bar{y}' = \bar{z}' = 0$. Thus, we write

$$I_x = \bar{I}_{x'} + m(\bar{y}^2 + \bar{z}^2) \qquad (7.31)$$

and similarly,

$$I_y = \bar{I}_{y'} + m(\bar{z}^2 + \bar{x}^2) \qquad I_z = \bar{I}_{z'} + m(\bar{x}^2 + \bar{y}^2) \qquad (7.32)$$

From Fig. 7.27, we see that the sum $\bar{y}^2 + \bar{z}^2$ in Eq. (7.31) represents the square of the distance between the x and x' axes. Similarly, the sums $\bar{z}^2 + \bar{x}^2$ and $\bar{x}^2 + \bar{y}^2$ represent the squares of the distances between the y and y' axes, and the z and z' axes, respectively.

The formulas in Eqs. (7.31) and (7.32) are called the *parallel-axis theorem* for mass moments of inertia, which may be stated as follows: *The moment of inertia I of a body of mass m about any given axis is equal to the moment of inertia $\bar{I}$ of the body about a central axis parallel to the given axis plus the product md^2, where d is the distance between those two axes, as shown in Fig. 7.28; i.e.,*

$$I = \bar{I} + md^2 \qquad (7.33)$$

Fig. 7.28

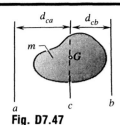

Fig. D7.47

Developmental Exercise

D7.47   If a, b, and c are three parallel axes and the axis c passes through the center of mass G of a body of mass m as shown, show that

$$I_a = I_b + m(d_{ca}^2 - d_{cb}^2)$$

EXAMPLE 7.21

Determine I_x of the circular cone in Example 7.17.

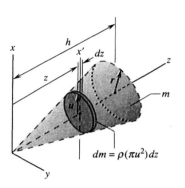

Solution.   Let dm be a thin disk and x' be a central axis of dm as shown, where the radius of the disk is $u = rz/h$ and its mass density is $\rho = 3m/(\pi r^2 h)$. Applying the *parallel-axis theorem* and the solution for I_x in Example 7.16 as a formula, we write

$$dI_x = d\bar{I}_{x'} + (dm)\, z^2 = \frac{1}{4}\,(dm)\, u^2 + (dm)\, z^2$$

$$= \left(\frac{1}{4}\, u^2 + z^2\right)(\rho\pi u^2\, dz) = \frac{\rho\pi r^2}{4h^4}\,(r^2 + 4h^2)z^4\, dz$$

$$I_x = \int dI_x = \frac{\rho\pi r^2}{4h^4}(r^2 + 4h^2)\int_0^h z^4\, dz = \frac{\rho\pi r^2 h}{20}(r^2 + 4h^2)$$

$$= \frac{3m}{\pi r^2 h}\frac{\pi r^2 h}{20}(r^2 + 4h^2)$$

$$I_x = \frac{3}{20}\, m(r^2 + 4h^2) \quad \blacktriangleleft$$

Fig. D7.49

Developmental Exercises

D7.48   Describe the parallel-axis theorem for mass moments of inertia.

D7.49   The circular cone shown has a mass m. Determine I_x of the cone.

7.14   Composite Bodies

A composite body has several component bodies. The moment of inertia I of a composite body about an axis is equal to the sum of the moments of inertia $I_1, I_2, \ldots, I_n$ of its component bodies about that axis; i.e.,

$$I = I_1 + I_2 + \cdots + I_n \tag{7.34}$$

For use in determining the moments of inertia of composite bodies, the moments of inertia of homogeneous bodies of some common shapes are given in Fig. 7.29, where the plates, disks, rings, and shells are understood to be thin, and m is the total mass of the body shown.

Slender rod

$$I_x = \frac{1}{3}mL^2\sin^2\theta \qquad I_y = \frac{1}{3}mL^2\cos^2\theta$$

$$I_z = \frac{1}{3}mL^2 \qquad \bar{I}_{x'} = \frac{1}{12}mL^2\sin^2\theta$$

$$\bar{I}_{y'} = \frac{1}{12}mL^2\cos^2\theta \qquad \bar{I}_{z'} = \frac{1}{12}mL^2$$

Circular ring

$$I_x = I_z = \frac{1}{2}mr^2$$

$$I_y = mr^2$$

Rectangular parallelepiped

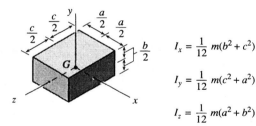

$$I_x = \frac{1}{12}m(b^2 + c^2)$$

$$I_y = \frac{1}{12}m(c^2 + a^2)$$

$$I_z = \frac{1}{12}m(a^2 + b^2)$$

Semicircular ring

$$\bar{z} = \frac{2r}{\pi} \qquad I_x = I_z = \frac{1}{2}mr^2 \qquad I_y = mr^2$$

$$\bar{I}_{x'} = \frac{1}{2\pi^2}mr^2(\pi^2 - 8)$$

$$\bar{I}_{y'} = \frac{1}{\pi^2}mr^2(\pi^2 - 4)$$

Rectangular plate

$$I_x = \frac{1}{12}mc^2 \qquad I_y = \frac{1}{12}m(c^2 + a^2) \qquad I_z = \frac{1}{12}ma^2$$

Circular disk

$$I_y = \frac{1}{2}mr^2$$

$$I_x = I_z = \frac{1}{4}mr^2$$

Circular cylinder

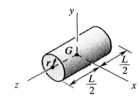

$$I_z = \frac{1}{2}mr^2$$

$$I_x = I_y = \frac{1}{12}m(3r^2 + L^2)$$

Semicircular disk

$$\bar{z} = \frac{4r}{3\pi}$$

$$I_y = \frac{1}{2}mr^2$$

$$I_x = I_z = \frac{1}{4}mr^2$$

$$\bar{I}_{x'} = \frac{1}{36\pi^2}mr^2(9\pi^2 - 64)$$

$$\bar{I}_{y'} = \frac{1}{18\pi^2}mr^2(9\pi^2 - 32)$$

Fig. 7.29 (a) Moments of inertia of bodies of common shapes.

Circular conic shell

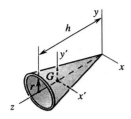

$$\bar{z} = \frac{2}{3} h$$

$$I_z = \frac{1}{2} mr^2$$

$$I_x = I_y = \frac{1}{4} m(r^2 + 2h^2)$$

$$\bar{I}_{x'} = \bar{I}_{y'} = \frac{1}{36} m(9r^2 + 2h^2)$$

Circular cone

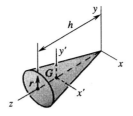

$$\bar{z} = \frac{3}{4} h$$

$$I_z = \frac{3}{10} mr^2$$

$$I_x = I_y = \frac{3}{20} m(r^2 + 4h^2)$$

$$\bar{I}_{x'} = \bar{I}_{y'} = \frac{3}{80} m(4r^2 + h^2)$$

Circular cylindrical shell

$$I_z = mr^2$$

$$I_x = I_y = \frac{1}{12} m(6r^2 + L^2)$$

Sphere

$$I_x = I_y = I_z = \frac{2}{5} mr^2$$

Triangular plate

$$\bar{I}_{x'} = \frac{1}{72} mh^2$$

$$I_x = \frac{1}{8} mh^2$$

Hemispherical shell

$$\bar{y} = -\frac{1}{2} r$$

$$I_x = I_y = I_z = \frac{2}{3} mr^2$$

$$\bar{I}_{x'} = \bar{I}_{z'} = \frac{5}{12} mr^2$$

Hemisphere

$$\bar{y} = -\frac{3}{8} r$$

$$I_x = I_y = I_z = \frac{2}{5} mr^2$$

$$\bar{I}_{x'} = \bar{I}_{z'} = \frac{83}{320} mr^2$$

Spherical shell

$$I_x = I_y = I_z = \frac{2}{3} mr^2$$

Half torus

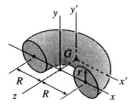

$$\bar{z} = -\frac{r^2 + 4R^2}{2\pi R}$$

$$I_x = I_z = \frac{1}{8} m(5r^2 + 4R^2)$$

$$I_y = \frac{1}{4} m(3r^2 + 4R^2)$$

Torus

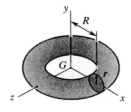

$$I_x = I_z = \frac{1}{8} m(5r^2 + 4R^2)$$

$$I_y = \frac{1}{4} m(3r^2 + 4R^2)$$

Fig. 7.29 (b) Moments of inertia of bodies of common shapes.

EXAMPLE 7.22

The mass density of the slender bent rod shown is $\rho_L = 3.75$ kg/m. Determine I_y and k_y of the rod.

Solution. The rod may be treated as a composite body consisting of the segments AB, BC, and CD as the component bodies. Thus, we write

$$I_y = (I_y)_{AB} + (I_y)_{BC} + (I_y)_{CD}$$

Since the segment AB lies along the y axis, we have $(I_y)_{AB} = 0$. To find $(I_y)_{BC}$ and $(I_y)_{CD}$ of the segments BC and CD, we first establish their central axes y' and y'' which are parallel to the y axis. Then, we apply the *parallel-axis theorem* and the formulas for a slender rod to write

$$(I_y)_{BC} = (\bar{I}_{y'})_{BC} + m_{BC}\,\overline{BE}^2 = \frac{1}{12} m_{BC}\,\overline{BC}^2 + m_{BC}\,\overline{BE}^2$$

$$= \frac{1}{12}\Big[3.75(1.2)\Big](1.2)^2 + \Big[3.75(1.2)\Big](0.6)^2 = 2.16$$

$$(I_y)_{CD} = (\bar{I}_{y''})_{CD} + m_{CD}\,\overline{BF}^2 = \frac{1}{12} m_{CD}\,\overline{CD}^2 + m_{CD}(\overline{BC}^2 + \overline{CF}^2)$$

$$= \frac{1}{12}\Big[3.75(1)\Big](1)^2 + \Big[3.75(1)\Big]\Big[(1.2)^2 + (0.5)^2\Big] = 6.65$$

$$I_y = 0 + 2.16 + 6.65 \qquad\qquad I_y = 8.81 \text{ kg}\cdot\text{m}^2 \blacktriangleleft$$

$$I_y = k_y^2 m: \quad 8.81 = k_y^2(3.75)(0.8 + 1.2 + 1)$$

$$k_y = 0.885 \text{ m} \blacktriangleleft$$

Developmental Exercises

D7.50 Determine I_x and k_x of the bent rod in Example 7.22.

D7.51 Determine I_z and k_z of the bent rod in Example 7.22.

EXAMPLE 7.23

A thin machine part of mass density $\rho_A = 4.83$ lbm/ft^2 is shown, where $a = 6$ in. Determine I_y of the machine part.

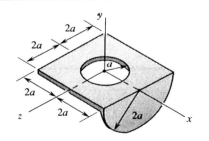

Solution. Let the machine part be a composite body consisting of a square plate of mass m_1, a semicircular plate of mass m_2, and an *empty* circular plate of mass m_3. Then, we have

$$m_1 = \rho_A(4a)^2 = 16\rho_A a^2$$

$$m_2 = \rho_A\Big[\frac{1}{2}\pi(2a)^2\Big] = 2\pi\rho_A a^2 \qquad m_3 = \rho_A(\pi a^2) = \pi\rho_A a^2$$

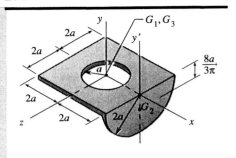

where $\rho_A = 0.15$ slug/ft^2 and $a = 0.5$ ft. Applying the *parallel-axis theorem* and appropriate formulas in Fig. 7.29, we write

$$I_y = (I_y)_1 + (I_y)_2 - (I_y)_3 = (I_y)_1 + \left[(\bar{I}_{y'})_2 + m_2 d_2^2\right] - (I_y)_3$$

$$= \frac{1}{12} m_1 \left[(4a)^2 + (4a)^2\right] + \left[\frac{1}{4} m_2 (2a)^2 + m_2 (2a)^2\right] - \frac{1}{2} m_3 a^2$$

$$= \frac{1}{6}(256 + 57\pi)\rho_A a^4 \qquad\qquad I_y = 0.680 \text{ slug} \cdot \text{ft}^2 \blacktriangleleft$$

Developmental Exercises

D7.52 Determine I_x and k_x of the machine part in Example 7.23.

D7.53 Determine I_z and k_z of the machine part in Example 7.23.

EXAMPLE 7.24

A body consists of a rectangular parallelepiped and a circular cylinder as shown, where $a = 20$ mm and the mass density is $\rho = 6$ Mg/m^3. Determine I_z and k_z of the body.

Solution. Let the component bodies be the rectangular parallelepiped of mass m_1 and the circular cylinder of mass m_2. Then, we have

$$m_1 = \rho(4a)(6a)(2a) = 48\rho a^3$$

$$m_2 = \rho(\pi a^2)(4a) = 4\pi\rho a^3$$

$$m = m_1 + m_2 = (48 + 4\pi)\rho a^3$$

where $\rho = 6000$ kg/m^3 and $a = 0.02$ m. Applying the *parallel-axis theorem* and appropriate formulas in Fig. 7.29, we write

$$I_z = (I_z)_1 + (I_z)_2 = \left[(\bar{I}_{z'})_1 + m_1 d_1^2\right] + (I_z)_2$$

$$= \left\{\frac{1}{12} m_1 \left[(4a)^2 + (6a)^2\right] + m_1 (2a)^2\right\} + \frac{1}{2} m_2 a^2$$

$$= (400 + 2\pi)\rho a^5 = (400 + 2\pi)(6000)(0.02)^5$$

$$I_z = 7.80 \times 10^{-3} \text{ kg} \cdot \text{m}^2 \blacktriangleleft$$

$$I_z = k_z^2 m: \quad (400 + 2\pi)\rho a^5 = k_z^2 (48 + 4\pi)\rho a^3$$

$$k_z = 51.8 \text{ mm} \blacktriangleleft$$

Developmental Exercises

D7.54 Determine I_x and k_x of the body in Example 7.24.

D7.55 Determine I_y and k_y of the body in Example 7.24.

PROBLEMS

NOTE. Unless otherwise specified, the formulas given in Fig. 7.29 as well as the parallel-axis theorem may be applied in solving the following problems of mass moments of inertia.

7.126 through 7.128 Determine I_x and I_y of the body made of slender rods as shown.

Fig. P7.126 and P7.129*

Fig. P7.127* and P7.130

Fig. P7.128 and P7.131

7.129* through 7.131 Determine I_z and k_z of the body made of slender rods as shown.

7.132 through 7.134 Determine I_x and I_y of the sheet metal part shown.

Fig. P7.132 and P7.135*

Fig. P7.133* and P7.136

Fig. P7.134 and P7.137

7.135* through 7.137 Determine I_z and k_z of the sheet metal part shown.

7.138 through 7.140 Determine I_y and k_y of the thin shell shown.

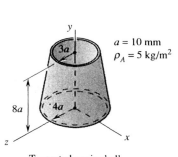

Truncated conic shell
Fig. P7.138 and P7.141*

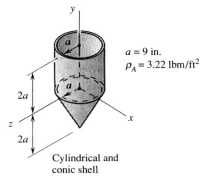

Cylindrical and conic shell
Fig. P7.139* and P7.142

Cylindrical and hemispherical shell
Fig. P7.140 and P7.143

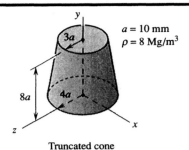

a = 10 mm
ρ = 8 Mg/m³

Truncated cone
Fig. P7.144 and P7.147*

7.141* through **7.143**   Determine I_x and k_x of the thin shell shown.

7.144 through **7.146**   Determine I_y and k_y of the body of revolution shown.

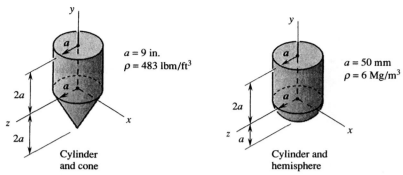

a = 9 in.
ρ = 483 lbm/ft³

Cylinder
and cone
Fig. P7.145* and P7.148

a = 50 mm
ρ = 6 Mg/m³

Cylinder and
hemisphere
Fig. P7.146 and P7.149

7.147* through **7.149**   Determine I_x and k_x of the body of revolution shown.

7.150   A horizontal hole of radius 1 in. is drilled through a body as shown. If $\rho = 14$ slugs/ft³, determine I_z and k_z of the body.

7.151   A flywheel is of the same shape and size as that obtained by revolving the area shown about the x axis one revolution, where $r_1 = 250$ mm, $r_2 = 200$ mm, $r_3 = 50$ mm, $r_4 = 25$ mm, $b_1 = 150$ mm, and $b_2 = 25$ mm. If $\rho = 7$ Mg/m³, determine I_x and k_x of the flywheel.

Fig. P7.150

Fig. P7.151

4a

4a

a = 20 mm

Fig. P7.152

7.152   A link consists of a cylinder and a half torus as shown. It is made of an alloy with $\rho = 6$ Mg/m³. Determine I_y and I_z of the link.

7.153*   Using single integration, verify the formula $I_x = \frac{1}{12}m(3r^2 + L^2)$ for the circular cylinder in Fig. 7.29.

7.154*   Using single integration, verify the formula $I_x = \frac{1}{12}m(6r^2 + L^2)$ for the circular cylindrical shell in Fig. 7.29.

7.155*   Using single integration, verify the formula $I_x = \frac{1}{8}m(5r^2 + 4R^2)$ for the torus in Fig. 7.29. (*Hint.* Refer to the *hint* in Prob. 7.119*.)

7.15   Concluding Remarks

The *moment of the first moment* of a quantity is called the *second moment* of the quantity. The *moment of inertia of an area* about a given axis may be thought of as the integral sum of the second moments of its differential areas about the given axis. Similarly, the *moment of inertia of a mass* about a given axis may be thought of as the integral sum of the second moments of its differential masses about the given axis.

The moments of inertia of an area are primarily the mathematical properties of the area with respect to the given axes. The prime motivation to study the moments of inertia of areas comes from the fact that such quantities have many practical applications in studying certain situations in engineering, such as those mentioned in Sec. 7.1. The necessity to find the principal moments and principal axes of inertia of an area associated with a certain point (e.g., the centroid of the area) in solving some problems (e.g., the buckling of columns) leads naturally to the introduction and study of the products of inertia of areas. Additionally, one will find in fluid mechanics that both the moments and the product of inertia of an area are used in expressing the coordinates of the *center of pressure* if neither of the coordinate axes is an axis of symmetry of a nonhorizontal submerged plane surface.

As a graphical scheme, Mohr's circle is shown to provide a convenient means for studies involving the rotation of axes for moments and products of inertia. Furthermore, it is a good geometric aid in determining the principal moments and principal axes of inertia.

The second part of this chapter presents the moments of inertia of a mass, which is mathematically similar to the moments of inertia of an area presented in the first part of the chapter. This material is intended to provide students with the prerequisite background for the study of plane kinetics of rigid bodies in dynamics. However, at the instructor's discretion, this material may either be covered along with area moments of inertia or be skipped in a course in statics. For convenience of those who prefer to postpone covering mass moments of inertia until the study of dynamics, this material is repeated in the first part of Chap. 16, just before it is needed in the study of plane kinetics of rigid bodies. The concepts of products and principal moments of inertia of a mass are needed in the study of kinetics of rigid bodies in three dimensions. Therefore, they are timely covered in Chap. 19.

REVIEW PROBLEMS

7.156 For the area and axes shown, determine the centroidal moments of inertia $\bar{I}_x$ and $\bar{I}_y$, and the centroidal radii of gyration $\bar{k}_x$ and $\bar{k}_y$.

7.157 and 7.158* For the area and axes shown, determine the centroidal moments of inertia $\bar{I}_x$, $\bar{I}_y$, $\bar{I}_u$, and $\bar{I}_v$, and the centroidal products of inertia $\bar{I}_{xy}$ and $\bar{I}_{uv}$.

Fig. P7.156

Fig. P7.157 and P7.159

Fig. P7.158* and P7.160*

7.159 and 7.160* For the area shown, determine (a) the centroidal principal moments of inertia $\bar{I}_b$ and $\bar{I}_s$, (b) the angles β and ψ locating the b and s axes.

7.161 For the area shown, determine the moments of inertia I_x and I_y, and the product of inertia I_{xy}. (*Hint*. Cf. the footnote to Prob. 7.67.)

Fig. P7.161

Dimensions in millimeters

Fig. P7.162

7.162 A slender steel rod with one *fixed end* and one free end has a length of 1 m and a cross-sectional area A as shown. The magnitude of the critical force $\mathbf{P}_{cr}$ for the buckling of this cantilever rod is given by the formula

$$P_{cr} = \frac{\pi^2 E\bar{I}_{min}}{4L^2}$$

where L is the length of the rod, E is the elastic modulus of the rod and is equal to 200 GPa, and $\bar{I}_{min}$ is the minimum moment of inertia of A about an axis through its centroid C as shown. Determine the value of P_{cr}.

7.163* For a given area in the xy plane, it is known that $I_y = 30 \times 10^6$ mm^4, $I_{xy} = 15 \times 10^6$ mm^4, and $I_{min} = 5 \times 10^6$ mm^4. Using Mohr's circle, determine (a) I_x, (b) I_{max}, (c) the angles β and ψ locating the principal axes b and s, respectively. Indicate the b and s axes on the xy plane.

7.164* Using Eqs. (7.19), prove the invariance property that

$$I_u I_v - I_{uv}^2 = I_x I_y - I_{xy}^2$$

7.165* Using Eqs. (7.19), (7.23), and (7.24), prove the invariance property that

$$I_{max} I_{min} = I_x I_y - I_{xy}^2 = I_u I_v - I_{uv}^2$$

7.166 A slender rod weighing 1.61 lb/ft is used to form a composite body as shown. Determine the moment of inertia I_x and the radius of gyration k_x of the body.

7.167* Refer to Prob. 7.166. Determine I_y and k_y of the body.

Fig. P7.166

7.168 A sheet of metal is cut and bent into a machine component as shown. If its mass density is $\rho_A = 16 \text{ kg/m}^2$, determine I_x and k_x of the component.

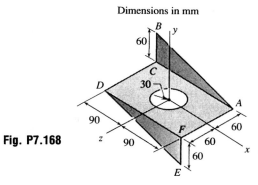

Dimensions in mm

Fig. P7.168

7.169 Refer to Prob. 7.168. Determine I_y and k_y of the component.

7.170* The mass density of the steel machine element is $\rho = 7.85 \text{ Mg/m}^3$. Determine I_x and k_x of the element.

7.171 Refer to Prob. 7.170*. Determine I_y and k_y of the element.

Dimensions in mm

Fig. P7.170*

Chapter 8

Structures

KEY CONCEPTS

- Truss, rigidly connected truss, pin-connected truss, idealized truss, simple truss, compound truss, complex truss, frame, machine, beam, and cable.
- Methods of joints and sections for analyzing plane trusses.
- Method of joints for analyzing space trusses.
- Equilibrium of frames and machines in a plane.
- Shear and bending-moment diagrams for beams.
- Parabolic and catenary cables.

A structure is a body, or an assemblage of component parts, which withstands applied loads. When the component parts are all straight two-force bodies, the assemblage is called an *idealized truss*. When one or more of the component parts are multiforce members, the assemblage is called a *frame* if all of the members are immovable, and a *machine* if some or all of its members are movable. The concepts developed previously in studying the equilibrium of particles and rigid bodies are adapted, in this chapter, to the study of the conditions of equilibrium of trusses, frames, machines, beams, and cables. In particular, useful methods are presented for the study of plane and space trusses, and a general procedure is given for solving equilibrium problems of frames and machines acted on by coplanar but nonconcurrent forces. The latter part of the chapter is devoted to the study of shears and bending moments in beams as well as to the analysis of cables. Since trusses, frames, machines, beams, and cables are basic structures, their analysis is fundamental and important.

TRUSSES

8.1   Trusses and Their Uses

A *joint* of a structure is a connection where two or more members of the structure are fastened together. A *truss* is a structure composed of a number of straight members connected (e.g., pinned, nailed, riveted, bolted, or welded) together at their ends by the joints of the structure where the loads are directly applied to the centers of the various joints, not to the portions of the members between the joints. Note that a joint may be a pin plus the

end portions of the members being connected by the pin. The pin by itself
is not a joint, although it is the center of the joint.

In modern engineering practice, wooden truss members are often nailed
or bolted together at the joints and metal truss members are often riveted,
bolted, or welded to gusset plates at the joints to achieve a greater degree
of rigidity for the structure as illustrated in Fig. 8.1(a). A truss whose members cannot freely rotate at the joints is called a *rigidly connected truss*. In
elementary studies, truss members are, however, taken as being pin-
connected at the joints as illustrated in Fig. 8.1(b). A truss that uses a
smooth pin at each joint to connect the truss members is called a *pin-
connected truss*.

(a)   (b)

Fig. 8.1

Trusses are frequently used in roofs, bridges, stadiums, towers, cranes,
etc. Some illustrations of trusses are shown in Fig. 8.2. Note in many draw-
ings that a solid line is used to denote a member and a small circle is used
to denote a joint. If a small circle is absent at the intersection of two solid
lines, there is no connection of the members at that intersection.

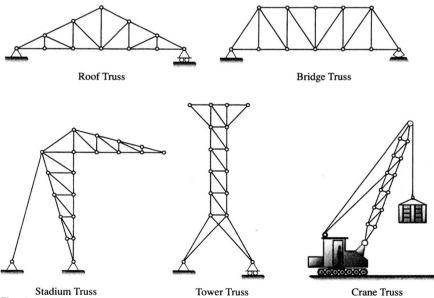

Roof Truss   Bridge Truss

Stadium Truss   Tower Truss   Crane Truss

Fig. 8.2

Developmental Exercises

D8.1 Define the terms: (a) joint, (b) truss, (c) rigidly connected truss, (d) pin-connected truss.

D8.2 Give three examples where trusses are used.

8.2 Usual Assumptions for Trusses

The *usual assumptions for trusses* in elementary studies are as follows:

1. Every member of a truss is straight and is connected at each end to a joint by a smooth circular pin on the longitudinal centroidal axis of the member. (No member is continuous through a joint, and deformations of members due to induced internal strains are taken as negligible.)

2. The weights of members of a truss are taken as being negligible compared with the applied loads. (If the weights of truss members were to be taken into account, they would be assumed to be applied to the joints, half of the weight of the member being applied to each of the two joints to which the member is connected.)

3. All loads are applied or transmitted to the joints only. (This assumption is usually met by providing a floor system to transmit loads between the joints. The manner in which floor beams and stringers are arranged to transmit loads to the joints of an actual bridge truss is illustrated in Fig. 8.3.)

Fig. 8.3

A truss satisfying the above usual assumptions is called an *idealized truss*. Clearly, *members of an idealized truss are all two-force bodies*. It is to be pointed out that results of studies based on the idealized version of the rigidly connected truss are relatively easy to obtain and are useful for practical purposes. In this text, all trusses are to be taken as idealized trusses.

Developmental Exercises

D8.3 What are the usual assumptions for trusses?

D8.4 What is an idealized truss?

8.3   Rigid Truss, Plane Truss, and Simple Truss

A truss is said to have collapsed if its original configuration has undergone a great change due to loading. A *rigid truss* is a truss which does not collapse under the action of loads. The trusses shown in Fig. 8.4(a) and (b) are nonrigid. They may be made rigid by providing them with one more member and one more component of external constraint, respectively, as shown in Fig. 8.4(c) and (d).

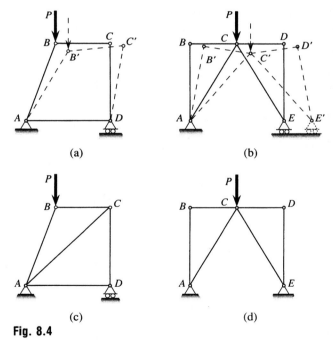

Fig. 8.4

The most basic two-dimensional rigid truss is a *triangular truss* which has three members and three joints as shown in Fig. 8.5(a). A *plane truss* is a two-dimensional truss; its members and the applied forces all lie in a plane. A *simple truss*[†] is a rigid plane truss which can be built from a triangular truss by repeating appropriately the addition of *two new members*

[†]The term *simple truss* denotes a plane truss. A *simple space truss* is defined in Sec. 8.10.

and *one new joint* at a time. This is sequentially illustrated in parts (a), (b), and (c) in Fig. 8.5.

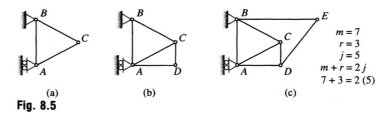

(a) (b) (c)

$$m = 7$$
$$r = 3$$
$$j = 5$$
$$m + r = 2j$$
$$7 + 3 = 2 (5)$$

Fig. 8.5

Note that in building a simple truss the two new members are fastened to two separate existing joints, and the new joint connecting these two new members does not lie on the line containing those two existing joints. Suppose that m is the number of members and j is the number of joints in a simple plane truss. Then, the total number of new members and the total number of new joints, which are added to the original triangular truss, are $m - 3$ and $j - 3$, respectively. By the definition of a *simple truss*, we must, therefore, have $m - 3 = 2(j - 3)$, or

$$m + 3 = 2j \qquad (8.1)$$

In associating Eq. (8.1) with a simple truss, it is necessary to define a *truss member* as a straight member which has *each of its ends connected to a joint* (which joins two or more members), rather than just a *hinge*. If one end of a member is connected to a joint but its other end is connected to a hinge, then that member is to be regarded as a *link* support, not as a truss member, of a simple truss. For example, the member AB in Fig. 8.6 needs to be regarded as a *link* support of the truss, not as a truss member, if the truss is to be classified as a simple truss; otherwise, Eq. (8.1) will not be satisfied.

Fig. 8.6

Since truss members are two-force bodies, the forces in the free-body diagram of each joint must form a system of concurrent forces. The analysis of the equilibrium of *each joint* is, therefore, similar to the analysis of the equilibrium of a *particle*. For equilibrium of a truss lying in the xy plane, there are *two* scalar equations of equilibrium to be satisfied by the forces in the free-body diagram of *each* joint. They are

$$\Sigma F_x = 0 \qquad \Sigma F_y = 0 \qquad \begin{array}{l} (3.4)^\dagger \\ \text{(repeated)} \end{array}$$

This means that we may write $2j$ equilibrium equations from the j free-body diagrams of the j joints of a plane truss.

Suppose that a plane truss has j joints, m unknown forces in the m members, and r unknown components of reactions from the supports. The equilibrium of all of the j joints of the truss implies and ensures that the entire truss as a rigid body is also in equilibrium. We readily see that when the $(m + r)$ unknowns satisfy those $2j$ equilibrium equations of the joints, they will also automatically satisfy the equations of equilibrium of the entire truss

---

†Note that the x and y axes may be chosen to be any pair of orthogonal axes having any desirable orientations. Furthermore, either of these two force equilibrium equations may be replaced with a moment equilibrium equation about any point other than the pin in the joint. (Cf. Example 8.1.)

or any portion of the truss. Thus, the *necessary condition* for a plane truss to be statically determinate is

$$m + r = 2j \tag{8.2}$$

This equation reduces to Eq. (8.1) when $r = 3$ (e.g., a truss with a hinge support and a roller support).

Developmental Exercises

D8.5 Define (a) a rigid truss, (b) a plane truss, (c) a simple truss.

D8.6 A simple truss has 15 members. How many joints does it have?

D8.7 A statically determinate plane truss has 5 joints and 4 components of reaction from its supports. How many members does it have?

8.4 Compound and Complex Trusses

A *compound truss* is a rigid plane truss which consists of two or more simple trusses directly connected together by linking joints, or linking members, or both, but cannot be built with the same procedure as that used in building a simple truss. Some illustrations of compound trusses are shown in Fig. 8.7, where component simple trusses are shaded and *LJ* denotes a linking joint and *LM* denotes a linking member.

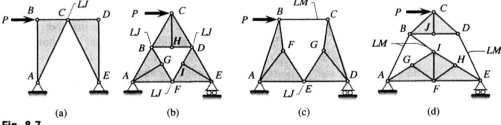

Fig. 8.7 (a) (b) (c) (d)

A *complex truss* is a rigid plane truss which cannot be classified as a simple truss or a compound truss. Some complex trusses are illustrated in Fig. 8.8. Note that there is no joint at the middle of each truss shown in

Fig. 8.8 (a) (b)

Fig. D8.10

this figure. The compound and complex trusses shown in Figs. 8.7 and 8.8 are statically determinate and can be shown to satisfy Eq. (8.2). We recall that the member AB in Fig. 8.6 needs to be regarded as a *link* support if the truss shown is to be classified as a *simple truss*. On the other hand, we can now regard the member AB as a truss member by classifying that truss as a *complex truss*, which satisfies Eq. (8.2) and is statically determinate.

Developmental Exercises

D8.8 Do the two trusses shown in Fig. 8.8 contain any simple trusses?

D8.9 Define (a) a compound truss, (b) a complex truss.

D8.10 The truss shown contains simple trusses. (a) Name those simple trusses. (b) Name the linking members connecting those simple trusses. (c) Is the truss shown a compound or a complex truss?

8.5 Forces in Truss Members

Since all loads are applied or transmitted to a truss at its joints only, any force induced in a truss member must come from the two pins at the two joints to which the member is connected. Thus, *each truss member is a two-force body*, as pointed out earlier. The *orientation* of the force in each truss member is the same as the axis of that member; only its *magnitude* and its *sense* need to be specified or determined. Therefore, the force in a truss member is also called an *axial force*.

If the axial force in a truss member tends to *stretch* the member or *pull* the end part of the member in an isolated joint away from its connecting pin, such as the force $\mathbf{F}_{AE}$ in Fig. 8.9, the axial force is a *tensile force* and the member is said to be in tension. A member that is in tension is called a *tensile member*. On the other hand, if the axial force in a truss member tends to *compress* the member or *push* the end part of the member in an isolated joint toward its connecting pin, such as the force $\mathbf{F}_{AB}$ or $\mathbf{F}_{AD}$ in Fig. 8.9, the axial force is a *compressive force* and the member is said to be in compression. A member that is in compression is called a *compressive member*.

In the analysis of trusses, each unknown axial force may initially be assumed to be either a tensile or a compressive force. If the value of the magnitude of the unknown axial force is algebraically found to be positive in the subsequent solution, the assumption is correct; if negative, we first drop the negative sign in the obtained value for the *magnitude* and then reverse the assumed *sense* of the axial force (i.e., change from tensile to compressive or vice versa) in the final assertion of the axial force.

The design criteria for tensile and compressive members of a truss are different. Therefore, each nonzero answer for the axial force in the solution of each truss problem should be labeled with a T when it is a tensile force and with a C when it is a compressive force. Note that the names of the joints to which a truss member is connected are used as the subscripts of the letter F to *label the magnitude of the axial force* in the truss member. The order of the subscripts is immaterial; e.g., $F_{AB} = F_{BA} =$ magnitude of axial force in the member which is connected to the joints A and B.

T = Tension

C = Compression

Fig. 8.9

Developmental Exercises

D8.11   Define the terms: (a) tensile force, (b) tensile member, (c) compressive force, (d) compressive member.

D8.12   How is the magnitude of the axial force in the member CD of a truss labeled in this chapter?

D8.13   Is $F_{GH} = F_{HG}$?

8.6   Zero-Force Members at T and V Joints

A *T joint* is a joint which connects three members where only two of them are collinear as shown in Fig. 8.10. *If no external load is applied to a T joint which is in equilibrium, the force in the noncollinear member must be zero.* This must be true; otherwise, the summation of the components of forces acting on the free body of the T joint in the direction perpendicular to the axis along the two collinear members cannot be zero. For example, if $F_{AD} \neq 0$ in Fig. 8.10, then $\Sigma F_y = -F_{AD} \sin\theta \neq 0$ because $\sin\theta \neq 0$. Since $\Sigma F_y = 0$ for equilibrium, we see that we must have $F_{AD} = 0$.

Fig. 8.10   **Fig. 8.11**

A *V joint* is a joint which connects two noncollinear members as shown in Fig. 8.11. *If no external load is applied to a V joint which is in equilibrium, the forces in those two members must be zero.* This is illustrated in Fig. 8.11(a) and must be true; otherwise, the summation of the components of forces acting on the free body of the V joint in the direction perpendicular to the axis of one of the two members, say ΣF_y, cannot be zero. Furthermore, when a V joint is acted on by a single force which is collinear with one of its two members, the force in the other member must be zero if equilibrium exists. This is illustrated in Fig. 8.11(b) and can be verified in a similar manner.

Developmental Exercises

D8.14   For equilibrium, verify that $F_{AC} = F_{AB}$ in Fig. 8.10 and $F_{AC} = P$ in Fig. 8.11(b).

D8.15   Which of the members in the truss loaded as shown are zero-force members?

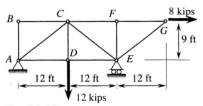

Fig. D8.15

8.7 Method of Joints for Plane Trusses

The equilibrium of each joint of a plane truss is similar to the equilibrium of a particle in a plane because the force system acting at each joint is either zero or coplanar and concurrent. The method used to determine the unknown axial forces or other parameters of a truss by considering the equilibrium of the individual joints, one at a time, is called the *method of joints*. This is one of the two basic methods for analyzing plane trusses; the other method is presented in Sec. 8.8.

According to Eq. (8.2), the *m* unknown axial forces and the *r* unknown components of reactions from the supports of a statically determinate plane truss with *j* joints may be determined by solving a set of 2*j* simultaneous equations of equilibrium written for those *j* joints. However, it is generally preferable to avoid solving a large set of simultaneous equations if possible. We see from Eqs. (3.4) that, for a plane truss, no more than two unknowns can be solved from the consideration of equilibrium of just one joint. Thus, whenever possible, we choose for each step of the solution an appropriate joint whose free-body diagram contains *no more than two unknowns*. This process is continued until all desired unknowns have been determined and checked. To facilitate such a process, we *sometimes* (but not always) need to determine beforehand the components of reactions from the supports of the truss by treating the entire truss as a rigid body in equilibrium.

We know that *when all joints of a truss are in equilibrium, the truss as a whole must be in equilibrium*. Therefore, redundant (or dependent) equations will be present in the set of equilibrium equations which are written for each joint of a truss *and* for the entire truss. However, the redundant equations may serve to check the solutions for the unknowns.

When the method of joints is applied to analyze a compound or complex truss, we may *sometimes* need to solve a set of more than two simultaneous equations which are coupled. For example, even if the *three* reaction components of the compound truss shown in Fig. 8.7(c) have been determined beforehand, the *method of joints* as applied to this truss will still lead to *eleven* unknown axial forces to be determined from the coupled equilibrium equations written for the various joints. A similar situation exists for the complex truss shown in Fig. 8.8(b). (Cf. Probs. 8.24 and 8.25*.)

EXAMPLE 8.1

Using the method of joints, determine the forces in members *AB* and *AE* of the simple cantilever truss shown.

Solution. We do not need to find the reactions from the supports because the free-body diagram of the joint *A* as shown contains only two unknowns which are being sought. Note that the unknown forces are here arbitrarily assumed to be tensile forces. For equilibrium, we write

$$+\uparrow \Sigma F_y = 0: \quad -\frac{3}{5} F_{AB} - 15 = 0 \quad\quad F_{AB} = -25$$

$$\xrightarrow{+} \Sigma F_x = 0: \quad -F_{AE} - \frac{4}{5} F_{AB} = 0 \quad\quad F_{AE} = 20$$

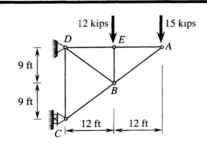

Since F_{AE} is positive but F_{AB} is negative, the answers are

$$F_{AB} = 25 \text{ kips } C \qquad F_{AE} = 20 \text{ kips } T \blacktriangleleft$$

where C denotes *compression* and T denotes *tension*.

REMARK. The value of F_{AE} may directly be obtained with a moment equilibrium equation as follows:

$$+\circlearrowleft \Sigma M_B = 0: \quad 9F_{AE} - 12(15) = 0 \quad F_{AE} = 20 \text{ kips } T \blacktriangleleft$$

EXAMPLE 8.2

Using the *method of joints*, determine the force in each member of the simple truss shown.

Solution. We need to determine the reactions from the supports before we can choose for study an appropriate joint whose free-body diagram contains no more than two unknowns. From the free-body diagram of the entire truss in equilibrium as shown, we write

$$+\circlearrowleft \Sigma M_A = 0: \quad 9\left(\frac{4}{5}\right)F - 6(120) = 0 \qquad F = 100$$

$$\xrightarrow{+} \Sigma F_x = 0: \quad A_x + \frac{3}{5}F = 0 \qquad A_x = -60$$

$$+\uparrow \Sigma F_y = 0: \quad A_y - 120 + \frac{4}{5}F = 0 \qquad A_y = 40$$

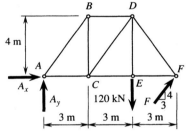

Thus, the reactions are $\mathbf{F} = 100 \text{ kN} \nearrow$, $\mathbf{A}_x = 60 \text{ kN} \leftarrow$, and $\mathbf{A}_y = 40 \text{ kN} \uparrow$.

The force in each member of the truss may now be determined by considering the equilibrium of the various joints as follows:

Joint A. Using the solutions above, we draw the free-body diagram of the joint A as shown. For equilibrium, we write

$$+\uparrow \Sigma F_y = 0: \quad \frac{4}{5}F_{AB} + 40 = 0 \qquad F_{AB} = -50$$

$$\xrightarrow{+} \Sigma F_x = 0: \quad \frac{3}{5}F_{AB} + F_{AC} - 60 = 0 \qquad F_{AC} = 90$$

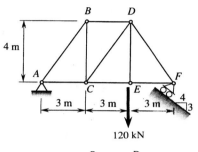

Noting the assumed senses of the unknown axial forces in the free-body diagram and the signs of the determined values of F_{AB} and F_{AC}, we write

$$F_{AB} = 50 \text{ kN } C \qquad F_{AC} = 90 \text{ kN } T \blacktriangleleft$$

Joint B. Knowing that the force in the member AB is a 50-kN compressive force, we draw the free-body diagram of the joint B as shown, where the force in the member BD is assumed to be compressive. For equilibrium, we write

$$\xrightarrow{+} \Sigma F_x = 0: \quad \frac{3}{5}(50) - F_{BD} = 0 \qquad F_{BD} = 30$$

$$+\uparrow \Sigma F_y = 0: \quad \frac{4}{5}(50) - F_{BC} = 0 \qquad F_{BC} = 40$$

$$F_{BC} = 40 \text{ kN } T \qquad F_{BD} = 30 \text{ kN } C \blacktriangleleft$$

Joint C. Knowing the forces in members *BC* and *AC*, we draw the free-body diagram of the joint *C* as shown. For equilibrium, we write

$$+\uparrow \Sigma F_y = 0: \quad 40 - \frac{4}{5} F_{CD} = 0 \qquad F_{CD} = 50$$

$$\xrightarrow{+} \Sigma F_x = 0: \quad F_{CE} - \frac{3}{5} F_{CD} - 90 = 0 \qquad F_{CE} = 120$$

$$F_{CD} = 50 \text{ kN } C \qquad F_{CE} = 120 \text{ kN } T \blacktriangleleft$$

Joint D. Knowing the forces in members *BD* and *CD*, we draw the free-body diagram of the joint *D* as shown. For equilibrium, we write

$$\xrightarrow{+} \Sigma F_x = 0: \quad 30 + \frac{3}{5}(50) - \frac{3}{5} F_{DF} = 0 \qquad F_{DF} = 100$$

$$+\uparrow \Sigma F_y = 0: \quad \frac{4}{5}(50) - F_{DE} + \frac{4}{5} F_{DF} = 0 \qquad F_{DE} = 120$$

$$F_{DE} = 120 \text{ kN } T \qquad F_{DF} = 100 \text{ kN } C \blacktriangleleft$$

Joint E. The free-body diagram of this joint is drawn as shown. For equilibrium, we write

$$\xrightarrow{+} \Sigma F_x = 0: \quad F_{EF} - 120 = 0 \qquad F_{EF} = 120 \text{ kN } T \blacktriangleleft$$

$$+\uparrow \Sigma F_y = 0: \quad 120 - 120 = 0$$

Joint F. The free-body diagram of this joint is drawn as shown. For equilibrium, we write

$$\xrightarrow{+} \Sigma F_x = 0: \quad \frac{3}{5}(100) + \frac{3}{5}(100) - 120 = 0$$

$$+\uparrow \Sigma F_y = 0: \quad \frac{4}{5}(100) - \frac{4}{5}(100) = 0$$

REMARK. We have written three equilibrium equations for the entire truss as a rigid body to determine the three unknown components of reactions (i.e., A_x, A_y, and F) in the beginning of the solution. This is why we have the last three extra equations which serve to check the results already obtained. If we considered only the equilibrium of the six joints but not that of the entire truss, we would have to solve a set of *twelve* simultaneous equations for the *nine* unknown axial forces and the *three* unknown components of reactions.

Developmental Exercises

D8.16 Using the *method of joints*, determine the force in each member of the simple truss shown.

D8.17 A complex truss is shown. Using the *method of joints*, express F_{CF} and F_{CD} in terms of F_{EF} with the assumption that the members CF, CD, and EF are in tension.

Fig. D8.16

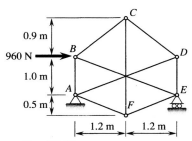

Fig. D8.17

PROBLEMS

8.1 through 8.12* Using the *method of joints*, determine the force in each member of the truss shown.

Fig. P8.1

Fig. P8.2

Fig. P8.3*

Fig. P8.4

Fig. P8.5

Fig. P8.6*

Fig. P8.7

Fig. P8.8

Fig. P8.9*

Fig. P8.10

Fig. P8.11

Fig. P8.12* and P8.13*

Fig. P8.14 and P8.18

8.13* through 8.17* Determine whether the truss shown is a simple truss, a compound truss, or a complex truss.

Fig. P8.16* and P8.20*

Fig. P8.17* and P8.21

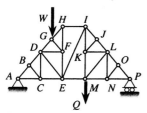

Fig. P8.15 and P8.19*

8.18 through 8.20* Determine the zero-force members in the truss shown.

8.21 Using the *method of joints*, determine the force in each member of the truss shown. (*Hint.* Determine the reactions first.)

8.22 and 8.23* Using the *method of joints*, determine the force in each member of the complex truss shown.

Fig. P8.22

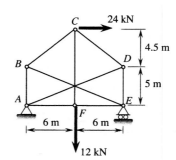

Fig. P8.23*

8.24 and 8.25* Using the *method of joints*, determine the force in each member of the compound truss shown.[†]

Fig. P8.24

Fig. P8.25*

8.8 Method of Sections for Plane Trusses

The method used to determine the unknown axial forces or other parameters of a truss by considering the equilibrium of a portion of the truss as a rigid body which contains two or more joints is called the *method of sections*. This method often provides a shortcut in determining the forces in certain truss members. Furthermore, this method can be used in conjunction with the method of joints to eliminate the need of solving a large number of simultaneous equations in determining the forces in members of some compound and complex trusses alluded to in Sec. 8.7. In using the method of sections, we *sometimes* (but not always) need to determine beforehand some of the components of reactions from the supports of the truss. We shall describe the *moment* and *shear approaches* in this method.[††]

I. MOMENT APPROACH. Oftentimes, we can isolate a portion of a plane truss by cutting along a section through certain members *whose axes all intersect at a common point*, except for the axes of the members whose axial forces are already known and for the axis of the member whose axial force is to be determined. If such is the case, we may proceed to consider the equilibrium of the free body of the isolated portion of the truss where pertinent components of reactions, if any, are determined beforehand from the free body of the entire truss. The desired axial force may then be determined by *summing to zero the moments of all forces acting on the free body of the isolated portion about the common point of intersection*.

[†]*Suggestion.* Do not determine beforehand the components of reactions from the supports. Treat them as undetermined unknowns so that a set of 2*j* equations in 2*j* unknowns will be obtained, where *j* is the number of joints. A computer may then be used to solve these equations (cf. App. D). Afterward, use the equilibrium equations for the entire truss to check the solutions for the reaction components.

[††]These two approaches simply pinpoint the shortest ways possible to solve for the unknowns in the subtruss as a rigid body.

EXAMPLE 8.3

Determine the forces in members *BD* and *CE* of the truss in Example 8.2.

Solution. We first determine **F** and then isolate the portion *DEF* of the truss to determine F_{BD} and F_{CE} by using the *moment approach* as follows:

Entire truss

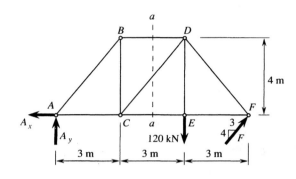

$$+ \circlearrowleft \ \Sigma M_A = 0: \quad 9\left(\frac{4}{5}\right)F - 6(120) = 0 \qquad \mathbf{F} = 100 \text{ kN } \nearrow$$

Portion DEF

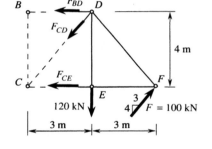

$$+ \circlearrowleft \ \Sigma M_C = 0: \quad 4F_{BD} - 3(120) + 6\left(\frac{4}{5}\right)(100) = 0$$

$$+ \circlearrowleft \ \Sigma M_D = 0: \quad -4F_{CE} + 4\left(\frac{3}{5}\right)(100) + 3\left(\frac{4}{5}\right)(100) = 0$$

These two equations yield $F_{BD} = -30$ and $F_{CE} = 120$. Thus, we write

$$F_{BD} = 30 \text{ kN } C \qquad F_{CE} = 120 \text{ kN } T \ \blacktriangleleft$$

REMARK. These answers check those obtained in Example 8.2. Note that the portion *ABC* could have been used in the solution if the reaction components $\mathbf{A}_x$ and $\mathbf{A}_y$ had been determined beforehand.

Developmental Exercise

D8.18   Determine F_{BD} in the truss in Example 8.1.

EXAMPLE 8.4

Determine the forces in members *AF*, *FN*, and *JK* of the compound truss shown.

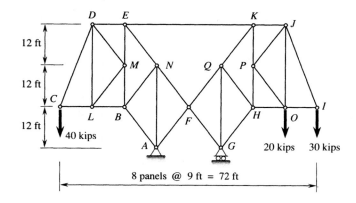

Solution. We first determine **G** and then isolate the portions *KQFGHOIJ* and *JPOI* of the truss to determine F_{AF}, F_{FN}, and F_{JK} by using the *moment approach* as follows:

Entire truss

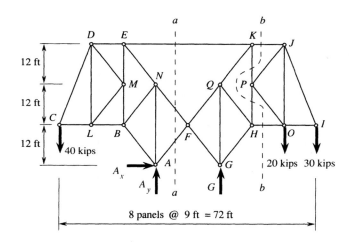

$$+ \circlearrowleft \ \Sigma M_A = 0: \quad 18G + 27(40) - 36(20) - 45(30) = 0$$

$$G = 55 \qquad \mathbf{G} = 55 \text{ kips} \uparrow$$

Portion KQFGHOIJ

We transmit $\mathbf{F}_{AF}$ to R, as shown on next page, and then write

$$+ \circlearrowleft \ \Sigma M_E = 0: \quad -48\left(\frac{3}{5}\right)F_{AF} + 27(55) - 45(20) - 54(30) = 0$$

$$F_{AF} = -35.9 \qquad F_{AF} = 35.9 \text{ kips } C \quad \blacktriangleleft$$

Furthermore, we transmit $\mathbf{F}_{FN}$ to E and then write

$$+ \circlearrowleft \ \Sigma M_K = 0: \quad -36\left(\frac{4}{5}\right)F_{FN} - 9(55) - 9(20) - 18(30) = 0$$

$$F_{FN} = -42.2 \qquad F_{FN} = 42.2 \text{ kips } C \quad \blacktriangleleft$$

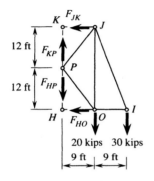

Portion JPOI

$$+ \circlearrowleft \ \Sigma M_H = 0: \quad 24F_{JK} - 9(20) - 18(30) = 0$$

$$F_{JK} = 30 \qquad F_{JK} = 30 \text{ kips } T \ \blacktriangleleft$$

REMARK. The same equation for determining the value of F_{JK} could similarly be obtained if we had chosen to isolate the portion *JPHOI* of the truss for consideration and set $\Sigma M_H = 0$.

Developmental Exercise

D8.19 Determine F_{EK}, F_{HO}, and F_{DE} in the truss in Example 8.4.

II. SHEAR APPROACH. At some other time, we can isolate a portion of a plane truss by cutting along a section through certain members *whose axes are all parallel to each other*, except for the members whose axial forces are already known and for the member whose axial force is to be determined. If such is the case, we may proceed to consider the equilibrium of the free body of the isolated portion of the truss where pertinent components of reactions, if any, are determined beforehand from the free body of the entire truss. The desired axial force may then be determined by *summing to zero the components of all forces acting on the free body of the isolated portion in the direction perpendicular to those parallel members*. (Note here that reaction components that are parallel to those parallel members do not need to be determined beforehand.)

EXAMPLE 8.5

Determine the force F_{CD} in the truss in Example 8.2.

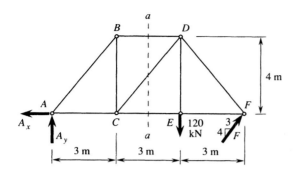

Solution. We first determine $\mathbf{A}_y$ from the entire truss and then isolate the portion ABC of the truss to determine F_{CD} by using the *shear approach* as follows:

Entire truss

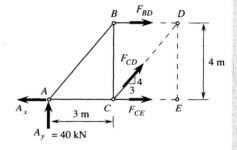

$$+\;\circlearrowright\; \Sigma M_F = 0: \quad -9A_y + 3(120) = 0$$

$$A_y = 40 \qquad \mathbf{A}_y = 40 \text{ kN } \uparrow$$

Portion ABC

$$+\;\circlearrowright\; \Sigma F_y = 0: \quad \frac{4}{5} F_{CD} + 40 = 0$$

$$F_{CD} = -50 \qquad F_{CD} = 50 \text{ kN } C \;\blacktriangleleft$$

REMARK. We do not need to find the value of $\mathbf{A}_x$ because $\mathbf{A}_x$ is parallel to the horizontal members BD and CE.

Developmental Exercises

D8.20 Determine F_{CN} in the truss shown.

D8.21 Which of the following best describes the value of F_{OK} in the truss shown? (a) 12 kips T, (b) 12 kips C, (c) 15 kips T, (d) 15 kips C, (e) statically indeterminate.

Fig. D8.20 and D8.21

EXAMPLE 8.6

Determine the force in the member *EJ* of the compound truss shown.

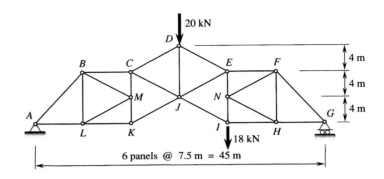

Solution. We first determine **G** and then isolate the portion *ENIHGF* of the truss to determine F_{EJ} by using the *shear approach* as follows:

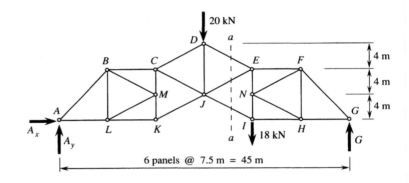

Entire truss

$$+ \circlearrowleft \ \Sigma M_A = 0: \quad 45G - 4(7.5)(18) - 3(7.5)(20) = 0$$

$$G = 22 \qquad \mathbf{G} = 22 \text{ kN} \uparrow$$

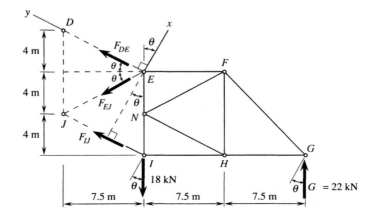

Portion ENIHGF

$$\tan\theta = \frac{4}{7.5} \qquad \theta = 28.07°$$

$$+\nearrow \Sigma F_x = 0: \quad -F_{EJ}\sin 2\theta - 18\cos\theta + 22\cos\theta = 0$$

$$F_{EJ} = 4.25 \qquad F_{EJ} = 4.25 \text{ kN } T \blacktriangleleft$$

Developmental Exercise

D8.22 Determine F_{CJ} in the truss in Example 8.6.

8.9 X Joints and Successive Method

An *X joint*, as shown in Fig. 8.12, is a joint which connects four members
in two collinear pairs lying in two intersecting straight lines. *If there is no
external force applied to an X joint which is in equilibrium, the tensile or
compressive forces in members of a collinear pair of members must be
equal*: however, the forces in this pair may not be equal to the forces in the
other pair. This must be true; otherwise, the summation of the components
of forces acting on the free body of the X joint in the direction perpendicular
to either of the said intersecting straight lines cannot be zero. This property
is useful in analyzing trusses containing X joints.

At times, the determination of the axial force in a specified member of a
truss cannot be achieved by using either the method of joints or the method
of sections only once. We may need to use the method of sections, or the
method of joints, or both, to determine the axial forces in certain other
nearby members before we can finally determine the specified axial force.
The method which successively employs the method of sections, or the
method of joints, or both, in determining a specified unknown axial force or
parameter of a truss is called the *successive method*.

Fig. 8.12

EXAMPLE 8.7

Determine the force in the member *DE* of the compound truss shown.

8.4 kN

F_{DI}

8 m

F_{DE}

8 m

Solution. The value of F_{DE} can be determined by applying the *successive method* to the portion *BCIJFGH* and the joint *D*, which are isolated as shown on the preceding page and this page, respectively.

Portion BCIJFGH

Recalling the property associated with X joints, we see that $F_{AH} = F_{HI} = F_{DI}$ and $F_{AG} = F_{GJ} = F_{EJ}$. Since there is *no net effect* on a rigid body from any pair of collinear forces which are equal in magnitude and opposite in sense, we may *disregard* the unknown forces $\mathbf{F}_{AH}$, $\mathbf{F}_{DI}$, $\mathbf{F}_{AG}$, and $\mathbf{F}_{EJ}$ acting on the free body of the isolated portion shown.

$$\xrightarrow{+} \Sigma F_x = 0: \quad F_{CD} + 8.4 = 0 \qquad F_{CD} = -8.4 \qquad F_{CD} = 8.4 \text{ kN } C$$

Joint D

$$+ \circlearrowleft \ \Sigma M_A = 0: \quad -8\left(\frac{4}{5}\right)F_{DE} - 8\left(\frac{3}{5}\right)F_{DE} - 8(8.4) = 0$$

$$F_{DE} = -6 \qquad\qquad F_{DE} = 6 \text{ kN } C \ \blacktriangleleft^\dagger$$

Developmental Exercises

D8.23 Determine F_{BH} in the truss in Example 8.7.

D8.24 Determine F_{AG} in the truss in Example 8.7.

EXAMPLE 8.8

Determine the force F_{FN} in the truss in Example 8.6.

Solution. Having previously determined that $\mathbf{G} = 22$ kN $\uparrow$, we may apply the *successive method* to determine F_{EF} first and then F_{FN}.

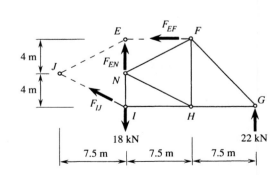

---

†Without using the property associated with X joints, as described in Sec. 8.9, much more effort would be needed to obtain the solution.

Portion FNIHG

$$+ \circlearrowright\ \Sigma M_I = 0: \quad 8F_{EF} + 15(22) = 0 \qquad F_{EF} = -41.25$$

$$F_{EF} = 41.25 \text{ kN } C$$

Portion FHG

$$+ \circlearrowright\ \Sigma M_H = 0: \quad 8\left(\frac{15}{17}\right)F_{FN} - 8(41.25) + 7.5(22) = 0$$

$$F_{FN} = 23.4 \qquad F_{FN} = 23.4 \text{ kN } T \ \blacktriangleleft$$

Developmental Exercises

D8.25 Determine F_{DJ} in the truss in Example 8.6.

D8.26 Determine F_{LM} in the truss in Example 8.6.

*8.10 Space Trusses

A *space truss* is a truss whose members do not all lie in a plane. The most basic rigid space truss is a *tetrahedron truss*, which has six members and four joints as shown in Fig. 8.13(a). In elementary studies of space trusses, the ends of truss members are assumed to be held by ball-and-socket joints, as illustrated in Fig. 8.13(b). Thus, each member of a space truss is treated as a two-force body.

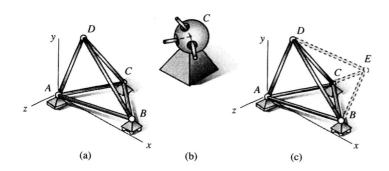

(a)   (b)   (c)

Fig. 8.13

A *simple space truss* is a rigid space truss which can be constructed from a tetrahedron truss by repeating appropriately the addition of *three new members* and *one new joint* at a time, as shown in Fig. 8.13(c). Note that the three new members are first fastened at their ends to three separate existing joints and then connected together at their other ends by the new joint. If *m* and *j* are the number of members and the number of joints in a simple space truss, respectively, we must have $m - 6 = 3(j - 4)$, or

$$m + 6 = 3j \tag{8.3}$$

Similar to Eq. (8.2), the *necessary condition* for a space truss to be statically determinate can be shown to be

$$m + r = 3j \tag{8.4}$$

where r is the number of unknown components of reaction from the supports. Space trusses considered in this chapter are statically determinate.

Developmental Exercises

D8.27 Define (a) a space truss, (b) a simple space truss.

D8.28 A simple space truss has 10 joints. How many truss members does it have?

★8.11   Method of Joints for Simple Space Trusses

The equilibrium of each joint of a space truss is similar to the equilibrium of a particle in space because the force system acting at each joint is either zero or concurrent and noncoplanar. In other words, the equilibrium of each joint of a space truss is described by Eq. (3.3) or its scalar version, Eqs. (3.5). Thus, we have

$$\Sigma \mathbf{F} = \mathbf{0} \tag{3.3}$$
(repeated)

$$\Sigma F_x = 0 \quad \Sigma F_y = 0 \quad \Sigma F_z = 0 \tag{3.5}$$
(repeated)

for the equilibrium of each joint of a space truss. From Eqs. (3.5), we infer that there are $3j$ equations describing the equilibrium of the j joints. From Eq. (8.3), we know in a simple space truss that $m + 6 = 3j$. Thus, those $3j$ equations may be used to determine the m unknown axial forces in the m members plus (up to) six unknown components of reactions from the supports of a simple space truss. This is the logic which forms the basis of the *method of joints for analyzing simple space trusses.*

Frequently, the effort required to solve the $3j$ equations may considerably be reduced by first determining the pertinent unknown reactions from the supports and then considering the equilibrium of the joints in such an order that each time no more than three new unknowns are involved. In this manner, we can avoid solving a large system of simultaneous equations.

EXAMPLE 8.9

A force $\mathbf{P} = 20\mathbf{i} - 60\mathbf{j} + 30\mathbf{k}$ kN is applied to the joint D of the space truss shown. Determine (a) the reactions from the supports, (b) the force in each member.

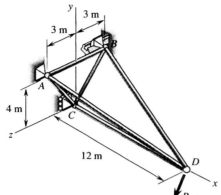

Solution. We first determine the reactions in the free-body diagram as shown.

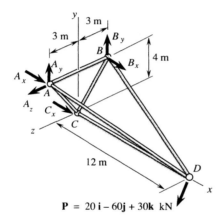

We write

$$\Sigma\mathbf{F} = 0: \quad (A_x + B_x + C_x + 20)\mathbf{i}$$
$$+ (A_y + B_y - 60)\mathbf{j} + (A_z + 30)\mathbf{k} = 0 \qquad (1)$$

$$\Sigma\mathbf{M}_A = 0: \quad (-6\mathbf{k}) \times (B_x\mathbf{i} + B_y\mathbf{j}) + (-4\mathbf{j} - 3\mathbf{k}) \times (C_x\mathbf{i})$$
$$+ (12\mathbf{i} - 4\mathbf{j} - 3\mathbf{k}) \times (20\mathbf{i} - 60\mathbf{j} + 30\mathbf{k}) = 0$$

$$(6B_y - 300)\mathbf{i} - (6B_x + 3C_x + 420)\mathbf{j} + (4C_x - 640)\mathbf{k} = 0 \qquad (2)$$

Equations (1) and (2) give a set of six scalar equations, which yield $A_x = -30$, $A_y = 10$, $A_z = -30$, $B_x = -150$, $B_y = 50$, and $C_x = 160$; i.e.,

$$\mathbf{A} = -30\mathbf{i} + 10\mathbf{j} - 30\mathbf{k} \text{ kN} \quad \blacktriangleleft$$

$$\mathbf{B} = -150\mathbf{i} + 50\mathbf{j} \text{ kN} \qquad \mathbf{C} = 160\mathbf{i} \text{ kN} \quad \blacktriangleleft$$

For use in the method of joints, we have

$$\vec{AB} = -6\mathbf{k} \qquad\qquad \lambda_{AB} = -\mathbf{k} = -\lambda_{BA}$$

$$\vec{AC} = -4\mathbf{j} - 3\mathbf{k} \qquad\qquad \lambda_{AC} = -\tfrac{1}{5}(4\mathbf{j} + 3\mathbf{k}) = -\lambda_{CA}$$

$$\vec{AD} = 12\mathbf{i} - 4\mathbf{j} - 3\mathbf{k} \qquad\qquad \lambda_{AD} = \tfrac{1}{13}(12\mathbf{i} - 4\mathbf{j} - 3\mathbf{k}) = -\lambda_{DA}$$

$$\vec{BC} = -4\mathbf{j} + 3\mathbf{k} \qquad\qquad \lambda_{BC} = \tfrac{1}{5}(-4\mathbf{j} + 3\mathbf{k}) = -\lambda_{CB}$$

$$\vec{BD} = 12\mathbf{i} - 4\mathbf{j} + 3\mathbf{k} \qquad\qquad \lambda_{BD} = \tfrac{1}{13}(12\mathbf{i} - 4\mathbf{j} + 3\mathbf{k}) = -\lambda_{DB}$$

$$\vec{CD} = 12\mathbf{i} \qquad\qquad \lambda_{CD} = \mathbf{i} = -\lambda_{DC}$$

The unknown axial forces may be determined as follows:

Joint D

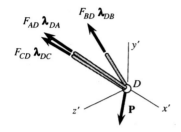

$$\Sigma \mathbf{F} = \mathbf{0}: \quad \left(-\frac{12}{13}F_{AD} - \frac{12}{13}F_{BD} - F_{CD} + 20\right)\mathbf{i}$$

$$+ \left(\frac{4}{13}F_{AD} + \frac{4}{13}F_{BD} - 60\right)\mathbf{j}$$

$$+ \left(\frac{3}{13}F_{AD} - \frac{3}{13}F_{BD} + 30\right)\mathbf{k} = \mathbf{0}$$

This vector equilibrium equation gives a set of three scalar equations, which yield $F_{AD} = 32.5$, $F_{BD} = 162.5$, and $F_{CD} = -160$. Thus, we have

$$F_{AD} = 32.5 \text{ kN } T \qquad F_{BD} = 162.5 \text{ kN } T \qquad F_{CD} = 160.0 \text{ kN } C \quad \blacktriangleleft$$

Joint C

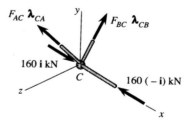

$$\Sigma \mathbf{F} = \mathbf{0}: \quad (160 - 160)\mathbf{i} + \left(\frac{4}{5}F_{AC} + \frac{4}{5}F_{BC}\right)\mathbf{j}$$

$$+ \left(\frac{3}{5}F_{AC} - \frac{3}{5}F_{BC}\right)\mathbf{k} = \mathbf{0}$$

This vector equilibrium equation gives *one trivial* and *two nontrivial* scalar equations, which yield

$$F_{AC} = F_{BC} = 0 \quad \blacktriangleleft$$

Joint B

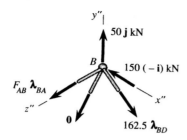

$$\Sigma \mathbf{F} = \mathbf{0}: \quad \left[\frac{12}{13}(162.5) - 150\right]\mathbf{i} + \left[-\frac{4}{13}(162.5) + 50\right]\mathbf{j}$$

$$+ \left[F_{AB} + \frac{3}{13}(162.5)\right]\mathbf{k} = \mathbf{0}$$

This vector equilibrium equation gives *two trivial* and *one nontrivial* scalar equations, which yield

$$F_{AB} = -37.5 \qquad F_{AB} = 37.5 \text{ kN } C \blacktriangleleft$$

Besides the above trivial scalar equations, the equilibrium of the joint *A* may be used to further check the obtained answers as follows:

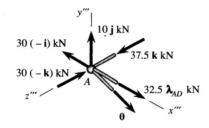

$$\Sigma \mathbf{F} = \left[\frac{12}{13}(32.5) - 30\right]\mathbf{i} + \left[-\frac{4}{13}(32.5) + 10\right]\mathbf{j}$$

$$+ \left[37.5 - \frac{3}{13}(32.5) - 30\right]\mathbf{k}$$

$$= 0\mathbf{i} + 0\mathbf{j} + 0\mathbf{k} = \mathbf{0}$$

REMARK. If we did not find the reactions from the supports first, we would need to solve simultaneous equations in determining the unknown axial forces in the truss.

Developmental Exercise

D8.29 Each member of the truss shown is 1 m long. (a) Determine the lengths of $\overline{AM}$, $\overline{OA}$, and $\overline{OD}$. (b) By symmetry, what should be the value of the y component of $\mathbf{F}_{DA}$ so that the joint D is in equilibrium? (c) Determine F_{AB}, F_{AC}, and F_{BC}.

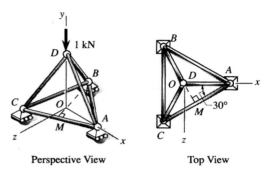

Perspective View   Top View

Fig. D8.29

PROBLEMS

8.26 Determine the forces in members BK and CJ of the truss shown.

Fig. P8.26 and P8.27

8.27 Determine the forces in members EI and DJ of the truss shown.

8.28* Determine the forces in members AB and BF of the truss shown.

Fig. P8.28* and P8.29

8.29 Determine the forces in members AF and BE of the truss shown.

8.30 Determine the forces in members *BL* and *CL* of the truss shown.

Fig. P8.30 and P8.31*

8.31* Determine the forces in members *DK* and *EK* of the truss shown.

8.32 Determine the forces in members *CO* and *EP* of the Baltimore bridge truss shown.

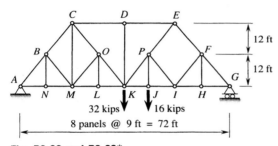

Fig. P8.32 and P8.33*

8.33* Determine the forces in members *CM* and *EI* of the truss shown.

8.34 Determine the forces in members *AB* and *FG* of the truss shown.

Fig. P8.34 and P8.35

8.35 Determine the forces in members *GI* and *DL* of the truss shown.

8.36* Determine the forces in members *DK* and *BJ* of the truss shown.

8.37 Determine the forces in members *FG* and *CJ* of the truss shown.

Fig. P8.36* and P8.37

Fig. P8.38

Fig. P8.39

8.38 Determine the forces in members *EN*, *LM*, and *GK* of the Fink roof truss shown.

8.39 Determine the forces in members *BC*, *BI*, and *DH* of the truss shown.

8.40* through 8.43* Determine the forces in members *AB*, *CD*, and *EF* of the truss shown.

Fig. P8.40*

Fig. P8.41

Fig. P8.42

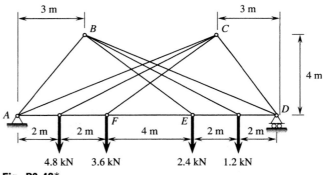

Fig. P8.43*

8.44 Determine the forces in members *BH* and *AI* of the truss when the crate is in equilibrium as shown.

Fig. P8.44 and P8.45*

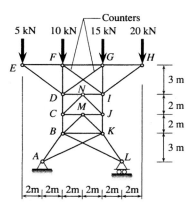

Fig. P8.46

8.45* Determine the forces in members *GH* and *CN* of the truss when the crate is in equilibrium as shown.

8.46 The members *FI* and *DG* of the tower truss shown are very slender and can act only in tension, and only one of them acts at a time; such members are known as *counters*. Determine the forces in the member *IJ* and the counter which is acting under the given loading. (*Hint.* First determine F_{IJ}, then isolate the portion *GIH* of the truss and consider $\Sigma F_y = 0$.)

8.47* and **8.48** Determine the zero-force members in the space truss shown.

Fig. P8.47*
and P8.49

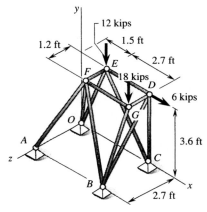

Fig. P8.48 and P8.50*

8.49 Determine the forces in members *BD*, *BE*, *BA*, and *BC* of the space truss shown.

8.50* Determine the forces in all nonzero-force members of the space truss shown.

FRAMES AND MACHINES

8.12 Structures Containing Multiforce Members

A *multiforce member* is a member or body which carries a force system and cannot be classified as a two-force body or as a particle. A *frame* is a structure which contains one or more multiforce members and is generally fully constrained. Frames are designed to support loads and are usually stationary. A *machine* is a structure which contains one or more movable multiforce members. Machines, which may be in the form of simple tools or complicated mechanisms, are designed to transmit or modify forces and motions. Machine problems in statics are usually associated with the transmission or modification of forces.

In studying frames and machines, we have the following *usual assumptions*: (a) the weights of members are negligible compared with the applied loads; (b) the connecting pin in each joint is circular and frictionless; (c) if a force acts at a joint, it is applied to the pin in the joint.

Members of a frame or machine are frequently detached for the analysis. For clarity in the sketch, we shall adopt a graphical convention that (1) *a pin or a pinhole keeping a pin* is indicated by a *solid circular dot*, (2) *a pinhole without a pin* in it is indicated by a small *circular curve*. By letting appropriate detached members keep certain pins in their pinholes, we often have the advantage of doing away with the need to consider separately the equilibrium of a pin connecting *three or more* members. However, the interacting forces between a pin and the walls of the pinholes in a joint can be studied by disassembling the joint, as illustrated in perspective views in Fig. 8.14 and in front views using the adopted graphical convention in Fig. 8.15, where Newton's third law is obeyed.

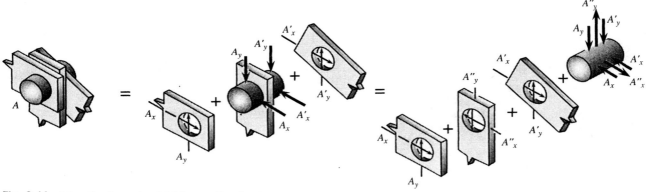

Fig. 8.14 Interacting forces in a joint (perspective views).

Fig. 8.15 Interacting forces in a joint (front views).

In modern structures, members of frames are seldom pin-connected. They are generally fastened at the joints with nails, bolts, rivets, or welding to achieve a greater degree of rigidity for the structure. However, results of studies based on the pin-connected version of the frame are relatively easy to obtain and are useful for practical purposes.

Developmental Exercises

D8.30 Define the terms: (a) multiforce member, (b) frame, (c) machine.

D8.31 What are the usual assumptions for frames and machines?

D8.32 As a graphical convention, what do we use to indicate (a) a pin in a pinhole, (b) a pinhole without a pin in it?

8.13 Force System Acting on a Detached Member

The force system in the free-body diagram of a detached member of a frame or machine generally includes the following:

1. the reactions to this detached member from its supports, if any;
2. the reactions to the walls of the pinholes of this detached member from the pins which this detached member does not keep;
3. the reactions from the walls of the pinholes of other members to the pins which this detached member keeps;
4. the external concentrated forces, if any, that are applied to the pins which this detached member keeps;
5. the external loads, if any, which are applied to this detached member.

Although the directions of unknown reaction components are initially assumed, it is of paramount importance to note that the free-body diagrams of the detached members of a frame or machine must be drawn in a *coordinated* manner consistent with *Newton's third law*. The letter designating the joint or support is generally used with appropriate subscripts (e.g., x or y) to label the unknown components of action and reaction at that joint or support. Furthermore, when a joint connects three or more multiforce members, superscripts (e.g., primes) in addition to the subscripts may be used along with the letter to name different unknown components of actions and reactions at that joint.

EXAMPLE 8.10

A frame is loaded as shown. Letting the member ABC keep the pin at C, draw the free-body diagram of each member separately.

Solution. According to Sec. 8.13 and Newton's third law, we draw the free-body diagram of each member as shown below. Note that (1) the member BD is a *two-force body* while the other members are multiforce members, (2) the *directions* of the unknown force components are assumed and indicated by the arrow signs consistent with *Newton's third law* (e.g., the action and reaction between the pinhole at B in the member ABC and the pin at B in the member BD are equal in magnitude and opposite in direction), (3) the *magnitudes* of the unknown force components are labeled with appropriate letters and subscripts, and (4) the *pin* at C kept by the member ABC is acted on by the external concentrated force of 240 lb as well as the reaction components from the member CDE.

EXAMPLE 8.11

A frame is loaded as shown. Letting the member BEF keep the pin at B, draw the free-body diagram of each member separately.

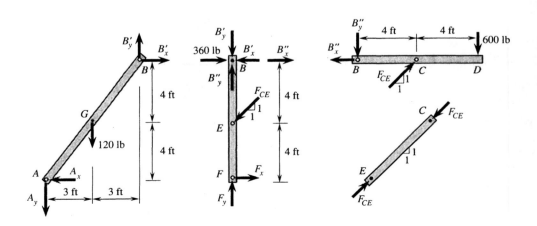

Solution. According to Sec. 8.13 and Newton's third law, we draw the free-body diagram of each detached member as shown. Note here that (1) the member CE is a *two-force body* while the others are all multiforce members; (2) the *directions* of the unknown force components are assumed and *Newton's third law* is obeyed; (3) the *magnitudes* of the unknown force components are labeled with appropriate letters and subscripts or both subscripts and superscripts; (4) the *magnitudes* of the unknown force components exerted between the pin at B and the member AB are labeled as B'_x and B'_y, while the magnitudes of those exerted between the pin at B and the member BCD are labeled as B''_x and B''_y; and (5) the *pin* at B kept by the member BEF carries the 360-lb force as well as the reactions from the members AB and BCD.

Developmental Exercises

D8.33 and 8.34 A frame is shown. Draw an appropriate free-body diagram of the member BCD if (a) it keeps no pins, (b) it keeps all the pins it can keep.

Fig. D8.34

Fig. D8.33

8.14   Method of Solution for Frames and Machines in a Plane

In studying a frame or machine which is in *equilibrium in a plane*, it is helpful to keep in mind that the forces acting on it are coplanar forces and the *maximum number of independent scalar equilibrium equations* that can be written from its free-body diagram under consideration is:

■ *three* if the unknown forces[†] are nonconcurrent and nonparallel;
■ *two* if the unknown forces are concurrent but noncollinear;
■ *two*, again, if the unknown forces are all parallel to each other and noncollinear but do not all form couples;
■ *one* if the unknown forces all pair up to form couples;
■ *one*, again, if there is only one unknown force or moment;
■ *zero* if there is no unknown force or moment.

[†]If any of the three characteristics of a force is unknown, the force is an unknown force.

Naturally, the equilibrium of all of the members of a frame or machine implies and ensures the equilibrium of the entire frame or machine as a rigid body, and vice versa. The scalar equilibrium equations written from the free-body diagrams of all individual members *and* the entire frame or machine are, therefore, not independent. Some of them may, however, serve to check the solutions for the unknowns.

Although certain problems can best be solved by a special set of steps, the following procedure may serve as a guide in solving *equilibrium problems of frames and machines in a plane*:

1. *Identify two-force bodies*, if any, in the system.
2. *Draw appropriate free-body diagrams* which involve all of the unknowns to be determined.
3. *Count* the number of *unknowns* and the maximum number of independent scalar equilibrium *equations* that can be written from the free-body diagrams drawn.
4. If necessary, *draw additional free-body diagrams* of detached members *until* the number of unknowns is equal to the number of independent scalar equilibrium equations.
5. *Write* the set of *simultaneous* scalar equilibrium *equations* from the free-body diagrams drawn.
6. *Solve* the simultaneous equations for the *unknowns* to be determined.

Developmental Exercise

D8.35 A frame is in equilibrium in a plane. Under what condition(s) can we write from its free-body diagram the following maximum number of independent scalar equilibrium equations: (a) three, (b) only two, (c) only one, (d) zero?

EXAMPLE 8.12

A frame is loaded as shown. Determine the reactions from the pins at C, D, and E to the horizontal member CDE.

Solution. The free-body diagram of the member CDE with no pins contains six unknowns: C_x, C_y, D_x, D_y, E_x, and E_y. So, we need to draw additional free-body diagrams and seek six independent scalar equilibrium equations in these six unknowns, if possible. We find that it is possible to do so as follows:

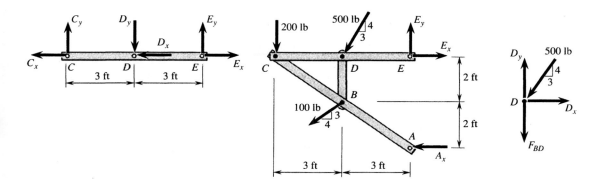

Entire frame

$$+\uparrow \Sigma F_y = 0: \quad E_y - \frac{4}{5}(500) - 200 - \frac{3}{5}(100) = 0$$

$$+\circlearrowleft \Sigma M_A = 0: \quad -4E_x + 4\left(\frac{3}{5}\right)(500) + 3\left(\frac{4}{5}\right)(500) + 6(200)$$

$$+2\left(\frac{4}{5}\right)(100) + 3\left(\frac{3}{5}\right)(100) = 0$$

These two equations yield $E_y = 660$ and $E_x = 985$.

Pin at D

$$\xrightarrow{+} \Sigma F_x = 0: \quad D_x - \frac{3}{5}(500) = 0 \qquad D_x = 300$$

Member CDE

$$+\circlearrowleft \Sigma M_D = 0: \quad 3E_y - 3C_y = 0 \qquad C_y = 660$$

$$\xrightarrow{+} \Sigma F_x = 0: \quad E_x - D_x - C_x = 0 \qquad C_x = 685$$

$$+\uparrow \Sigma F_y = 0: \quad E_y - D_y + C_y = 0 \qquad D_y = 1320$$

Thus, the reactions from the pins at C, D, and E to the member CDE are

$$\mathbf{C} = -685\mathbf{i} + 660\mathbf{j} \text{ lb} \qquad \mathbf{D} = -300\mathbf{j} - 1320\mathbf{j} \text{ lb} \quad \blacktriangleleft$$

$$\mathbf{E} = 985\mathbf{i} + 660\mathbf{j} \text{ lb} \quad \blacktriangleleft$$

REMARK. Note that the 500-lb concentrated force is applied to the pin at D. Furthermore, this pin receives a reaction from the two-force member BD as well as two reaction components from the multiforce member CDE, according to *Newton's third law*, as shown.

Developmental Exercises

D8.36   Determine the reactions from the pins to the member *ABC* of the frame in Example 8.10.

D8.37   Determine the reactions from the pins to the member *AGB* of the frame in Example 8.11.

EXAMPLE 8.13

If the tension in the spring *HI* of the vise grip shown is 17 N, determine the magnitude of the gripping forces at *G*.

Solution.   Let the magnitude of the gripping forces be *G*. To find *G*, we need to draw free-body diagrams of some detached members. Note that the member *CD* is a *two-force body* because the release lever is not exerting any force on it. The free-body diagram of the upper jaw as shown contains four unknowns: G, F_x, F_y, and F_{CD}. Thus, we need to draw at least one more free-body diagram and seek four independent scalar equilibrium equations in these four unknowns, if possible. We find that it is possible to do so as follows:

Upper jaw

$$+ \circlearrowleft \; \Sigma M_C = 0: \quad 145G - 95F_y - 45\left(\frac{24}{51}\right)(17) + 35(300) = 0 \quad (1)$$

$$\overset{+}{\rightarrow} \; \Sigma F_x = 0: \quad F_x - \frac{60}{68}F_{CD} + \frac{45}{51}(17) = 0 \quad (2)$$

$$+\uparrow \; \Sigma F_y = 0: \quad G - F_y + \frac{32}{68}F_{CD} - \frac{24}{51}(17) - 300 = 0 \quad (3)$$

Lower jaw

$$+ \circlearrowleft \; \Sigma M_E = 0: \quad -55G + 44F_x + 5F_y + 20\left(\frac{45}{51}\right)(17) = 0 \quad (4)$$

Solving these four simultaneous equations, we obtain

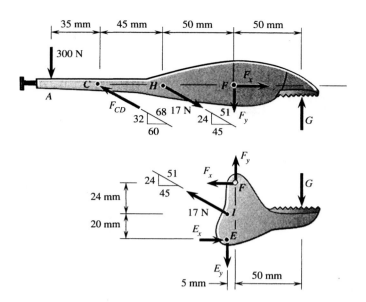

$$G = 8712 \text{ N} \qquad F_x = 9360 \text{ N} \qquad F_y = 13404 \text{ N} \qquad F_{CD} = 10625 \text{ N}$$

$$G = 8.71 \text{ kN} \quad \blacktriangleleft$$

Thus, we see that the two 300-N applied forces have been magnified about 29 times when they are transmitted to the rod at the jaws of the vise grip.

EXAMPLE 8.14

The loader holds a 2.4-kip load which acts through the point G as shown. The machine is symmetrical about a vertical plane passing through the boom $ABCDE$ and has two sets of linkages and hydraulic cylinders. Determine the compressive forces in the cylinders KL and BM.

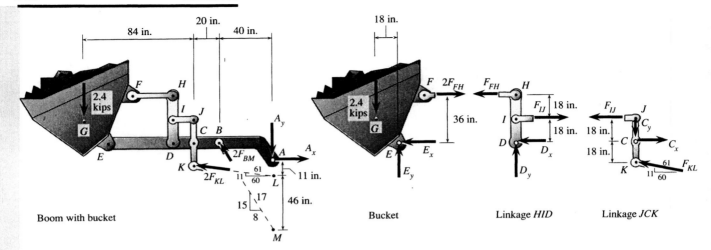

Boom with bucket   Bucket   Linkage *HID*   Linkage *JCK*

Solution. We first draw the free-body diagrams needed to solve for the unknowns F_{KL} and F_{BM}. Note that we indicate $2F_{BM}$, $2F_{KL}$, and $2F_{FH}$ in the first two free-body diagrams because there are two sets of linkages and hydraulic cylinders to hold the load. We may solve this problem in the following sequence:

Bucket

$$+ \circlearrowleft \Sigma M_E = 0: \quad 18(2.4) - 36(2F_{FH}) = 0 \qquad F_{FH} = 0.6 \text{ kips}$$

Linkage HID

$$+ \circlearrowleft \Sigma M_D = 0: \quad 36(F_{FH}) - 18F_{IJ} = 0 \qquad F_{IJ} = 1.2 \text{ kips}$$

Linkage JCK

$$+ \circlearrowleft \Sigma M_C = 0: \quad 18F_{IJ} - 18\left(\frac{60}{61}\right)F_{KL} = 0 \qquad F_{KL} = 1.22 \text{ kips} \quad \blacktriangleleft$$

Boom with bucket

We first transmit $2\mathbf{F}_{KL}$ to the point L and $2\mathbf{F}_{BM}$ to the point M; then, we apply Varignon's theorem to write

$$+ \circlearrowleft \Sigma M_A = 0: \quad 144(2.4) - 11\left(\frac{60}{61}\right)(2F_{KL}) - 57\left(\frac{8}{17}\right)(2F_{BM}) = 0$$

$$F_{BM} = 5.95 \text{ kips} \quad \blacktriangleleft$$

Developmental Exercise

D8.38 Determine the total shearing force (i.e., the resultant reaction) supported by the pin at A of the loader in Example 8.14.

PROBLEMS

8.51 through 8.54* Letting the member *ABC* of the frame shown keep all the pins it can keep, draw the free-body diagram of each member as detached.

Fig. P8.51 and P8.55

Fig. P8.52* and P8.56*

Fig. P8.53* and P8.57

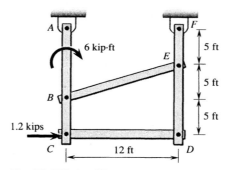

Fig. P8.54* and P8.58

8.55 through 8.58   Determine the reactions *from* the pins *to* the member *ABC* of the frame shown.

8.59   Two bent pipes are shown. Neglecting the friction between them in the portion *CD*, determine the reactions from the supports at *A* and *F*.

Fig. P8.59

Fig. P8.60

8.60   Determine the reactions from the pins to the member *ABC* as shown.

Fig. P8.61

8.61 and 8.62* Determine the reactions from the supports at *A* and *G* of the hinged beam shown.[†]

Fig. P8.62*

8.63* and 8.64 Neglecting friction, determine the reactions from the supports at *A*, *B*, and *C* if the weight of each pipe shown is 100 lb.

Fig. P8.63*   Fig. P8.64

8.65 and 8.66* Determine the magnitude of the forces exerted on the rod at *A* as shown.

Fig. P8.65

Fig. P8.66*

[†]A beam of this type is also known as a *Gerber beam (Gerberbalken)*, after the German engineer Heinrich Gerber (1832–1912).

8.67 Determine the forces exerted at C and D on the member BCD of the tongs shown.

8.68 Two shafts are connected by a universal joint at D and supported by narrow bearings without stops at B and C and by a narrow bearing with stop at E as shown. When the arm of the crosspiece attached to the shaft AD is vertical and the system is in equilibrium, determine (a) the magnitude of the torque $\mathbf{M}_F$, (b) the reactions from the three bearings.

Fig. P8.67

Fig. P8.68

8.69* Solve Prob. 8.68 for the case in which the crosspiece attached to the shaft DF is vertical.

8.70 Determine the magnitude of the force $\mathbf{F}$ required to release the self-locking brace of a shelf which supports a 60-lb crate as shown.

Fig. P8.70

Fig. P8.71

8.71 The pair of meshing bevel gears shown is in equilibrium. If the pitch angle $\beta = 30°$, determine the magnitude of the torque $\mathbf{M}$.

(a)

(b)

Fig. P8.72

8.72 A safety device to prevent overloading is shown in (a). The breaking strength of the shear pin at S is 240 lb. When $P = P_{max}$, the shear pin at S will break and thus release the load as shown in (b). If $L = 1.5$ in., determine the value of P_{max}. (*Hint*. Note symmetry in the system.)

8.73* Solve Prob. 8.72 if $L = 1.25$ in.

8.74 If the gear train shown is in equilibrium, determine the tension T.

Dimensions in millimeters

Fig. P8.74

8.75 A 99-N force **P** is exerted on the handle of the crusher as shown. For $a/b = 40/9$, determine the magnitude of the crushing force **Q**.

Fig. P8.75

8.76* Solve Prob. 8.75 for $a/b = 60/11$.

8.77 A bar *DE* is to be cut by the cutter shown. Determine the magnitude of the cutting force in terms of the applied force **P**.

Dimensions in millimeters

Fig. P8.77

8.78 Half of the weight *W* of an automobile hood is held in equilibrium with the hinge counterbalance arrangement as shown. Find the force in the spring *CH*.

Fig. P8.78

8.79 Determine the forces in the hydraulic cylinders *DE* and *CF* of the dump truck in the position shown.

Fig. P8.79

8.80 Three hydraulic cylinders are used to control the action of the backhoe bucket. For a 1170-N load acting as shown, determine (a) the force developed in each cylinder, (b) the reactions from the pins at *A* and *D* to the member *ABCD*.

Fig. P8.80

BEAMS AND CABLES

★8.15 Internal Forces and Moments in Multiforce Members

When loads are applied to a structure, internal forces and moments are developed in the cross sections of its multiforce members which act to withstand the effects of the applied loads. We shall here study such forces and moments in the cross sections of multiforce members.

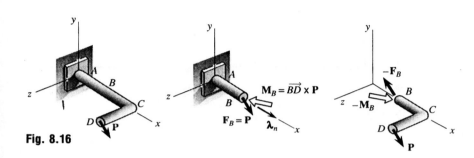

Fig. 8.16

In Fig. 8.16(a), a cantilever bent rod $ABCD$ carrying a force $\mathbf{P}$ at its free end D is shown. Using the procedure of *resolving a given force into a force-moment system at another point*, as presented in Sec. 5.1, we find that the force $\mathbf{F}_B$ and the moment $\mathbf{M}_B$ acting in the cross section at B on the end of the portion AB of the bent rod are

$$\mathbf{F}_B = \mathbf{P} \qquad \mathbf{M}_B = \overrightarrow{BD} \times \mathbf{P} \qquad (8.5)$$

which are shown in Fig. 8.16(b).[†] By Newton's third law, the force and the moment acting in the cross section at B on the end of the portion BCD must be $-\mathbf{F}_B$ and $-\mathbf{M}_B$, respectively, and are shown in Fig. 8.16(c). Thus, the force and the moment acting in a cross section of a member may, equivalently, be determined by *considering the equilibrium* of a portion of the member as divided by the cross section.[††]

Unless otherwise stated, all cross sections are to be understood as right cross sections. Let the *unit vector* perpendicular to the cross section at B and pointing outward at the end of the chosen portion AB be $\boldsymbol{\lambda}_n$ as shown in

---

[†]If there are other forces as well as moments acting on the portion BCD of the bent rod, each of the forces may similarly be resolved into a force at B and a moment. The vector sum of the forces brought to B by the resolutions will give the value of $\mathbf{F}_B$. Similarly, the vector sum of the moments acting on the portion BCD and the moments brought forth by the resolutions will give the value of $\mathbf{M}_B$.

[††]If the member is not in equilibrium, the forces and moments in its cross sections should be computed from the kinetics of the member.

Fig. 8.16(b). By resolving the force $\mathbf{F}_B$ as well as the moment $\mathbf{M}_B$ in the cross section at B into two components, one perpendicular and the other parallel to the cross section, we write

$$\mathbf{F}_B = \mathbf{F}_\perp + \mathbf{F}_\parallel \qquad \mathbf{M}_B = \mathbf{M}_\perp + \mathbf{M}_\parallel \qquad (8.6)$$

where

$$\mathbf{F}_\perp = (\boldsymbol{\lambda}_n \cdot \mathbf{F}_B)\boldsymbol{\lambda}_n \qquad \mathbf{F}_\parallel = \mathbf{F}_B - \mathbf{F}_\perp$$

$$\mathbf{M}_\perp = (\boldsymbol{\lambda}_n \cdot \mathbf{M}_B)\boldsymbol{\lambda}_n \qquad \mathbf{M}_\parallel = \mathbf{M}_B - \mathbf{M}_\perp \qquad (8.7)$$

The perpendicular component $\mathbf{F}_\perp$ is called the *normal force* or the *axial force*, while the parallel component $\mathbf{F}_\parallel$ is called the *shear* or the *shearing force* in the cross section. On the other hand, the perpendicular component $\mathbf{M}_\perp$ is called the *torque*, while the parallel component $\mathbf{M}_\parallel$ is called the *bending moment* in the cross section. (Cf. Sec. 4.18.)

EXAMPLE 8.15

In the cross section at B on the end of the portion AB of the cantilever bent rod shown, determine (a) the internal force $\mathbf{F}_B$, (b) the normal force $\mathbf{F}_\perp$, (c) the shear $\mathbf{F}_\parallel$, (d) the internal moment $\mathbf{M}_B$, (e) the torque $\mathbf{M}_\perp$, (f) the bending moment $\mathbf{M}_\parallel$.

Solution. Applying Eqs. (8.5) and (8.7), we write

$$\mathbf{P} = 8\mathbf{i} - 6\mathbf{j} \text{ kN} \qquad \boldsymbol{\lambda}_n = \mathbf{i}$$

$$\mathbf{F}_B = \mathbf{P} \qquad\qquad \mathbf{F}_B = 8\mathbf{i} - 6\mathbf{j} \text{ kN} \qquad \blacktriangleleft$$

$$\mathbf{F}_\perp = (\boldsymbol{\lambda}_n \cdot \mathbf{F}_B)\boldsymbol{\lambda}_n \qquad\qquad \mathbf{F}_\perp = 8\mathbf{i} \text{ kN} \qquad \blacktriangleleft$$

$$\mathbf{F}_\parallel = \mathbf{F}_B - \mathbf{F}_\perp \qquad\qquad \mathbf{F}_\parallel = -6\mathbf{j} \text{ kN} \qquad \blacktriangleleft$$

$$\overrightarrow{BD} = 2\mathbf{i} + 3\mathbf{k} \text{ m} \qquad \mathbf{M}_B = \overrightarrow{BD} \times \mathbf{P}$$

$$\mathbf{M}_B = 18\mathbf{i} + 24\mathbf{j} - 12\mathbf{k} \text{ kN·m} \qquad \blacktriangleleft$$

$$\mathbf{M}_\perp = (\boldsymbol{\lambda}_n \cdot \mathbf{M}_B)\boldsymbol{\lambda}_n \qquad\qquad \mathbf{M}_\perp = 18\mathbf{i} \text{ kN·m} \qquad \blacktriangleleft$$

$$\mathbf{M}_\parallel = \mathbf{M}_B - \mathbf{M}_\perp \qquad \mathbf{M}_\parallel = 24\mathbf{j} - 12\mathbf{k} \text{ kN·m} \qquad \blacktriangleleft$$

Developmental Exercise

D8.39 A short section of semicircular shell having a mass of 10 kg rests on a smooth horizontal surface as shown. In the cross section at B on the end of the portion AB of the shell, determine (a) the outward unit vector λ_n which is perpendicular to the cross section, (b) the internal force $\mathbf{F}_B$, (c) the normal force $\mathbf{F}_\perp$, (d) the shear $\mathbf{F}_\parallel$, (e) the internal moment $\mathbf{M}_B$, (f) the torque $\mathbf{M}_\perp$, (g) the bending moment $\mathbf{M}_\parallel$.

Fig. D8.39

★8.16   Shears and Bending Moments in Beams

A *beam* is a structural member designed to support mainly transverse loads. In general, beams are *long, straight, prismatic,* and *horizontal*. Since transverse loads are perpendicular to the axes of beams, they will induce shears and bending moments in the cross sections of beams.

A *load* may be a force or a moment, which may be distributed or concentrated on a beam. The basic types of *loads carried by beams*, as illustrated in Fig. 8.17, are usually expressed in (1) N/m or lb/ft for *distributed forces*, (2) N or lb for *concentrated forces*, (3) N·m/m or lb·ft/ft for *distributed moments*, and (4) N·m or lb·ft for *concentrated moments*. Note that the symbols w and $\tilde{m}$ in Fig. 8.17 denote the intensities of the distributed force and the distributed moment, respectively. Furthermore, the distributed moment $\tilde{m}$ and the concentrated moment T may graphically be represented as shown in parts (b) and (c) of Fig. 8.17.

Distributed and concentrated forces	Distributed and concentrated moments	Distributed and concentrated moments
(a)	(b)	(c)

Fig. 8.17

For studying shears and bending moments in the cross sections of beams which are horizontal and under the action of force systems in the vertical plane, we adopt the following *sign convention*:

1. *The shear in a cross section of a beam is positive when it acts upward on the left end of the right portion* (or downward on the right end of the left portion) *of the beam as divided by the cross section.*
2. *The bending moment in a cross section of a beam is positive when it tends to cause compression at the top* (or tension at the bottom) *of the beam at that cross section.*

Positive shear and positive bending moment	Effect of positive shear	Effect of positive bending moment
(a)	(b)	(c)

Fig. 8.18

The positive shear and positive bending moment in a cross section are illustrated in Fig. 8.18(a), where Newton's third law is observed. Note that the effects of the positive shear and positive bending moment on a beam as illustrated in Fig. 8.18(b) and (c) are somewhat exaggerated. If the shear or bending moment is not zero and cannot be classified as positive, it is negative.

According to the above sign convention and Sec. 8.15, two *general rules* for computing shears and bending moments may be stated as follows:

1. The *shear in a cross section* of a beam is equal to the algebraic sum of all forces which lie *on a chosen side* of the cross section of the free body of the beam. If the *left side* is chosen, *upward forces are positive* while downward forces are negative. If the *right side* is chosen, the reverse is true.
2. The *bending moment in a cross section* of a beam is equal to the algebraic sum of moments, about the centroid of the cross section, of all forces and couples which lie *on a chosen side* of the cross section of the free body of the beam. The moment is positive if its action tends to cause *compression at the top* of the beam at that cross section.

Developmental Exercises

D8.40   Are beams usually vertical, horizontal, or inclined?

D8.41   Name the four basic types of loads carried by beams.

EXAMPLE 8.16

Determine the shears and the bending moments in the cross sections at *E* and *F* of the beam shown.

Solution. The reactions from the supports are determined by considering the equilibrium of the free body of the entire beam; they are

$$\mathbf{B} = 34 \text{ kN} \uparrow \qquad \mathbf{D} = 7 \text{ kN} \uparrow$$

Suppose that we choose the *left side* of E to compute the shear V_E and the bending moment M_E. Noting the replacement of the distributed load with its resultant force of 18 kN at G as shown, we write

$$V_E = -18 + 34 \qquad V_E = +16 \text{ kN} \blacktriangleleft$$

$$M_E = -4(18) + 2(34) \qquad M_E = -4 \text{ kN·m} \blacktriangleleft$$

To find the shear V_F and the moment M_F in the cross section at F, we note that it is simpler to work with the *right side* of F. Thus, we write

$$V_F = -7 \qquad V_F = -7 \text{ kN} \blacktriangleleft$$

$$M_F = 2(7) \qquad M_F = +14 \text{ kN·m} \blacktriangleleft$$

Note that these same results may be obtained by considering the *equilibrium* of a portion of the member.

Developmental Exercises

D8.42 Refer to Example 8.16. Compute V_E and M_E by working with the *right side* of E.

D8.43 Refer to Example 8.16. Compute V_F and M_F by working with the *left side* of F.

EXAMPLE 8.17

A beam with a bracket is loaded as shown. Determine the equations for the shear and bending moment in the cross sections of the portions AB and BC.

Solution. The 1.2-kip force at D is resolved into a 1.2-kip force and a moment of 7.2 kip·ft at B as shown. The reactions from the supports as shown are determined by considering the equilibrium of the free body of the entire beam.

Portion AB. Let an arbitrary cross section in this portion be located at E which is at a distance x from A. Summing the forces as well as the moments about E of the forces which lie on the *left side* of E, we write

$$V_{AB} = 2.7 - 0.2x \qquad\qquad V_{AB} = 2.7 - 0.2x \text{ kips} \blacktriangleleft$$

$$M_{AB} = x(2.7) - \frac{x}{2}(0.2x) \qquad M_{AB} = 2.7x - 0.1x^2 \text{ kip·ft} \blacktriangleleft$$

Portion BC. Let an arbitrary cross section in this portion be located at F which is at a distance x from A. Noting that it is simpler to work with the *right side* of F, we write

$$V_{BC} = -0.9 \qquad\qquad V_{BC} = -0.9 \text{ kips} \blacktriangleleft$$

$$M_{BC} = (24 - x)(0.9) \qquad M_{BC} = 21.6 - 0.9x \text{ kip·ft} \blacktriangleleft$$

Developmental Exercises

D8.44 Describe the general rule for computing the shear in a cross section of a beam.

D8.45 Describe the general rule for computing the bending moment in a cross section of a beam.

D8.46 Is it true that an upward force on a beam always has a positive contribution to the bending moment in any cross section of the beam?

★8.17 Shear and Bending-Moment Diagrams

A graph showing the shear V in the cross section of a beam as a function of the position coordinate x of the cross section is called a *shear diagram*. Similarly, a graph showing the bending moment M in the cross section of a beam as a function of the position coordinate x of the cross section is called a *bending-moment diagram*. The curves in these diagrams are called the *shear* and *bending-moment curves*, respectively.

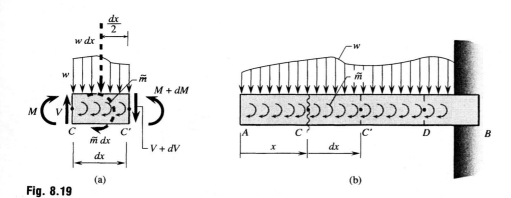

Fig. 8.19

Note that a concentrated force of magnitude P on a beam may be considered as a distributed force of intensity $w = P/\Delta x$ over a very short length Δx of the beam. Similarly, a concentrated moment of magnitude T on a beam may be considered as a distributed moment of intensity $\tilde{m} = T/\Delta x$ over a very short length Δx of the beam. Thus, let us consider the free body shown in Fig. 8.19(a) for the differential beam element CC' of the beam in Fig. 8.19(b), where the beam is subjected to a *downward* distributed force of intensity w and a *clockwise* distributed moment of intensity $\tilde{m}$. Since the beam element as well as the beam is in equilibrium, we refer to Fig. 8.19(a) and write

$$+\uparrow \Sigma F_y = 0: \quad V - w\, dx - (V + dV) = 0$$

$$\frac{dV}{dx} = -w \tag{8.8}$$

$$+\circlearrowleft \Sigma M_{C'} = 0: \quad -M - (dx)V + \frac{dx}{2}(w\, dx) - \tilde{m}\, dx + (M + dM) = 0$$

$$dM = V\, dx + \tilde{m}\, dx - \frac{w}{2}(dx)^2 \approx V\, dx + \tilde{m}\, dx$$

$$\frac{dM}{dx} = V + \tilde{m} \tag{8.9}^\dagger$$

Let $x = x_C$, $V = V_C$, $M = M_C$ at the point C and $x = x_D$, $V = V_D$, $M = M_D$ at the point D of the beam. Integrating Eqs. (8.8) and (8.9) from

---

[†]This equation reduces to $dM/dx = V$ if there is no distributed moment acting on the beam. However, without this equation, the slopes of the lines in the bending-moment diagram in Example 8.21 would be inexplicable. Further discussions on distributed moments and couple stresses may be found in texts on continuum mechanics.

the point C to the point D of the beam in Fig. 8.19(b), we obtain

$$V_D - V_C = -\int_{x_C}^{x_D} w \, dx \tag{8.10}$$

$$M_D - M_C = \int_{x_C}^{x_D} V \, dx + \int_{x_C}^{x_D} \tilde{m} \, dx \tag{8.11}$$

From Eqs. (8.8) and (8.9), we can show that

$$\frac{d^2V}{dx^2} = -\frac{dw}{dx} \tag{8.12}$$

$$\frac{d^2M}{dx^2} = -w + \frac{d\tilde{m}}{dx} \tag{8.13}$$

Notice in Eqs. (8.8) through (8.13) that w and $\tilde{m}$ are *positive* when the distributed force and the distributed moment act in the *downward* and *clockwise* directions, respectively; otherwise, they are *negative*. From these six equations, we state the following six *relations*, which are useful in plotting shear and bending-moment curves:

1. The *slope of the shear curve* at any point is equal to negative of the intensity of the distributed force on the beam at that point. Thus, the shear curve has a vertical slope at the point where a concentrated force is applied.
2. The *slope of the bending-moment curve* at any point is equal to the sum of the shear in the cross section and the intensity of the distributed moment on the beam at that point. Furthermore, the bending-moment curve has a vertical slope at the point where a concentrated moment is applied.
3. The *change in shear* from C to D of the beam is equal to negative of the sum of all forces on the beam between C and D.
4. The *change in bending moment* from C to D of the beam is equal to the "area" bounded by the shear curve and the x axis between C and D plus the sum of all moments on the beam between C and D.
5. The *concavity of the shear curve* in a certain portion is upward if $dw/dx < 0$ in that portion of the beam; and downward, if $dw/dx > 0$. Of course, the portion of the shear curve is a straight line if $dw/dx = 0$ in that portion.
6. The *concavity of the bending-moment curve* in a certain portion is upward if $-w + d\tilde{m}/dx > 0$ in that portion of the beam; and downward, if $-w + d\tilde{m}/dx < 0$. Of course, the portion of the bending-moment curve is a straight line if $-w + d\tilde{m}/dx = 0$ in that portion.

From the six relations, we infer that the shear and bending-moment diagrams have the following special *features*: (a) a shear curve has vertical jumps *at and only at* the points where concentrated forces are applied to the beam, (b) *each vertical jump in the shear curve* at a point is equal to the concentrated force at that point, (c) a bending-moment curve has vertical jumps *at and only at* the points where concentrated moments are applied to the beam, (d) *each vertical jump in the bending-moment curve* at a point is equal to the concentrated moment at that point, (e) *each point of inflection* (if any) *in the shear curve* is located at the point where the value of dw/dx

on the beam changes its sign, and (f) *each point of inflection* (if any) *in
the bending-moment curve* is located at the point where the value of $-w +
d\tilde{m}/dx$ on the beam changes its sign.

Developmental Exercise

D8.47 In proceeding from the left to the right along a shear curve of a beam,
does the vertical jump in the curve at a certain point go in the same direction and
have the same magnitude as the concentrated force there?

EXAMPLE 8.18

Draw the shear and bending-moment diagrams for the beam shown.

Solution. The reactions are first determined. The shear and bending-
moment diagrams are then drawn according to Secs. 8.16 and 8.17. In particular, note that the vertical jumps in the shear curve are equal to the corresponding concentrated forces.

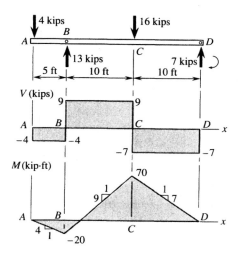

Developmental Exercise

D8.48 Describe the six *relations* which are useful in drawing shear and bending-
moment diagrams.

EXAMPLE 8.19

Draw the shear and bending-moment diagrams for the beam in Example 8.16.

Solution. We first draw the free-body diagram of the beam as shown. The shear and bending-moment diagrams are then drawn according to Secs. 8.16 and 8.17. Note that, instead of computing the *area* bounded by the shear curve and the x axis in the range $0 \leq x \leq 3$ m, we may compute the value of the bending moment at B directly from the free body of the beam. Thus, replacing the distributed force with its resultant force at G, we write

$$M_B = -2(18) \qquad M_B = -36 \text{ kN·m}$$

Furthermore, note that (1) $dV/dx = 0$ at $x = 3$ m, (2) $dM/dx = 0$ at $x = 0$, (3) the shear curve between A and B is concave upward, and (4) the bending-moment curve between A and B is concave downward.

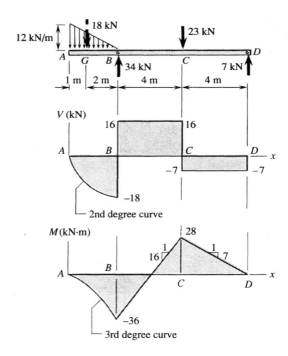

Developmental Exercise

D8.49 Refer to the solution in Example 8.19. Give the values of the slopes of
(a) the shear curve at $x = 0$, (b) the bending-moment curve at the point just to
the left of $x = 3$ m, (c) the bending-moment curve at the point just to the right
of $x = 3$ m.

EXAMPLE 8.20

Draw the shear and bending-moment diagrams for the beam in Example
8.17.

Solution. The 1.2-kip force at D is replaced with a 1.2-kip force and a
moment of 7.2 kip·ft at B as shown. After drawing the free-body diagram
of the beam as shown, the shear and bending-moment diagrams are drawn
according to Secs. 8.16 and 8.17. In particular, note that the vertical jumps
in the shear and bending-moment curves at $x = 12$ ft are equal to the concentrated force and concentrated moment on the beam at B, respectively.
Furthermore, note that the bending-moment curve between A and B is concave downward.

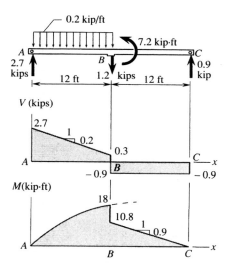

Developmental Exercises

D8.50 In proceeding from the left to the right along a bending-moment curve,
do the vertical jump in the curve and the corresponding concentrated moment on
the beam have the same magnitude?

D8.51 In proceeding from the left to the right along a bending-moment curve,
does the vertical jump in the curve at a certain point go upward or downward if
the concentrated moment at that point is (a) clockwise, (b) counterclockwise?

EXAMPLE 8.21

A beam is subjected to a uniformly distributed moment of intensity 10 kN·m/m as shown. Draw the shear and bending-moment diagrams for the beam.

Solution. We first draw the free-body diagram as shown. The shear and bending-moment diagrams are then drawn according to Secs. 8.16 and 8.17. Note that the beam experiences a constant shear of -15 kN in the portion BC. In particular, note from Eq. (8.9) that the slopes of the bending-moment curve are

$$\left(\frac{dM}{dx}\right)_{AB} = (V + \tilde{m})_{AB} = -15 + 10 = -5 \ (\text{kN·m/m})$$

$$\left(\frac{dM}{dx}\right)_{BC} = (V + \tilde{m})_{BC} = 0 + 10 = 10 \ (\text{kN·m/m})$$

Furthermore, note from Eq. (8.11) that the changes in bending moment between key points are

$$M_B - M_A = \int_{x_A}^{x_B} V \, dx + \int_{x_A}^{x_B} \tilde{m} \, dx$$

$$= -15(2) + 10(2) = -10 \ (\text{kN·m})$$

$$M_C - M_B = \int_{x_C}^{x_B} V \, dx + \int_{x_C}^{x_B} \tilde{m} \, dx$$

$$= 0 + 10(1) = 10 \ (\text{kN·m})$$

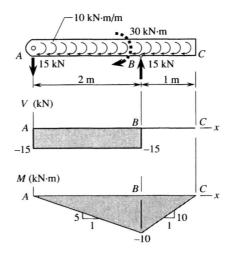

Developmental Exercise

D8.52   Work Example 8.21 if the beam is cantilevered at A and the roller support at B is removed.

PROBLEMS

8.81   In the cross section at A of the cantilever bent rod shown, determine (a) the normal force $\mathbf{F}_\perp$, (b) the shear $\mathbf{F}_\parallel$, (c) the torque $\mathbf{M}_\perp$, (d) the bending moment $\mathbf{M}_\parallel$.

Fig. P8.81

Fig. P8.82*

8.82*   A short section of semicircular shell having a weight W and a radius R is shown. In the cross section at B on the end of the portion BC of the shell, determine, in terms of W, R, and θ, (a) the normal force $\mathbf{F}_\perp$, (b) the shear $\mathbf{F}_\parallel$, (c) the torque $\mathbf{M}_\perp$, (d) the bending moment $\mathbf{M}_\parallel$.

8.83*   Refer to Prob. 8.82*. Determine (a) the value of θ for which the normal force in the cross section at B is maximum, (b) the magnitude of the maximum normal force.

8.84* through **8.98***   Draw the shear and bending-moment diagrams for the beam shown.

Fig. P8.84*

Fig. P8.85

Fig. P8.86*

Fig. P8.87*

Fig. P8.88

Fig. P8.89*

Fig. P8.90*

Fig. P8.91

Fig. P8.92*

Fig. P8.93*

Fig. P8.94

Fig. P8.95*

Fig. P8.96*

Fig. P8.97

Fig. P8.98*

8.99 Determine the equations for the shear and bending moment in the beam shown.

8.100* The shear diagram for a beam, which has no moments applied between the ends, is shown. Draw for the beam (a) the free-body diagram, (b) the bending-moment diagram.

Fig. P8.99

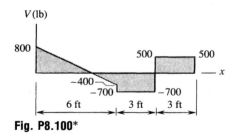

Fig. P8.100*

★8.18   Cables and the Governing Differential Equation

A *cable* is a cordlike element which is assumed to have great tensile strength and negligible resistance to bending. In other words, a cable is usually assumed to be a flexible structural member and the *force developed* anywhere in the cable is always a *tensile force tangent to the centerline of the cable there*. In elementary study, the elongation of the cable under loading is

assumed to be negligible; i.e., all cables are assumed to be *inextensible* as
well as *flexible*.

Cables are commonly used in suspension bridges, transmission lines,
telephone lines, guy wires, aerial tramways, cranes, derricks, and many
other applications. The applied load, tension, span, sag, and length are important parameters in the design of cables.

A cable may carry *vertical concentrated loads* as shown in Fig. 8.20(a),
or it may carry a *vertical distributed load* as shown in Fig. 8.20(b). In some
cases, the weight of the cable is taken as negligible compared with the applied loads. In other cases, the weight of the cable may be an appreciable
load or the sole load and cannot be neglected. For instance, the weight of
the cable in Fig. 8.20(c) is the sole load carried by the cable.

<table>
<tr><td>(a)</td><td>(b)</td><td>(c)</td></tr>
</table>

Fig. 8.20

When a cable carries only concentrated loads, such as that illustrated in
Fig. 8.20(a), the weight of the cable is usually taken as negligible. In such
a case, any portion of the cable between successive loads is considered as a
two-force body, and the force of tension developed in the cable is directed
along the cable. Thus, the method of solution outlined in Sec. 3.10 is applicable to the solution of problems involving cables carrying only concentrated loads. Furthermore, the concept of rigid-body equilibrium is also applicable as illustrated in Example 5.10.

The equilibrium of a cable carrying a *vertical distributed load w* may be
studied by considering the free body of a typical differential element of the
cable as shown in Fig. 8.21, where T is the tension in the cable and w is
the intensity of the vertical distributed load *measured in force per unit of*

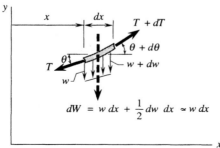

$$dW = w\,dx + \frac{1}{2}\,dw\,dx \approx w\,dx$$

Fig. 8.21

horizontal length. Note here that *the positive direction of w is downward.* For equilibrium of the element as shown, we write

$$\xrightarrow{+} \Sigma F_x = 0: \quad (T + dT)\cos(\theta + d\theta) - T\cos\theta = 0$$

$$+\uparrow \Sigma F_y = 0: \quad (T + dT)\sin(\theta + d\theta) - T\sin\theta - w\,dx = 0$$

Expanding the sine and cosine of the sum of two angles, recognizing that $\cos d\theta \approx 1$ and $\sin d\theta \approx d\theta$, and neglecting higher-order terms in these two equations, we obtain

$$dT\cos\theta - T\sin\theta\,d\theta = 0$$

$$dT\sin\theta + T\cos\theta\,d\theta - w\,dx = 0$$

which may be written as

$$d(T\cos\theta) = 0 \qquad (8.14)$$

$$d(T\sin\theta) - w\,dx = 0 \qquad (8.15)$$

Equation (8.14) indicates the property that

$$T\cos\theta = T_o \qquad (8.16)$$

where T_o is a constant equal to the tension in the cable at the point where θ and hence the slope of the cable are zero. This means that the horizontal component of the tension in the cable is a constant T_o, which is clear because *the cable carries only vertical loads.* Since $-1 \le \cos\theta \le 1$, it is obvious that the *minimum tension* in the cable is T_o. Noting that $\tan\theta = dy/dx$, we may obtain from Eqs. (8.15) and (8.16) that

$$d\left(T_o\frac{dy}{dx}\right) - w\,dx = 0$$

$$\boxed{\frac{d^2y}{dx^2} = \frac{w}{T_o}} \qquad (8.17)$$

Equation (8.17) is the *governing differential equation* for the cable carrying a vertical distributed load. The conditions at the fixed ends of the cable are called the *boundary conditions* of the cable. The shape of the cable is defined by the functional relation $y = f(x)$ which satisfies Eq. (8.17) and the boundary conditions of the cable.

EXAMPLE 8.22

The cable AE loaded as shown is in equilibrium. Determine (a) the elevations y_B and y_D, (b) the tension T_{AB} in the portion AB.

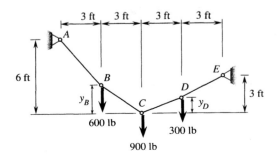

Solution. We first note that there are six unknowns involved in the equilibrium of the cable. They are the two unknown elevations y_B and y_D and the four unknown tensions T_{AB}, T_{BC}, T_{CD}, and T_{DE}. The equilibrium of each of the three loading points B, C, and D yields two independent equilibrium equations. Thus, there are six independent equations for the solution of the six unknowns. However, this problem may more directly be solved by applying the concept of the equilibrium of rigid bodies as follows:

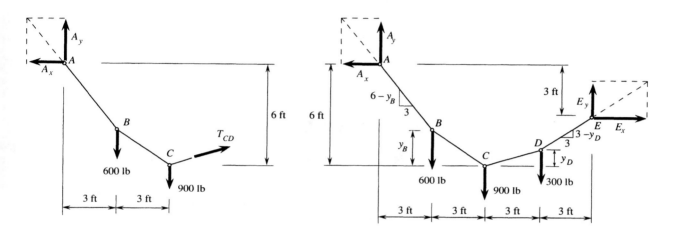

Portion ABC

$$+\ \text{\reflectbox{$\circlearrowright$}}\ \Sigma M_C = 0: \quad 6A_x - 6A_y + 3(600) = 0$$

Entire cable

$$+\ \text{\reflectbox{$\circlearrowright$}}\ \Sigma M_E = 0: \quad 3A_x - 12A_y + 9(600) + 6(900) + 3(300) = 0$$

$$\xrightarrow{+}\ \Sigma F_x = 0: \quad -A_x + E_x = 0$$

$$+\uparrow\ \Sigma F_y = 0: \quad A_y + E_y - 600 - 900 - 300 = 0$$

Solving these four equations, we obtain

$$A_x = 900 \text{ lb} \qquad A_y = 1200 \text{ lb} \qquad E_x = 900 \text{ lb} \qquad E_y = 600 \text{ lb}$$

Since portions AB and DE are two-force bodies, the resultant reaction from A must be equal to $-\mathbf{T}_{AB}$, and the resultant reaction from E must be equal to $-\mathbf{T}_{ED}$. Noting the slopes of the portions AB and ED as indicated in the free-body diagram of the entire cable, we write

$$\frac{6 - y_B}{3} = \frac{A_y}{A_x} = \frac{1200}{900} \qquad\qquad y_B = 2 \text{ ft} \ \blacktriangleleft$$

$$\frac{3 - y_D}{3} = \frac{E_y}{E_x} = \frac{600}{900} \qquad\qquad y_D = 1 \text{ ft} \ \blacktriangleleft$$

$$T_{AB}^2 = A_x^2 + A_y^2 = (900)^2 + (1200)^2 \qquad T_{AB} = 1500 \text{ lb} \ \blacktriangleleft$$

Developmental Exercises

D8.53 What are the usual assumptions for cables?

D8.54 Refer to Example 8.22. Determine the tension T_{DE} in the portion DE of the cable.

D8.55 Refer to Example 8.22. Determine the tensions T_{BC} and T_{CD} in the portions BC and CD of the cable, respectively.

D8.56 Refer to the cable in Fig. 8.20(b). Let T_A, T_B, and T_C be the tensions in the cable at A, B, and C, respectively. Using Eq. (8.16) as a guide, determine which of the following is true: (a) $T_A > T_B > T_C$, (b) $T_B > T_C > T_A$, (c) $T_C > T_A > T_B$, (d) $T_B > T_A > T_C$, (e) $T_A = T_B = T_C$.

D8.57 In Eq. (8.17), is the positive direction of the distributed load w the same as that of the ordinate y?

D8.58 In Eq. (8.17), is w measured along the cable or along the horizontal?

★8.19 Parabolic Cable

Let us here consider a cable AB carrying a *vertical load distributed with a constant intensity w along the horizontal* as shown in Fig. 8.22(a). The weight of the cable is not uniformly distributed along the horizontal but is assumed to be negligibly small compared with the load carried. Such a cable closely simulates the cables of a suspension bridge where the weight of the roadway is the dominant load and may be assumed to be uniformly distributed along the horizontal.

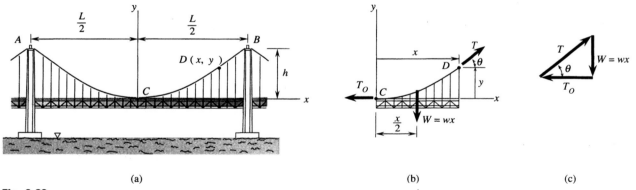

(a)   (b)   (c)

Fig. 8.22

We shall show that the aforementioned cable AB hangs in a *parabolic arc*. As shown in Fig. 8.22(a), the cable has a span L and sag h, and the origin of coordinates is at the *lowest point C* of the span where the slope dy/dx of the cable is zero. Knowing that w and T_o are not functions of x, we may integrate Eq. (8.17) to yield

$$\frac{dy}{dx} = \frac{w}{T_o}x + C_1$$

$$y = \frac{w}{2T_o}x^2 + C_1 x + C_2$$

where C_1 and C_2 are constants of integration. Since $dy/dx = y = 0$ at $x = 0$, we find that $C_1 = C_2 = 0$. Thus, we have

$$y = \frac{w}{2T_o}x^2 \qquad (8.18)$$

which we see is a *vertical parabola*. Note that this equation may also be obtained by considering the equilibrium of the portion CD, shown in Fig. 8.22(b), as follows:

$$+\circlearrowleft \; \Sigma M_D = 0: \quad \frac{x}{2}(wx) - yT_o = 0$$

Solving for y from this equation, we obtain Eq. (8.18). The boundary conditions of the cable at A and B require that $y = h$ when $x = \pm L/2$. Substitution of these values of y and x into Eq. (8.18) yields

$$T_o = \frac{wL^2}{8h} \qquad y = \frac{4h}{L^2}x^2 \qquad (8.19)$$

The equilibrium of the portion CD shown in Fig. 8.22(b) requires that the three force vectors there form a right triangle as shown in Fig. 8.22(c). Thus, the tension T is given by

$$T^2 = T_o^2 + (wx)^2 = \left(\frac{wL^2}{8h}\right)^2 + (wx)^2$$

$$T = \frac{w}{8h}(64h^2x^2 + L^4)^{1/2} \qquad (8.20)$$

The tension T is maximum when $x = L/2$. Hence, the maximum tension is

$$T_{\max} = \frac{wL}{8h}(16h^2 + L^2)^{1/2} \qquad (8.21)$$

The length S of the complete cable AB may be studied from the relation

$$ds = [(dx)^2 + (dy)^2]^{1/2} = \left[1 + \left(\frac{dy}{dx}\right)^2\right]^{1/2} dx$$

We find from Eqs. (8.19) that $dy/dx = 8hx/L^2$. Thus, we write

$$S = 2\int_0^{L/2} ds = 2\int_0^{L/2}\left(1 + \frac{64h^2x^2}{L^4}\right)^{1/2} dx$$

$$= 2\int_0^{L/2}\left(1 + \frac{32h^2x^2}{L^4} - \frac{512h^4x^4}{L^8} + \cdots\right) dx$$

$$S = L\left[1 + \frac{8}{3}\left(\frac{h}{L}\right)^2 - \frac{32}{5}\left(\frac{h}{L}\right)^4 + \cdots\right] \qquad (8.22)$$

The series in Eq. (8.22) can be shown to converge for values of $h/L \le \frac{1}{4}$. In most cases, h is much smaller than $L/4$, and the first three terms of the series will give a sufficiently good approximation. If the supports of a parabolic cable have *different elevations*, the relations expressed in the preceding equations may be applied to the part of the cable on one side of the *lowest point* of the cable. The other side is studied separately.

We know that a cable hanging under its own weight is not loaded uniformly along the horizontal. However, the shape of such a cable may be approximated by a parabola when the cable is sufficiently taut. Cables hanging under their own weight are studied in Sec. 8.20.

EXAMPLE 8.23

The mass density of the pipe shown is 10 kg/m. Determine (a) the minimum and maximum tensions in the cable, (b) the length S_{AC} of the segment AC of the cable.

Solution. Let the lowest point C of the cable be chosen as the origin of the coordinates as shown. The intensity of the load is $w = 10g$ N/m $= 98.1$ N/m measured along the horizontal, and the shape of the cable is a *parabola* defined by Eq. (8.18); i.e.,

$$y = \frac{98.1}{2T_o} x^2$$

where T_o is the minimum tension in the cable at C. Imposing the boundary conditions of the cable at A and B on this equation and noting the relation between a and b, we write

$$4 = \frac{98.1}{2T_o}(-a)^2 \qquad 8 = \frac{98.1}{2T_o}(b)^2 \qquad a + b = 50$$

These three equations yield

$$a = 20.711 \text{ m} \qquad b = 29.289 \text{ m} \qquad T_o = 5.26 \times 10^3 \text{ N}$$

$$T_{min} = 5.26 \text{ kN} \quad \blacktriangleleft$$

We note from Eq. (8.16) that T is maximum when θ is as close to 90° as possible. Since θ is closest to 90° at B, we write

$$\tan\theta_B = \left.\frac{dy}{dx}\right|_{x=b} = \frac{98.1}{T_o}b = \frac{98.1(29.289)}{5.26 \times 10^3} \qquad \theta_B = 28.6°$$

$$T_{max} = T_B = T_o \sec\theta_B = 5.99 \times 10^3 \text{ N}$$

$$T_{max} = 5.99 \text{ kN} \quad \blacktriangleleft$$

Note that this answer may alternatively be obtained by substituting $L = 2b = 58.58$ m, $h = 8$ m, and $w = 98.1$ N/m into Eq. (8.21).

The length S_{AC} of the segment AC of the cable may be computed by treating it as the left half of a parabolic cable of span $2a = 41.42$ m. Thus, substituting $S = 2S_{AC}$, $L = 41.42$ m, and $h = 4$ m into Eq. (8.22), we write

$$2S_{AC} = 41.42\left[1 + \frac{8}{3}\left(\frac{4}{41.42}\right)^2 - \frac{32}{5}\left(\frac{4}{41.42}\right)^4 + \cdots\right]$$

$$S_{AC} = 21.2 \text{ m} \blacktriangleleft$$

Developmental Exercises

D8.59 In using Eq. (8.18) to define the shape of a parabolic cable, how is the origin of coordinates chosen?

D8.60 Refer to Example 8.23. Determine the tension T_A in the cable at the support A.

D8.61 Refer to Example 8.23. Determine (a) the length S_{CB} of the segment CB of the cable, (b) the total length S_{AB} of the cable.

*8.20 Catenary Cable

Let us now consider a cable AB carrying a vertical load distributed with a constant intensity γ_L along the cable itself as shown in Fig. 8.23(a). The load intensity γ_L is *measured in force per unit of length along the cable*. A cable hanging under its own weight is loaded in this way. We shall show that such a cable hangs in the shape of a *hyperbolic cosine curve*—a *catenary*.

Choosing the lowest point C of the cable as the origin of coordinates, we draw the free-body diagram of an arbitrary segment CD of the cable as shown in Fig. 8.23(b), where T_o is still the minimum tension in the cable at

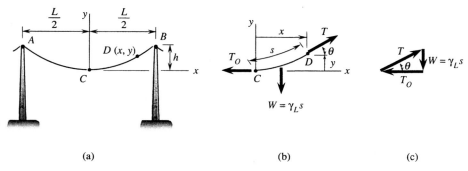

(a)   (b)   (c)

Fig. 8.23

C, s is the arc length of the cable measured from C to D, and the present load intensity parameter γ_L is related to the load intensity parameter w in Eq. (8.17) as follows:

$$\gamma_L \, ds = w \, dx \qquad w = \gamma_L \frac{ds}{dx}$$

$$w = \gamma_L \left[1 + \left(\frac{dy}{dx} \right)^2 \right]^{1/2} \tag{8.23}$$

For simplicity in the subsequent studies, we introduce the constant

$$c = \frac{T_o}{\gamma_L} \tag{8.24}$$

Substituting Eq. (8.23) into Eq. (8.17) and using Eq. (8.24), we write

$$\frac{d^2y}{dx^2} = \frac{1}{c} \left[1 + \left(\frac{dy}{dx} \right)^2 \right]^{1/2}$$

Substitution of $u = dy/dx$ into this equation yields

$$\frac{du}{dx} = \frac{1}{c}(1 + u^2)^{1/2} \qquad (1 + u^2)^{-1/2} \, du = \frac{1}{c} \, dx$$

Integrating from the point C (where $x = 0$ and $dy/dx = u = 0$) to the point D of the cable in Fig. 8.23(a), we write

$$\int_0^u (1 + u^2)^{-1/2} \, du = \frac{1}{c} \int_0^x dx$$

$$\ln[u + (1 + u^2)^{1/2}] = \frac{x}{c} \qquad e^{x/c} = u + (1 + u^2)^{1/2}$$

Solving for u, we find

$$u = \frac{dy}{dx} = \frac{1}{2}(e^{x/c} - e^{-x/c}) = \sinh \frac{x}{c}$$

This slope of the cable may be integrated from $C(0, 0)$ to $D(x, y)$ of the cable in Fig. 8.23(a) to yield

$$y = c \left(\cosh \frac{x}{c} - 1 \right) \tag{8.25}$$

Thus, the shape of the cable AB hanging under its own weight is defined by Eq. (8.25), which is a vertical *hyperbolic cosine curve* shifted a distance c downward and is commonly called a *catenary*.

The equilibrium of the portion CD shown in Fig. 8.23(b) requires that the three force vectors there form a right triangle as shown in Fig. 8.23(c). From this figure and Eq. (8.24), we write

$$\frac{dy}{dx} = \tan \theta = \frac{\gamma_L s}{T_o} = \frac{s}{c} \qquad s = c \frac{dy}{dx}$$

which together with Eq. (8.25) yields

$$s = c \sinh \frac{x}{c} \tag{8.26}$$

The tension T in the cable may be obtained from Fig. 8.23(c), Eq. (8.24), and Eq. (8.26) as follows:

$$T^2 = T_o^2 + (\gamma_L s)^2 = T_o^2 \left[1 + \left(\frac{s}{c} \right)^2 \right] = T_o^2 \left(1 + \sinh^2 \frac{x}{c} \right)$$

$$= T_o^2 \cosh^2 \frac{x}{c}$$

$$T = T_o \cosh \frac{x}{c} \qquad (8.27)$$

From Eqs. (8.24), (8.25), and (8.27), we also obtain that

$$T = T_o + \gamma_L y \qquad (8.28)$$

If the supports of a catenary cable have *different elevations*, the relations expressed in the preceding equations may be applied to the part of the cable on one side of the *lowest point* of the cable. The other side is studied separately. However, the analysis of such a catenary cable often leads to simultaneous transcendental equations whose solution would require a numerical method and a computer to be practical.

EXAMPLE 8.24

The transmission wire AB as shown has a mass density of 1 kg/m measured along its own length. Determine (a) the minimum and maximum tensions in the wire, (b) the total length of the wire.

Solution.   Let the lowest point C of the wire be chosen as the origin of the coordinates as shown. The intensity of the load is $\gamma_L = 1g$ N/m $= 9.81$ N/m measured along the wire. Since the slope and the coordinates of the wire at C are all zero, the shape of the wire is a *catenary* defined by

$$y = c \left(\cosh \frac{x}{c} - 1 \right)$$

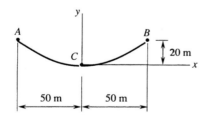

Imposing the boundary conditions at B (or A) on this equation, we write

$$20 = c \left(\cosh \frac{50}{c} - 1 \right)$$

This transcendental equation may be solved by trial and error, a numerical method (cf. Sec. B5.8 of App. B), or with a *digital root finder* to yield[†]

$$c = 65.59$$

[†]The solution of that transcendental equation with a *digital root finder* is illustrated in App. E.

From Eq. (8.24), we write

$$T_o = c\gamma_L = 65.59(9.81) = 643.4$$

$$T_{min} = T_o \qquad\qquad T_{min} = 643 \text{ N} \blacktriangleleft$$

Since the maximum tension occurs at B (or A), we apply Eq. (8.27) to obtain

$$T_{max} = T_B = T_o \cosh \frac{x_B}{c} = 643.4 \cosh \frac{50}{65.59} = 839.6$$

$$T_{max} = 840 \text{ N} \blacktriangleleft$$

Note that this answer may also be obtained from Eq. (8.28). Applying Eq. (8.26), we find the total length S_{AB} of the wire as follows:

$$S_{AB} = 2s_B = 2c \sinh \frac{x_B}{c} = 2(65.59) \sinh \frac{50}{65.59} = 109.97$$

$$S_{AB} = 110.0 \text{ m} \blacktriangleleft$$

Developmental Exercises

D8.62   Refer to Example 8.24. If the sag h is reduced to 10 m, determine (a) the constant c as defined in Eq. (8.24), (b) the maximum tension in the wire.

D8.63   Refer to Example 8.24. Determine S_{AB} using Eq. (8.22) as an approximation.

PROBLEMS

8.101* The cable AE is loaded as shown. If $h_C = 25$ ft, determine (a) the horizontal component of the reaction from the support at E, (b) the maximum tension in the cable.

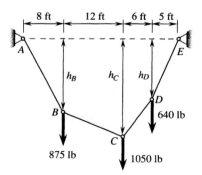

Fig. P8.101* and P8.102

8.102   If the cable shown has a maximum tension of 1560 lb, determine the values of h_B, h_C, and h_D.

8.103 If $a = 5.76$ m in the cable shown, determine (a) the load P, (b) the maximum tension in the cable.

2.4 m 2.4 m 2.4 m 2.4 m

a

6.3 m

E

A

D

B C

P

540 N

360 N

Fig. P8.103 and P8.104

8.104 If $P = 200$ N in the cable shown, determine the distance a.

8.105* The main span of a suspension bridge consists of a uniform roadway suspended from two cables. The load carried by *each* cable is 15 Mg/m along the horizontal. If the span is 1300 m and the sag is 150 m, determine the minimum and maximum tensions in each cable.

8.106 Refer to Prob. 8.105*. Determine the length of each cable used in the main span of the suspension bridge.

8.107* Applying Formula (b) in Sec. B5.6 of App. B, show that Eq. (8.18) for the parabolic cable can be obtained from Eq. (8.25) for the catenary cable by taking the first two terms in the series expansion of cosh x/c if the sag is very small and the cable is sufficiently taut.

8.108 and 8.109 A cable supports a uniformly distributed load with respect to the horizontal as shown. Knowing that the slope of the cable at C is zero, determine the minimum and maximum tensions in the cable.

8.110 and 8.111 A cable supports a uniformly distributed load along the horizontal as shown. Knowing that the slope of the cable at C is zero, determine the total length of the cable.

8.112* A cable supports a distributed load along the horizontal as shown. The intensity of the load at B is w_o per unit length and decreases uniformly to zero at C. The slope of the cable at C is zero. Determine (a) the equation of the curve assumed by the cable, (b) the maximum tension in the cable.

100 ft

B

20 ft C

$w = 10$ lb/ft

Fig. P8.108 and P8.110

100 m

A

5 m B

10 m C

$\dfrac{w}{g} = 10$ kg/m

Fig. P8.109 and P8.111

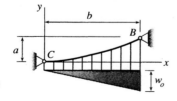

y

b

B

a

C

x

w_o

Fig. P8.112*

8.113 A 10-m cord hangs under its own weight as shown. If the span is $L = 2$ m, determine the sag h. [*Hint*. Use Eqs. (8.25) and (8.26). Start with $c \approx 0.3$ as a trial solution.][†]

L

h

Fig. P8.113 and P8.114*

[†]Cf. App. E or Sec. B5.8 of App. B for the solution of a transcendental equation.

8.114* A 10-m cord hangs under its own weight as shown. If the sag is $h = 4$ m, determine the span L.

8.115 Cables are hung under their own weight between transmission towers as shown. If each cable has a weight density of 10 lb/ft and the resultant force **R** exerted by the cables at B is vertical, determine the sag h of the cable BC.

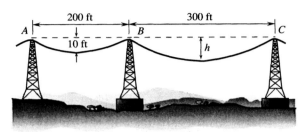

Fig. P8.115

8.116 Refer to Prob. 8.115. Determine (a) the angle which the cable just to the left of B makes with the horizontal, (b) the angle which the cable just to the right of B makes with the horizontal, (c) the magnitude of **R**.

8.117* Refer to Prob. 8.115. Determine the total length of each of the cables AB and BC.

8.118 Solve Prob. 8.115 by assuming $w \approx \gamma_L$ in Eq. (8.23) and using the relations for the parabolic cable as an approximation.

8.119 Solve Prob 8.116 by assuming $w \approx \gamma_L$ in Eq. (8.23) and using the relations for the parabolic cable as an approximation.

8.120* Solve Prob. 8.117* by assuming $w \approx \gamma_L$ in Eq. (8.23) and using the relations for the parabolic cable as an approximation.

8.21   Concluding Remarks

This chapter covers the study of the conditions of equilibrium of trusses, frames, machines, beams, and cables. Such a study entails applications of all the concepts of equilibrium of particles and rigid bodies in a plane and in space, which were treated in Chaps. 1 through 5.

Specifically, the method of joints for analyzing plane trusses and simple space trusses is established from the *concept of equilibrium of particles in a plane and in space.* The method of sections for analyzing the forces in members of plane trusses and the determination of reactions from the supports of trusses, frames, machines, beams, and cables under the action of coplanar force systems are based on the *concept of equilibrium of rigid bodies in a plane.* Furthermore, the determination of reactions from the supports of space trusses is based on the *concept of equilibrium of rigid bodies in space.*

In retrospect, it is clear that the study of the conditions of equilibrium of frames and machines in a plane is a direct extension of the analysis of simple rigid-body equilibrium problems covered in Chap. 5. Such a study is more advanced in nature. However, it is greatly facilitated by the drawing of pertinent free-body diagrams with strict observance of Newton's third law.

Although the shears and bending moments in beams can be determined from the concept of equilibrium, we have emphasized the use of the concept of resolving a given force into a force-moment system at another point to determine those quantities. The two general rules in Sec. 8.16 and the six relations listed in Sec. 8.17 are intended to offer helpful guides in drawing the shear and bending-moment diagrams of beams.

Cables carrying concentrated loads have, in fact, been studied to some extent in Chap. 3 as well as in Example 5.8. They are only briefly discussed in this chapter. The preceding three sections are mainly devoted to the study of cables carrying distributed loads. It is true that many of the equations derived for parabolic and catenary cables are frequently used as formulas if problems involving such cables are to be efficiently solved. Furthermore, the analysis of catenary cables often leads to transcendental equations, which offer students the opportunity of finding roots by trial and error, a numerical method, or a *digital root finder*.[†]

REVIEW PROBLEMS

8.121* Refer to the compound truss in Prob. 8.21. Using the method of sections, determine the forces in members CH, DE, and AE.

8.122* Refer to the compound truss in Prob. 8.24. Using the method of sections, determine the forces in members BF, CD, and AD.

8.123 Determine the forces in members GH, EI, BK, KJ, and CL of the truss shown.

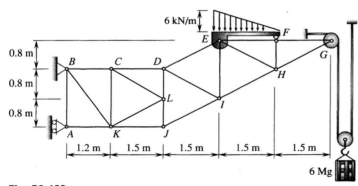

Fig. P8.123

8.124 Determine the reactions from the pins to the member BEG of the frame shown.

8.125 A pair of compound-lever snips is used to cut a wire at M as shown. For the gripping forces of 200 N applied at A and B, determine the magnitude of the cutting forces exerted by the blades on the wire at M.

8.126* Draw the shear and bending-moment diagrams for the beam shown.

[†]Cf. App. E.

Fig. P8.124

Fig. P8.125

Fig. P8.126*

Fig. P8.127

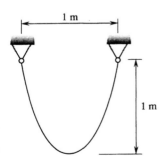

Fig. P8.128

8.127   A cable supports a section of pipeline which weighs 200 lb/ft as shown. Determine (a) the maximum tension in the cable, (b) the total length of the cable.

8.128   A rope hanging under its own weight has a span of 1 m and a sag of 1 m as shown. Determine the total length of the rope.

Friction

Chapter 9

KEY CONCEPTS

- Full film lubrication, boundary lubrication, dry friction, friction force, coefficient of static friction, coefficient of kinetic friction, angle of static friction, and angle of kinetic friction.
- Laws of dry friction.
- Types of friction problems and the methods for their solution.
- Friction problems involving rigid bodies, square-threaded screws, journal bearings, and thrust bearings.
- Friction problems involving flat belts and V belts.

The study of the conditions of physical systems often requires that the effects of friction be taken into account if more accurate results are to be obtained. This chapter is concerned with the conditions of equilibrium of systems involving dry friction between contacting surfaces. In other words, friction problems considered in this chapter are essentially equilibrium problems which may partially be governed by the laws of dry friction. Thus, it is important to be able to appropriately introduce the equations representing the laws of dry friction when the circumstances warrant. Although the effects of dry friction are sometimes undesirable, there are numerous situations in which such effects are not only desirable but also necessary. By thinking of the friction needed to walk and run, the braking needed to stop, the transmission of power through belt drives, and the like, we readily see that friction is ever-present in our daily activities and in many engineering systems.

FRICTION BETWEEN RIGID BODIES

9.1 State of Sliding Surfaces

Friction is a phenomenon in which the surfaces of contact of two bodies resist relative sliding, or the tendency toward relative sliding. We assumed in preceding chapters that surfaces in contact were either *frictionless* or *rough*. This assumption was an idealization. Most surfaces of contact have some asperities which interlock to produce resistance to relative mo-

Fig. 9.1

tion between the surfaces. Such asperities may be magnified as illustrated in Fig. 9.1.

The state of two sliding surfaces is said to be in *full film lubrication* when sufficient lubricant is present to completely separate the asperities of the two surfaces. It is said to be in *solid film lubrication* if a solid lubricant, such as graphite, is used. It is said to be in *boundary lubrication* when a significant amount of fluid lubricant is present, but opposing asperities penetrate the fluid film and come in contact. When no significant amount of lubricant covers the interface, the state of two sliding surfaces is said to be in *dry friction* or *Coulomb friction*, after the French engineer and physicist Charles Augustin Coulomb (1736–1806). In this chapter, we shall be concerned with systems involving dry friction.

Developmental Exercise

D9.1   Define the terms: (a) friction, (b) full film lubrication, (c) solid film lubrication, (d) boundary lubrication, (e) dry friction.

9.2   Friction Force

A *friction force* is the force component which is tangent to the contact surfaces of two bodies and *opposes* the sliding motion, or the tendency toward sliding motion, of one contact surface relative to the other. Like other forces, friction forces always occur in pairs, one on each of the contact surfaces. The *sense of the friction force* acting on one contact surface is always opposite to the sense or tendency of its relative sliding motion with respect to the other contact surface. When sliding between the contact surfaces occurs, the friction force is called a *kinetic friction force*, which will be denoted by $\mathbf{F}_k$; otherwise, it is called a *static friction force*, which will be denoted by $\mathbf{F}$.

As an illustration, let us consider a block of weight W which is initially at rest on a rough surface as shown in Fig. 9.2(a). Upon application of the horizontal force $\mathbf{P}$ as shown, the block will either move or tend to move to the right. Thus, the friction force $\mathbf{F}$ acting on the block, as shown in Fig. 9.2(b), is directed to the left to oppose the motion, or tendency toward motion, of the block to the right. By Newton's third law, we note that the friction force $\mathbf{F}$ and the normal force $\mathbf{N}$ acting on the block are related to the friction force $\mathbf{F}'$ and the normal force $\mathbf{N}'$ acting on the support as follows: $F' = F$, $\mathbf{F}' = -\mathbf{F}$; $N' = N$, $\mathbf{N}' = -\mathbf{N}$.

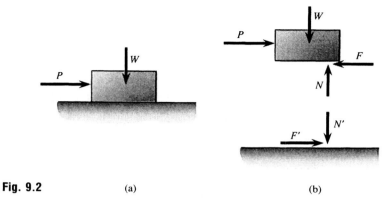

Fig. 9.2   (a)   (b)

Developmental Exercises

D9.2   Define the terms: (a) friction force, (b) kinetic friction force, (c) static friction force.

D9.3[†]   The wedge A having a weight W_A is driven by the horizontal force **P** to raise the body B having a weight W_B as shown. All contact surfaces are rough. Draw the free-body diagrams of (a) the wedge A, (b) the body B, (c) the wedge A and the body B together. (*Hint.* The senses of the forces must be correctly indicated.)

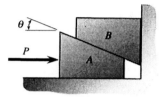

Fig. D9.3

9.3   Laws of Dry Friction

The elementary *laws of dry friction* are applicable to friction surfaces with no significant amount of lubricant. They are, to a first-order approximation, applicable to *boundary lubrication* and *solid film lubrication*. We may state these laws as follows:[††]

1. For bodies not in relative motion, the magnitude of the friction force **F** reaches a maximum value F_m *when slipping impends*; i.e.,

$$F \leq F_m \tag{9.1}$$

$$F_m = \mu_s N \tag{9.2}^{†††}$$

[†]Continued in D9.16.

[††]These laws are attributed to Charles Augustin Coulomb, whose friction memoir *Théorie des Machines Simples* was declared in the spring of 1781 by the Academy of Sciences in Paris, France, to be the winner of the double prize for the solution of "problems of friction of sliding and rolling surfaces, the resistance to bending in cords, and the application of these solutions to simple machines used in the navy." One of Coulomb's two-term formulas for friction between sliding surfaces is, in his notation, written as $F = A + (P/\mu)$, where F is the maximum friction force, P is the normal force, μ is the *inverse* of the coefficient of friction, and A is a constant force accounting for adhesive or cohesive effects. However, we shall neglect the adhesive or cohesive effects in our study. For a detailed account, see C. S. Gillmor, *Coulomb and the Evolution of Physics and Engineering in Eighteenth-Century France* (Princeton, N.J.: Princeton University Press, 1971).

[†††]In some mechanics literature, Eqs. (9.2) and (9.3) have been written merely as $F = \mu N$, where F represents either F_m or F_k and μ represents either μ_s or μ_k as appropriate. Note that the F in $F = \mu N$ is *not* the same as the F we use here to denote the magnitude of the general friction force **F** where slipping does not necessarily occur or impend. [Cf. Eqs. (9.1) and (9.2).]

where N is the magnitude of the normal force **N** acting on the surface of contact, and μ_s is called the *coefficient of static friction*.

2. For bodies in relative motion, the magnitude of the kinetic friction force **F**$_k$ is given by

$$F_k = \mu_k N \qquad (9.3)$$

where μ_k is called the *coefficient of kinetic friction*.

3. The values of μ_s and μ_k are dependent on the materials involved as well as the nature of the surfaces in contact, but independent of the sliding speed and size of the area of contact.

4. In general, we have $\mu_s > \mu_k$.

According to these laws, a horizontal force **P** applied to a block of weight W resting on a horizontal dry surface, as shown in Fig. 9.2(a), will cause a friction force **F** to be developed on the block as shown in Fig. 9.3(a). The relation between F and P is given by the graph shown in Fig. 9.3(b) where the same scale is used for both F and P axes.

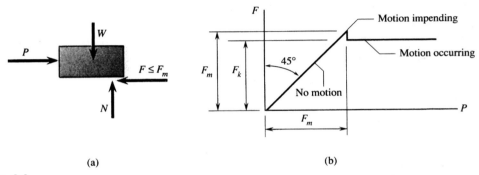

(a)   (b)

Fig. 9.3

Note in Fig. 9.3 that the maximum value of the friction force which can be developed is designated by F_m; i.e., $F_m = F_{\max}$. As P is increased from 0 to F_m, we see that $F = P \leq F_m$ and the block, which is originally at rest, remains at rest. At $F = P = F_m$, slipping of the block impends. When $P > F_m$, we see that $F = F_k < F_m$ and the block is being accelerated to the right by the net unbalanced force of magnitude $(P - F_k)$. However, if P is then decreased to F_k, the block will move with a constant velocity. The approximate ranges of values of μ_s for several dry surfaces are given in Table 9.1.

Table 9.1   Approximate Ranges of Values of Coefficients of Static Friction

Dry Surfaces	μ_s	Dry Surfaces	μ_s
Metal on metal	0.2–1.2	Wood on leather	0.2–0.5
Metal on wood	0.2–0.6	Stone on stone	0.4–0.7
Metal on stone	0.3–0.7	Rubber on asphalt	0.7–1.1
Metal on leather	0.3–0.6	Rubber on concrete	0.6–0.9
Wood on wood	0.2–0.6	Rubber on ice	0.05–0.2

Note. The values of μ_s are mostly less than 1, but they can exceed 1 as indicated. For metal on metal, the wide variation in the value of μ_s depends on the metals involved (e.g., copper on copper, mild steel on aluminum, hard steel on lead).

Developmental Exercises

D9.4 Define the terms: (a) coefficient of static friction, (b) coefficient of kinetic friction.

D9.5 Is it correct to say that the friction force is *always* equal to the product of the coefficient of friction and the normal force acting on the surface of contact?

D9.6 A man pushes an 80-lb crate with a horizontal force **P** as shown. He finds that the crate is on the verge of moving when $P = 32$ lb. When the crate is moving, he finds that he needs to keep P at 28 lb if the crate is to move at a constant velocity. Determine the values of (a) μ_s, (b) μ_k, (c) **F** (the friction force) acting on the crate at rest if $P = 25$ lb, (d) **F** acting on the crate which is moving to the left but is slowing down when a 25-lb force **P** as shown is acting on the crate.

D9.7 Give a brief summary of the laws of dry friction.

Fig. D9.6

9.4 Angles of Friction

The *angle of static friction*, denoted by ϕ_s, is the angle between the normal force **N** and the reaction **R**, where **R** is the resultant force of **N** and the maximum static friction force $\mathbf{F}_m$ developed on the surface of contact. Similarly the *angle of kinetic friction*, denoted by ϕ_k, is the angle between the normal force **N′** and the reaction **R′**, where **R′** is the resultant force of **N′** and the kinetic friction force $\mathbf{F}_k$ developed on the surface of contact. For two given contact surfaces, we generally have

$$\phi_s > \phi_k \tag{9.4}$$

The angles ϕ_s and ϕ_k are shown in Fig. 9.4, where **R** is vertical, but **R′** is not. We write

$$\tan\phi_s = \frac{F_m}{N} = \frac{\mu_s N}{N} \qquad \tan\phi_k = \frac{F_k}{N'} = \frac{\mu_k N'}{N'}$$

$$\boxed{\tan\phi_s = \mu_s} \qquad \boxed{\tan\phi_k = \mu_k} \tag{9.5}$$

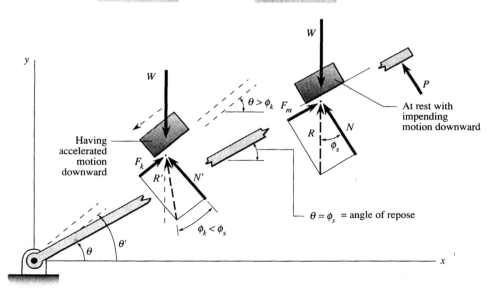

Fig. 9.4

In Fig. 9.4, if the block is originally at rest and the angle θ between the inclined plank and the horizontal is slowly increased from relatively small values to ϕ_s, slipping of the block down the inclined plank will impend at $\theta = \phi_s$. The angle of inclination of the plank corresponding to impending motion is sometimes called the *angle of repose*. Clearly, the angle of repose is equal to ϕ_s. If the block is originally at rest, it will remain at rest for all $\theta < \phi_s$; but it will have motion down the inclined plank if $\theta > \phi_s$.

Developmental Exercises

D9.8 Define the terms: (a) angle of static friction, (b) angle of kinetic friction, (c) angle of repose.

D9.9 The vase on the table starts to slide when the table a young boy lifts makes an angle of 20° with the horizontal as shown. Between the vase and the table, what are the values of (a) ϕ_s, (b) μ_s?

D9.10 The brick shown moves with a constant velocity if $\theta = 30°$, but it will accelerate down the plank when θ exceeds 30°. What are the values of ϕ_k and μ_k between the brick and the plank?

Fig. D9.9

Fig. D9.10

9.5   Types of Friction Problems

Friction problems considered in statics are mainly *equilibrium problems involving dry friction*. For developing systematic solutions, we may classify most friction problems into the following three types:

- *Type I. Relative motion (slipping, tipping, or rolling) in the system is not assured to impend or occur.* In problems of this type, it is usually necessary to ascertain whether the relative motion in the system will impend or occur.
- *Type II. The type and location of relative motion which impends or occurs in the system are known.* In problems of this type, we know that (1) friction forces have magnitudes equal to F_m or F_k, as appropriate, on surfaces where slipping impends or occurs, and (2) the resultant reaction from the support acts through a corner or along an edge of the body if tipping or rolling impends or occurs at that support.
- *Type III. Relative motion in the system is known to impend or occur, but the type and location of relative motion which impends or occurs are not specified.* In problems of this type, the various possible ways for relative motion to start should be analyzed and checked.

Developmental Exercise

D9.11 Describe briefly each of the three types of friction problems.

9.6 Method of Solution for Type I Problems

The following suggested steps are generally helpful in *analyzing and solving type I friction problems* described in Sec. 9.5:

1. *Assume* that the system is at *rest*.
2. *Solve the problem as an equilibrium problem* without using laws of dry friction. (The general procedure for solving equilibrium problems in previous chapters may be followed.)
3. If appropriate, *compute the maximum value F_m of the friction force* which may be developed on each contact surface.
4. *Check the initial assumption* in step (1) by appropriately comparing the magnitude F of the friction force, obtained under that assumption, with its maximum value F_m obtained in step (3). If $F \leq F_m$, the initial assumption is correct, and the problem is solved. If $F > F_m$, equilibrium does not exist and the body must have kinetic friction; therefore, the correct value of F is F_k.

EXAMPLE 9.1

A 100-lb block is initially at rest on the horizontal support before a horizontal force **P** is applied as shown. Knowing that $\mu_s = 0.30$ and $\mu_k = 0.25$ between the surfaces of contact, determine the magnitude F of the friction force acting on the block if the magnitude of **P** is (a) 20 lb, (b) 27 lb, (c) 30 lb, (d) 40 lb, (e) 45 lb.

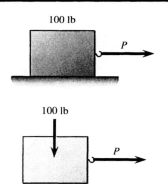

Solution. From the statement of the problem, relative motion of the block is not assured to impend or occur. Thus, this is a *type I friction problem*. Since the block cannot move in the vertical direction, we may draw the free-body diagram of the block as shown and write

$$+\uparrow \Sigma F_y = 0: \quad N - 100 = 0 \qquad N = 100 \text{ lb}$$

Although the block may be moved to the right by **P**, we know that $F \leq P$. By the laws of dry friction, we write

$$F_m = \mu_s N = 0.30(100) \qquad F_m = 30 \text{ lb}$$

$$F_k = \mu_k N = 0.25(100) \qquad F_k = 25 \text{ lb}$$

(a) In this case, $P = 20 \text{ lb} < F_m$. Thus, the block is in equilibrium. We write

$$\overset{+}{\rightarrow} \Sigma F_x = 0: \quad P - F = 0 \qquad F = 20 \text{ lb} \blacktriangleleft$$

(b) In this case, $P = 27 \text{ lb} < F_m$. Thus, the block is in equilibrium even though $P > F_k$. This is so because the block is initially at rest. We write

$$\overset{+}{\rightarrow} \Sigma F_x = 0: \quad P - F = 0 \qquad F = 27 \text{ lb} \blacktriangleleft$$

(c) In this case, $P = 30$ lb $= F_m$. Thus, slipping impends but the block is still in equilibrium. We write

$$\xrightarrow{+} \Sigma F_x = 0: \quad P - F = 0 \qquad\qquad F = 30 \text{ lb} \quad \blacktriangleleft$$

(d) In this case, $P = 40$ lb $> F_m$. Thus, slipping occurs and the block has kinetic friction. We write

$$F = F_k \qquad\qquad\qquad F = 25 \text{ lb} \quad \blacktriangleleft$$

(e) In this case, $P = 45$ lb $> F_m$. Thus, slipping occurs and the block has kinetic friction. We write

$$F = F_k \qquad\qquad\qquad F = 25 \text{ lb} \quad \blacktriangleleft$$

Developmental Exercise

26 lb
5
12

100 lb

$\mu_s = 0.22$

Fig. D9.12

D9.12 A 26-lb force acts on a 100-lb weight which is initially at rest as shown. If $\mu_s = 0.22$, determine (a) N, (b) F, (c) F_m, (d) whether the weight will move.

EXAMPLE 9.2

A 10-lb horizontal force **P** is applied to a homogeneous body of weight W which is initially at rest as shown. If $W = 35$ lb and $\mu_s = 0.3$ between the body and the support, determine whether or not motion of the body will impend or occur.

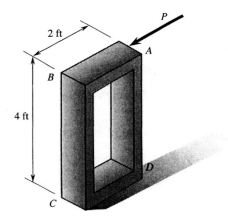

Solution. We first note that, under the action of **P**, the body could slide, tip, or simply remain at rest. Thus, this is a *type I friction problem*. We may analyze and solve this problem as follows:

(a) *Force* **P** *required to cause sliding of the body to impend.* From the free-body diagram shown below, where we set $F = 0.3\,N$, we write

$$+\uparrow\ \Sigma F_y = 0:\quad N - 35 = 0$$

$$\xrightarrow{+}\ \Sigma F_x = 0:\quad 0.3N - P = 0$$

$$N = 35\ \text{lb}\qquad P = 10.5\ \text{lb} > 10\ \text{lb}$$

Therefore, sliding does not impend or occur.

(b) *Force* **P** *required to cause tipping of the body to impend.* In this case, the resultant reaction from the floor must act through the corner C as shown above. We write

$$+\ \circlearrowright\ \Sigma M_C = 0:\quad 4P - 1(35) = 0$$

$$P = 8.75\ \text{lb} < 10\ \text{lb}$$

Conclusion: Motion of the body will occur by tipping. ◀

Developmental Exercise

D9.13 Work Example 9.2 if (a) $W = 41$ lb and $\mu_s = 0.24$, (b) $W = 41$ lb and $\mu_s = 0.25$, (c) $W = 39$ lb and $\mu_s = 0.26$.

9.7 Method of Solution for Type II Problems

The following suggested steps are generally helpful in *analyzing and solving type II friction problems* described in Sec. 9.5:

1. *Write the equations governing the impending slipping or kinetic friction* between all contact surfaces where impending slipping or kinetic friction is known to exist.
2. *Solve the problem as an equilibrium problem* by incorporating appropriately the friction equations in step (1) in the solution. (The general pro-

cedure for solving equilibrium problems in previous chapters may be followed.)

3. *Decide*, if required, *the range of a parameter* by checking the obtained values with the equations governing the impending slipping or kinetic friction.

EXAMPLE 9.3

Determine the range of the magnitude of the horizontal force **P** which will keep the 200-lb block in equilibrium as shown, where $\mu_s = 0.25$ between the surfaces of contact.

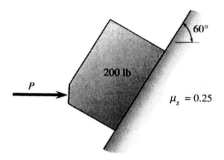

Solution.   We first find that the angle of static friction is

$$\phi_s = \tan^{-1}\mu_s = \tan^{-1} 0.25 = 14.04° < 60°$$

This reveals that the block will slide down the incline without the application of the force **P**. Clearly, slipping of the block downward or upward on the incline will impend when the magnitude of **P** reaches its lower bound or upper bound, respectively. Thus, this is a *type II friction problem*.

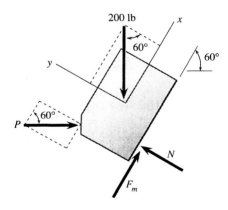

Case (a). Impending motion down the incline. The free-body diagram in this case is drawn as shown. We write

Friction law: $F_m = 0.25N$

$+\nearrow \Sigma F_x = 0$: $P \cos 60° + F_m - 200 \sin 60° = 0$

$+\nwarrow \Sigma F_y = 0$: $-P \sin 60° + N - 200 \cos 60° = 0$

Solution of these three equations yields

$P = 206.8$ lb $\qquad N = 279.1$ lb $\qquad F_m = 69.8$ lb

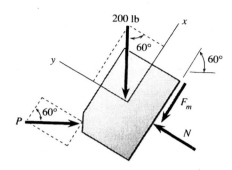

Case (b). Impending motion up the incline. The free-body diagram in this case is drawn as shown. We write

Friction law: $F_m = 0.25N$

$+\nearrow \Sigma F_x = 0$: $P \cos 60° - F_m - 200 \sin 60° = 0$

$+\nwarrow \Sigma F_y = 0$: $-P \sin 60° + N - 200 \cos 60° = 0$

Solution of these three equations yields

$P = 699.2$ lb $\qquad N = 705.5$ lb $\qquad F_m = 176.4$ lb

From the results in the above two cases, we conclude that the solution to the problem is

$$207 \text{ lb} \le P \le 699 \text{ lb} \quad \blacktriangleleft$$

Developmental Exercise

D9.14 Refer to the block in Example 9.3. Determine the magnitude of **P** for which the friction force between the block and the incline is zero.

EXAMPLE 9.4

A 400-kg machine is mounted on an 80-kg horizontal beam as shown. In order to raise the left end of the beam somewhat, a wedge is driven under the base at A by a horizontal force **P**. Knowing that $\mu_s = 0.3$ between all surfaces of contact, determine (a) the maximum value of θ of the wedge for which the base at B will not move, (b) the magnitude of **P** required to drive the wedge when θ is maximum.

Solution.   When the wedge has the desired maximum value of θ and is under the action of the required force **P**, the base at A will have an *impending slipping* on the wedge while the base at B will have an *impending slipping* on the horizontal support. Thus, this is a *type II friction problem*. The free-body diagrams of the wedge and the body AB at impending slipping are drawn as shown. Note that the friction law $F_m = \mu_s N$ has been employed

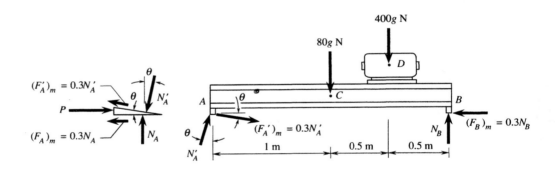

in writing friction forces in the free-body diagrams. In the two free-body diagrams, there are five unknowns: N_A, N'_A, N_B, P, and θ. They may be determined as follows:

Wedge

$$\xrightarrow{+} \Sigma F_x \; = \; 0: \quad P - 0.3N_A - N'_A \sin\theta - 0.3N'_A \cos\theta = 0 \tag{1}$$

$$+\uparrow \Sigma F_y \; = \; 0: \quad N_A - N'_A \cos\theta + 0.3N'_A \sin\theta = 0 \tag{2}$$

Body AB

$$\xrightarrow{+} \Sigma F_x \; = \; 0: \quad N'_A \sin\theta + 0.3N'_A \cos\theta - 0.3N_B = 0 \tag{3}$$

$$+\uparrow \Sigma F_y \; = \; 0: \quad N'_A \cos\theta - 0.3N'_A \sin\theta + N_B - 80g - 400g = 0 \tag{4}$$

$$+\circlearrowright \Sigma M_A \; = \; 0: \quad 2N_B - 1(80g) - 1.5(400g) = 0 \tag{5}$$

Solving these five simultaneous equations, we obtain

$$N_A = 1373 \text{ N} \qquad N_B = 3335 \text{ N} \qquad \theta = 19.38°$$
$$N_A' = 1628 \text{ N} \qquad R = 1413 \text{ N}$$
$$\theta_{max} = 19.38° \qquad P = 1.413 \text{ kN} \blacktriangleleft$$

Developmental Exercises

D9.15 Work Example 9.4 if $\mu_s = 0.2$.

D9.16 Refer to the system in D9.3. Determine the magnitude of **P** required to raise the body B if $\theta = 30°$, $W_A = W_B = 100$ lb, and $\mu_s = 0.35$ between all surfaces of contact.

D9.17 A force **P** pulls on a 10-kg block which rests on a horizontal support as shown. For $\mu_s = 0.22$, determine the minimum value of P for which the block will start to move.

Fig. D9.17

PROBLEMS

9.1 through 9.3* The body shown is homogeneous and is tilted on the rough surface by the horizontal force **P**. The angle α is equal to 30° when sliding impends. Determine the value of μ_s between the surfaces of contact.

Hemispherical shell

Fig. P9.1 and P9.4

Semicircular cylinder

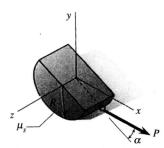

Fig. P9.2 and P9.5*

Semicircular pipe

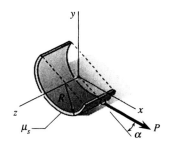

Fig. P9.3* and P9.6

9.4 through 9.6 The body shown is homogeneous and is tilted on the rough surface by the horizontal force **P**. If $\mu_s = 0.5$ between the surfaces of contact, determine the angle α when sliding impends.

9.7 through 9.12* A 58-N horizontal force **P** is applied to a 10-kg homogeneous body which is initially at rest as shown. If $\mu_s = 0.6$ between the body and the support, determine whether or not motion of the body will impend or occur.

9.13 through 9.18 An adequate horizontal force **P** is applied to move the homogeneous body which is initially at rest as shown. Determine the maximum allowable value of μ_s for which tipping will not occur.

Fig. P9.7, P9.13 and P9.44*

Fig. P9.8, P9.14*, and P9.45

Fig. P9.9*, P9.15 and P9.46

Fig. P9.10, P9.16 and P9.47*

Fig. P9.11, P9.17* and P9.48

Fig. P9.12*, P9.18 and P9.49

9.19* A section of pipe weighing 40 lb rests against a curb of height $h = 0.2$ ft. It is known that $\mu_s = 0.45$ between the pipe and the curb. Can an adequate horizontal force **P** roll the pipe over the curb as shown?

Fig. P9.19* and P9.20

Fig. P9.21

9.20 A section of pipe weighing 60 lb rests against a curb of height $h = 0.1$ ft. It is desired to apply a horizontal force **P** to roll the pipe over the curb as shown. Determine the required minimum values of (a) μ_s between the pipe and the curb, (b) the magnitude of **P**.

9.21 The values of μ_s between the track AB and the 8-kg board shown are 0.2 at C and 0.25 at D. If the board is to be moved to the right, determine the horizontal force **P** which must be applied to the handle at E.

9.22* Solve Prob. 9.21 if the board is to be moved to the left.

9.23 Two crates on the ramp are released from rest as shown. It is known that $\mu_s = 0.4$ and $\mu_k = 0.35$ between the ramp and the 60-kg crate A, and $\mu_s = 0.3$ and $\mu_k = 0.25$ between the ramp and the 40-kg crate B. For $\theta = 20°$, determine (a) whether the crates will move, (b) the friction force acting on each crate.

Fig. P9.23

Fig. P9.25

9.24* Solve Prob. 9.23 if $\theta = 19°$.

9.25 A cant hook is used to lift a log as shown. It is known that $\mu_s = 0.4$ between the log and the device at A. Determine the minimum value of θ for which the log may be lifted by an adequate force **P** as shown.

9.26 The weights of the blocks A and B, as shown, are 200 lb and 52 lb, respectively. The system is initially at rest and $\mu_s = 0.4$ between all surfaces of contact. Determine the magnitude of the horizontal force **P** that will cause motion of the block A to impend.

9.27 A 20-lb body A rests on a wedge B of negligible weight and against a smooth vertical wall as shown. If $\mu_s = 0.25$ between A and B as well as between B and the horizontal support, determine whether the wedge B is self-locking in the position shown when (a) $\theta = 25°$, (b) $\theta = 30°$, (c) $\theta = 35°$.

Fig. P9.26

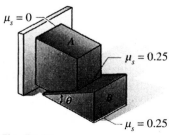

Fig. P9.27

9.28   It is known that the 20.8-lb body A and the 21.7-lb body B are initially at rest as shown and $\mu_s = 0.1$ between all surfaces of contact. Determine the magnitude of **P** that will cause the body A to have an impending motion down the incline.

Fig. P9.28

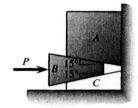

Fig. P9.29

9.29   If the bodies A and B as shown have masses $m_A = 100$ kg and $m_B \approx 0$ and $\mu_s = 0.25$ between all surfaces of contact, determine the range of the magnitudes of **P** if equilibrium exists.

9.30*   The body A shown weighs 60 lb and $\mu_s = 0.25$ between all surfaces of contact. Neglecting the weight of the wedge B, determine the minimum magnitude of **P** which will cause an upward motion of the body A to impend.

Fig. P9.30*

9.31   During a manufacturing process, a piece of plywood with a thickness of h is to be further compressed to a final thickness of 20 mm by steel rollers driven as shown. It is known that $\mu_s = 0.15$ between the plywood and either of the rollers. Determine the maximum value of h for which the plywood will be pulled into the rollers by the friction forces.

Fig. P9.31

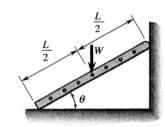

Fig. P9.32 and P9.33*

9.32   A ladder of weight W and length L is placed against a vertical wall as shown, where $\mu_s = 0.5$ between all surfaces of contact. What is the minimum value of θ for which the ladder will not slide down?

9.33*   A ladder is shown, where $\mu_s = 0.4$ between all surfaces of contact. Determine whether the ladder will be in equilibrium for $\theta = 45°$.

9.34   A 10-kg movable chalkboard $ABCD$ is normally supported by two 5-kg counterweights E and F by means of cords and pulleys as shown. One day, the left cord is broken and the chalkboard is found to remain in the position shown due to friction. Assume that the chalkboard is slightly smaller than the fixed guiding frame and it will bind only at the points A and C. For equal values of μ_s between the surfaces of contact at A and C, determine the smallest possible value of μ_s.

Fig. P9.34

Fig. P9.35

Fig. P9.37

Fig. P9.39

9.35 A 600-mm uniform slender rod is placed in a hemispherical bowl which has an interior radius of 200 mm as shown. If the achievable maximum value of the angle θ is 30°, determine μ_s between the rod and the bowl.

9.36* Refer to the system in Prob. 9.35. Determine the value of μ_s between the rod and the bowl if the achievable minimum value of θ is 10°.

9.37 A uniform slender rod ABC of length L is held by a horizontal pipe which has an interior diameter of 15 in. as shown. Determine the minimum value of μ_s existing between the rod and the pipe if $L = 75$ in.

9.38* Refer to the system in Prob. 9.37. Determine the minimum value of μ_s existing between the rod and the pipe if $L = 84$ in.

9.39 Two uniform rods, each of weight W, are held by frictionless pins at A and B as shown. For $\mu_s = 0.4$ between the surfaces of contact at C, determine whether the rods will remain in equilibrium if $\theta = 50°$.

9.40* Refer to the system in Prob. 9.39. Determine the required minimum value of μ_s in terms of the angle θ if the system is to remain at rest.

9.41 For $\mu_s = 0.25$ between the body A of mass m_A and the incline as shown, determine the range of values of m_A for which the 10-kg body B will remain in equilibrium.

9.42 It is known that the mass of the crate A is 10 kg, the mass of the cylinder B is 6 kg, and $\mu_s = 0.55$ between the crate and the horizontal support as shown. For $0 \le \theta \le 90°$, determine the range of values of θ for which the crate will not move.

9.43 If the mass of the collar A is $m_A = 16$ kg and $\mu_s = 0.25$ between the collars and the rods, determine the range of values of the mass m_B of the collar B for which the collars as shown will not move.

Fig. P9.41

Fig. P9.42

Fig. P9.43

9.8   Method of Solution for Type III Problems

The following suggested steps are generally helpful in *analyzing and solving type III friction problems* described in Sec. 9.5:

1. *List all of the possible ways* in which the system can have the stated relative motion.
2. *Find the solution for each of the possible ways* listed in step (1). Each of these possible ways now corresponds to a type II friction problem or a simple equilibrium problem.
3. *Select the correct solution* which satisfies the situation or condition stated in the problem.

EXAMPLE 9.5

A horizontal force **P** is applied to a 100-lb homogeneous quarter circular disk which is initially at rest as shown. If $\mu_s = 0.6$ between the disk and the support, determine the minimum magnitude of **P** required to move the disk.

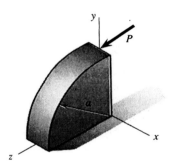

Solution.   When the magnitude of **P** is large enough, the disk will be caused to have impending slipping or tipping. However, it is not yet known whether slipping or tipping impends first. Thus, this is a *type III friction problem*.

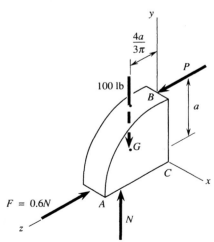

Case (a). Assume that slipping impends first. From the free-body diagram shown on the preceding page, we write

$$+\uparrow \Sigma F_y = 0: \quad N - 100 = 0 \qquad N = 100 \text{ lb}$$

$$+\swarrow \Sigma F_z = 0: \quad P - 0.6N = 0 \qquad P = 60 \text{ lb}$$

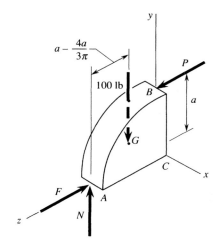

Case (b). Assume that tipping impends first. In this case, the components of reaction from the support pass through the corner at A as shown in the free-body diagram above. Thus, we write

$$+ \,\circlearrowleft\, \Sigma M_A = 0: \quad aP - [a - 4a/(3\pi)](100) = 0$$

$$P = 57.6 \text{ lb}$$

Decision. The smaller of the two values of P obtained in the above two cases is the minimum magnitude of **P** required. Thus, we have

$$P_{\min} = 57.6 \text{ lb} \quad \blacktriangleleft$$

Developmental Exercise

D9.18 A horizontal force **P** acts on a 120-lb cabinet which rests on a floor as shown. It is known that $\mu_s = 0.21$. (a) What is the magnitude of **P** if slipping impends? (b) Where will the resultant floor reaction act if tipping impends? (c) What is the magnitude of **P** if tipping impends? (d) What is the smallest magnitude of **P** which will cause the cabinet to move?

Fig. D9.18

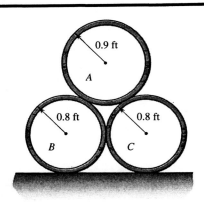

EXAMPLE 9.6

The cylindrical pipe A is to be stacked on top of the cylindrical pipes B and C as shown. The pipes have weights $W_A = 270$ lb, $W_B = W_C = 240$ lb, respectively. Assume that μ_s is the same between all contact surfaces. Determine the smallest value of μ_s for which the pipe A will not fall down.

Solution. We first recognize that the motion or collapse of the pipe A downward simultaneously with the motion of the pipes B and C outward will impend when μ_s equals the smallest value. However, the type and location of such motion (slipping or rolling) are not yet known. Thus, this is a *type III friction problem*.

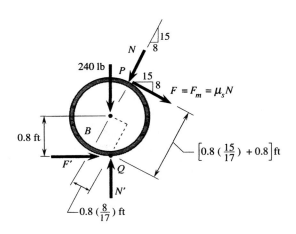

Case (a). Assume that slipping of the pipe A on the pipes B and C impends, while the pipes B and C tend to move outward by rolling (and possibly slipping) on the horizontal support. The free-body diagram of the pipe B in this case is drawn as shown above. Note that slipping is now assumed to impend at the point P. (Slipping does not necessarily impend at the point Q.) The friction law, $F_m = \mu_s N$, is indicated at P in the free-body diagram. Thus, we write

$$+\,\text{℥}\,\Sigma M_Q = 0: \quad 0.8(\tfrac{8}{17})(N) - [0.8(\tfrac{15}{17}) + 0.8](\mu_s N) = 0$$

$$\mu_s = \tfrac{1}{4} = 0.25$$

Case (b). Assume that slipping of the pipes B and C on the horizontal support impends, while the pipe A tends to move downward by either rolling or slipping on the pipes B and C. In this case, let us first find the normal component of the reaction from the support to the pipe B. From the free-body diagram of the system shown on the next page, we write

$$+\,\text{℥}\,\Sigma M_R = 0: \quad -1.6(N') + 1.6(240) + 0.8(270) = 0$$

$$N' = 375 \text{ lb}$$

Having determined N', we may draw the free-body diagram of the pipe B in this case as shown above. The friction law, $F'_m = \mu_s N' = 375\mu_s$, has been indicated in this diagram for the assumed impending slipping at the point Q. (Slipping does not necessarily impend at the point P.) We write

$$+ \circlearrowleft \ \Sigma M_P = 0: \quad -0.8(\tfrac{8}{17})(375 - 240)$$

$$+ \ [0.8 + 0.8(\tfrac{15}{17})](375\mu_s) = 0$$

$$\mu_s = 0.09 < 0.25$$

Decision. The larger of the two values of μ_s obtained in the above two cases is the smallest μ_s required. Thus, the answer to this problem is

$$\mu_s = 0.25 \quad \blacktriangleleft$$

REMARK. If this problem asked for the minimum value of μ_s between the pipes and the horizontal support as well as the minimum value of μ_s between the pipes themselves so that the pipe A could be stacked, then this problem would become a type II friction problem. On the other hand, if the values of μ_s between all surfaces of contact were given and the problem asked whether the pipes as stacked would stay together, then this problem would become a type I friction problem.

Developmental Exercise

D9.19 If μ_s is the same between all surfaces of contact, determine its minimum value for which the three identical pipes may be placed as shown.

Fig. D9.19

PROBLEMS

9.44* through 9.49 (*The figures for these problems are shown and specified along with those for Probs. 9.7 through 9.12* on p. 364.*) The body shown is homogeneous and weighs 18 lb. It is initially at rest when the horizontal force **P** acts on it. If $\mu_s = 0.6$ between the body and the support, determine the minimum magnitude of **P** required to cause the body to have impending motion.

9.50* If $\mu_s = 0.4$ between the 14-kg uniform plank *ABC* and either of the two joists shown, determine the minimum magnitude of the horizontal force **P** required to move the plank.

Fig. P9.50*

Fig. P9.51

Fig. P9.52

Fig. P9.53

9.51 Bodies *A* and *B* are of the same weight *W*. If $\mu_s = 0.3$ between *A* and *B* and between *B* and the adjustable frame shown, determine the value of θ at which relative motion between *A* and *B* impends.

9.52 If $\mu_s = 0.5$ between the 40-lb crate and the floor shown, determine the minimum magnitude of **P** required to cause an impending motion of the crate.

9.53 The value of μ_s is 0.35 between the 10-kg crate *A* and the 20-kg crate *B*, and is 0.2 between the crate *B* and the floor as shown. Determine the minimum value of the magnitude of **P** required to cause impending motion of the system.

9.54 A 20-kg concrete block rests on an incline as shown. If $\mu_s = 0.2$, determine the maximum magnitude of the horizontal force **P** for which the block remains in equilibrium.

9.55* Solve Prob. 9.54 if $\mu_s = 0.3$.

9.56 It is desired to stack the 20-kg cylinder A on top of the 45-kg cylinders B and C as shown. Assume that μ_s is the same between all surfaces of contact. Determine the minimum value of μ_s for which the cylinders may remain in equilibrium as shown.

Fig. P9.54

Fig. P9.56

Fig. P9.57

9.57 A cylinder C having a small weight W_C and a vertical bar B having negligible weight are placed between an adjustable guide A and a vertical wall D as shown. Assume that $\mu_s = 0.5$ between all surfaces of contact. Determine the minimum value of the angle θ if the system is to be self-locking for any magnitude of the vertical force **P** applied to the bar B.

9.58* Refer to the system in Prob. 9.57. Determine the required minimum value of μ_s in terms of the angle θ if the system is to be self-locking for any magnitude of **P**.

9.59* Two links are used to connect the 20-kg block A and the 10-kg block B as shown. If $\mu_s = 0.25$ between all contact surfaces, determine the maximum magnitude of the vertical force **P** which will not cause motion of the system to impend.

9.60 A section of pipe having a mass of 26 kg is held on an incline as shown. It is known that $\mu_s = 0.3$ between all contact surfaces. If $\theta = 40°$, determine the minimum magnitude of **P** required to prevent downward motion from occurring.

Fig. P9.59*

Fig. P9.60

9.61 Solve Prob. 9.60 if $\theta = 50°$.

Fig. P9.62

9.62 Two identical sections of pipes A and B, each weighing 72 lb, are held on an incline as shown. If $\mu_s = 0.2$ between all contact surfaces and $\theta = 30°$, determine the minimum magnitude of $\mathbf{P}$ required to prevent downward motion from occurring.

9.63* Solve Prob. 9.62 if $\theta = 20°$.

★9.9 Dry Friction in Square-Threaded Screws

A *screw* is a machine which consists of a head and a cylindrical or tapered shank that has at least one helical thread built on its surface. Screws are used for fastenings and for transmitting power or motion. A *square-threaded screw* is a screw whose shank is cylindrical and whose thread has a square or rectangular cross section. Square-threaded screws are generally more efficient than screws of other types in transmitting power or motion.

A dry or partially lubricated square-threaded jack is shown in Fig. 9.5(a), where an axial load $\mathbf{W}$ is to be raised by applying a torque $\mathbf{M}$ about the axis

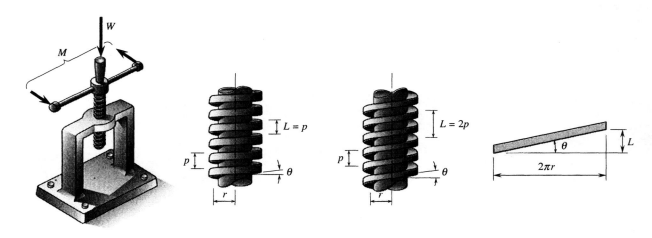

Fig. 9.5   (a)   (b)   (c)   (d)

of the screw. The *lead*, denoted by L, is the distance through which the screw advances in one turn; the *pitch*, denoted by p, is the distance measured between two consecutive threads. These are illustrated in Fig. 9.5(b) for a *single-threaded screw* and in Fig. 9.5(c) for a *double-threaded screw*. For an *n-threaded screw*, we have

$$L = np \qquad (9.6)$$

If one turn of the thread is unwrapped from the screw and stretched as shown in Fig. 9.5(d), we see that the *lead angle* θ is given by

$$\tan\theta = \frac{L}{2\pi r} \qquad (9.7)$$

where r is the mean radius of the thread.

The action of a screw is closely related to the friction developed in the threads. When the applied torque $\mathbf{M}$ in Fig. 9.5(a) is sufficiently large so that raising of the axial load $\mathbf{W}$ impends, the reaction $\Delta\mathbf{R}$ from the thread of the jack frame to a small representative portion of the thread of the screw may be depicted as shown in Fig. 9.6, where ϕ_s is the angle of static friction. Similar reactions exist on all portions of the screw thread where contact with the thread of the jack frame occurs. Let the component of $\Delta\mathbf{R}$ perpendicular to the axis of the screw be $\Delta\mathbf{R}_\perp$. Then, the magnitude of $\Delta\mathbf{R}_\perp$ is $\Delta R_\perp = \Delta R \sin(\theta + \phi_s)$. The torque about the axis of the screw produced by $\Delta\mathbf{R}$ is $r(\Delta R_\perp) = [r\sin(\theta + \phi_s)]\Delta R$. The total torque produced by all reactions is $\Sigma(r\sin(\theta + \phi_s)\Delta R] = [r\sin(\theta + \phi_s)]\Sigma(\Delta R)$. Therefore, equilibrium of moments in the axial direction of the screw in Fig. 9.6 yields

$$M = [r\sin(\theta + \phi_s)]\,\Sigma(\Delta R)$$

Fig. 9.6

On the other hand, equilibrium of forces in the axial direction of the screw in Fig. 9.6 requires that

$$W = \Sigma[\Delta R\cos(\theta + \phi_s)] = [\cos(\theta + \phi_s)]\Sigma(\Delta R)$$

Dividing M by W and rearranging, we obtain the magnitude of the torque required to *raise* the axial load $\mathbf{W}$ on the jack (i.e., to *advance* the screw) as

$$M = Wr\tan(\theta + \phi_s) \qquad (9.8)$$

The action of the entire screw may be simulated by using the unwrapped thread of the screw on an incline as a model as shown in Fig. 9.7. The

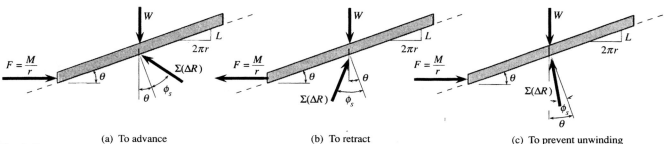

(a) To advance

(b) To retract
$(\theta \leq \phi_s)$

(c) To prevent unwinding
$(\theta \geq \phi_s)$

Fig. 9.7

equivalent force required to move this unwrapped thread on the fixed incline is $F = M/r$. By considering the equilibrium of the unwrapped thread in Fig. 9.7(a), we can also obtain Eq. (9.8). When the torque is removed, the screw will remain in place and be self-locking if $\theta \leq \phi_s$.

If $\theta \leq \phi_s$, the torque **M** required to *lower* the axial load **W** on the jack (i.e., to *retract* the screw) can readily be obtained as

$$M = Wr \tan(\phi_s - \theta) \tag{9.9}$$

by considering the equilibrium of the free body in Fig. 9.7(b). If $\theta > \phi_s$, the screw will unwind by itself, and we see from Fig.9.7(c) that the torque required to *prevent unwinding* is given by

$$M = Wr \tan(\theta - \phi_s) \tag{9.10}$$

The important relations which are used in studying situations involving the impending motion of square-threaded screws are summarized in Fig. 9.8, where **W** and **M** form a wrench.

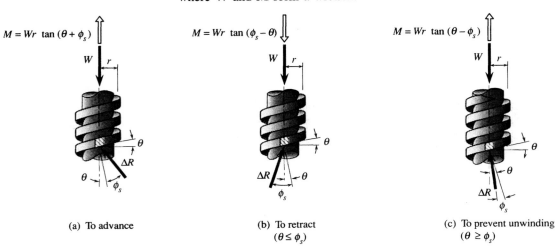

| (a) To advance | (b) To retract $(\theta \leq \phi_s)$ | (c) To prevent unwinding $(\theta \geq \phi_s)$ |

Fig. 9.8

Developmental Exercises

D9.20 Define the *lead*, the *pitch*, and the *lead angle* of a square-threaded screw with n threads.

D9.21 For a square-threaded screw with n threads, show that Eq. (9.8) may equivalently be written as $M = Wr(np + 2\pi\mu_s r)/(2\pi r - \mu_s np)$.

EXAMPLE 9.7

The scissors jack shown supports a load **P** from a car. The screw has a double square thread with a mean diameter of 10 mm and a pitch of 2 mm. The static coefficient of friction between threads is 0.2. The jack is symmetrical about a vertical plane containing the axis of the screw and has two sets of linkages. If a torque $M = 10$ N·m is required to raise **P** as shown, determine (a) the magnitude of **P**, (b) the torque M' required to lower **P**.

Solution. We note that the pairs of linkages *AB*, *AD*, *BC*, and *CD* are all
two-force members. Although the application of the torque **M** will cause the
front and back members in the pairs of linkages to develop different axial
forces, we shall combine the axial forces in each pair into a single force and
denote the combined axial forces in those pairs of linkages by $\mathbf{F}_{AB}$, $\mathbf{F}_{AD}$,
$\mathbf{F}_{BC}$, and $\mathbf{F}_{CD}$, respectively.

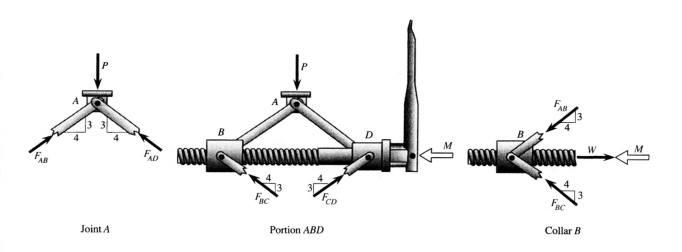

| Joint *A* | Portion *ABD* | Collar *B* |

Joint A

$$\xrightarrow{+} \Sigma F_x = 0: \quad \frac{4}{5} F_{AB} - \frac{4}{5} F_{AD} = 0 \qquad F_{AB} = F_{AD}$$

$$+\uparrow \Sigma F_y = 0: \quad \frac{3}{5} F_{AB} + \frac{3}{5} F_{AD} - P = 0 \qquad F_{AB} = F_{AD} = \frac{5}{6} P$$

Portion ABD

$$\xrightarrow{+}\ \Sigma F_x = 0: \quad -\frac{4}{5}F_{BC} + \frac{4}{5}F_{CD} = 0 \qquad F_{BC} = F_{CD}$$

$$+\uparrow \Sigma F_y = 0: \quad \frac{3}{5}F_{BC} + \frac{3}{5}F_{CD} - P = 0 \qquad F_{BC} = F_{CD} = \frac{5}{6}P$$

Collar B

$$\xrightarrow{+}\ \Sigma F_x = 0: \quad W - \frac{4}{5}F_{AB} - \frac{4}{5}F_{BC} = 0 \qquad W = \frac{4}{3}P$$

In the free-body diagram of the collar B, we may move the force **W** and torque **M** to the left end of the screw as shown, where $r = 5$ mm, $p = 2$ mm, $L = 2p = 4$ mm, and

$$\tan\theta = \frac{L}{2\pi r} = \frac{4}{2\pi(5)} \qquad \theta = 7.256°$$

$$\tan\phi_s = \mu_s = 0.2 \qquad \phi_s = 11.310°$$

$$M = 10 \text{ N·m} \qquad W = \tfrac{4}{3}P$$

From the relation summarized in Fig. 9.8(a), we write

$$10 = \tfrac{4}{3}P(\tfrac{5}{1000})\tan(7.256° + 11.310°)$$

$$P = 4.466 \times 10^3 \qquad\qquad P = 4.47 \text{ kN} \blacktriangleleft$$

Since $\phi_s > \theta$, the screw will not unwind by itself. Thus, the torque M' required to lower the load **P** may be computed by applying the relation summarized in Fig. 9.8(b) as follows:

$$M' = \tfrac{4}{3}(4.466 \times 10^3)(\tfrac{5}{1000})\tan(11.310° - 7.256°)$$

$$M' = 2.11 \text{ N·m} \blacktriangleleft$$

Developmental Exercise

D9.22   Work Example 9.7 for the case in which the screw has a single thread.

★9.10 Axle Friction in Journal Bearings

A *journal bearing* is a bearing that provides lateral support to a rotating axle or shaft in contrast to axial or thrust support. The friction between an axle and its journal bearings is referred to as *axle friction*.

Let a dry or partially lubricated journal bearing be shown in Fig. 9.9 where the clearance between the axle and bearing is greatly exaggerated. As the axle begins to turn in the direction shown, it rolls up the inner surface of the bearing until slippage occurs. The point of contact remains more or less at a fixed position A as rotation of the axle continues. To keep the axle rotating at a constant speed, we find that a torque $\mathbf{M}$ must be applied as shown. The torque $\mathbf{M}$ and the lateral load $\mathbf{W}$ on the axle at the bearing cause a reaction $\mathbf{R}$ at A. For equilibrium, $\mathbf{W}$ and $\mathbf{R}$ must be *equal in magnitude but not collinear*. Note that $\mathbf{R}$ is tangent to a small circle of radius r_f. This circle is called the *circle of friction* of the axle and bearing and is generally independent of $\mathbf{W}$. The angle between $\mathbf{R}$ and its normal component $\mathbf{N}$ is the angle of kinetic friction ϕ_k. For the axle rotating at a constant speed, we write

$$+\uparrow \Sigma F_y = 0: \quad R - W = 0 \qquad R = W$$

$$+\circlearrowleft \Sigma M_A = 0: \quad r_f W - M = 0 \qquad M = r_f W \qquad (9.11)$$

Noting that $\mu_k = \tan\phi_k$ and $\sin\phi_k \approx \tan\phi_k$ for small values of ϕ_k, we write

$$r_f = r \sin\phi_k \approx r \tan\phi_k \qquad r_f \approx r\mu_k \qquad (9.12)$$

In situations involving impending slipping in the journal bearing, we replace μ_k with μ_s; i.e.,

$$r_f \approx r\mu_s \qquad (9.13)$$

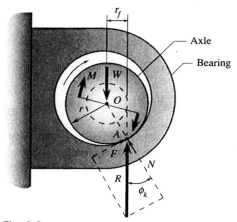

Fig. 9.9

EXAMPLE 9.8

A 100-lb crate is suspended from a pulley of diameter 10 in. as shown. For $\mu_s = 0.2$ between the pulley and the axle of diameter 1 in., determine the minimum force **P** required to maintain equilibrium of the system.

Solution. When the magnitude of **P** has the minimum value required to maintain equilibrium of the system, the pulley tends to rotate counterclockwise about the axle or, equivalently, the axle tends to rotate clockwise relative to the pulley and to climb from A to B. Thus, the free-body diagram of the pulley may be drawn as shown, where Q denotes the cable tension on the right pulley. As this problem involves impending slipping, rather than kinetic friction, we write

$$r_f \approx r\mu_s = (0.5 \text{ in.})(0.2) = 0.1 \text{ in.}$$

$$+\circlearrowleft \ \Sigma M_C = 0: \quad -10P + (5 - 0.1)(100) = 0$$

$$\mathbf{P} = 49 \text{ lb} \uparrow \ \blacktriangleleft$$

Developmental Exercises

D9.23 Refer to Example 9.8. Determine the maximum force **P** consistent with equilibrium.

D9.24 The kinetic coefficient of friction between a 1-in.-diameter axle and its bearing is 0.15. The load on the axle at the bearing is 300 lb. Determine (a) the torque required to keep the axle rotating at a constant speed, (b) the radius of the friction circle.

★9.11   Disk Friction in Thrust Bearings

A *thrust bearing* is a bearing that gives axial or thrust support to a rotating shaft or disk. Examples of thrust bearings include pivot bearings, collar bearings, disk brakes, and disk clutches. The friction between a disk or a shaft and its thrust bearing is referred to as *disk friction*.

Let us consider a hollow shaft rotating against a pivot bearing as shown in Fig. 9.10. The hollow shaft of inside radius R_i and outside radius R_o is acted on by a *wrench* consisting of an axial force **P** and a torque **M** which keeps the shaft rotating at a constant speed. Assuming that the normal pres-

Fig. 9.10

sure p between the contact surfaces is uniform, we find that

$$p = \frac{P}{\pi(R_o^2 - R_i^2)} \qquad (9.14)$$

The normal force $d\mathbf{N}$ and the friction force $d\mathbf{F}$ acting on the differential area dA have magnitudes $p\,dA$ and $\mu_k p\,dA$, respectively, where μ_k is the coefficient of kinetic friction. The moment of $d\mathbf{F}$ about the axis of the shaft is $u\,dF = \mu_k pu\,dA = \mu_k pu(u\,du\,d\theta) = \mu_k pu^2\,du\,d\theta$, and the total moment is

$$M = \mu_k p \int_{\theta=0}^{2\pi} \int_{u=R_i}^{R_o} u^2\,du\,d\theta = \frac{2}{3}\pi\mu_k p(R_o^3 - R_i^3)$$

$$M = \frac{2\mu_k P(R_o^3 - R_i^3)}{3(R_o^2 - R_i^2)} \qquad (9.15)$$

If the shaft is solid and has a radius R, we let $R_i = 0$ and $R_o = R$ in Eq. (9.15) to obtain

$$M = \tfrac{2}{3}\mu_k PR \qquad (9.16)$$

Thus, the largest torque which can be transmitted by a disk clutch without causing slippage is

$$M = \tfrac{2}{3}\mu_s PR \qquad (9.17)$$

Developmental Exercise

D9.25   When a disk clutch passes its wearing-in period, the normal pressure p may be assumed to be inversely proportional to the radial distance u from its center; i.e., $p = c/u$ where c is a constant. For the disk clutch shown, show that (a) $c = P/(2\pi R)$, (b) the largest torque which may be transmitted after the wearing-in period is given by $M = \tfrac{1}{2}\mu_s PR$.

Fig. D9.25

Fig. P9.64

Fig. P9.65

PROBLEMS

9.64 Two plates are fastened with a bolt and nut as shown. Assume that the bolt has a single square thread of mean diameter 0.5 in. and a lead of 0.1 in. For $\mu_s = 0.35$ between threads, determine the torque which must be applied to the bolt and nut to induce a bolt tension of 1 kip.

9.65 The screw of the vise shown has a double square thread of mean diameter 20 mm and a pitch of 4 mm. A clamping force of magnitude F is induced at C when a horizontal force of magnitude Q is applied normal to the handle at D to tighten the vise. For $\mu_s = 0.3$ between threads and neglecting effects of friction at B, determine the ratio of F/Q.

9.66* Solve Prob. 9.65 for $\mu_s = 0.2$.

9.67 The worm gear shown consists of a worm AB (i.e., a threaded segment of shaft) and a gear wheel C which is rigidly attached to a drum D. The worm AB has a single square thread with a mean radius of 30 mm and a pitch of 10 mm. For $\mu_s = 0.2$ between the thread of AB and the teeth of C, determine the torque M which must be applied to the shaft EF in order to raise the 200-kg crate shown.

Fig. P9.67

9.68* In Prob. 9.67, determine the torque which must be applied to the shaft EF in order to lower the crate.

9.69 The jack shown supports a load **P** of 6 kN from a car. The screw has a single square thread with a mean radius of 5 mm and a pitch of 2 mm. The jack is symmetrical about a vertical plane containing the axis of the screw and has two sets of linkages. For $\mu_s = 0.25$ between threads, determine the torque M required (a) to raise the load **P**, (b) to lower the load **P**.

40 mm | 60 mm | P | 60 mm | 40 mm

45 mm

45 mm

Smooth collar

75 mm

M

60 mm

Smooth washer

Threaded collar

80 mm | 40 mm

Fig. P9.69

9.70 A bushing of outside radius 1.5 in. fits loosely on a horizontal shaft of radius 1 in. as shown. No slippage occurs between the rope and the bushing when the force **P** starts to raise the 100-lb crate. For $\mu_s = 0.1$ between the shaft and the bushing, determine the magnitude of **P** if (a) $\theta = 0$, (b) $\theta = 90°$.

1 in.

θ

P

100 lb

Fig. P9.70

9.71 A 100-kg crate is to be raised as shown. Each of the 100-mm-diameter pulleys rotates on a 20-mm-diameter axle. For $\mu_s = 0.15$ between the pulley and axle, determine the magnitude of **P** required to raise the crate.

20 mm

E D A

F | 100 mm | 100 mm

P B

C 20 mm

100 kg

Fig. P9.71

9.72* A cart having four steel wheels of 30-in. diameter is to be moved on rails as shown. Each steel wheel supports a 4-in.-diameter axle and carries an axle load of 2 kips. For $\mu_k = 0.02$ between the wheel and axle, determine the magnitude of the horizontal force **P** required to move the cart at a constant velocity.

Fig. P9.72*

9.73 A collared shaft and its cross section in a thrust bearing are shown. If a wrench consisting of a 1500-lb thrust **P** and a 100-lb·ft torque **M** will cause rotation of the shaft to impend, determine the value of μ_s between the collar and the bearing.

Fig. P9.73   **Fig. P9.74***

9.74* A shaft and its cross section in a conical pivot bearing are shown. The coefficient of kinetic friction is μ_k and the bearing pressure is uniform. If a wrench consisting of a thrust **P** and a torque **M** will keep the shaft rotating at a constant speed, show that

$$M = \frac{(d_1^2 + d_1 d_2 + d_2^2)\mu_k P}{3(d_1 + d_2)\sin\theta}$$

9.75 A shaft and its cross section in a spherical thrust bearing are shown, where the coefficient of static friction is μ_s. The bearing pressure p may be assumed to be equal to $K \sin\theta$. The shaft carries a wrench consisting of **P** and **M**. Determine the expressions of (a) K, (b) the magnitude of **M** required to start rotating the shaft.

9.76 A 20-kg electric floor polisher is shown. Assume that the normal pressure between the floor and the ring-shaped area of the attached polishing disk is uniform and $\mu_k = 0.2$. Determine the magnitude Q of the forces in the couple required to prevent rotation of the handle during polishing.

Fig. P9.75

Bottom view

Fig. P9.76

BELT FRICTION

9.12   Flat Belts

Friction forces developed between a belt and a drum have various engineering applications (e.g., belt drives and brakes). The relationship between the two tensile forces T_1 and T_2 in the two parts of a flat belt (or rope), which

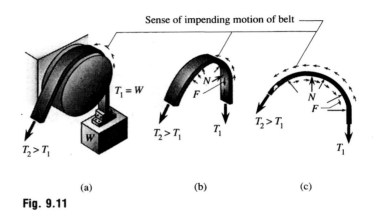

(a)   (b)   (c)

Fig. 9.11

are not in contact with the drum as shown in Fig. 9.11(a), can be established when *slipping* of the belt around the drum *impends* (and the conditions of equilibrium still hold true). A perspective view and a front view of the free-body diagram of the belt are shown in Fig. 9.11(b) and (c), respectively. The drum may have either constant or varying curvature as shown in Fig. 9.12; however, the coefficient of friction is assumed to be the same at all points of contact between the belt and the drum.

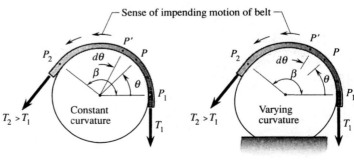

Fig. 9.12   (a)   (b)

The free-body diagram of the infinitesimal belt element PP' in Fig. 9.12 is shown in Fig. 9.13. For equilibrium of this element, we write

$$+\searrow \Sigma F_x = 0: \quad T \cos \frac{d\theta}{2} + dF - (T + dT) \cos \frac{d\theta}{2} = 0 \quad (9.18)^\dagger$$

$$+\nearrow \Sigma F_y = 0: \quad dN - T \sin \frac{d\theta}{2} - (T + dT) \sin \frac{d\theta}{2} = 0 \quad (9.19)$$

$$\text{Friction law:} \quad dF = \mu_s \, dN \quad (9.20)$$

Recognizing that $\cos(d\theta/2) \approx 1$, $\sin(d\theta/2) \approx d\theta/2$, and neglecting second-order infinitesimals in comparison with first-order infinitesimals, we obtain, from the above equations,

$$\frac{dT}{T} = \mu_s \, d\theta \quad (9.21)$$

By integrating both sides of Eq. (9.21) from P_1 to P_2 of the flat belt, as shown in Fig. 9.12, we write

$$\int_{T_1}^{T_2} \frac{dT}{T} = \mu_s \int_0^\beta d\theta \qquad \ln \frac{T_2}{T_1} = \mu_s \beta \quad (9.22)$$

Thus, the equation governing the impending slipping of a flat belt may be written as

$$T_2 = T_1 e^{\mu_s \beta} \quad (9.23)$$

where e (equal to $2.718...$) is the base of natural logarithms, μ_s is the coefficient of static friction between the belt and the drum, β is the *angle of contact* (i.e., the angle between the radii of curvature of the drum, which are drawn at the two points P_1 and P_2 where the belt leaves the drum), and T_1 and T_2 are, respectively, the smaller tensile force and the larger tensile force in the two parts of the belt not in contact with the drum.

In using the above formula for flat-belt friction, it is to be noted that the angle of contact β must be in radians and the belt must have impending relative motion with respect to the drum. If a belt or rope is wrapped n times around a post, β is equal to $2n\pi$. Furthermore, this belt friction formula is *valid for any shape of surface of the drum* provided that the coefficient of friction is the same at all points of contact. In kinetic friction, if the mass of the belt is negligible, we may simply replace μ_s with μ_k and write

$$T_2 = T_1 e^{\mu_k \beta} \quad (9.24)$$

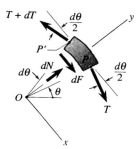

Fig. 9.13

Developmental Exercise

D9.26 Write the equation governing the impending slipping of a flat belt around a drum and define the symbols used in the equation.

---

†This equation or its equivalent can also be obtained from Fig. 9.13 by applying $\Sigma M_O = 0$.

EXAMPLE 9.9

If $\mu_s = 0.25$ between all contact surfaces and the body A shown has a mass of 80 kg, determine the maximum mass of the body B for which the system will remain in equilibrium.

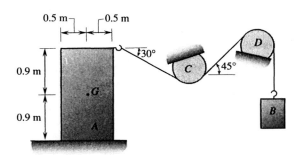

Solution. When the body B has the said maximum mass, its weight will cause the body A to have either *impending slipping* or *impending tipping*. Thus, this is a *type III friction problem*. We may solve this problem by first determining the tension T_{AC} between A and C that will cause A to have impending motion (slipping or tipping).

Case (a). Assume that slipping of the body A impends. (See above.)

Friction law: $(F_A)_m = 0.25N_A$

$$\xrightarrow{+} \Sigma F_x = 0: \quad T_{AC} \cos 30° - (F_A)_m = 0$$

$$+\uparrow \Sigma F_y = 0: \quad N_A - T_{AC} \sin 30° - 80g = 0$$

Solution of these three equations yields

$$T_{AC} = 264.8 \text{ N} \qquad N_A = 917.2 \text{ N} \qquad (F_A)_m = 229.3 \text{ N}$$

Case (b). Assume that tipping of the body A impends. In this case, the normal reaction component is $\mathbf{N}_P$ and the friction force is $\mathbf{F}_P$. They act through the edge P of the body A as shown. We have

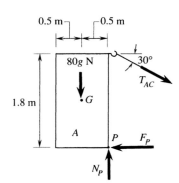

$$+ \, \text{\Large \curvearrowright} \, \Sigma M_P = 0: \quad 0.5(80g) - 1.8(T_{AC} \cos 30°) = 0$$

$$T_{AC} = 251.7 \text{ N} < 264.8 \text{ N}$$

Decision. The value of T_{AC} at which the motion of the body A impends is the smaller value of T_{AC} obtained in the above two cases. This means that the impending motion is in the form of *impending tipping*; so, we take

$$T_{AC} = 251.7 \text{ N}$$

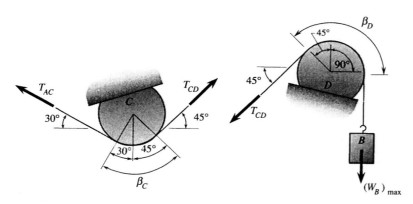

As indicated, the angles of contact of the cord around the drums are

$$\beta_C = 30° + 45° = 75° = \frac{5\pi}{12} \text{ rad}$$

$$\beta_D = 45° + 90° = 135° = \frac{3\pi}{4} \text{ rad}$$

By repeated applications of Eq. (9.23), we write

$$(W_B)_{max} = T_{CD} \, e^{\mu_s \beta_D} = (T_{AC} \, e^{\mu_s \beta_C}) \, e^{\mu_s \beta_D}$$

$$= 251.7 \, e^{0.25(5\pi/12)} \, e^{0.25(3\pi/4)}$$

$$= 629.3 = (m_B)_{max} \, (9.81)$$

$$(m_B)_{max} = 64.1 \text{ kg} \quad \blacktriangleleft$$

Developmental Exercises

D9.27   A boy and a girl pull on the ends of a rope which is wrapped around a post as shown. When the girl pulls with a 5-lb force on one end of the rope, the boy finds that he needs to exert a 50-lb force on the other end of the rope if he wants to pull her toward him. Determine (a) the angle of contact β, (b) the value of μ_s between the rope and the post.

D9.28   It is known that $\mu_s = 0.25$ between the elliptic drum and the rope which carries a 10-kg block on one side and is held by a force P on the other side as shown. Determine (a) the angle of contact β, (b) the minimum and the maximum values of P if equilibrium exists.

Fig. D9.27

Fig. D9.28

9.13   V Belts

In automobile engines and other machines, the belts used for belt drives are frequently V-shaped. Let α be the angle between the two friction sides of the V belt as shown in Fig. 9.14(a). The V belt is in contact with the sides of the groove of the pulley; it does not touch the bottom of the groove. The condition of impending slipping of the V belt in the groove of the pulley will be studied.

The perspective view of the free-body diagram of an infinitesimal V-belt element subtending an angle $d\theta$ in Fig. 9.14(a) is shown in Fig. 9.14(b). The orthographic projections of this free-body diagram are shown in Fig. 9.14(c). We write

$$\xrightarrow{+} \Sigma F_x = 0: \quad T \cos \frac{d\theta}{2} + 2\,dF - (T + dT) \cos \frac{d\theta}{2} = 0 \tag{9.25}$$

$$+\uparrow \Sigma F_y = 0: \quad 2(dN) \sin \frac{\alpha}{2} - T \sin \frac{d\theta}{2} - (T + dT) \sin \frac{d\theta}{2} = 0 \tag{9.26}$$

Friction law: $dF = \mu_s\,dN$ \tag{9.27}

Fig. 9.14

From Eqs. (9.25) through (9.27), we can similarly obtain the equation governing the impending slipping of a V belt as

$$T_2 = T_1 \, e^{\mu_s \, \beta / \sin \, (\alpha/2)} \qquad (9.28)$$

where T_1, T_2, μ_s, and β have the same meaning as those used in Sec. 9.12. If kinetic friction occurs, μ_s is replaced by μ_k and we get

$$T_2 = T_1 \, e^{\mu_k \, \beta / \sin \, (\alpha/2)} \qquad (9.29)$$

It is readily seen from Eqs. (9.23), (9.24), (9.28), (9.29), and Fig. 9.14 that when $\alpha \rightarrow \pi$ rad, the formulas governing the V-belt friction do reduce to the formulas governing the flat-belt friction.

Developmental Exercise

D9.29 Write the equation governing the impending slipping of a V belt around a pulley and define the symbols used in the equation.

EXAMPLE 9.10

A V belt with its cross section shown is to deliver a torque of 100 N·m to a lathe pulley at A from a motor pulley at B rotating counterclockwise at a constant speed. If $R = 400$ mm, $r = 40$ mm, and $\mu_s = 0.3$ between the belt and the pulleys, determine (a) the minimum tensile strength required of the V belt, (b) the driving torque of the motor.

Solution. In solving this problem, we need to investigate the condition under which the *difference* between the magnitudes of the tensile forces in the portions of the V belt not in contact with the pulleys is maximum. When such a difference in maximum, impending slipping exists between the V belt and the motor pulley, rather than the lathe pulley, because $\beta_B < \beta_A$ as shown. Thus, the present problem is a *type II friction problem*.

Note that the reaction torque $\mathbf{M}_A$ of 100 N·m from the shaft of the lathe pulley and the driving torque $\mathbf{M}_B$ of the motor have been indicated in the free-body diagrams of the pulleys on the next page. The tensile forces T_2 and T_1 are related to each other by the condition of impending slipping of the V belt around the motor pulley. at B. From the given data, we have

$$\mu_s = 0.3 \qquad \alpha = 38° \qquad \beta_B = 180° - 30° = 150° = \frac{5\pi}{6} \text{ rad}$$

$$\text{V-belt friction:} \quad T_2 = T_1 \, e^{0.3(5\pi/6)/\sin \, (38°/2)} \qquad (1)$$

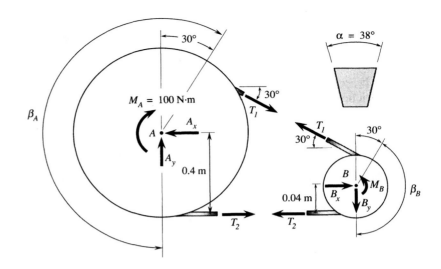

Since the lathe pulley and the motor pulley have no angular acceleration, they are in rotational equilibrium. Thus, we write

$$\textit{Lathe pulley,} \ +\!\!\!\curvearrowright \ \Sigma M_A = 0: \quad 0.4T_2 - 0.4T_1 - 100 = 0 \quad (2)$$

$$\textit{Motor pulley,} \ +\!\!\!\curvearrowright \ \Sigma M_B = 0: \quad M_B + 0.04T_1 - 0.04T_2 = 0 \quad (3)$$

Solution of the above three simultaneous equations yields

$$T_1 = 24.60 \qquad T_2 = 274.60 \qquad M_B = 10$$

Minimum tensile strength required = 275 N ◀

Driving torque of the motor = 10 N·m ◀

Developmental Exercise

D9.30 In a car engine as shown, the crankshaft pulley C, which rotates clockwise, delivers power to pulleys A and F. When the car is accelerated, a screeching noise is heard due to slipping of the V belt around the pulley C. Determine the ratio T_{CF}/T_{CA} of belt tensions in portions CF and CA when (a) the screeching impends, (b) the screeching noise is being heard.

Fig. D9.30

PROBLEMS

9.77 A rope is fastened to a crate of mass m and passed over a drum as shown. If $m = 100$ kg and $\mu_s = 0.1$ between all surfaces of contact, what is the smallest magnitude of the vertical force $\mathbf{Q}$ which will move the crate?

9.78* A man fastens a rope to a crate of weight W and passes it over a drum as shown. The force $\mathbf{Q}$ exerted by the man on the rope may reach a maximum value of 150 lb. If $\mu_s = 0.1$ between all surfaces of contact, determine the maximum weight of the crate which can be moved by the man.

Fig. P9.77and P9.78*

9.79 A ship pulls on a hawser with a force of 10 kips as shown. For $\mu_s = 0.25$ between the hawser and the bollard, determine the number of full turns a dock worker has to wrap the hawser around the bollard if he is to hold the ship by exerting on his side of the hawser a force $\mathbf{P}$ of magnitude not greater than 20 lb.[†]

9.80 If $\mu_s = 0.3$ between the rope and the pipes shown, determine the minimum horizontal force $\mathbf{P}$ required to hold the 10-kg mass in equilibrium.

Fig. P9.79

Fig. P9.80 and P9.81*

9.81* If $\mu_s = 0.3$ between the rope and the pipes shown, determine the minimum horizontal force $\mathbf{P}$ required to pull the 10-kg mass upward from rest.

9.82 through 9.84* A force $\mathbf{P}$ is applied to a horizontal bar AB which has a negligible weight and is supported by a rope passing around cantilever pipes as shown. For $a = L/4$, determine the minimum value of μ_s between the rope and the pipes for which the bar will remain horizontal.

Fig. P9.82 and P9.85

Fig. P9.83 and P9.86*

Fig. P9.84* and P9.87

9.85 through 9.87 A force $\mathbf{P}$ is applied to a horizontal bar AB which has a negligible weight and is supported by a rope passing around cantilever pipes as shown. For $\mu_s = 0.2$ between the rope and the pipes, determine the minimum ratio of a/L for which the bar will remain horizontal.

[†]Note that a *bollard* is a thick post mounted on a wharf and cannot be rotated, while a *capstan* is a type of windlass which can be rotated manually or by motor.

9.88 through 9.90* A vertical force **P** is applied to a rope which passes around cantilever pipes and is connected to a 100-kg homogeneous body *B* as shown. The values of μ_s between the rope and the pipes and between the body and the horizontal support are 0.2 and 0.27, respectively. Determine the minimum magnitude of **P** required to cause impending motion.

Fig. P9.88

Fig. P9.89   **Fig. P9.90***

9.91 The two drums *C* and *D* are rigidly fastened together and are free to rotate about a frictionless bearing at *O* as shown. The body *A* weighs 100 lb. The value of μ_s is 0.1 between the belt and the small drum *D* and 0.2 between the belt and the large drum *C*. Determine the minimum magnitude of the force **P** which will lift the body *A*.

9.92* Solve Prob. 9.91 if the value of μ_s is 0.15 between the belt and the small drum *D* and 0.3 between the belt and the large drum *C*.

9.93 A constant torque of 100 N·m is applied to the axle of a flywheel to keep it rotating at a constant speed. The coefficients of friction between the brake band and the drum of the flywheel as shown are $\mu_s = 0.3$ and $\mu_k = 0.25$. Determine the magnitude of the force **P** acting at the hand bar if the direction of the applied torque and the rotation of the flywheel are clockwise.

Fig. P9.91

Fig. P9.93

9.94 Solve Prob. 9.93 if the direction of the applied torque and the rotation of the flywheel are counterclockwise.

9.95 A 20-lb force is applied to the lever at *A* and $\mu_k = 0.5$ between the flat belt and the drum which is attached to a flywheel as shown. A torque **M** is applied to the axle of the flywheel. Determine the torque **M** required to keep the flywheel rotating counterclockwise at a constant speed.

Fig. P9.95

9.96* Refer to the system in Prob. 9.95. Determine the torque **M** required to keep the flywheel rotating clockwise at a constant speed.

9.97 A 100-lb weight C and a body D of weight W are connected by a flat belt which passes around a shaft A and a fixed cylindrical drum B as shown. For $\mu_s = 0.3$ and $\mu_k = 0.2$ between all surfaces of contact, determine the maximum weight W which can be raised if the shaft A is rotated from rest.

9.98 A 10-kg body A and a body B of mass m_B are connected by a rope which passes around a smooth pulley C and a fixed drum D as shown. If $\mu_s = 0.2$ between all surfaces of contact and $m_B = 17$ kg, determine the minimum magnitude of the force **P** which will cause impending motion between the rope and the drum D.

Fig. P9.97

Fig. P9.98

9.99* Solve Prob. 9.98 for $m_B = 18$ kg.

9.100 A flat belt is used to deliver a torque from the pulley D to a load at the pulley C. It passes around two smooth idler pulleys A and B as shown. The value of μ_s between the belt and the pulley D is 0.3. A block E of mass m is attached to the axle of the pulley D by a cable to keep the pulley D from slipping on the belt. What is the minimum mass m of the block E which will prevent slippage from occurring when a 50 N·m torque is applied to the pulley D?

Fig. P9.100

9.101* Refer to the system in Prob. 9.100. Determine the minimum value of μ_s between the belt and the pulley C if the belt is not to slip around the pulley C.

9.102 The tape of a recorder is driven by the pulley A with a torque of 0.3 N·m as shown. The maximum tension which is allowed to be developed in the tape is 40 N. Determine the minimum value of μ_s required between the tape and the pulley A.

9.103* Solve Prob. 9.102 if the maximum tension which is allowed to be developed in the tape is 30 N.

Fig. P9.102

Fig. P9.104

9.104 A V belt passes around a grooved wheel and is connected to form a brake as shown. For $\mu_k = 0.3$ between the belt and the wheel, determine the resisting torque of the brake if the wheel is turning clockwise.

9.105* Solve Prob. 9.104 if the wheel is turning counterclockwise.

9.106 A V belt of negligible mass is used to deliver a torque from the motor at B to a load at the pulley A as shown. If the maximum allowable belt tension is 900 lb and $\mu_s = 0.3$, determine the largest torque which can be delivered to the pulley A.

Fig. P9.106

9.107* Solve Prob. 9.106 if the maximum allowable belt tension is 700 lb and $\mu_s = 0.2$.

9.14 Concluding Remarks

A dry friction force is simply the tangential component of the reaction between two nonlubricated or partially lubricated surfaces of contact of two bodies. This tangential component of the reaction is found, through experiments, to be proportional to the normal component of the reaction *when and only when* slipping between the two contact surfaces impends or is occurring. The constant of proportionality is called the *coefficient of friction*.

Many equilibrium problems involving dry friction are really statically indeterminate problems because they cannot be solved by using only the equations of equilibrium. These statically indeterminate problems may be solved if the equations representing the laws of dry friction are appropriately introduced. There are many other classes of statically indeterminate equilibrium problems which are not treated in this text. They may be found in texts on mechanics of materials, advanced structural analysis, elasticity, etc.

The classification of friction problems into certain types is very helpful in developing methods for solving them. Such a classification is similar to the classification of problems in other subjects (e.g., differential equations) into certain types for the study of their solutions. Thus, a careful review of Secs. 9.5 through 9.8 is recommended.

The resistance to rolling of a wheel arising from the deformation between the wheel and its supporting surface is referred to as the *rolling resistance*. This resistance is not due to tangential friction forces and therefore is a phenomenon entirely different from that of dry friction. As illustrated in Fig. 9.15 where θ is usually small, the force **P** necessary to initiate and maintain

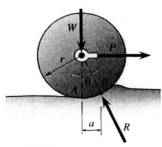

Fig. 9.15

rolling at constant velocity can be found by writing

$$+ \circlearrowright \; \Sigma M_B = 0: \quad aW - (r\cos\theta)P \approx aW - rP = 0$$

$$\mu_r = \frac{a}{r} \qquad P = \mu_r W \qquad (9.30)$$

where the moment arm of **P** is taken as r and **W** is the load on the wheel. The ratio $\mu_r = a/r$ is referred to as the *coefficient of rolling resistance*, which may vary with the speed of travel, the radius of the wheel, the elastic and plastic properties of the wheel and its support, and the roughness of the surface of support. Some tests indicate that the distance a varies only slightly with the radius r. Thus, the distance a has also been referred to as the coefficient of rolling resistance. Note that a has the dimension of length, while μ_r is dimensionless.

REVIEW PROBLEMS

9.108 If $\mu_s = 0.5$ between the 26-lb homogeneous body and the support as shown, determine the minimum magnitude of the horizontal force **P** required to cause the body to have impending motion.

9.109 Two horizontal forces **F** and $-\mathbf{F}$ are applied to a 20-kg horizontal cylinder A which rests on a 10-kg vertical cylinder B as shown. The value of μ_s between the cylinder A and its surfaces of contact is 0.35, and that between the cylinder B and the horizontal support is 0.25. Determine the smallest magnitude of **F** for which the cylinder A will start to move.

Fig. P9.108

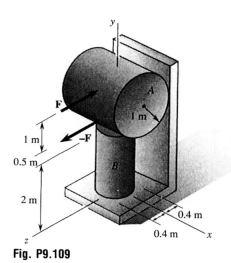

Fig. P9.109

9.110 The screw of the clamp shown has a single square thread, a mean diameter of 12 mm, and a pitch of 3 mm. If the clamp is tightened with a torque of 1 N·m and $\mu_s = 0.25$ between the threads, determine (a) the magnitude of the force pressing the two pieces of wood together, (b) the torque required to loosen the clamp if self-locking exists.

Fig. P9.110

Fig. P9.111* and P9.112*

9.111* If the weight of the bar AB is negligible and $a = L/4$, determine the minimum value of μ_s between the rope and the pipes for which the bar under the action of the force **P** will remain horizontal as shown.

9.112* If the weight of the bar AB is negligible and $\mu_s = 0.2$ between the rope and the pipes, determine the minimum ratio of a/L for which the bar under the action of the force **P** will remain horizontal as shown.

9.113 A strap wrench is used to grip a pipe as shown. If $a = 50$ mm and $\mu_s = 0.25$ between the strap and the pipe, determine the maximum value of b for which the wrench will be self-locking.[†]

Fig. P9.113 and P9.114*

9.114* Show that the strap wrench shown will be self-locking if the value of μ_s between the strap and the pipe exceeds that given by the equation

$$3\pi\mu_s + 2 \ln \mu_s = 0$$

Determine that minimum value of μ_s.[††]

Fig. P9.115

[†]Note that the strap is not to slip on the pipe, however large the force **F** may be. At impending slipping, we have $T_2 > T_1 = \mu_s N_w$ where N_w is the normal force exerted by the wrench on the strap at the top of the pipe.

[††]Cf. App. E or Sec. B5.8 of App. B for the solution of a transcendental equation.

9.115 The mass of body A is 10 kg and $\mu_s = 0.5$ between all surfaces of contact. For $\theta = 35°$ as shown, determine the range of values of the mass of body B for which the system will remain in equilibrium.

A

L

a

30°

B

Fig. P9.116

9.116 A man sits on a horizontal bar of negligible weight as shown. For $a = L/3$, determine the angle θ_A and the minimum value of μ_s between the rope and the pipe for which the bar will remain horizontal.

9.117 The car engine shown is at rest. Determine the ratio T_{AF}/T_{FC} of the tensions in the portions AF and FC of the V belt when a mechanic manually rotates the fan clockwise and the belt slips around the fan pulley F.

$\alpha = 32°$

$\mu_s = 0.35$

$\mu_k = 0.3$

A

F

3

4

C

Fig. P9.117

Chapter 10   Virtual Work

In mechanics, *work* is done by a force or a moment to (or on) a body during a displacement of the body; and no work can be done to a body which remains stationary. Having carefully defined work, we proceed to derive the *principle of virtual work*, which is then applied to solve equilibrium problems of *frames* as well as *machines*. Here, by giving a free body in equilibrium an appropriate compatible virtual displacement and setting the corresponding total virtual work to zero, we can frequently determine a specific unknown quantity in a seemingly complex equilibrium problem without having to solve a system of simultaneous equations. This is an outstanding feature of the principle of virtual work. Later, a *quantitative definition* for the potential energy of a body is given, and the *principle of potential energy* is deduced. In the last part, the potential energy and its expansion into a *Taylor's series* are employed to establish the criteria for gauging the *stability of equilibrium* of conservative systems.

PRINCIPLE OF VIRTUAL WORK

10.1 Displacements

The *displacement* of a body is the change of position of the body.[†] When the body is in translation or being treated as a particle, the displacement of the body is simply a linear displacement. The *linear displacement* of a par-

---

[†]Unless otherwise stated, the displacement of a body is to be understood as being measured from a fixed position in an inertial reference frame. Virtual displacements are discussed in Sec. 10.4.

ticle moving from the point A along *any* path (e.g., the path C) to the point B is a vector $\mathbf{q} = \overrightarrow{AB}$ which is directed from the point A to the point B and has a magnitude equal to the shortest distance between the points A and B as shown in Fig. 10.1. In terms of the position vectors $\mathbf{r}_A$ and $\mathbf{r}_B$ of A and B, we write

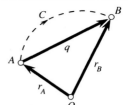

Fig. 10.1

$$\mathbf{q} = \mathbf{r}_B - \mathbf{r}_A \qquad (10.1)$$

The *angular displacement* of a body is the change of orientation of the body; it is produced by the rotation of the body. The *direction of the angular displacement* $\Delta\theta$ of a rigid body, as shown in Fig. 10.2, is defined to be parallel to the extended thumb of the right hand when the other fingers of the right hand are curled in the direction of rotation of the body. Note that finite angular displacements of a rigid body are *not* vector quantities because they do not follow the commutative law as well as the parallelogram law for the addition of vectors.[†] However, *infinitesimal angular displacements* $\overrightarrow{d\theta}_1$ and $\overrightarrow{d\theta}_2$ (as well as their time rates of change) of a rigid body are vectors because they do obey the laws for the addition of vectors.[††] We have

Fig. 10.2

$$\overrightarrow{d\theta}_1 + \overrightarrow{d\theta}_2 = \overrightarrow{d\theta}_2 + \overrightarrow{d\theta}_1 \qquad (10.2)$$

The *total distance traveled* by a body is the total accumulated length of path traversed by the body. Similarly, the *total angle turned through* by a body is the total accumulated angle rotated through by the body.

Developmental Exercises

D10.1 Define the terms: (a) displacement, (b) linear displacement, (c) angular displacement, (d) total distance traveled, (e) total angle turned through.

D10.2 A particle travels along the x axis from the point A first to the point B, then back to the point C, as shown. (a) What is the displacement of the particle? (b) What is the total distance traveled by the particle?

Fig. D10.2

D10.3 A rod is pinned to a support at A. The rod rotates from the position AB first to the position AC, then back to the position AD, as shown. (a) What is the angular displacement of the rod? (b) What is the total angle turned through by the rod?

Fig. D10.3

10.2 Work of a Force[†††]

The *work of a force* $\mathbf{F}$ *acting on a particle during a differential displacement* $d\mathbf{r}$ *of the particle* from the point A to a neighboring point A', as shown in Fig. 10.3, is defined as an infinitesimal scalar quantity dU equal to the dot product of $\mathbf{F}$ and $d\mathbf{r}$; i.e.,

$$dU = \mathbf{F} \cdot d\mathbf{r} \qquad (10.3)$$

[†]Cf. D1.3.

[††]Cf. App. C for the proof that they are vectors.

[†††]Note that *work* is done *by* a force *to* (or *on*) a body. It is the *force* (not the *body*) which does work; a body *receives* work from the force acting on it during its displacement from one position to another. If the action of the force on a particle and the displacement of the particle do not occur simultaneously, no work is done by the force to the particle. The *work of a force* acting on a body is the *work done by a force* to a body.

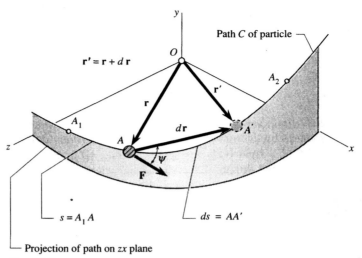

Fig. 10.3

Alternatively, we may write $dU = F \cos\psi\, ds$, where $ds = |d\mathbf{r}|$ and ψ is the angle between $\mathbf{F}$ and $d\mathbf{r}$. Anyway, the *work $U_{1\to2}$ of a force $\mathbf{F}$ acting on a particle during a finite displacement of the particle* from the point A_1 along a path C to the point A_2, as shown in Fig. 10.3, is

$$U_{1\to2} = \int_C \mathbf{F} \cdot d\mathbf{r} \tag{10.4}$$

where the label C at the integral sign signifies that the work $U_{1\to2}$ done by $\mathbf{F}$ to the particle is generally dependent on the path C along which the particle travels from A_1 to A_2. If $U_{1\to2}$ is dependent on the coordinates of A_1 and A_2, but not on those of C, then $\mathbf{F}$ is referred to as a *conservative force* (which is discussed in Sec. 10.9) and we may write

$$U_{1\to2} = \int_{A_1}^{A_2} \mathbf{F} \cdot d\mathbf{r} \tag{10.5}$$

For example, if $\mathbf{F}$ is a friction force, which is a nonconservative force, Eq. (10.4) will generally yield a different work for each different path C. By definition, the work of a force is a scalar and has the dimensions of *force × length*. The primary units for work are joules (1 joule = 1 J = 1 N·m) in the SI and foot-pounds (ft·lb) in the U.S. customary system.[†]

The *work of a force $\mathbf{F}$ acting on a rigid body* is equal to the sum of works done by $\mathbf{F}$ to the particles of the rigid body. When work is done to a body by a force, *the particle* (or particles) *of the body*, on which the force acts, *must have a nonzero component of displacement in the direction of the force during the action of the force*; otherwise, no work is done. Note that

---

[†]The joule (J) is the primary unit of *work* and *energy* in the SI. Note that *moment* is not a form of energy. Although 1 J = 1 N·m, the primary units of *moment* of a force are N·m, *not* J, in the SI. In the U.S. customary system, the primary units are usually lb·ft for *moment*, but ft·lb for *work*. [However, some do not strictly follow the conventions for lb·ft and ft·lb, which are in reverse order to those shown in Eqs. (4.18) and (10.4).]

the work done by any force to a stationary rigid body by sliding on its surface is zero. (Cf. D10.6 through D10.8.)

The work as defined above may be referred to as a *kinetic work*, which is shown in dynamics to be directly related to the *change of kinetic energy* of the body under consideration. Since velocity is the time rate of increase of displacement (or the time rate of change of position), it should be noted that *the work of a force acting on a rigid body is zero if the particle of the rigid body, on which the force acts, has, during the instant of action of the force, a zero velocity.*

If the force **F** is *constant* in magnitude and direction, the work $U_{1 \to 2}$ of **F** acting on a particle during its motion from the point A_1 along a path C to the point A_2, as shown in Fig. 10.4, is

$$U_{1 \to 2} = \int_C \mathbf{F} \cdot d\mathbf{r} = \mathbf{F} \cdot \int_C d\mathbf{r} = \mathbf{F} \cdot \mathbf{q}$$

$$U_{1 \to 2} = Fq \cos\psi \tag{10.6}$$

where **q** is the displacement of the particle from A_1 to A_2 and ψ is the angle between the positive directions of **F** and **q**. We readily observe that this work $U_{1 \to 2}$ is positive if ψ is acute and negative if ψ is obtuse. There are three cases which are of special interest: (1) if **F** and **q** have the same direction, then $U_{1 \to 2} = Fq$; (2) if **F** and **q** have opposite directions, then $U_{1 \to 2} = -Fq$; (3) if **F** is perpendicular to **q**, then $U_{1 \to 2} = 0$. Using $\mathbf{q}_{\parallel}$ to denote the component of the displacement of the particle parallel to the constant force **F**, we may write Eq. (10.6) as

$$U_{1 \to 2} = Fq_{\parallel} \tag{10.7}$$

where

$$q_{\parallel} = q \cos\psi \tag{10.8}$$

Note that the value of $q_{\parallel}$ is negative when $90° < \psi \le 180°$.

Furthermore, *the work of a force is equal to the sum of the works of its component forces, and conversely.* This property is analogous to *Varignon's theorem* and can be proved by applying the distributive property of dot products of vectors to the definition of the work of a force and the relation between the force and its component forces.[†]

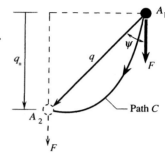

Fig. 10.4

EXAMPLE 10.1

If $\mu_k = 0.2$ between the 100-lb body and the horizontal support as shown, determine the works U_P, U_W, U_N, and U_F done to the body by the 50-lb

---

[†]Some refer to this property as the *theorem of work*.

pushing force **P**, the 100-lb weight force **W** of the body, the normal reaction **N**, and the friction force **F**, respectively, during the 3-ft displacement of the body from A to B.[†]

$P = 50$ lb $W = 100$ lb $q = 3$ ft

$30°$ F_k N

Solution. Since the body can move only in the horizontal direction, it is in equilibrium in the vertical direction. We write

$$+\uparrow \Sigma F_y = 0: \quad N - 100 - 50\sin 30° = 0 \qquad N = 125 \text{ lb}$$

$$\text{Friction law:} \quad F_k = \mu_k N \qquad\qquad F_k = 25 \text{ lb}$$

The angles ψ_P, ψ_W, ψ_N, and ψ_F between the displacement **q** of the body and the forces **P**, **W**, **N**, and F_k, respectively, are

$$\psi_P = 30° \qquad \psi_W = 90° \qquad \psi_N = 90° \qquad \psi_F = 180°$$

Noting that the forces considered are constant and $U_{1\rightarrow 2} = Fq\cos\psi$, we write

$$U_P = Pq\cos\psi_P = 50(3)(\sqrt{3}/2) \qquad U_W = Wq\cos\psi_W = 100(3)(0)$$

$$U_N = Nq\cos\psi_N = 125(3)(0) \qquad U_F = F_k q\cos\psi_F = 25(3)(-1)$$

$$U_P = 129.9 \text{ ft·lb} \qquad U_W = U_N = 0 \qquad U_F = -75 \text{ ft·lb} \quad \blacktriangleleft$$

Alternative solution for U_P. Resolving the 50-lb force **P** into two rectangular components $P_x = 50\cos 30°$ lb $\rightarrow$ and $P_y = 50\sin 30°$ lb $\downarrow$, we write

$$U_P = P_x q\cos 0° + P_y q\cos 90° = (50\cos 30°)(3)(1) + 0$$

$$U_P = 129.9 \text{ ft·lb} \quad \blacktriangleleft$$

[†]Cf. the third footnote on p. 401.

Developmental Exercises

D10.4 Define the work of a force acting on a (a) particle, (b) rigid body.

D10.5 A 100-N force **F** acts on a particle P during a 2-m displacement **q** of the particle as shown. Compute the work of **F**.

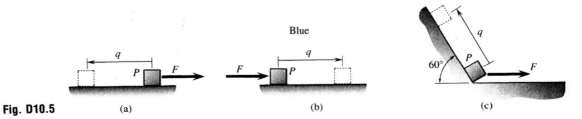

Blue

q P F F P q q $60°$ P F

Fig. D10.5   (a)   (b)   (c)

EXAMPLE 10.2

The tension in the cord which holds a spring of modulus $k = 300$ N/m in compression as shown is 60 N. A package slides from the position A down the incline and displaces the right end of the spring from B to C by an amount $\delta = 0.6$ m before it comes to a stop. Determine the work U_s of the spring force $\mathbf{F}_s$ acting on the package during its displacement from A to C.

Solution. We first draw the free-body diagram of the package as shown. Note that the spring force $\mathbf{F}_s$ will act on the package only when the package comes to the region between B and C. The other forces are ever present during the displacement.

Initially, the tensile force in the cord and the compressive force in the spring must have the same magnitude; i.e., $F_{s1} = 60$ N. For a spring modulus $k = 300$ N/m, the magnitude of the initial deflection x_1 of the spring is given by

$$x_1 = F_{s1}/k = 60/300 \qquad x_1 = 0.2 \text{ m}$$

The final deflection x_2 of the spring is given by

$$x_2 = x_1 + 0.6 \qquad x_2 = 0.8 \text{ m}$$

The magnitude of the spring force $\mathbf{F}_s$ in terms of the deflection x of the spring is $F_s = kx = 300x$. The differential displacement $d\mathbf{r}$ of the package and $\mathbf{F}_s$ acting on the package are opposite in sense. Thus, the angle ψ_s between $d\mathbf{r}$ and $\mathbf{F}_s$ is 180°. Noting that $|d\mathbf{r}| = dx$, we write

$$U_s = \int_C \mathbf{F}_s \cdot d\mathbf{r} = -\int_{x_1}^{x_2} F_s \, dx = -\int_{0.2}^{0.8} 300x \, dx$$

$$U_s = -90 \text{ J} \blacktriangleleft$$

REMARK. By the concept of "area" in calculus, the value of U_s may also be obtained from the "area" of the trapezoid in the spring-force diagram as shown. Using the formula for the area of a trapezoid (cf. Sec. B2.3 of App. B), we write

$$U_s = -\tfrac{1}{2}(60 + 240)(0.6)$$

$$U_s = -90 \text{ J} \blacktriangleleft$$

Developmental Exercises

D10.6 A line is drawn with a piece of chalk on a fixed chalkboard. Is the work *greater than, equal to,* or *less than* zero when it is done by the friction force to (a) the chalkboard, (b) the chalk?

D10.7 A 13-kg wheel *rolls without slipping* down an incline through a 2-m displacement as shown. The free-body diagram of the wheel and the normal and friction forces acting on the incline are also shown. If $\mu_s = 0.6$ and $\mu_k = 0.5$, determine during that displacement the work done by (a) **W** to the wheel, (b) **N** to the wheel, (c) **F** to the wheel, (d) **N'** to the incline, (e) **F'** to the incline.

D10.8 Is the work of a constant force **F** during the displacement **q** *of its point of application* on a stationary rigid body equal to **F·q**?

Fig. D10.7

10.3 Work of a Couple†

If a rigid body is in a translational motion, its orientation must remain unchanged, and the displacements of all particles of the rigid body are equal. Thus, *the work of a couple acting on a rigid body in translation is always equal to zero.* This is obviously true because, as shown in Fig. 10.5, the total work of the two forces **F** and −**F** in the couple acting on the rigid body through the common displacement *d***r** is always zero; i.e.,

$$dU = \mathbf{F} \cdot d\mathbf{r} + (-\mathbf{F}) \cdot d\mathbf{r} = 0 \qquad (10.9)$$

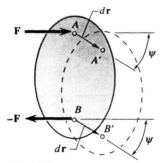

Fig. 10.5

Now, consider the case in which a rigid body is acted on by a couple of moment **M** and, at the same time, undergoes an infinitesimal angular displacement $\overrightarrow{d\theta}$ which is parallel to **M**. Let the couple be represented by the forces **F** and −**F** which are tangent to the arcs of rotation at the points A and B as shown in Fig. 10.6. By Eq. (10.7), the work of the couple acting on the rigid body during the angular displacement $d\theta$ is

$$dU = |-\mathbf{F}|(-R\, d\theta) + |\mathbf{F}|(R + r)\, d\theta = Fr\, d\theta$$

Since the product Fr is equal to the magnitude of the moment **M** of the couple, we write

$$dU = M\, d\theta = \mathbf{M} \cdot \overrightarrow{d\theta} \qquad (10.10)$$

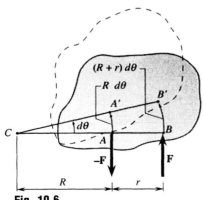

Fig. 10.6

---

†The work of a couple is the same as the work of the *moment* of the couple. (Cf. the footnote on p. 90.)

Note that $dU = 0$ if $\mathbf{M} \perp \overrightarrow{d\theta}$. Thus, *the work $U_{1 \to 2}$ of a moment* $\mathbf{M}$, *or a couple with moment* $\mathbf{M}$, acting on a rigid body during its rotation from the angular position θ_1 to the angular position θ_2 is

$$U_{1 \to 2} = \int_{\theta_1}^{\theta_2} \mathbf{M} \cdot \overrightarrow{d\theta} \qquad (10.11)$$

If the moment $\mathbf{M}$ is *constant*, the work $U_{1 \to 2}$ of $\mathbf{M}$ acting on a rigid body during its finite angular displacement $\Delta\theta$, which is expressed in radians and is in the same direction as that of $\mathbf{M}$, is

$$\boxed{U_{1 \to 2} = M \, \Delta\theta} \qquad (10.12)$$

Clearly, $U_{1 \to 2}$ is negative if M and $\Delta\theta$ have opposite senses. Note that the work of a moment, or a couple, is a scalar whose dimensions are also *force × length* because angles are taken as dimensionless.

Developmental Exercise

D10.9   Define the work of a couple acting on a rigid body.

EXAMPLE 10.3

Determine the total work dU of the force system acting on the bent bar shown if it undergoes an infinitesimal angular displacement $\overrightarrow{d\theta} = d\theta \, \circlearrowleft$ about the point B.

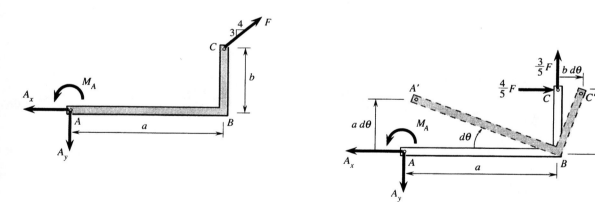

Solution.   The angular displacement of the bent bar is drawn as shown. By resolving $\mathbf{F}$ at C into two components and applying Eqs. (10.7) and (10.12), we write

$$dU = A_x(0) + A_y(-a \, d\theta) + M(-d\theta) + \frac{4}{5}F(b \, d\theta) + \frac{3}{5}F(0)$$

$$dU = \left(\frac{4}{5}Fb - A_ya - M \right) d\theta \quad \blacktriangleleft$$

Developmental Exercises

D10.10   Work Example 10.3 if the bent bar undergoes an infinitesimal linear displacement $\overrightarrow{dx} = dx \rightarrow$.

D10.11   Work Example 10.3 if the bent bar undergoes an infinitesimal linear displacement $\overrightarrow{dy} = dy \uparrow$.

PROBLEMS

10.1   A 10-lb force **P** is applied to a 40-lb crate to move it through a 30-ft displacement **q** on a smooth horizontal surface as shown. Determine the work of (a) **P**, (b) all external forces.

Fig. P10.1

10.2   Solve Prob. 10.1 if $P = 20$ lb, $q = 18$ ft, and the surfaces of contact are rough with $\mu_k = 0.2$.

10.3 through 10.6   The crate undergoes a displacement **q** on a ramp as shown. With the coefficient of friction between the crate and the ramp as indicated, determine the total work of all external forces acting on the crate during this displacement.

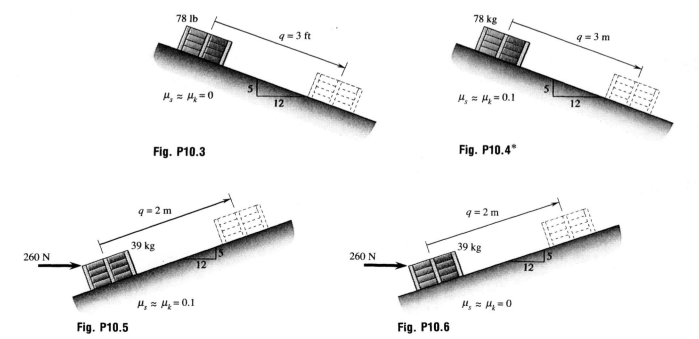

Fig. P10.3   **Fig. P10.4*

Fig. P10.5   **Fig. P10.6**

10.7   A cord is used to tie a box to a wall with a compressed spring separating (but not connecting) them as shown. The spring has a free length $L = 1.5$ m and a modulus $k = 20$ N/m. Neglecting friction, determine the work of the spring force acting on the box after the cord is severed.

10.8*   Solve Prob. 10.7 if $k = 15$ N/m.

10.9 and 10.10   The spring BC has a free length L and a modulus k as shown. As the bar is released to rotate from its horizontal position AB to its vertical position AB', the total work done to the bar is known to be zero. Determine the mass of the bar.

$\mu_s \approx \mu_k = 0$

Fig. P10.7

Fig. P10.9

Fig. P10.10

10.11*   Solve Prob. 10.9 if the total work done to the bar is known to be 9 J.

10.12   Determine the total work dU of the forces and moments acting on the frame shown if it undergoes an infinitesimal linear displacement of $\overrightarrow{dy} = dy$ ft $\uparrow$.

10.13*   Solve Prob. 10.12 if the frame undergoes an infinitesimal linear displacement $\overrightarrow{dx} = dx$ ft $\rightarrow$.

10.14*   Solve Prob. 10.12 if the frame rotates about the point A through a differential angle $\overrightarrow{d\theta} = d\theta \uparrow$.

Fig. P10.12

10.4   Virtual Displacements

An *actual displacement* of a body, as the name implies, gives the actual change of position of the body. A *virtual displacement* of a body is a given or imaginary differential displacement, which is possible but does not necessarily take place under actual motion, of the body. There are *linear* and *angular virtual displacements*; they are vector quantities.[†] The symbol $\delta\mathbf{r}$ is used to denote the virtual displacement of a particle at the position $\mathbf{r}$; it distinguishes the virtual displacement from the actual differential displacement $d\mathbf{r}$ of the particle. For a particle located at the point (x, y), the components of its virtual displacement are usually denoted by $\overrightarrow{\delta x}$ and $\overrightarrow{\delta y}$. Similarly, the angular virtual displacement of a line or body, whose angular position coordinate is θ, is denoted by $\overrightarrow{\delta\theta}$.

A *compatible virtual displacement* of a body is a set of imaginary *first-order* differential displacements, which conforms to the integrity (i.e., no breakage or rupture) of its free body within the framework of first-order accuracy, where the body may be a particle, a rigid body, or a set of connected particles or rigid bodies.[††] Note that a compatible virtual displacement

---

[†]Cf. App. C.

[††]Second- and higher-order virtual displacements are considered in studying the *stability of equilibrium* later in this chapter.

of a body does not necessarily conform to the constraints at the supports in the space diagram. On the other hand, a *constrained virtual displacement* of a body is a compatible virtual displacement of the body which conforms to the constraints at the supports in the space diagram, as illustrated in Fig. 10.7.

(a) A linkage

(b) Free body

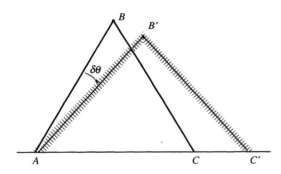

(c) This shows a *constrained*, as well as compatible, *virtual displacement* of the linkage.

(d) This shows a *compatible*, but not a *constrained*, *virtual displacement* of the linkage.

Fig. 10.7

EXAMPLE 10.4

A beam AB is given an angular virtual displacement $\overrightarrow{\delta\theta}$ about its end A so that its end B moves to the point B' as shown. The line segment $\overline{AB'B''}$ is the hypotenuse of the right triangle ABB''. In terms of a Maclaurin's series (cf. Sec. B5.5 of App. B), we have

$$\sec(\delta\theta) = 1 + \frac{1}{2!}(\delta\theta)^2 + \text{(higher-order terms of } \delta\theta)$$

Determine (a) the approximate length of the line segment $\overline{B'B''}$ in terms of L and $\delta\theta$, (b) whether it is permissible for the length and position of the displaced beam to be represented by the hypotenuse $\overline{AB''}$ in considering questions of equilibrium, (c) the virtual displacement $\boldsymbol{\delta}_B$ of the end B.

Solution.   Since $\delta\theta$ is infinitesimal, we write

$$\sec(\delta\theta) \approx 1 + \frac{1}{2!}(\delta\theta)^2 = 1 + \frac{1}{2}(\delta\theta)^2$$

Thus, the approximate length of $\overline{B'B''}$ is

$$\overline{B'B''} = L\sec(\delta\theta) - L \approx L\left[1 + \frac{1}{2}(\delta\theta)^2\right] - L$$

$$\overline{B'B''} \approx \frac{L}{2}(\delta\theta)^2 \quad \blacktriangleleft$$

This shows that $\overline{B'B''}$ is a second-order quantity in $\delta\theta$. Therefore, in considering questions of equilibrium, we may neglect $\overline{B'B''}$ and take $\overline{AB'}$ as being equal to $\overline{AB''}$. In other words, *it is permissible for the length and the position of the displaced beam to be represented by the hypotenuse $\overline{AB''}$.* The arc BB' and the line segment $\overline{BB''}$ are taken to be equal because $\delta\theta$ is infinitesimal. Thus, the virtual displacement of the end B is

$$\delta_B = \overline{BB''} = L\,\delta\theta \qquad \boldsymbol{\delta}_B = L\,\delta\theta\;\uparrow \quad \blacktriangleleft$$

Developmental Exercises

D10.12   Define the terms: (a) virtual displacement, (b) compatible virtual displacement, (c) constrained virtual displacement.

D10.13   A hinge B connects two beams AB and BC as shown. An angular virtual displacement $\vec{\delta\theta}$ is given to the beam AB about its end A, and the end C of the beam BC is to remain on the x axis. The exact configuration of the displaced beams is indicated by the line $AB'C'$ as shown. In terms of a Maclaurin's series, we have

$$\cos(\delta\theta) = 1 - \frac{1}{2!}(\delta\theta)^2 + \frac{1}{4!}(\delta\theta)^4 - \cdots$$

For considerations of equilibrium, determine (a) the approximate length of the line segment $\overline{CC'}$ in terms of L and $\delta\theta$, (b) whether it is permissible for the lengths and the positions of the displaced beams to be represented by the lines $\overline{AB''}$ and $\overline{B''C}$ as shown, (c) the virtual displacement $\boldsymbol{\delta}_B$ of the hinge B.

Fig. D10.13

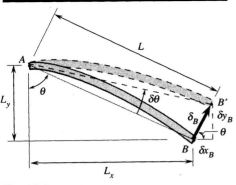

Fig. 10.8

10.5 Components of a Linear Virtual Displacement

From time to time, we need to compute the rectangular components of a linear virtual displacement of a point on a rigid body which is given an angular virtual displacement about a fixed point. Let $\vec{\delta x_B}$ and $\vec{\delta y_B}$ be the x and y components of the linear virtual displacement $\boldsymbol{\delta}_B$ of the point B on the rigid body AB which is given an angular virtual displacement $\vec{\delta\theta}$ about the fixed point A as shown in Fig. 10.8. From this figure, we write

$$\delta x_B = \delta_B \cos\theta = (L\,\delta\theta)\cos\theta = (L\cos\theta)\,\delta\theta = L_y\,\delta\theta$$

$$\delta y_B = \delta_B \sin\theta = (L\,\delta\theta)\sin\theta = (L\sin\theta)\,\delta\theta = L_x\,\delta\theta$$

$$\boxed{\delta x_B = L_y\,\delta\theta \qquad \delta y_B = L_x\,\delta\theta} \qquad (10.13)$$

The formulas in Eqs. (10.13) show that *the x (or y) component of the linear virtual displacement of the point B on a rigid body which is given an angular virtual displacement $\vec{\delta\theta}$ about a fixed point A is equal to the projection of the line segment $\overline{AB}$ in the y (or x) direction times the infinitesimal angle $\delta\theta$.* Alternatively, we may use the cross product to write

$$\boxed{\boldsymbol{\delta}_B = \vec{\delta\theta} \times \overrightarrow{AB}} \qquad (10.14)$$

where $\boldsymbol{\delta}_B = \delta x_B\mathbf{i} + \delta y_B\mathbf{j}$, $\vec{\delta\theta} = \delta\theta\mathbf{k}$, and $\overrightarrow{AB} = L_x\mathbf{i} - L_y\mathbf{j}$.

EXAMPLE 10.5

It is desired to give the member $ABCDE$ of the free body shown a clockwise angular virtual displacement $\vec{\delta\theta}$ about the pinhole at A with the pinhole at F allowed to move only along the x axis. Determine in terms of $\delta\theta$ the linear virtual displacements $\vec{\delta x_F}$, $\vec{\delta y_C}$, $\vec{\delta x_D}$, and $\vec{\delta y_E}$ of the referred points on the free body.

Solution. The desired virtual displacements of the free body are sketched to first-order accuracy as shown. Using Eqs. (10.13) and noting that $\delta x_F = \delta x_B$, we write

$$\delta x_B = (\overline{AB})_y \, \delta\theta = \overline{AB} \, \delta\theta = 6 \, \delta\theta = \delta x_F \qquad \overrightarrow{\delta x_F} = 6 \, \delta\theta \text{ ft} \rightarrow \blacktriangleleft$$

$$\delta y_C = (\overline{AC})_x \, \delta\theta = \overline{AA} \, \delta\theta = 0 \, (\delta\theta) \qquad \overrightarrow{\delta y_C} = \mathbf{0} \blacktriangleleft$$

$$\delta x_D = (\overline{AD})_y \, \delta\theta = \overline{AC} \, \delta\theta = 10 \, \delta\theta \qquad \overrightarrow{\delta x_D} = 10 \, \delta\theta \text{ ft} \rightarrow \blacktriangleleft$$

$$\delta y_E = (\overline{AE})_x \, \delta\theta = \overline{AP} \, \delta\theta = 4 \, \delta\theta \qquad \overrightarrow{\delta y_E} = 4 \, \delta\theta \text{ ft} \downarrow \blacktriangleleft$$

REMARK. The preceding virtual displacements may also be obtained by using the cross product. For instance, we write

$$\boldsymbol{\delta}_D = \overrightarrow{\delta\theta} \times \overrightarrow{AD} = (-\delta\theta\mathbf{k}) \times (4\mathbf{i} + 10\mathbf{j})$$
$$= -4 \, \delta\theta\mathbf{j} + 10 \, \delta\theta\mathbf{i} = \delta x_D\mathbf{i} + \delta y_D\mathbf{j}$$

Thus, $\delta x_D = 10 \, \delta\theta$ and $\delta y_D = -4 \, \delta\theta$.

Developmental Exercises

D10.14   Let the entire body in Example 10.5 be rigidly rotated about F through $\delta\theta \, \circlearrowleft$. Determine in terms of $\delta\theta$ the linear virtual displacements $\overrightarrow{\delta y_C}$, $\overrightarrow{\delta x_D}$, $\overrightarrow{\delta y_E}$, $\overrightarrow{\delta x_A}$, and $\overrightarrow{\delta y_A}$ of the referred points on the free body.

D10.15   Using the cross product, determine $\overrightarrow{\delta x_E}$ and $\overrightarrow{\delta y_E}$ in Example 10.5.

10.6 Differential Calculus and Displacement Center[†]

The relations among the virtual displacements of certain particles or members of a system need to be determined in many applications. They can frequently be found by using *differential calculus* or the *displacement center*, or both, which are explained in the following paragraphs.

[†]For a detailed treatment, see Jong, I. C., and C. W. Crook, "Introducing the Concept of Displacement Center in Statics," *Engineering Education*, ASEE, vol. 80, no. 4, May/June 1990, pp. 477-479.

For a body or system allowed to have only a single degree of freedom, the use of *differential calculus* may be implemented as follows: (a) study the manner in which the virtual displacements of the system are to be specified; (b) select a governing position variable whose differential is the key parameter in specifying the virtual displacements; (c) write the expressions or equations which relate the relevant position variables of the particles or members to the governing position variable selected in step (b); (d) take the differentials of the expressions or equations established in step (c); (e) solve for the differentials of the desired relevant position variables in terms of the differential of the governing position variable. For illustrations, see Examples 10.6 and 10.11.

The *displacement center* of a body (or a member of a system) is the center of rotation of the body when it undergoes an angular virtual displacement and the points on it incur linear virtual displacements. Such a center is located at the intersection of two straight lines which are perpendicular to two linear virtual displacement vectors of two different points of that body which pass through those two points, respectively. For example, if a body AB undergoes an angular virtual displacement $\vec{\delta\theta}$ so that the points A and B have linear virtual displacements $\delta\mathbf{r}_A$ and $\delta\mathbf{r}_B$, respectively, as shown in Fig. 10.9, the displacement center D of the body is located at the intersection of the two straight lines AD and BD which are perpendicular to $\delta\mathbf{r}_A$ and $\delta\mathbf{r}_B$, respectively.

(a)

(b)

(c)

Fig. 10.9

Note in Fig. 10.9 that the angular virtual displacements of the lines AD, BD, and any other lines drawn from the body to the point D must all be equal to $\delta\theta$. If a system consists of two or more members, *each member will generally have a different displacement center, and each displacement center has zero virtual displacement*. In Fig. 10.9(c), note that the displacement center of the member BC is at C.

EXAMPLE 10.6[†]

It is desired to give the member AB of the free body shown a counterclockwise angular virtual displacement $\vec{\delta\theta}$ about the pinhole at A with the pinhole at C moving only along the x axis. Determine in terms of $\delta\theta$ (a) the angular

---

[†]Continued in D10.24.

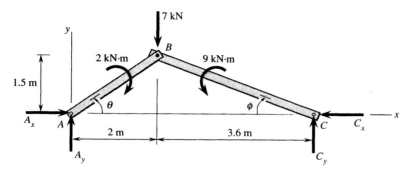

virtual displacement $\overrightarrow{\delta\phi}$ of the member BC, (b) the linear virtual displacement $\overrightarrow{\delta x_C}$ of the pinhole at C.

Solution. For illustrative purposes, let the problem in this example be solved in the following two ways:

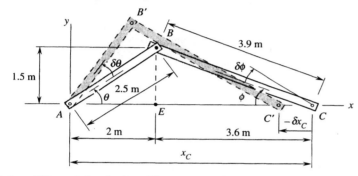

(a) *Using differential calculus*. The two angular position variables θ and ϕ may be related as follows:

$$\overline{BE} = 2.5 \sin\theta = 3.9 \sin\phi$$

Taking the differentials, we write

$$2.5 \,(\cos\theta)\, \delta\theta = 3.9 \,(\cos\phi)\delta\phi$$

Since $\cos\theta = \dfrac{2}{2.5}$ and $\cos\phi = \dfrac{3.6}{3.9}$, we obtain

$$\delta\phi = \frac{2.5 \cos\theta}{3.9 \cos\phi}\, \delta\theta = \frac{5}{9}\, \delta\theta \qquad\qquad \overrightarrow{\delta\phi} = \frac{5}{9}\, \delta\theta \;\curvearrowright \quad \blacktriangleleft$$

The variable x_C defining the position of the pinhole at C is given by $x_C = 2.5 \cos\theta + 3.9 \cos\phi$. Taking the differentials, we write

$$\delta x_C = -2.5 \sin\theta\, \delta\theta - 3.9 \sin\phi\, \delta\phi$$

Since $\sin\theta = \dfrac{1.5}{2.5}$, $\sin\phi = \dfrac{1.5}{3.9}$, and $\delta\phi = \dfrac{5}{9}\, \delta\theta$, we have

$$\delta x_C = -\frac{7}{3}\, \delta\theta \qquad\qquad \overrightarrow{\delta x_C} = \frac{7}{3}\, \delta\theta \; \text{m} \leftarrow \quad \blacktriangleleft$$

Note that the negative sign in the value of δx_C indicates that the direction of $\overrightarrow{\delta x_C}$ is opposite to the increasing direction of x_C.

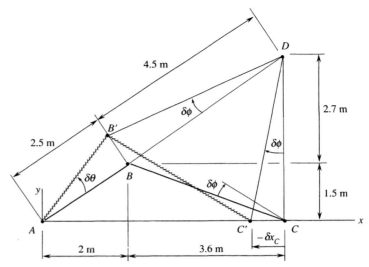

(b) *Using displacement center.* Based on the specified virtual displacements of the system, we see that the displacement centers of members *AB* and *BC* are at *A* and *D*, respectively, as shown in the sketch for the centerlines of the members. We note that $\overline{BB'} = \overline{AB}\,\delta\theta = \overline{BD}\,\delta\phi$. Thus, we write

$$2.5\,\delta\theta = 4.5\,\delta\phi \qquad \delta\phi = \frac{2.5}{4.5}\,\delta\theta \qquad \vec{\delta\phi} = \frac{5}{9}\,\delta\theta \;\downdownarrows \quad \blacktriangleleft$$

$$\delta x_C = -\overline{CD}\,\delta\phi = -4.2\left(\frac{5}{9}\,\delta\theta\right) \qquad \vec{\delta x}_C = \frac{7}{3}\,\delta\theta \text{ m} \leftarrow \quad \blacktriangleleft$$

Developmental Exercise

D10.16† It is desired to give the member *ABC* of the free body shown a counterclockwise angular virtual displacement $\vec{\delta\theta}$ about the pinhole at *A* with the pinhole at *E* moving only along the *x* axis. (a) Relate the variables θ and ϕ which define the angular positions of members *ABC* and *BDE*. (b) Determine in terms of $\delta\theta$ (1) the angular virtual displacement $\vec{\delta\phi}$, (2) the linear virtual displacements $\vec{\delta x}_E$ and $\vec{\delta y}_C$. (c) Using displacement center, verify the answers obtained in (b).

Fig. D10.16

---

†Continued in D10.25.

10.7   Principle of Virtual Work†

The work of a force or a moment acting on a body during a virtual displacement of the body is called a *virtual work*. Let δU be the *total virtual work* of the forces $\mathbf{F}_1, \mathbf{F}_2, \ldots, \mathbf{F}_n$ acting on a particle during a virtual displacement $\delta\mathbf{r}$ of the particle from the point A to the point A' as shown in Fig. 10.10. Then, we write

$$\begin{aligned} \delta U &= \mathbf{F}_1 \cdot \delta\mathbf{r} + \mathbf{F}_2 \cdot \delta\mathbf{r} + \cdots + \mathbf{F}_n \cdot \delta\mathbf{r} \\ &= (\mathbf{F}_1 + \mathbf{F}_2 + \cdots + \mathbf{F}_n) \cdot \delta\mathbf{r} \\ &= \mathbf{R} \cdot \delta\mathbf{r} \end{aligned}$$

Fig. 10.10

where $\mathbf{R}$ is the resultant force. If equilibrium exists, we have $\mathbf{R} = \mathbf{0}$; therefore, $\delta U = \mathbf{0} \cdot \delta\mathbf{r} = 0$ is a necessary condition. At the same time, $\delta U = 0$ for any $\delta\mathbf{r}$ is a sufficient condition for equilibrium because we must have $\mathbf{R} = \mathbf{0}$ if $\delta U = \mathbf{R} \cdot \delta\mathbf{r} = 0$ for any $\delta\mathbf{r}$. Thus, *if a particle is in equilibrium, the total virtual work of all forces acting on the particle is zero for any virtual displacement of the particle, and conversely.*

By Newton's third law, all internal forces acting on the particles of a rigid body or a system of pin-connected rigid bodies must exist in pairs, each of which consists of two collinear forces that are equal in magnitude and opposite in direction; and the collinear forces in each pair remain at a fixed distance from each other within a rigid body. Since pin connections are taken as frictionless, we see that *the total virtual work of all internal forces must be zero during any compatible virtual displacement of a rigid body or a system of pin-connected rigid bodies.*

Furthermore, all particles forming a body must be in equilibrium when the body is in equilibrium. Thus, the total virtual work of all external and internal forces acting on all particles of the body in equilibrium must be zero during any compatible virtual displacement of the body. Since the total virtual work of all *internal forces* is zero, the total virtual work of all *external forces* acting on that body must also be zero; i.e.,

$$\delta U = 0 \qquad\qquad (10.15)$$

The properties stated above lead to the *principle of virtual work*†† in statics, which may be stated as follows: *If a body is in equilibrium,* the total virtual work of the external force system acting on its free body during any compatible virtual displacement of its free body is equal to zero, and conversely, *where the body may be a particle, a set of connected particles, a rigid body, a frame, or a machine.*

---

†Recent historical studies show that on February 26, 1715, the Swiss mathematician Johann (Jean) Bernoulli (1667–1748) communicated to Pierre Varignon (1654–1722) the principle of virtual velocities in analytical form for the first time. That was the forerunner of the principle of virtual work today. The approach to mechanics based on the principle of virtual work was formally treated by Joseph Louis Lagrange (1736–1813) in his *Mécanique Analytique* published in 1788.

††Cf. the *energy criterion* in Sec. 10.15.

Fig. D10.19

Developmental Exercises

D10.17 What is a virtual work?

D10.18 Describe the *principle of virtual work* in statics.

D10.19 A particle at A is in equilibrium as shown, where $W = 280$ N. (a) Write the analytical expressions of **P** and **Q**. (b) If the particle is given a virtual displacement $\delta\mathbf{r} = (\delta r_1)(\tfrac{4}{5}\mathbf{i} + \tfrac{3}{5}\mathbf{j})$, which is perpendicular to **P**, compute the total virtual work δU of **P**, **Q**, and **W**. (c) Solve for Q by setting $\delta U = 0$. (d) Having solved for Q, determine P by letting $\delta\mathbf{r} = \delta r_2\mathbf{i}$ and setting $\delta U = 0$.

D10.20 The free body of a bent rod and its virtual displacement are shown. (a) Determine the virtual displacements $\overrightarrow{\delta x_C}$, $\overrightarrow{\delta y_C}$, and $\overrightarrow{\delta\phi}$ in terms of $\delta\theta$. (b) Determine $\mathbf{M}_A$ if the bent rod is in equilibrium.

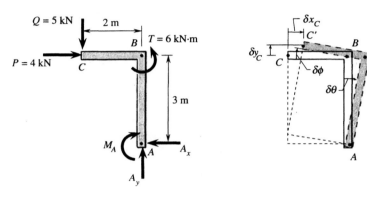

Fig. D10.20

10.8 Applications to Equilibrium Problems

The principle of virtual work in statics frequently enables us to solve for unknown quantities in seemingly complex equilibrium problems of frames and machines without having to solve simultaneous equations. The following suggested steps are generally helpful in applying the *principle of virtual work* to the solution of many equilibrium problems:

1. *Draw* the appropriate *free-body diagram* of the system.
2. *Give*, if possible, the free body or its centerlines *a compatible virtual displacement* in such a manner that only one unknown parameter, besides the given loading, will be involved in doing the virtual work. (If more than one unknown parameter is involved, the need to solve simultaneous equations will arise.)
3. *Mark and compute* the pertinent components of the *virtual displacements* at the various locations where forces and moments are acting. (This step may involve considerations of geometry, trigonometry, and differential calculus or displacement center.)
4. *Write* the expression of the *total virtual work* in terms of the unknown parameter and the virtual displacement.
5. *Set the total virtual work to zero* and solve for the unknown parameter in the resulting equation.
6. *Repeat steps* (2) *through* (5) until all required unknown parameters have been determined.

EXAMPLE 10.7

A frame is loaded as shown. Determine the reactions from the supports at A and C.

Solution. The free-body diagram of the entire system is drawn as shown, where the original 300-lb force at D is replaced by its rectangular components. The unknown reaction components are A_x, A_y, C_x, and C_y.

 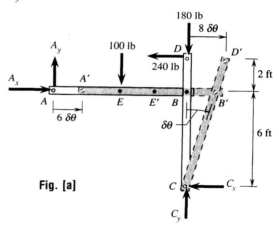

Fig. [a]

(a) To determine A_x, let the member CD be given a clockwise angular virtual displacement $\overrightarrow{\delta\theta}$ about the pinhole at C and the member AB remain horizontal as shown in Fig. [a].[†] From the sketch, we write

$$\overline{AA'} = \overline{BB'} = 6\,\delta\theta \qquad \overline{DD'} = 8\,\delta\theta$$

Applying the *principle of virtual work*, we write

$$\delta U = A_x(6\,\delta\theta) + 240(-8\,\delta\theta) = 0$$

$$A_x = 320 \qquad\qquad \mathbf{A_x} = 320\ \text{lb} \rightarrow \blacktriangleleft$$

(b) To determine A_y, let the member AB be given a counterclockwise angular virtual displacement $\overrightarrow{\delta\theta}$ about the pin at B and the member DBC remain unmoved as shown in Fig. [b].[†] From the sketch and by the

---

[†]This is just a *compatible* (not constrained) *virtual displacement*.

Fig. [b]

principle of virtual work, we write

$$\delta U = A_y(-8\ \delta\theta) + 100(4\ \delta\theta) = 0$$

$$A_y = 50 \qquad\qquad \mathbf{A}_y = 50\ \text{lb}\ \uparrow \quad \blacktriangleleft$$

(c) To determine $\mathbf{C}_x$, let the member DBC be given a clockwise angular virtual displacement $\overrightarrow{\delta\theta}$ about the pin at B and the member AB remain unmoved as shown in Fig. [c] (see footnote on p. 419). From the sketch and by the *principle of virtual work*, we write

$$\delta U = C_x(6\ \delta\theta) + 240(-2\ \delta\theta) = 0$$

$$C_x = 80 \qquad\qquad \mathbf{C}_x = 80\ \text{lb}\ \leftarrow \quad \blacktriangleleft$$

Fig. [c]

Fig. [d]

(d) To determine $\mathbf{C}_y$, let the member AB be given a clockwise angular virtual displacement $\overrightarrow{\delta\theta}$ about the end at A and the member DBC remain vertical as shown in Fig. [d] (see footnote on p. 419). From the sketch and by the *principle of virtual work*, we write

$$\delta U = C_y(-8\ \delta\theta) + 180(8\ \delta\theta) = 100(4\ \delta\theta) = 0$$

$$C_y = 230 \qquad\qquad \mathbf{C}_y = 230\ \text{lb}\ \uparrow \quad \blacktriangleleft$$

REMARK. The above solutions for A_x, A_y, C_x, and C_y did not involve the solution of simultaneous equations; i.e., these unknowns have been found independently of each other. Moreover, the same results would be obtained if the directions of the virtual displacements were reversed.

Developmental Exercise

D10.21 If the free body shown in Example 10.5 is in equilibrium, determine the unknown force $\mathbf{F}_x$ acting on that free body.

EXAMPLE 10.8

A frame is shown. Determine the reaction from the support at C.

Solution. We first draw the free-body diagram of the entire system as shown. Note that the distributed load over the portion AB of the frame has been replaced in the free-body diagram by its resultant force which is equal to the area of the loading diagram and acts through its centroid. The unknown reaction components are $\mathbf{C}_x$, $\mathbf{C}_y$, $\mathbf{D}_x$, and $\mathbf{D}_y$. We need to determine $\mathbf{C}_x$ and $\mathbf{C}_y$.

(a) To determine C_x, let the portion $\underline{DEFGH}$ of the frame be given a clockwise angular virtual displacement $\overrightarrow{\delta\theta}$ about the pinhole at D and the orientation of the portion $ABCE$ of the frame be unchanged.[†] Note that $ABCE$ of the frame has only a linear virtual displacement to the right equal to the length of $\overline{EE'}$. From the sketch and by the *principle of virtual work*, we write

$$\delta U = C_x(-8\,\delta\theta) + 1200(\delta\theta) + 600(12\,\delta\theta) = 0$$

$$C_x = 1050 \qquad\qquad\qquad \mathbf{C}_x = 1050 \text{ lb} \leftarrow \quad \blacktriangleleft$$

(b) To determine C_y, let the portion $ABCE$ be given a clockwise angular virtual displacement $\overrightarrow{\delta\theta}$ about the pin E and the other portion be unmoved.[†] This is satisfactory because C_x is already known. From the sketch and by the *principle of virtual work*, we write

$$\delta U = 300(-10\,\delta\theta) + C_x(4\,\delta\theta) + C_y(6\,\delta\theta) = 0$$

$$C_y = 500 - \tfrac{2}{3}C_x = 500 - \tfrac{2}{3}(1050) = -200$$

$$\mathbf{C}_y = 200 \text{ lb} \downarrow \quad \blacktriangleleft$$

[†]This is just a *compatible* (not constrained) *virtual displacement*.

Developmental Exercise

D10.22   Refer to the frame in Example 10.8. Determine D_x and D_y.

EXAMPLE 10.9

Determine the reactions from the supports at A and D of the hinged beam shown.[†]

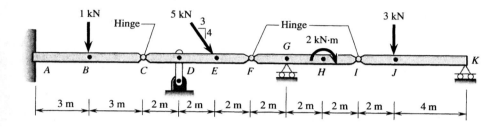

Solution.   The free-body diagram of the entire system is drawn as shown. Note that the 5-kN force in the space diagram has been replaced by its rectangular components at E in the free-body diagram. The unknown reaction components are $\mathbf{M}_A$, $\mathbf{A}_x$, $\mathbf{A}_y$, $\mathbf{D}$, $\mathbf{G}$, and $\mathbf{K}$.

To use the *principle of virtual work* to determine $\mathbf{M}_A$, $\mathbf{A}_x$, $\mathbf{A}_y$, and $\mathbf{D}$, we choose four appropriate *compatible* (not constrained) *virtual displacements* of the centerline of the hinged beam as shown. From the virtual displacement shown in Fig. (a), we write

$$\delta U = M_A\,(\delta\theta) + 1(-3\,\delta\theta) + 4(6\,\delta\theta)$$
$$+ 2(-6\,\delta\theta) + 3(-16\,\delta\theta) = 0$$

$$M_A = 39 \qquad \mathbf{M}_A = 39 \text{ kN·m} \; \curvearrowright \; \blacktriangleleft$$

From the virtual displacement shown in Fig. (b) on the next page, we write

$$\delta U = A_x(-\delta x) + 3(\delta x) = 0$$

$$A_x = 3 \qquad \mathbf{A}_x = 3 \text{ kN} \leftarrow \; \blacktriangleleft$$

[†]Cf. the footnote on p. 318.

Fig. [a]

Fig. [b]

Fig. [c]

Fig. [d]

From the virtual displacement shown in Fig. [c], we write

$$\delta U = A_y(2\ \delta\theta) + 1(-2\ \delta\theta) + 4(2\ \delta\theta)$$
$$+ 2(-2\ \delta\theta) + 3(-\tfrac{16}{3}\ \delta\theta) = 0$$

$$A_y = 7 \qquad\qquad \mathbf{A}_y = 7 \text{ kN} \uparrow \quad \blacktriangleleft$$

From the virtual displacement shown in Fig. [d], we write

$$\delta U = D(\delta\theta) + 4(-2\ \delta\theta) + 2(\tfrac{3}{2}\ \delta\theta) + 3(4\ \delta\theta) = 0$$

$$D = -7 \qquad\qquad \mathbf{D} = 7 \text{ kN} \downarrow \quad \blacktriangleleft$$

Developmental Exercise

D10.23   Determine the reactions from the supports at G and K of the beam in
Example 10.9.

EXAMPLE 10.10

Determine the moment **M** for which the system shown is in equilibrium.

Solution.   The free-body diagram and a *constrained virtual displacement*
of the centerlines of the members of the system are sketched as shown,
where the *displacement centers* of the members AB and BC are at A and D,
respectively.[†] Note that the angular virtual displacements for AB and BC are

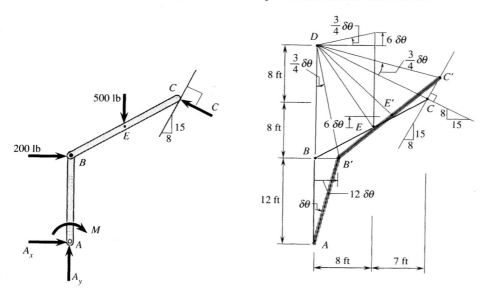

[†]Note that a *constrained virtual displacement* is also a *compatible virtual displacement*. (Cf.
Sec. 10.4.)

$\delta\theta \, \text{\Large$\circlearrowright$}$ and $\frac{3}{4}\delta\theta \, \text{\Large$\circlearrowleft$}$, respectively. From the sketches and by the *principle of virtual work*, we write

$$\delta U = M(\delta\theta) + 200(12 \, \delta\theta) + 500(-6 \, \delta\theta)$$
$$= (M - 600) \, \delta\theta = 0$$
$$M = 600 \qquad \mathbf{M} = 600 \text{ lb·ft } \text{\Large$\circlearrowright$} \quad \blacktriangleleft$$

Developmental Exercises

D10.24 If the free body shown in Example 10.6 is in equilibrium, determine the unknown force $\mathbf{C}_x$ acting on that free body.

D10.25 If the free body shown in D10.16 is in equilibrium, determine the unknown force $\mathbf{E}_x$ acting on that free body.

EXAMPLE 10.11

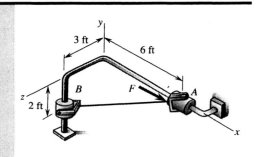

Collars A and B, each weighing 20 lb, are connected by a thin wire AB and may slide freely on the smooth bent rod as shown. Determine the magnitude of $\mathbf{F}$ consistent with equilibrium.

Solution. Since the collar A slides along the x axis and the collar B slides along the vertical portion of the rod, the coordinates of A and B may be written as $A(x_A, 0, 0)$ ft and $B(0, y_B, 3)$ ft, respectively. In the position shown, we have

$$x_A = 6 \text{ ft} \qquad y_B = -2 \text{ ft}$$

The length of the wire AB can readily be found to be 7 ft. The free-body diagram of the collars and the wire is sketched as shown. Note that $\mathbf{B}_{zx}$ is the reaction from the bent rod to the collar B and is parallel to the zx plane. In order to prevent B_{zx}, A_y, and A_z from entering the expression of δU, we give the collar A a virtual displacement $\overrightarrow{\delta x_A}$ and the collar B a corresponding virtual displacement $\overrightarrow{\delta y_B}$ as shown. Using the Pythagorean theorem, we have

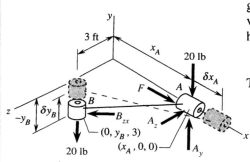

$$x_A^2 + y_B^2 + 3^2 = 7^2$$

Taking the differentials, we write

$$2x_A \, \delta x_A + 2y_B \, \delta y_B = 0$$
$$\delta y_B = -\frac{x_A}{y_B} \delta x_A = -\left(\frac{6}{-2}\right) \delta x_A = 3\delta x_A$$

From the sketch and by the *principle of virtual work*, we write

$$\delta U = F(\delta x_A) + 20(-\delta y_B) = [F - 20(3)] \, \delta x_A = 0$$
$$F = 60 \text{ lb} \quad \blacktriangleleft$$

Developmental Exercise

D10.26 Using the *principle of virtual work*, do Example 3.11. (*Hint.* Let the coordinates of A and B be $A(0, y_A, 12)$ ft and $B(x_B, y_B, 0)$ ft, where $y_B = 12 - \frac{3}{4}x_B$.)

PROBLEMS

NOTE. The following problems are to be solved using the *principle of virtual work*.

10.15 and 10.16 Determine the reaction from the support at A as shown.

Fig. P10.15

Fig. P10.16

10.17* A 24-kg ladder, with its center of gravity at its midpoint G, is shown. Neglecting friction forces at the supports, determine the magnitude of the horizontal force **P** necessary to keep the ladder from sliding.

10.18 A 96-lb ladder with its center of gravity at its midpoint G is kept from sliding by a 14-lb horizontal force at its bottom as shown. Neglecting friction forces at the supports, determine the distance x.

Fig. P10.17*

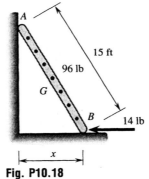

Fig. P10.18

10.19 through 10.23* Determine the magnitude of the force **P** for which the linkage as shown is in equilibrium.

Fig. P10.19

Fig. 10.20*

Fig. P10.21

Fig. P10.22

Fig. P10.23*

Fig. P10.24

10.24   Two 20-kg uniform bars AB and BC are connected and supported as shown. The spring AC has a free length of 0.4 m and a modulus of 400 N/m. Determine the values of x consistent with equilibrium.[†]

10.25 and 10.26* Two 10-kg uniform bars AB and BC are connected and supported as shown. The spring DE can develop either tension or compression and has a free length of 0.6 m as well as a modulus of 900 N/m. Determine the values of θ consistent with equilibrium.[†]

Fig. P10.25

Fig. P10.26*

10.27   Determine the reactions from the supports at A and H of the hinged beam shown.

Fig. P10.27 and P10.28*

10.28* Determine the reactions from the supports at C and I of the hinged beam shown.

10.29   Determine the reactions from the supports at A and G of the hinged beam shown.

[†]The solution of this problem leads to a quartic equation. (Cf. App. E or Sec. B5.8 of App. B for the solution of a nonlinear equation.)

Fig. P10.29 and P10.30*

10.30* Determine the reactions from the supports at D and I of the hinged beam shown.

10.31 Determine the reactions from the supports at A and B of the frame shown.

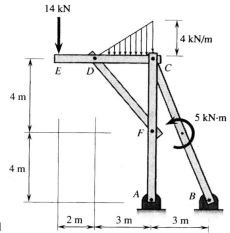

Fig. P10.31

10.32 Determine the magnitude of the force exerted by the cutter on the bolt at E as shown.

Fig. P10.32

10.33 A body of weight W is suspended at B of the system as shown. The spring has a modulus k of 72 lb/ft and is neither stretched nor compressed when AB and BC are horizontal. Determine the value of W if the system is in equilibrium when $\theta = 30°$.

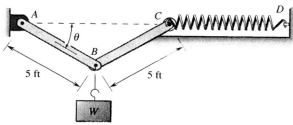

Fig. P10.33

10.34* Refer to Prob. 10.33. Determine the angle θ if $W = 400$ lb when the system is in equilibrium.[†]

10.35 and 10.36* Determine the magnitude of the force **P** required to hold the structure as shown in equilibrium.

Fig. P10.35

Fig. P10.36*

10.37 A hydraulic cylinder FG is used to adjust the height of a mobile platform which supports a crate of 1 Mg as shown. The point C is the midpoint of the supporting members AE and BD, each of which is 2 m long. For $\theta = 60°$, determine the magnitude of the compressive force $\mathbf{F}_{GF}$ in the hydraulic cylinder. (*Hint.* The virtual work of $\mathbf{F}_{GF}$ may be calculated by first computing δx_F and δy_F and then applying the *theorem of work* alluded to on p. 403.)

10.38 Collars A and B, each weighing 35 lb, are connected by a thin wire AB and may slide freely on the smooth bent rod. Determine the magnitude of **F** consistent with equilibrium as shown.

Fig. P10.37

Fig. P10.38

POTENTIAL ENERGY AND STABILITY

★10.9 Conservative System and Potential Energy

An *analytic force field* is a region in space where a body experiences a determinable single-valued force or force system at every position in the region. Since the kinetic friction force developed on a body at a given position may have different values in magnitude and direction, a kinetic friction force field is not an analytic force field.

[†]The footnote on p. 428 applies.

A *conservative force field* is an analytic force field in which the work of the force in the field acting on a body traveling in the field is dependent on the initial and final positions of the body, not on the path traveled by the body. The force of a conservative force field is called a *conservative force*. In the rest of this chapter, we shall deal with topics related to the equilibrium as well as the stability of equilibrium of structures subjected to the following conservative forces: gravitational forces, elastic spring forces, and applied constant loads.

Any force which acts on a body but does no work to the body during a displacement of the body is called a *nonworking force*. For instance, when a body moves on a surface, the normal force exerted by the surface on the body does no work to the body and is a nonworking force. A *conservative system* is a system where the body or assemblage of bodies is subjected to only conservative forces and nonworking forces, if any.

For practical use, the *potential energy of a body* in a conservative force field relative to an appropriate reference datum is defined as *the amount of work which the body will receive from the conservative force if the body travels from its present position to the reference datum.*[†] Thus, the potential energy has the same dimensions and units as the work of a force. The potential energy of a body situated at the reference datum is, of course, equal to zero.

Developmental Exercises

D10.27 Define (a) analytic force field, (b) conservative force field, (c) conservative force, (d) nonworking force, (e) conservative system.

D10.28 Define the potential energy of a body.

★10.10 Gravitational Potential Energy

We know that the force exerted on a body in a *gravitational force field* is the weight force of the body. As a body travels from one position to a nearby position in the vicinity of the surface of the earth, the weight force of the body is essentially a downward vertical force of constant magnitude. Thus, the *gravitational potential energy* V_g *of a body* with a weight W near the surface of the earth at the position A of altitude y above the reference datum, as shown in Fig. 10.11, is equal to the work which the body will receive from the constant weight force $\mathbf{W}$ of the body if the body travels

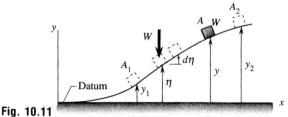

Fig. 10.11

[†]This is a *quantitative definition. Qualitatively*, the potential energy of a body is the energy possessed by the body by virtue of its position relative to the reference datum in the conservative force field.

from its present position at $\eta = y$ to the reference datum at $\eta = 0$. We write

$$V_g = \int_y^0 W(-d\eta) \qquad \boxed{V_g = Wy} \qquad (10.16)$$

where V_g is positive only if the datum chosen is situated between the body and the center of the earth. The work $U_{1\to2}$ of the weight force $\mathbf{W} = W\downarrow$ acting on the body traveling from A_1 to A_2 is found to be related to the *change in gravitational potential energy,* ΔV_g, as follows:

$$U_{1\to2} = \int_{y_1}^{y_2} W(-d\eta) = -(Wy_2 - Wy_1) = -\left[(V_g)_2 - (V_g)_1\right]$$

$$U_{1\to2} = -\Delta V_g \qquad (10.17)$$

EXAMPLE 10.12

A body of weight W is at an altitude y above the surface of the earth whose radius is R. Choosing the surface of the earth as the reference datum, show that Wy is a good approximation for the potential energy V_g of the body if $y \ll R$.

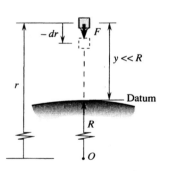

Solution. The weight W of a body is usually measured on the surface of the earth. Letting O be the center of the earth as shown, m be the mass of the body, and M be the mass of the earth, we may apply Newton's law of gravitation and the definition of V_g to write

$$F = \frac{GMm}{r^2} \qquad W = F|_{r=R} = \frac{GMm}{R^2}$$

$$V_g = \int_{R+y}^R F(-dr) = -GMm \int_{R+y}^R r^{-2}\, dr$$

$$= GMm\left(\frac{1}{R} - \frac{1}{R+y}\right)$$

$$= \frac{GMmy}{R(R+y)}$$

Since $y \ll R$, we have $R + y \approx R$ and

$$V_g \approx \frac{GMm}{R^2}\, y = Wy \qquad \text{Q.E.D.} \blacktriangleleft$$

Developmental Exercises

D10.29 Define the gravitational potential energy of a body near the surface of the earth.

D10.30 When will the gravitational potential energy be positive?

★10.11   Elastic Potential Energy

The force exerted by an elastic spring on a body attached to the spring is proportional to the deformation in length of the spring (cf. Sec. 3.5). A body subjected to the force of an elastic spring is said to be in an *elastic force field*. The *reference datum in an elastic force field* is generally chosen to be the position of the body where the spring attached to the body is neutral (i.e., neither stretched nor compressed). For a body at A connected by a spring with a modulus k and an elongation x as shown in Fig. 10.12, the *elastic potential energy V_e of the body* is equal to the work which the

Fig. 10.12

body will receive from the spring force $\mathbf{F} = k\xi \leftarrow$ if the body travels from its present position at $\xi = x$ to the reference datum at $\xi = 0$. Thus, we have

$$V_e = \int_x^0 k\xi(-d\xi) \qquad \boxed{V_e = \frac{1}{2}kx^2} \qquad (10.18)$$

Similarly, we find that the work $U_{1\rightarrow2}$ of the spring force $\mathbf{F}$ acting on the body traveling from A_1 to A_2 is related to the *change in elastic potential energy*, ΔV_e, as follows:

$$U_{1\rightarrow2} = \int_{x_1}^{x_2} k\xi(-d\xi) = -\left(\frac{1}{2}kx_2^2 - \frac{1}{2}kx_1^2\right) = -[(V_e)_2 - (V_e)_1]$$

$$U_{1\rightarrow2} = -\Delta V_e \qquad (10.19)$$

In the case where a torsional spring of modulus K is used to supply a restoring moment $M = K\theta$ to a body which undergoes an angular rotation θ from the neutral position of the spring as shown in Fig. 10.13, the *elastic potential energy V_e of the body* can similarly be shown to be

$$V_e = \int_\theta^0 K\theta(-d\theta) \qquad \boxed{V_e = \frac{1}{2}K\theta^2} \qquad (10.20)$$

Fig. 10.13

where the angle θ is measured in radians and the modulus K is measured in moment per radian (e.g., N·m/rad or lb·ft/rad). Of course, Eq. (10.19) is also true in the case of a torsional spring.

EXAMPLE 10.13

A bar of negligible weight is attached on one end to a spring of modulus k and acted on the other end by a constant force **P** as shown. The spring is neutral when $\theta = 0$. For the position shown, determine the elastic potential energy V_e of the bar in terms of θ.

Solution.   From the sketch shown, we see that the spring has an elongation of

$$x = \overline{CA} = L\sin\theta$$

Thus, we write

$$V_e = \frac{1}{2}kx^2 = \frac{1}{2}k(L\sin\theta)^2$$

$$V_e = \frac{1}{2}kL^2\sin^2\theta \quad \blacktriangleleft$$

Developmental Exercises

D10.31   Define the elastic potential energy of a body.

D10.32   How is the reference datum for the elastic potential energy of a body generally chosen?

★10.12   Applied Potential Energy

Besides gravitational and elastic force fields, there are other conservative force fields which exist to exert conservative forces on various bodies. A body deflected by an *applied conservative force* is said to possess an applied potential energy because the body will receive a definite amount of work from the applied conservative force if it returns to its undeflected state. The *applied potential energy V_a of a body* in a deflected state under the action of an applied conservative force is equal to the work which the body will receive from the applied conservative force if the body travels from its present deflected position to its undeflected position of equilibrium, which is the reference datum. However, we shall here focus our attention on applied conservative forces which are constant in magnitude and direction. Note that an *applied constant force* is a constant force which is suddenly or instantaneously applied to a body, and its magnitude as well as direction remains the same thereafter.

Let us now consider a body AB carrying an applied constant force **P** as shown in Fig. 10.14(a). The *applied potential energy V_a of the body* situated in the deflected position $A'B$ under the action of the applied constant force

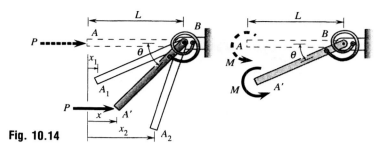

Fig. 10.14

P is equal to the work which the body will receive from **P** if the body returns from its present deflected position $A'B$ to its undeflected neutral position AB of equilibrium, which is the reference datum; i.e.,

$$V_a = \int_x^0 P\,dx \qquad \boxed{V_a = -Px} \qquad (10.21)$$

The work $U_{1\to2}$ of the applied constant force **P** on the member moving from the position where $x = x_1$ to the position where $x = x_2$ is related to the *change in the applied potential energy*, ΔV_a, as follows:

$$U_{1\to2} = \int_{x_1}^{x_2} P\,dx = -(-Px_2 + Px_1) = -[(V_a)_2 - (V_a)_1]$$

$$U_{1\to2} = -\Delta V_a \qquad (10.22)$$

In the case where an applied constant moment **M** acts on a body AB as shown in Fig. 10.14(b), the *applied potential energy V_a of the body* can similarly be shown to be

$$\boxed{V_a = -M\theta} \qquad (10.23)$$

Of course, Eq. (10.22) is also true in the case of an applied constant moment **M**.

EXAMPLE 10.14

Refer to the bar in Example 10.13. Determine in terms of θ the applied potential energy V_a of the bar under the action of the applied constant force **P**.

Solution.   From the sketch shown, we see that when the bar returns from its present deflected position AB to its original undeflected neutral position CD of equilibrium, the right end of the bar moves from the point B to the point D. The distance from B to D is

$$x = \overline{BD} = \overline{CD} - \overline{CB} = L - L\cos\theta$$

Thus, we write

$$V_a = -Px = -P(L - L\cos\theta)$$

$$V_a = -PL(1 - \cos\theta) \quad \blacktriangleleft$$

★10.13 Principle of Potential Energy

The *potential energy V* of a body in a conservative system is the sum of the gravitational, elastic, and applied potential energies of the body; i.e.,

$$V = V_g + V_e + V_a \qquad (10.24)$$

By its definition, we note that the body in a conservative system receives work only from the conservative forces when it undergoes a displacement. From Eqs. (10.16) through (10.24), we can say that any differential work dU done by the conservative forces to the body, which has a differential displacement, is equal to the negative of the corresponding differential increment dV in the potential energy of the body; i.e.,

$$dU = -dV \qquad (10.25)$$

In the case of a virtual displacement, Eq. (10.25) becomes

$$\delta U = -\delta V \qquad (10.26)$$

For a single-degree-of-freedom system, whose position can adequately be defined by a single variable θ, we write

$$V = V(\theta) \qquad \delta V = \frac{dV}{d\theta} \delta\theta$$

Since $\delta\theta \neq 0$, the principle of virtual work for the equilibrium of the system leads to $\delta U = -\delta V = 0$ and

$$\frac{dV}{d\theta} = 0 \qquad (10.27)$$

Thus, *the derivative of the potential energy of a conservative system in equilibrium with respect to its position variable is zero.*[†] This property, a *corollary* of the principle of virtual work, is referred to as the *principle of potential energy. If the position of a system depends on n independent variables* $\theta_1, \theta_2, \ldots, \theta_n$, *the system is said to possess n degrees of freedom.* These independent variables are called the *generalized coordinates* of the system. For equilibrium of an *n*-degree-of-freedom conservative system, we have $V = V(\theta_1, \theta_2, \ldots, \theta_n)$ and

$$\delta U = -\delta V = -\left(\frac{\partial V}{\partial \theta_1} \delta\theta_1 + \frac{\partial V}{\partial \theta_2} \delta\theta_2 + \cdots + \frac{\partial V}{\partial \theta_n} \delta\theta_n \right) = 0$$

[†]This means that the potential energy has an *extremum value* when the conservative system is in equilibrium.

Since the virtual displacements $\delta\theta_1$, $\delta\theta_2$, . . . , $\delta\theta_n$ are independent of one another, it is necessary that the partial derivatives of V with respect to each independent variable be equal to zero for equilibrium; i.e.,

$$\frac{\partial V}{\partial \theta_1} = 0 \qquad \frac{\partial V}{\partial \theta_2} = 0 \quad \cdots \quad \frac{\partial V}{\partial \theta_n} = 0 \qquad (10.28)$$

These equations represent the *principle of potential energy* for a conservative system with n degrees of freedom.

EXAMPLE 10.15

The spring shown has a modulus k and a free length L, and each of the two bars has a weight W, where $W < 4kb$. Determine the values of θ for which equilibrium exists. (Cf. Example 10.18 for stability analysis.)

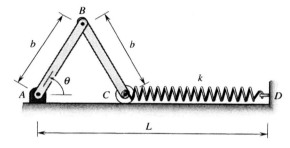

Solution. The system shown is a conservative system. The conservative forces are the resultant weight forces at the midpoints of the bars and the spring force at the point C as shown. Others are nonworking forces. Since the free length of the spring is L, the contraction of the spring is x as shown. Choosing the line AC and the point A as the reference datums for V_g and V_e of the system, respectively, we write

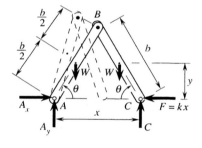

$$y = \frac{b}{2}\sin\theta \qquad V_g = 2Wy = Wb\sin\theta$$

$$x = 2b\cos\theta \qquad V_e = \frac{1}{2}kx^2 = 2kb^2\cos^2\theta$$

$$V = V_g + V_e = Wb\sin\theta + 2kb^2\cos^2\theta$$

By the *principle of potential energy*, we write

$$\frac{dV}{d\theta} = b\cos\theta(W - 4kb\sin\theta) = 0$$

Thus, the positions of equilibrium are given by

$$\cos\theta = 0 \qquad\qquad \theta = \frac{\pi}{2} \quad \blacktriangleleft$$

$$W - 4kb\sin\theta = 0 \qquad \theta = \sin^{-1}\frac{W}{4kb} \quad \blacktriangleleft$$

REMARK. In the first position, the points A and C coincide. Since $W < 4kb$, the second position does exist.

Developmental Exercise

D10.35   A uniform bent rod ABC has a total weight $3W$ and is hung by a wire at B as shown. (a) Choosing the horizontal line passing through B as the reference datum, express the potential energy V of the bent rod as a function of θ. (b) Determine the value of θ for which the bent rod may be in equilibrium.

Fig. D10.35

EXAMPLE 10.16

Refer to the system in Example 10.13. Determine the values of θ for which the bar may be in equilibrium if $\mathbf{P} = \frac{1}{2} kL$.

Solution. The system shown is a conservative system. The elastic and applied potential energies of the bar, as established in Examples 10.13 and 10.14, are

$$V_e = \frac{1}{2} kL^2 \sin^2\theta \qquad V_a = -PL(1 - \cos\theta)$$

Since $P = \frac{1}{2} kL$, we write the potential energy of the bar as follows:

$$V = V_e + V_a = \frac{1}{2} kL^2 \sin^2\theta - PL(1 - \cos\theta)$$

$$= \frac{1}{2} kL^2(\sin^2\theta + \cos\theta - 1)$$

For equilibrium, we apply the *principle of potential energy* to write

$$\frac{dV}{d\theta} = kL^2 \sin\theta\left(\cos\theta - \frac{1}{2}\right) = 0$$

$$\sin\theta = 0 \qquad\qquad\qquad \theta = 0 \blacktriangleleft$$

$$\cos\theta - \frac{1}{2} = 0 \qquad\qquad \theta = 60° \blacktriangleleft$$

Developmental Exercise

D10.36   Describe the *principle of potential energy*.

★10.14   Stability of Equilibrium†

Three types of equilibrium have been discussed in Sec. 5.6. They are *stable equilibrium*, *unstable equilibrium*, and *neutral equilibrium*. We may illustrate them again as shown in Fig. 10.15, where each of the three pin-supported uniform bars has a weight W and a length L. The corresponding potential energies of these bodies are plotted in Fig. 10.16.

| (a) Stable equilibrium at $\theta = \pi/2$ | (b) Unstable equilibrium at $\theta = \pi/2$ | (c) Neutral equilibrium at $\theta = \pi/2$ |

Fig. 10.15

We observe from Figs. 10.15 and 10.16 that (a) the potential energy of the bar at its position of *stable* equilibrium is *minimum*, (b) the potential energy of the bar at its position of *unstable* equilibrium is *maximum*, and (c) the potential energy of the bar at its position of *neutral* equilibrium is *constant*. These observed results are quite general for conservative systems because *the conservative force in the force field always tends to do positive work to the system and thus to decrease the potential energy of the system* [cf. Eq. (10.25)].

  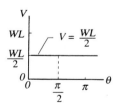

| (a) Stable equilibrium at $\theta = \pi/2$ | (b) Unstable equilibrium at $\theta = \pi/2$ | (c) Neutral equilibrium at $\theta = \pi/2$ |

Fig. 10.16

---

†This section deals only with the stability of equilibrium of *rigid bodies* subjected to conservative force fields where gravitational, elastic, and applied potential energies of the rigid body are definable. The loss of stability of *deformable bodies* through *buckling* or *flutter* is usually treated in mechanics of materials, structural stability, and theory of vibrations.

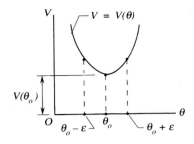

Fig. 10.17

For a single-degree-of-freedom conservative system with potential energy V and generalized coordinate θ, we may employ the conditions under which a function is minimum, maximum, or constant in calculus to summarize the conditions for the equilibrium of the system as follows:

$$\frac{dV}{d\theta} = 0 \quad \text{and} \quad \frac{d^2V}{d\theta^2} > 0: \quad \text{stable equilibrium}$$

$$\frac{dV}{d\theta} = 0 \quad \text{and} \quad \frac{d^2V}{d\theta^2} < 0: \quad \text{unstable equilibrium} \quad\quad (10.29)$$

$$\frac{dV}{d\theta} = \frac{d^2V}{d\theta^2} = \cdots = 0: \quad \text{neutral equilibrium}$$

If we expand the potential energy function $V(\theta)$ into a Taylor's series (cf. Sec. B5.5 of App. B) around an equilibrium position at $\theta = \theta_o$ and transpose the first term, we have

$$V(\theta) - V(\theta_o) = (\theta - \theta_o) V'(\theta_o) + \frac{1}{2!}(\theta - \theta_o)^2 V''(\theta_o)$$

$$+ \frac{1}{3!}(\theta - \theta_o)^3 V'''(\theta_o) + \frac{1}{4!}(\theta - \theta_o)^4 V^{iv}(\theta_o)$$

$$+ \cdots + \frac{1}{n!}(\theta - \theta_o)^n V^{(n)}(\theta_o) \quad\quad (10.30)$$

where the superscripts of V indicate derivatives of V with respect to θ. We see from Eq. (10.30) and Fig. 10.17 that the potential energy is a minimum at $\theta = \theta_o$ if $V(\theta) - V(\theta_o)$ is positive for all values of θ different from θ_o in the interval $\theta_o - \varepsilon \leq \theta \leq \theta_o + \varepsilon$, where ε is a positive number as small as we please so that the series converges rapidly. If $V'(\theta_o) = V''(\theta_o) = 0$ and $V'''(\theta_o) \neq 0$, then $V(\theta) - V(\theta_o)$ as well as $(\theta - \theta_o)^3$ changes its sign as θ increases through θ_o; and $V(\theta_o)$ is neither a minimum nor a maximum. If $V'(\theta_o) = V''(\theta_o) = V'''(\theta_o) = 0$ and $V^{iv}(\theta_o) \neq 0$, then the sign of $V(\theta) - V(\theta_o)$ is dependent on that of $V^{iv}(\theta_o)$. Therefore, *if the first and the second derivatives of V with respect to θ are zero while the first nonzero higher-order derivative of V with respect to θ is of even order and has a positive value, then the value of V at $\theta = \theta_o$ is a minimum and the equilibrium is still stable; otherwise, the equilibrium is unstable.*

For a two-degree-of-freedom conservative system with generalized coordinates θ_1 and θ_2, the potential energy $V(\theta_1, \theta_2)$ is minimum and hence the equilibrium is stable if the following are simultaneously satisfied:[†]

[†]Cf. Sec. 10.15 and the second footnote on p. 447.

$$\frac{\partial V}{\partial \theta_1} = \frac{\partial V}{\partial \theta_2} = 0$$

$$\frac{\partial^2 V}{\partial \theta_1^2}\frac{\partial^2 V}{\partial \theta_2^2} - \left(\frac{\partial^2 V}{\partial \theta_1 \partial \theta_2}\right)^2 > 0 \qquad (10.31)$$

$$\frac{\partial^2 V}{\partial \theta_1^2} > 0 \quad \text{or} \quad \frac{\partial^2 V}{\partial \theta_2^2} > 0$$

These criteria correspond to the special case of $n = 2$ in the criteria stated later in Sec. 10.15. Furthermore, note that

$$\frac{\partial^2 V}{\partial \theta_1^2} \quad \text{and} \quad \frac{\partial^2 V}{\partial \theta_2^2}$$

must be of the same sign whenever

$$\frac{\partial^2 V}{\partial \theta_1^2}\frac{\partial^2 V}{\partial \theta_2^2} - \left(\frac{\partial^2 V}{\partial \theta_1 \partial \theta_2}\right)^2 > 0$$

is satisfied.

EXAMPLE 10.17

A thin hemispherical shell of mean radius R and weight W is placed in equilibrium on the top of a fixed semicylindrical drum of radius R as shown. Assuming that slipping does not occur between the shell and the drum, determine the stability of equilibrium of the shell in the position shown.

Solution.   Since the shell does not slip on the drum, the arcs BA and BA' are equal and the slightly displaced position of the shell can be defined by the generalized coordinate θ as shown. From Fig. 6.10, we know that the center of gravity G of the shell is at a distance $R/2$ from its center C as shown. Choosing the fixed support as the reference datum, we may write the potential energy V of the shell as

$$V = Wy = W\left(2R\cos\theta - \frac{1}{2}R\cos 2\theta\right)$$

By differentiation, we write

$$\frac{dV}{d\theta} = WR(-2\sin\theta + \sin 2\theta) \qquad \frac{d^2V}{d\theta^2} = WR(-2\cos\theta + 2\cos 2\theta)$$

At $\theta = 0$, which is the position of equilibrium, we find that

$$\frac{dV}{d\theta} = \frac{d^2V}{d\theta^2} = 0$$

Thus, we continue to investigate the values of higher-order derivatives at $\theta = 0$:

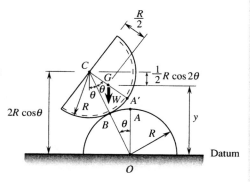

$$\left.\frac{d^3V}{d\theta^3}\right|_{\theta=0} = WR(2\sin\theta - 4\sin 2\theta)\Big|_{\theta=0} = 0$$

$$\left.\frac{d^4V}{d\theta^4}\right|_{\theta=0} = WR(2\cos\theta - 8\cos 2\theta)\Big|_{\theta=0} = -6WR < 0$$

Since the first nonzero higher-order derivative is of even order but has a *negative* value, we conclude that V is maximum.

The equilibrium of the hemispherical shell at $\theta = 0$ is unstable.   ◀

Fig. D10.37

Developmental Exercise

D10.37 A thin semicylindrical shell of meán radius R and weight W is placed in equilibrium on the top of a fixed semicylindrical drum of radius R as shown. Assuming that slipping does not occur between the shell and the drum, determine the stability of equilibrium of the shell in the position shown.

EXAMPLE 10.18

Refer to the system in Example 10.15, where it is found that the system with $W < 4kb$ may be in equilibrium in the positions $\theta = \pi/2$ and $\theta = \sin^{-1} W/(4kb)$. Assume that the bars cannot rotate as a single rigid body about A when A and C coincide at $\theta = \pi/2$. Determine the stability of the system in these two positions.

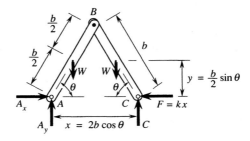

Solution. The potential energy of the bars, as established in Example 10.15, is

$$V = Wb \sin\theta + 2kb^2 \cos^2\theta$$

By differentiation, we write

$$\frac{dV}{d\theta} = Wb \cos\theta - 4kb^2 \cos\theta \sin\theta$$

$$\frac{d^2V}{d\theta^2} = -Wb \sin\theta - 4kb^2(-\sin^2\theta + \cos^2\theta)$$

Noting that $\cos^2\theta = 1 - \sin^2\theta$, we write

$$\frac{d^2V}{d\theta^2} = 8kb^2 \sin^2\theta - Wb \sin\theta - 4kb^2$$

(a) *Stability in the position* $\theta = \pi/2$. We have $\dfrac{dV}{d\theta} = 0$. Since $\sin\theta = 1$ and $W < 4kb$, we write

$$\frac{d^2V}{d\theta^2} = 8kb^2 - Wb - 4kb^2 = 4kb^2 - Wb = b(4kb - W) > 0$$

The equilibrium of the system at $\theta = \dfrac{\pi}{2}$ is stable. ◀

(b) *Stability in the position* $\theta = \sin^{-1} \dfrac{W}{4kb}$. We have $\dfrac{dV}{d\theta} = 0$. Since $\sin\theta = \dfrac{W}{4kb}$ and $W < 4kb$, we write

$$\frac{d^2V}{d\theta^2} = 8kb^2\left(\frac{W}{4kb}\right)^2 - Wb\left(\frac{W}{4kb}\right) - 4kb^2 = \frac{W^2}{4k} - 4kb^2$$

$$= \frac{1}{4k}(W + 4kb)(W - 4kb) < 0$$

The equilibrium of the system at $\theta = \sin^{-1} \dfrac{W}{4kb}$ is unstable. ◄

Developmental Exercise

D10.38 Refer to the system in Example 10.15. Determine the stability of equilibrium of the system in the position $\theta = \pi/2$ if $W = 4kb$.

EXAMPLE 10.19 ——————————

Refer to the system in Example 10.13. Determine the range of the magnitudes of **P** for which the equilibrium of the system is stable in the position $\theta = 0$.

Solution. The potential energy of the bar, as established in Example 10.16, is

$$V = \frac{1}{2} kL^2 \sin^2\theta - PL(1 - \cos\theta)$$

By differentiation, we write

$$\frac{dV}{d\theta} = kL^2 \sin\theta \cos\theta - PL \sin\theta$$

$$\frac{d^2V}{d\theta^2} = kL^2(\cos^2\theta - \sin^2\theta) - PL \cos\theta$$

We note that $\dfrac{dV}{d\theta} = 0$ at $\theta = 0$. For *stable* equilibrium in the position $\theta = 0$, we write

$$\left.\frac{d^2V}{d\theta^2}\right|_{\theta = 0} = kL^2 - PL > 0 \qquad P < kL \quad ◄$$

Developmental Exercise

D10.39 Refer to the system studied in Examples 10.13 and 10.19. Note that $d^2V/d\theta^2 = 0$ if $P = kL$. Determine the stability of equilibrium of the system in the position $\theta = 0$ if $P = kL$.

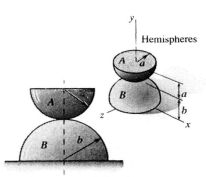

PROBLEMS

10.39* through 10.44* Using the *principle of potential energy*, solve the following problems:

10.39*	Prob. 10.24.	10.40*	Prob. 10.25
10.41*	Prob. 10.26*.	10.42*	Prob. 10.33.
10.43*	Prob. 10.34*.	10.44*	Prob. 10.35.

10.45 through 10.47* Determine the stability of equilibrium of the uniform bar *AB* of weight *W* supported by two vertical light links as shown.

Fig. P10.46   **Fig. P10.47***

10.48 through 10.51* The homogeneous body *A* is placed in equilibrium on the top of the body *B* as shown. Assuming that the radii *a* and *b* are equal and slipping does not occur between these two bodies, determine the stability of equilibrium of the body *A* in the position shown.

Fig. P10.45

Fig. P10.48 and P10.52*

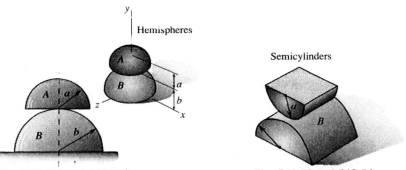

Fig. P10.49* and P10.53*   **Fig. P10.50 and P10.54**

Semicylinders

Fig. P10.51* and P10.55*

10.52* through 10.55* The homogeneous body *A* is placed in equilibrium on the top of the body *B* as shown. Assuming that slipping does not occur between these two bodies, determine the range of values of the ratio *a/b* of the radii for which this equilibrium of the body *A* is stable.

10.56 A homogeneous composite body is composed of a hemisphere of radius *b* an a cylinder of height *a* as shown. Determine the range of values of the ratio *a/b* for which the equilibrium of the body is stable in its upright position.

10.57 A container is composed of a hemispherical shell of radius *b* and a cylindrical shell of height *a*, both made from the same sheet metal, as shown. Determine the range of values of the ratio *a/b* for which the equilibrium of the body is stable in its upright position.

Fig. P10.56

Fig. P10.57

Fig. P10.58

10.58 A torsional spring with modulus $K = 100$ N·m/rad is installed at the base of an inverted pendulum which carries a mass $m = 10$ kg at a height h as shown. It is known that the pendulum in its upright position as shown is in equilibrium. Determine the range of values of h for which this equilibrium is stable.

10.59* A spring with modulus k is used to stabilize a bent bar which is under the action of a horizontal force **P** as shown. It is known that the bent bar is in equilibrium in the position shown. Determine the range of magnitudes of **P** for which this equilibrium is stable.

Fig. P10.59*

10.60 Determine the range of values of the modulus k of the spring for which the equilibrium of the platform AB of mass m as shown is stable.

Fig. P10.60

Fig. P10.61

10.61 A uniform bar of length L and mass m is supported as shown. The spring with modulus k is neutral when $\theta = 0$. Determine the angle θ for equilibrium and prove that this equilibrium is stable.

10.62 A uniform bar of length L and weight W is supported as shown. The spring with modulus k is neutral when $\theta = 0$. Determine all possible positions of equilibrium of the bar and their stability if $W = 2kL$.

10.63* Solve Prob. 10.62 if $W = 3kL$.

10.64 Solve Prob. 10.62 if $W = \frac{3}{2}kL$.

Fig. P10.62

Fig. P10.65

Fig. P10.69

Fig. P10.71

10.65　An eccentric weight W is carried at G by a wheel which is held by a spring with modulus k wrapping around its drum of radius r as shown. The spring is neutral when $\theta = 0$. For $r = 6$ in., $a = 10$ in., $k = 2$ lb/in., and $W = 10$ lb, determine the positions of equilibrium and their stability.

10.66　A body of mass m is hung at the end A of the mechanism shown. The rod CE can freely slide through the collar pivoted at D. The spring with modulus k is compressed by the stopper at E and the collar at D and is neutral when $\theta = 0$. It is known that $k = 98.1$ N/m and $L = 800$ mm. Determine the value of m for which the mechanism will be in *neutral equilibrium* for any value of θ.

Fig. P10.66

10.67*　Refer to the system in Prob. 10.66. Determine the range of values of m for which the equilibrium of the mechanism is stable in the position $\theta = 0$.

10.68　Refer to the system in Prob. 10.66. Determine the range of values of m for which the equilibrium of the mechanism is stable in the position $\theta = 180°$.

10.69　A horizontal force $\mathbf{P}$ is applied to the mechanism shown. The rod CE can freely slide through the collar D which rests on rollers. The spring with modulus k is compressed by the stopper at E and the collar D and is neutral when $\theta = 0$. Determine the positions of equilibrium of the mechanism and their stability if $P = kL$.

10.70*　Refer to the system in Prob. 10.69. Determine the range of magnitudes of $\mathbf{P}$ for which the equilibrium of the mechanism is stable in the position $\theta = 90°$.

10.71　A uniform slender rod AB of length L and mass m is supported by a ball and socket at B and rests on a smooth horizontal member of a frame as shown. Determine the stability of equilibrium of the rod AB in the position shown.

10.15　Concluding Remarks

The concept of *work* is both important and subtle in mechanics. It is the *force* (not the *body*) which does work. A body *receives* work from a force only when the particle (of the body) acted on by the force has a displacement during the action of the force. In other words, if a force *slides* on a *stationary* body, no work is done by the force to the stationary body. This is discussed in Sec. 10.2 and emphasized in Example 10.1 and D10.6 through D10.8, which involve a variety of situations.

As an extension, we may here employ the concept of virtual work in this chapter to alternatively define the concept of *equivalent systems of forces*, presented in Sec. 5.3, as follows: *If each of any two systems of forces will do the same amount of total virtual work to a body for the same given compatible virtual displacement of the body, then those two systems of forces are equivalent; otherwise, they are either equipollent or nonequivalent.* This alternative definition is referred to as the *energy criterion* for equivalent systems of forces.[†] We see that, when the compatible virtual displacement is a linear virtual displacement of the entire body, the energy criterion will yield the condition that the resultant force of each of the two equivalent systems of forces must be the same (i.e., $\mathbf{R} = \mathbf{R}'$). We also see that, when the compatible virtual displacement is an angular virtual displacement of the entire body about a given point, the energy criterion will yield the condition that the resultant moment of each of the two equivalent systems of forces about a given point must be the same [i.e., $\mathbf{M}_P^R = (\mathbf{M}_P^R)'$]. Thus, the conditions of equivalent systems of forces as defined in Eqs. (5.4) are fully covered by the *energy criterion*. Moreover, the energy criterion will clearly lead to the *principle of virtual work*, as presented in Sec. 10.7, when *the force system* in the free-body diagram of a body is *equivalent to a zero wrench*.

The *potential energy* of a body is a vital concept in mechanics. The quantitative definition for the potential energy of a body as given in Sec. 10.9 is a working definition devised to enable one to quickly acquire a firm grasp of that concept. Note that the potential energy is not defined for a body in a nonconservative force field. The *principle of potential energy* for conservative systems is a direct consequence of the *principle of virtual work*; the former is a corollary of the latter.

The criteria for the stability of equilibrium of a conservative system are expressed in terms of the potential energy of the system. Such criteria are discussed and summarized in Sec. 10.14 for single- and two-degree-of-freedom conservative systems. For an *n*-degree-of-freedom conservative system with generalized coordinates $\theta_1, \theta_2, \ldots, \theta_n$, it can be shown that the potential energy $V(\theta_1, \theta_2, \ldots, \theta_n)$ is minimum and hence the equilibrium is stable if the following criteria, which consist of *n* equations and *n* inequalities, are simultaneously satisfied:

$$\frac{\partial V}{\partial \theta_1} = \frac{\partial V}{\partial \theta_2} = \cdots = \frac{\partial V}{\partial \theta_n} = 0 \tag{10.32}$$

$$\begin{vmatrix} \dfrac{\partial^2 V}{\partial \theta_1^2} & \dfrac{\partial^2 V}{\partial \theta_1 \partial \theta_2} & \cdots & \dfrac{\partial^2 V}{\partial \theta_1 \partial \theta_i} \\[2mm] \dfrac{\partial^2 V}{\partial \theta_2 \partial \theta_1} & \dfrac{\partial^2 V}{\partial \theta_2^2} & \cdots & \dfrac{\partial^2 V}{\partial \theta_2 \partial \theta_i} \\[2mm] \cdots & \cdots & \cdots & \cdots \\[2mm] \dfrac{\partial^2 V}{\partial \theta_i \partial \theta_1} & \dfrac{\partial^2 V}{\partial \theta_i \partial \theta_2} & \cdots & \dfrac{\partial^2 V}{\partial \theta_i^2} \end{vmatrix} > 0 \tag{10.33}$$

where the left-hand side of the inequality sign is a determinant of the *i*th order and $i = 1, 2, \ldots, n$ to yield the *n* inequalities.[††] It is obvious that

---

[†]In this alternative definition, the body may be a particle, a rigid body, or a set of connected particles or rigid bodies.

[††]For a detailed treatment, see G. J. Simitses, *An Introduction to the Elastic Stability of Structures* (Englewood Cliffs, N.J.: Prentice-Hall, 1976).

the stability criteria become increasingly complex for systems with three or more degrees of freedom. The stability of equilibrium is a measure of the quality of equilibrium; hence, the stability analysis of equilibrium represents an in-depth analysis of equilibrium. Thus, this is logically the last chapter in the study of statics.

Fig. P10.72

REVIEW PROBLEMS

10.72   Determine the weight W which balances the 180-lb load as shown.[†] Does the value of W depend on the values of a and b?

10.73 and 10.74   Determine the reactions from the support at A of the structure shown.

Fig. P10.73

Fig. P10.74

Fig. P10.75

[†]This type of balance is also known as a *Roberval balance*, after the French mathematician Gilles Personne De Roberval (1602–1675).

10.75 Determine the moment **M** for which the linkage as shown is in equilibrium.

Fig. P10.76* and P10.77*

10.76* Uniform rods, each having a weight W and a length L, are connected as shown. For equilibrium, determine the angle θ_2.

10.77* Uniform rods, each having a weight W and a length L, are connected as shown. For equilibrium, determine the angle θ_1.

Fig. P10.78*

Fig. P10.79

10.78* and 10.79 Determine the reactions from the supports at A and B of the frame shown.

10.80 A bar of negligible weight is supported and loaded as shown. The free length of the spring with modulus k is L. Determine the range of magnitudes of the constant force **P** for which the equilibrium of the bar is stable in the position shown.

Fig. P10.80

10.81 A cylinder of radius r and weight Q is suspended from the ceiling to *stabilize* the equilibrium of a uniform slender rod of length L and weight W which is hinged at its base and leans against a vertical wall as shown. Determine (a) the value of the derivative $d\theta/d\phi$ as evaluated in the position $\theta = \phi = 0$, (b) the range of values of Q required.

Fig. P10.81

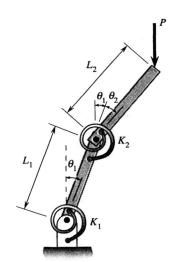

Fig. P10.82

10.82 A double pendulum lies in a smooth *horizontal* plane as shown. The two bars are hinged with torsional springs at the support and the joint. The torsional springs have moduli K_1 and K_2 as indicated. The system is in equilibrium when $\theta_1 = \theta_2 = 0$. For $L_1 = 2L$, $L_2 = L$, and $K_1 = K_2 = K$, determine the range of magnitudes of the horizontal force **P** for which the equilibrium position is stable.

10.83* Solve Prob. 10.82 if $L_1 = L_2 = L$, $K_1 = 2K$, and $K_2 = K$.

10.84* Using the *principle of virtual work*, solve Prob. 3.55.

10.85* Using the *principle of virtual work*, solve Prob. 3.56.

Prior Basic Mathematics   Appendix A

Prior mathematical skills needed for studying Chaps. 1 through 5 involve an ability to perform calculation with a calculator and a proficiency in basic algebra, geometry, and trigonometry. However, some prior background in basic analytic geometry and elementary calculus will be needed starting with Chap. 6. The following problems are designed to reflect the typical prior skills required. Answers to these problems are provided at the end of the text. Some review materials on the aforementioned basic mathematics are presented in App. B.

Choose the correct or best item to complete the sentence or answer the question in each of the following:

1. The solution for x in the equation $3x + 3 = -2$ is
   (a) 3/5.   (b) $-3/5$.   (c) 5/3.
   (d) $-5/3$.   (e) -1.

2. The simultaneous equations $2x + 5y = 21$ and $-3x + y = 11$ have a solution given by
   (a) $(-2, 5)$.   (b) $(-5, -2)$.   (c) $(2, 5)$.
   (d) $(-5, 2)$.   (e) $(-2, -5)$.

3. The roots of the quadratic equation $x^2 + x - 12 = 0$ are
   (a) -1 and 12.   (b) 1 and -12.   (c) -2 and 6.
   (d) 3 and -4.   (e) 3 and 4.

4. The roots of the quadratic equation $x^2 + x - 9 = 0$ are
   (a) -1 and 9.   (b) 1 and -9.   (c) 2.54 and -3.54.
   (d) 5.08 and -7.08.   (e) -5.08 and 7.08.

5. Expansion of $(a + b)^3$ yields
   (a) $a^3 + b^3$.   (b) $a^3 - 2a^2b + 2ab^2 - b^3$.
   (c) $a^3 + 2a^2b + 2ab^2 + b^3$.   (d) $a^3 - 3a^2b + 3ab^2 - b^3$.
   (e) $a^3 + 3a^2b + 3ab^2 + b^3$.

6. The term $(x^2)^3$ is equivalent to
   (a) x^5.   (b) x^2.   (c) x^6.
   (d) x^3.   (e) $x^{2.5}$.

7. The term $(a/b)^{-3/2}$ is equivalent to
   (a) $(b/a)^{3/2}$.   (b) $-a^3/b^2$.   (c) $a^{3/2}/b^{3/2}$.
   (d) $\sqrt{(a/b)^3}$.   (e) $-\sqrt{(a/b)^3}$.

8. The term $\frac{1}{2} \ln (a^{1/3})$ is equivalent to
   (a) $\ln (a^{5/6})$.   (b) $\ln (a^{6/5})$.   (c) $\frac{1}{3} \ln (a^{1/2})$.
   (d) $\ln (a^{2/3})$.   (e) $\ln (a^{3/2})$.

9. The term $\frac{1}{3}\ln(a/b)$ is equivalent to
   (a) $\ln(a^{1/3}b)$.   (b) $\ln(a^{1/3}) + \ln(b^{1/3})$.   (c) $\frac{1}{3}(\ln a + \ln b)$.
   (d) $\frac{1}{3}(\ln a)(\ln b)$.   (e) $\frac{1}{3}(\ln a - \ln b)$.

10. If $3.45^x = 8$, then the value of x is
    (a) 1.679.   (b) 1.583.   (c) 1.402.   (d) 1.338.
    (e) 1.238.

11. The value of the determinant $\begin{vmatrix} 5 & -2 \\ 8 & -2 \end{vmatrix}$ is
    (a) 26.   (b) -26.   (c) 9.   (d) 6.   (e) -6.

12. The value of the determinant $\begin{vmatrix} 5 & -7 & 9 \\ 2 & 0 & 0 \\ 6 & -4 & 1 \end{vmatrix}$ is
    (a) -14.   (b) -54.   (c) 54.   (d) 58.   (e) -58.

13. The value of $[6.44 + (4.75)^{1/3}(e^{0.4\pi})\ln 3.65 - 3\sin 35° \cot 215°]$ is
    (a) 7.30.   (b) 8.90.   (c) 11.63   (d) 12.88.   (e) 16.54.

14. The shortest distance from the point A to the line which is tangent to the circle at the point B, as shown, is
    (a) 9.5 m.   (b) 9 m.   (c) 8.5 m.   (d) 8 m.   (e) 7.5 m.

15. The length of the line segment $\overline{AB}$, as shown, is
    (a) 5.29 ft.   (b) 5.66 ft.   (c) 6.29 ft.   (d) 6.66 ft.
    (e) 6.92 ft.

16. The area of the triangle ABC shown is
    (a) 9 m².   (b) 12 m².   (c) 18 m².   (d) 21 m².   (e) 24 m².

17. The area of the trapezoid shown is
    (a) 28 in².   (b) 42 in².   (c) 48 in².   (c) 63 in².   (e) 84 in².

Fig. P14

Fig. P15

Fig. P16

Fig. P17   Fig. P18   Fig. P19

18. If $\theta = 3/5$ rad, the length of the arc AB, as shown, is
    (a) 0.6 m.   (b) 1.667 m.   (c) 9 m.   (d) 12 m.   (e) 8 m.

19. The circumference of the circle shown is
    (a) 3π m².   (b) 6π m².   (c) 9π m².   (d) 3π m.   (e) 6π m.

20. The area of the quarter circular sector shown is
    (a) 2π m².   (b) 4π m².   (c) 8π m².   (d) 2π m.   (e) 4π m.

21. In the figure shown, the angle α is
    (a) 44°.   (b) 45°.   (c) 46°.   (d) 47°.   (e) 60°.

Fig. P20

Fig. P21

22. The length of the side $\overline{BC}$ of the right triangle shown is
    (a) 8 ft.   (b) 6 ft.   (c) 7.5 ft.   (d) 4.12 ft.   (e) 8.5 ft.

23. In the triangle shown, $\tan\theta$ is equal to
    (a) 3/5.   (b) 5/3.   (c) 3/4.   (d) 4/3.   (e) 4/5.

Fig. P22

Fig. P23

Fig. P24

24. The length of the side $\overline{AB}$ of the triangle shown is
    (a) 13 m.   (b) 11.27 m.   (c) 9.22 m.   (d) 8.95 m.
    (e) 6.48 m.

25. The length of the side $\overline{AB}$ of the right triangle shown is
    (a) 12.6 m.   (b) 11.8 m.   (c) 10.5 m.   (d) 9.24 m.
    (e) 8.51 m.

26. The angle θ of the right triangle shown is
    (a) 35.0°.   (b) 36.9°.   (c) 44.4°.   (d) 45.6°.   (e) 53.1°.

Fig. P25

Fig. P26

Fig. P27

27. The angle θ of the triangle shown is
    (a) 0.785 rad.   (b) 0.813 rad.   (c) 1.05 rad.
    (d) 1.57 rad.   (e) 3.14 rad.

28. The length of the side $\overline{AC}$ of the triangle shown is
    (a) 5.00 in.   (b) 5.08 in.   (c) 5.32 in.   (d) 5.53 in.
    (e) 5.77 in.

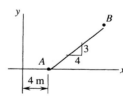

Fig. P28

29. The term $\cos^2\theta$ is equivalent to
    (a) $(1 - \sin 2\theta)/2$.   (b) $(1 - \cos 2\theta)/2$.   (c) $(1 + \sin 2\theta)/2$.
    (d) $(1 + \cos 2\theta)/2$.   (e) $-\sin 2\theta$.

30. The expression $1 - \sin^2\theta$ is equivalent to
    (a) $\tan^2\theta$.   (b) $\sec^2\theta$.   (c) $1 + \cot^2\theta$.   (d) $\cos^2\theta - 1$.
    (e) $\cos^2\theta$.

31. The equation of the line AB shown is
    (a) $3x - 4y = 0$.   (b) $4x - 3y = 0$.
    (c) $3x - 4y + 12 = 0$.   (d) $3x - 4y - 12 = 0$.
    (e) $4x - 3y + 12 = 0$.

32. The length of the line segment $\overline{AB}$ shown is
    (a) 22 m.   (b) 24 m.   (c) 26 m.   (d) 28 m.   (e) 38 m.

Fig. P31

Fig. P32

Fig. P33

Fig. P34

33. The equation of the parabola shown is
(a) $y^2 = 16x$.   (b) $x^2 = 2y$.   (c) $x^2 = y + 8$.
(d) $y^2 = x + 60$.   (e) $4y = 8x$.

34. The equation of the ellipse shown is
(a) $\dfrac{x}{4} + \dfrac{y}{3} = 1$.   (b) $\dfrac{x}{4} - \dfrac{y}{3} = 1$.   (c) $\dfrac{x^2}{4} + \dfrac{y^2}{3} = 1$.
(d) $\dfrac{x^2}{16} - \dfrac{y^2}{9} = 1$.   (e) $\dfrac{x^2}{16} + \dfrac{y^2}{9} = 1$.

35. Which of the following represents an ellipse in polar coordinates?
(a) $\dfrac{1}{r} = 3 \cos\theta$.   (b) $\dfrac{1}{r} = 3\theta$.   (c) $\dfrac{1}{r} = 3 + 3 \cos\theta$.
(d) $\dfrac{1}{r} = 3 + 4 \cos\theta$.   (e) $\dfrac{1}{r} = 4 + 3 \cos\theta$.

36. The differential area element dA in polar coordinates, as shown, is
(a) $u\, du\, d\theta$.   (b) $\theta\, du$.   (c) $du\, d\theta$.   (d) $u\, d\theta$.   (e) $dx\, dy$.

Fig. P36

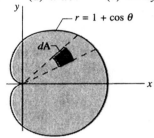

Fig. P37

37. The area enclosed by the cardioid shown is given by
(a) $\displaystyle\int_{0}^{2\pi} (1 + \cos\theta)\, d\theta$.   (b) $\displaystyle 2\int_{\theta=0}^{\pi}\int_{u=0}^{1+\cos\theta} du\, d\theta$.
(c) $\displaystyle\int_{\theta=0}^{2\pi}\int_{u=0}^{1+\cos\theta} du\, d\theta$.   (d) $\displaystyle 2\int_{\theta=0}^{\pi}\int_{u=0}^{1+\cos\theta} u\, du\, d\theta$.
(e) $\displaystyle\int_{\theta=0}^{2\pi}\int_{u=0}^{1} (1 + \cos\theta)\, du\, d\theta$.

38. The differential area element dA, as shown, is
(a) $(y^2/4)\, dy$.   (b) $2x^{1/2}\, dx$.   (c) $[(y^2 - 16)/4]\, dy$.
(d) $[(16 - y^2)/4]\, dy$.   (e) $(4 - y)\, dy$.

Fig. P38

39. If $\dfrac{dx}{dt} > 0$, which of the following statements is *always* true?
(a) The magnitude of x is increasing.
(b) The algebraic value of x is increasing.
(c) The magnitude of x is decreasing.
(d) The algebraic value of x is decreasing.
(e) The value of x is positive.

40. The slope of the normal to the curve $2x^2 + 9y = 0$ at the point $(3, -2)$ is
(a) 3/4.   (b) 4/3.   (c) $-4/3$.   (d) $-3/4$.   (e) 3/5.

41. The radius of curvature of the curve $2x^2 + 9y = 0$ at the point $(3, -2)$ is
    (a) 0.096.   (b) 1.25.   (c) 3.75.   (d) 10.42.   (e) 15.52.

42. The differential area element dA, as shown, is
    (a) $2x^{1/2}\, dx$.   (b) $\frac{1}{4} y^2 \, dy$.   (c) $(2x^{1/2} - 2)\, dx$.
    (d) $(4 - 2x^{1/2})\, dx$.   (e) $(2x^{1/2} - 4)\, dx$.

43. The area of the shaded region shown is given by
    (a) $\int_1^3 2y^{1/2}\, dy$.   (b) $\int_1^4 2y^{1/2}\, dy$.   (c) $\int_0^4 \frac{x^2}{4}\, dx$.
    (d) $\int_2^4 \frac{x^2}{4}\, dx$.   (e) $\int_1^3 (4 - y)\, dy$.

Fig. P42

Fig. P43

4y = x² y = x²/4

— 4 x = y²

4 m

4 y = x²

Fig. P44

44. The area of the shaded region shown is given by
    (a) $\int_0^4 \left(\frac{1}{4} x^2 - 2x^{1/2} \right) dx$.   (b) $\int_0^4 \frac{1}{4} x^2 \, dx$.
    (c) $\int_0^4 \left(2y^{1/2} - \frac{1}{4} y^2 \right) dy$.   (d) $\int_0^4 2y^{1/2}\, dy$.   (e) $\int_0^4 x \, dy$.

45. The volume of the right circular cone shown is given by
    (a) $\int_0^h (2\pi b/h)x \, dx$.   (b) $\int_0^h 2\pi x \, dx$.   (c) $\int_0^h \pi x^2 \, dx$.
    (d) $\int_0^h (\pi h^2/b^2)x^2 \, dx$.   (e) $\int_0^h (\pi b^2/h^2)x^2 \, dx$.

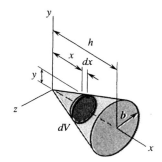

Fig. P45

46. The value of $0.234 \int_1^9 (2x - x^{-1} - x^{-1/2})\, dx$ is
    (a) 76.0.   (b) 73.8.   (c) 17.78.   (d) 17.27.   (e) -2.20.

47. Which of the following is *false*?
    (a) $\sin^2 x + \cos^2 x = 1$.   (b) $\sinh^2 x + \cosh^2 x = 1$.
    (c) $\sec^2 x - \tan^2 x = 1$.   (d) $\operatorname{sech}^2 x + \tanh^2 x = 1$.
    (e) $\csc^2 x - \cot^2 x = 1$.

48. Which of the following represents the shape of the curve defined by
    $y = \cosh x$:

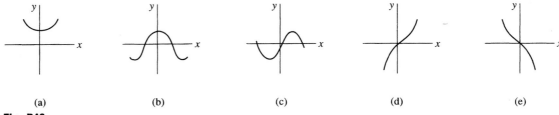

(a)   (b)   (c)   (d)   (e)

Fig. P48

49. If $e^x = y + (1 + y^2)^{1/2}$, then y is equal to
    (a) $e^x + e^{-x}$.   (b) $e^x - e^{-x}$.   (c) $\tanh x$.
    (d) $\cosh x$.   (e) $\sinh x$.

50. The value of $\displaystyle\int_0^1 \sinh x\, dx$ is
    (a) -1.543.   (b) -0.1752.   (c) 0.1752.   (d) 0.543.
    (e) 1.175.

Review of Basic Mathematics

Appendix B

B1 ALGEBRA

B1.1 Fundamental Principles

The following principles allow us to solve a great many equations:

1. *Addition principle.* If the equation $a = b$ is true, then $a + c = b + c$ is true, for any number or expression c.
2. *Multiplication principle.* If the equation $a = b$ is true, then $ac = bc$ is true for any number or expression c.
3. *Principle of zero products.* For any numbers or expressions a and b, if $ab = 0$, then $a = 0$ or $b = 0$, and if $a = 0$ or $b = 0$, then $ab = 0$.
4. *Principle of powers.* If the equation $a = b$ is true, then $a^n = b^n$ is true for any number n.

B1.2 Roots of a Quadratic Equation

If $ax^2 + bx + c = 0$ and $a \neq 0$, then

$$x = \frac{-b \pm \sqrt{b^2 - 4ac}}{2a}$$

B1.3 Solution of n Linear Simultaneous Equations

A set of n linear simultaneous equations in n unknowns $x_1, x_2, \ldots, x_n$ may be written in the form

$$
\begin{array}{ll}
a_{11}x_1 + a_{12}x_2 + \cdots + a_{1n}x_n = b_1 & (1) \\
a_{21}x_1 + a_{22}x_2 + \cdots + a_{2n}x_n = b_2 & (2) \\
\cdots \cdots \cdots \cdots \cdots \cdots \cdots \cdots & \cdots \\
a_{n1}x_1 + a_{n2}x_2 + \cdots + a_{nn}x_n = b_n & (n)
\end{array}
$$

If these n equations are independent, nonhomogeneous, and consistent, then they have a unique nontrivial solution. Such a solution may be obtained by the following steps:

1. From one of the n given equations, express the unknown x_1 in terms of the remaining $n - 1$ unknowns $x_2, x_3, \ldots, x_n$.
2. Substitute the expression of x_1 obtained in step (1) into the remaining $n - 1$ equations to obtain a set of $n - 1$ linear simultaneous equations in $n - 1$ unknowns $x_2, x_3, \ldots, x_n$.
3. From one of the $n - 1$ equations obtained in step (2), express the unknown x_2 in terms of the remaining $n - 2$ unknowns $x_3, x_4, \ldots, x_n$.
4. Substitute the expression of x_2 obtained in step (3) into the remaining $n - 2$ equations to obtain a set of $n - 2$ linear simultaneous equations in $n - 2$ unknowns $x_3, x_4, \ldots, x_n$.
5. Continue the reduction process similar to the above four steps until the size of the set is reduced to one linear equation in one unknown.
6. Solve for the single unknown in the last equation and perform the back substitutions of determined values into the expressions of other unknowns to obtain the values of all other unknowns.

B1.4   Second- and Third-Order Determinants

Second- and third-order determinants can be expanded by special schemes as follows:

$$\begin{vmatrix} a & b \\ c & d \end{vmatrix} = ad - bc$$

$$\begin{vmatrix} a & b & c \\ d & e & f \\ g & h & i \end{vmatrix}$$

$$= aei + bfg + cdh - ceg - bdi - afh$$

Note that the sign to be used with each individual product is indicated at the tail of each individual arrow. Furthermore, the above schematic methods of expansion *cannot* be used with determinants of order higher than three. However, any determinant may be expanded by using the cofactors of its row elements or column elements. Using the cofactors of the elements of the first row of a third-order determinant, we write

$$\begin{vmatrix} a & b & c \\ d & e & f \\ g & h & i \end{vmatrix} = a \begin{vmatrix} e & f \\ h & i \end{vmatrix} - b \begin{vmatrix} d & f \\ g & i \end{vmatrix} + c \begin{vmatrix} d & e \\ g & h \end{vmatrix}$$

$$= a(ei - fh) - b(di - fg) + c(dh - eg)$$

If *any* two rows or columns of a determinant are interchanged, the value of the determinant will change its sign. For example, we write

$$\begin{vmatrix} a & b & c \\ d & e & f \\ g & h & i \end{vmatrix} = - \begin{vmatrix} d & e & f \\ a & b & c \\ g & h & i \end{vmatrix} = \begin{vmatrix} d & e & f \\ g & h & i \\ a & b & c \end{vmatrix} = - \begin{vmatrix} g & h & i \\ d & e & f \\ a & b & c \end{vmatrix}$$

Thus, by interchanging the rows twice, we write

$$\begin{vmatrix} a & b & c \\ d & e & f \\ g & h & i \end{vmatrix} = \begin{vmatrix} d & e & f \\ g & h & i \\ a & b & c \end{vmatrix} = \begin{vmatrix} g & h & i \\ a & b & c \\ d & e & f \end{vmatrix}$$

This property is useful in studying the volume of a parallelepiped using the scalar triple product in Sec. 4.15.

B1.5 Binomials

(a) $a^2 - b^2 = (a + b)(a - b)$

(b) $a^3 - b^3 = (a - b)(a^2 + ab + b^2)$

(c) $(a + b)^2 = a^2 + 2ab + b^2$

(d) $(a - b)^2 = a^2 - 2ab^2 + b^2$

(e) $(a + b)^3 = a^3 + 3a^2b + 3ab^2 + b^3$

(f) $(a - b)^3 = a^3 - 3a^2b + 3ab^2 - b^3$

(g) $(a + b)^n = a^n + na^{n-1}b + \dfrac{n(n-1)}{2!} a^{n-2}b^2$

$$+ \dfrac{n(n-1)(n-2)}{3!} a^{n-3}b^3 + \cdots + b^n$$

(binomial theorem)

B1.6 Exponents

(a) $A^a A^b = A^{a+b}$

(b) $(A^a)^b = A^{ab}$

(c) $A^{-a} = \dfrac{1}{A^a}$

(d) $A^{a/b} = (A^a)^{1/b} = \sqrt[b]{A^a}$

B1.7 Logarithms

The logarithm of a number N to a base b is the exponent x indicating the power to which b must be raised to produce the number N. Thus, we write $\log_b N = x$ if $b^x = N$. The base of the common logarithm is 10, while that of the natural logarithm is e, where

$$e = 1 + \frac{1}{1!} + \frac{1}{2!} + \frac{1}{3!} + \frac{1}{4!} + \cdots = 2.718281828 \cdots$$

The natural logarithm is usually abbreviated as ln. We have

(a) $\ln (AB) = \ln A + \ln B$
(b) $\ln (A/B) = \ln A - \ln B$
(c) $\ln (A^a) = a \ln A$
(d) $A^a = e^{a \ln A}$

B2 GEOMETRY

B2.1 Useful Postulates and Theorems

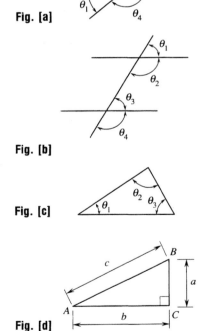

Fig. [a]

Fig. [b]

Fig. [c]

Fig. [d]

1. *Vertical angles*: Vertical angles formed by two intersecting lines are congruent. In Fig. [a], we have $\theta_1 = \theta_3$ and $\theta_2 = \theta_4$.
2. *Corresponding angles*: When a transversal intersects two parallel lines, corresponding angles are congruent, and conversely. In Fig. [b], we have $\theta_1 = \theta_3$ and $\theta_2 = \theta_4$.
3. *Angle sum for triangles*: The sum of the measures of the angles of a triangle is 180°. In Fig. [c], we have $\theta_1 + \theta_2 + \theta_3 = 180°$.
4. *Pythagorean theorem*: In a right triangle, the square of the length of its hypotenuse is equal to the sum of the squares of the lengths of its legs.[†] In Fig. [d], we have $c^2 = a^2 + b^2$.
5. *Congruent angles*: If the two lines including an angle are respectively perpendicular to the two lines including another angle, then these two angles must be either congruent angles or supplementary angles. In Fig. [e], we have $\theta_1 = \theta_2$ and $\theta_1 + \theta_3 = 180°$.

Fig. [e]

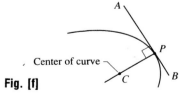

Center of curve

Fig. [f]

6. *Center of curve*: A line drawn perpendicular to the tangent to a curve and passing through the point of tangency will also pass through the

---

[†]The general formula $[\frac{1}{2}(m^2 + 1)]^2 = m^2 + [\frac{1}{2}(m^2 - 1)]^2$, where m is any odd integer, is attributed to the Greek philosopher and mathematician Pythagoras (ca. 580–500 B.C.).

center of the curve at that point. In Fig. [f], the point of tangency is P and $\overline{CP} \perp \overline{AB}$.

7. *Shortest distance*: The shortest distance from a point to a line is equal to the length of the perpendicular segment drawn from the point to the line. In Fig. [g], d_s is the shortest distance from the point P to the line AB.

Fig. [g]

B2.2   Radian Measure

One radian is defined as the angle which subtends a circular arc whose length is equal to the length of the radius of the circular arc. As shown in Fig. [h], we have

$$\theta = \frac{s}{R} \qquad s = R\theta$$

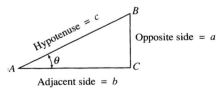

Fig. [h]

where θ is measured in radians.

B2.3   Mensuration Formulas for Some Common Shapes

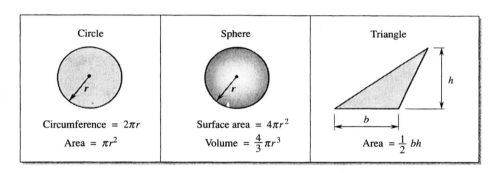

Circle	Sphere	Triangle
Circumference $= 2\pi r$	Surface area $= 4\pi r^2$	
Area $= \pi r^2$	Volume $= \frac{4}{3}\pi r^3$	Area $= \frac{1}{2}bh$

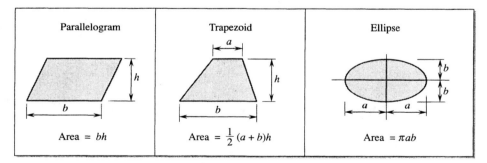

Parallelogram	Trapezoid	Ellipse
Area $= bh$	Area $= \frac{1}{2}(a+b)h$	Area $= \pi ab$

B3   TRIGONOMETRY

B3.1   Definitions with Mnemonics

Let a right triangle be oriented as shown in Fig. [a], where the angle θ is on the left and the right angle is on the right. Then, the values of $\sin\theta$,

Fig. [a]

Fig. [b]

Fig. [c]

Fig. [d]

Fig. [e]

Fig. [f]

$\cos\theta$, and $\tan\theta$ are, respectively, equal to those obtained by *dividing the first side touched into the second side touched* when the letters ▵, c, and t are written in *script* on the triangle as shown in Figs. [b], [c], and [d]. We have

$$\sin\theta = \frac{a}{c} = \frac{\text{Opposite side}}{\text{Hypotenuse}} = \theta - \frac{\theta^3}{3!} + \frac{\theta^5}{5!} - \frac{\theta^7}{7!} + \cdots \ (\theta \text{ in rad})$$

$$\cos\theta = \frac{b}{c} = \frac{\text{Adjacent side}}{\text{Hypotenuse}} = 1 - \frac{\theta^2}{2!} + \frac{\theta^4}{4!} - \frac{\theta^6}{6!} + \cdots \ (\theta \text{ in rad})$$

$$\tan\theta = \frac{a}{b} = \frac{\text{Opposite side}}{\text{Adjacent side}} = \theta + \frac{\theta^3}{3} + \frac{2\theta^5}{15} + \frac{17\theta^7}{315} + \cdots \ (\theta \text{ in rad})$$

B3.2 Some Useful Properties with Mnemonics

The quadrants in which the trigonometric functions have positive values may be indicated as shown in Fig. [e]. Furthermore, several trigonometric identities may readily be written from the hexagon shown in Fig. [f]. The mnemonic rules associated with Fig. [f] are as follows:[†]

1. The product of the functions at the ends of each diagonal is equal to 1. We have

$$\sin\theta \csc\theta = 1 \qquad \cos\theta \sec\theta = 1 \qquad \tan\theta \cot\theta = 1$$

2. Any function is equal to the product of its two neighboring functions. We have

$$\sin\theta = \cos\theta \tan\theta \qquad \csc\theta = \sec\theta \cot\theta$$
$$\cos\theta = \sin\theta \cot\theta \qquad \sec\theta = \tan\theta \csc\theta$$
$$\tan\theta = \sin\theta \sec\theta \qquad \cot\theta = \cos\theta \csc\theta$$

3. In each of the three shaded triangles, the square of the quantity at the lower angle is equal to the sum of the squares of the quantities at the upper two angles. We have

$$\sin^2\theta + \cos^2\theta = 1$$
$$\tan^2\theta + 1 = \sec^2\theta \qquad 1 + \cot^2\theta = \csc^2\theta$$

B3.3 Other Useful Trigonometric Identities

(a) $\sin(\alpha \pm \beta) = \sin\alpha \cos\beta \pm \cos\alpha \sin\beta$

(b) $\cos(\alpha \pm \beta) = \cos\alpha \cos\beta \mp \sin\alpha \sin\beta$

(c) $\tan(\alpha \pm \beta) = \dfrac{\tan\alpha \pm \tan\beta}{1 \mp \tan\alpha \tan\beta}$

(d) $\sin^2\theta = \frac{1}{2}(1 - \cos 2\theta)$

[†]Note that the order in which the trigonometric functions appear on the hexagon is that $\sin\theta$ and $\cos\theta$ are in the first row, $\tan\theta$ and $\cot\theta$ are in the second row, and $\sec\theta$ and $\csc\theta$ are in the third row as shown.

(e) $\cos^2\theta = \frac{1}{2}(1 + \cos 2\theta)$

(f) $\sin\alpha \pm \sin\beta = 2 \sin \frac{1}{2}(\alpha \pm \beta) \cos \frac{1}{2}(\alpha \mp \beta)$

(g) $\cos\alpha + \cos\beta = 2 \cos \frac{1}{2}(\alpha + \beta) \cos \frac{1}{2}(\alpha - \beta)$

(h) $\cos\alpha - \cos\beta = -2 \sin \frac{1}{2}(\alpha + \beta) \sin \frac{1}{2}(\alpha - \beta)$

(i) $\sin\alpha \cos\beta = \frac{1}{2}[\sin(\alpha + \beta) + \sin(\alpha - \beta)]$

(j) $\sin\alpha \sin\beta = \frac{1}{2}[\cos(\alpha - \beta) - \cos(\alpha + \beta)]$

(k) $\cos\alpha \cos\beta = \frac{1}{2}[\cos(\alpha + \beta) + \cos(\alpha - \beta)]$

B3.4   Law of Sines

For a triangle with angles and sides as shown in Fig. [g], we may write the law of sines as follows:

$$\frac{a}{\sin\alpha} = \frac{b}{\sin\beta} = \frac{c}{\sin\gamma}$$

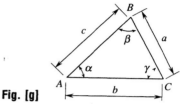

Fig. [g]

B3.5   Law of Cosines

For a triangle with angles and sides as shown in Fig. [g], we may write the law of cosines as follows:

$$a^2 = b^2 + c^2 - 2bc \cos\alpha \qquad b^2 = c^2 + a^2 - 2ca \cos\beta$$

$$c^2 = a^2 + b^2 - 2ab \cos\gamma$$

B3.6   Inverse Functions

(a) $\alpha = \sin^{-1} x$ if $x = \sin\alpha$

(b) $\theta = \csc^{-1} u$ if $u = \csc\theta$

(c) $\beta = \cos^{-1} y$ if $y = \cos\beta$

(d) $\phi = \sec^{-1} v$ if $v = \sec\phi$

(e) $\gamma = \tan^{-1} z$ if $z = \tan\gamma$

(f) $\psi = \cot^{-1} w$ if $w = \cot\psi$

B4   ANALYTIC GEOMETRY

B4.1   Equation of a Straight Line

Fig. [a]

Fig. [b]

B4.2 Equations of Conic Sections

A *conic section* is any curve which is the locus of a point that moves in such a way that the ratio of its distance from a fixed point to its distance from a fixed line is constant. This ratio is called the *eccentricity*, the fixed point the *focus*, and the fixed line the *directrix* of the curve. These are illustrated as shown, where $\varepsilon = eccentricity$.

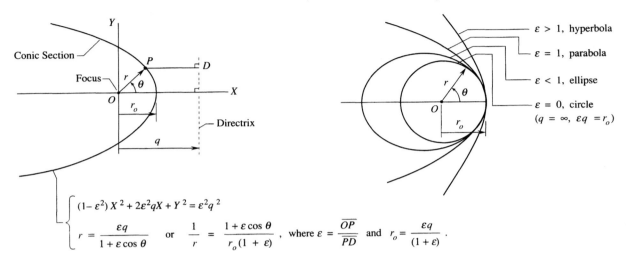

$$(1 - \varepsilon^2)\, X^2 + 2\varepsilon^2 qX + Y^2 = \varepsilon^2 q^2$$

$$r = \frac{\varepsilon q}{1 + \varepsilon \cos \theta} \quad \text{or} \quad \frac{1}{r} = \frac{1 + \varepsilon \cos \theta}{r_o (1 + \varepsilon)}, \quad \text{where } \varepsilon = \frac{\overline{OP}}{\overline{PD}} \text{ and } r_o = \frac{\varepsilon q}{(1 + \varepsilon)}.$$

- $\varepsilon > 1$, hyperbola
- $\varepsilon = 1$, parabola
- $\varepsilon < 1$, ellipse
- $\varepsilon = 0$, circle ($q = \infty$, $\varepsilon q = r_o$)

B4.3 Equation of a Circle

Note that $\varepsilon = 0$, $q = \infty$, and $\varepsilon q = R$.

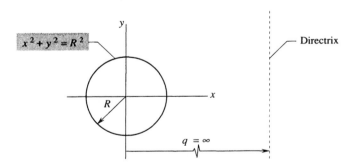

B4.4 Equation of an Ellipse[†]

Note that $a^2 - b^2 = c^2$, $\varepsilon = \dfrac{c}{a}$, $b = a(1 - \varepsilon^2)^{1/2}$, and $q = \pm\dfrac{b^2}{c}$.

[†]If a conic section has two foci, the geometric center of the conic section does not lie between a focus and its associated directrix.

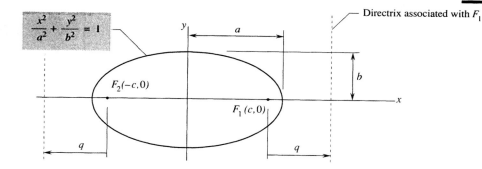

B4.5 Equation of a Parabola

Note that $\varepsilon = 1$ and $q = -2c$.

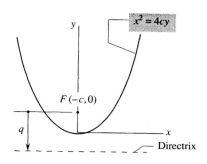

B4.6 Equation of a Hyperbola

Note that $a^2 + b^2 = c^2$, $\varepsilon = \dfrac{c}{a}$, $b = a(\varepsilon^2 - 1)^{1/2}$, and $q = \pm\dfrac{b^2}{c}$.

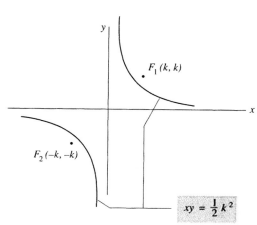

B5   CALCULUS

B5.1   Derivatives

(a) $\dfrac{dC}{dx} = 0$ (C is a constant.)

(b) $\dfrac{dx}{dx} = 1$

(c) $\dfrac{dy}{dt} = \dfrac{dy}{dx}\dfrac{dx}{dt}$

(d) $\dfrac{dx^n}{dx} = n\,x^{n-1}$

(e) $\dfrac{de^x}{dx} = e^x$

(f) $\dfrac{da^x}{dx} = (\ln a)\,a^x$

(g) $\dfrac{d}{dx}(\ln x) = \dfrac{1}{x}$

(h) $\dfrac{d}{dx}(\sin x) = \cos x$

(i) $\dfrac{d}{dx}(\cos x) = -\sin x$

(j) $\dfrac{d}{dx}(\tan x) = \sec^2 x$

(k) $\dfrac{d}{dx}(\cot x) = -\csc^2 x$

(l) $\dfrac{d}{dx}(\sec x) = \sec x \tan x$

(m) $\dfrac{d}{dx}(\csc x) = -\csc x \cot x$

(n) $\dfrac{d}{dx}(uv) = \dfrac{du}{dx}v + u\dfrac{dv}{dx}$

(o) $\dfrac{d}{dx}\left(\dfrac{u}{v}\right) = \dfrac{v\dfrac{du}{dx} - u\dfrac{dv}{dx}}{v^2}$

B5.2   Integrals

(a) $\int u\,dv = uv - \int v\,du$ (integration by parts)

(b) $\int x^n\,dx = \dfrac{1}{n+1}x^{n+1}$ $(n \neq -1)$

(c) $\int \dfrac{dx}{x} = \ln x$

(d) $\int \sin x\,dx = -\cos x$

(e) $\int \cos x\,dx = \sin x$

(f) $\int \tan x \, dx = -\ln(\cos x)$

(g) $\int \cot x \, dx = \ln(\sin x)$

(h) $\int \sec x \, dx = \ln(\sec x \tan x)$

(i) $\int \csc x \, dx = -\ln(\csc x + \cot x)$

(j) $\int \sin^n x \, dx = -\dfrac{1}{n} \sin^{n-1} x \cos x + \dfrac{1}{n}(n-1) \int \sin^{n-2} x \, dx$

(k) $\int \cos^n x \, dx = \dfrac{1}{n} \cos^{n-1} x \sin x + \dfrac{1}{n}(n-1) \int \cos^{n-2} x \, dx$

(l) $\int \sec^n x \, dx = \dfrac{1}{n-1} \sin x \sec^{n-1} x + \dfrac{n-2}{n-1} \int \sec^{n-2} x \, dx$

(m) $\int \csc^n x \, dx = -\dfrac{1}{n-1} \cos x \csc^{n-1} x + \dfrac{n-2}{n-1} \int \csc^{n-2} x \, dx$

(n) $\int e^x \, dx = e^x$

(o) $\int (a^2 - x^2)^{3/2} \, dx = \tfrac{1}{8}\Bigg[2x(a^2 - x^2)^{3/2}$
$$+ \, 3a^2 x(a^2 - x^2)^{1/2} + 3a^4 \sin^{-1}\frac{x}{a} \Bigg]$$

(p) $\int (a^2 - x^2)^{-1/2} \, dx = \sin^{-1}\dfrac{x}{a}$

(q) $\int (x^2 \pm a^2)^{-1/2} \, dx = \ln[x + (x^2 \pm a^2)^{1/2}]$

(r) $\int x^{-1}(x^2 - a^2)^{-1/2} \, dx = \dfrac{1}{a} \cos^{-1}\dfrac{a}{x}$

(s) $\int x^2(a^2 - x^2)^{-1/2} \, dx = \tfrac{1}{2}\Bigg[-x(a^2 - x^2)^{1/2} + a^2 \sin^{-1}\dfrac{x}{a} \Bigg]$

(t) $\int (x^2 \pm a^2)^{1/2} \, dx = \tfrac{1}{2} \{ x(x^2 \pm a^2)^{1/2} \pm a^2 \ln[x + (x^2 \pm a^2)^{1/2}] \}$

(u) $\int x^2(x^2 \pm a^2)^{1/2} \, dx = \tfrac{1}{8} \{ 2x(x^2 \pm a^2)^{3/2} \mp a^2 x(x^2 \pm a^2)^{1/2}$
$$- \, a^4 \ln[x + (x^2 \pm a^2)^{1/2}] \}$$

(v) $\int x^3(a^2 \pm x^2)^{1/2} \, dx = \tfrac{1}{15}(\pm 3x^2 - 2a^2)(a^2 \pm x^2)^{3/2}$

(w) $\int (a^2 - x^2)^{1/2} \, dx = \tfrac{1}{2}\Bigg[x(a^2 - x^2)^{1/2} + a^2 \sin^{-1}\dfrac{x}{a} \Bigg]$

(x) $\int x(a^2 - x^2)^{1/2} \, dx = -\tfrac{1}{3}(a^2 - x^2)^{3/2}$

(y) $\int x^2(a^2 - x^2)^{1/2} \, dx = \tfrac{1}{8}\Bigg\{ -2x(a^2 - x^2)^{3/2}$
$$+ \, a^2 \Bigg[x(a^2 - x^2)^{1/2} + a^2 \sin^{-1}\dfrac{x}{a} \Bigg] \Bigg\}$$

(z) $\int x^3(a^2 - x^2)^{1/2} \, dx = \tfrac{1}{15}(3x^2 + 2a^2)(a^2 - x^2)^{3/2}$

B5.3   Differential of a Function

For the purpose of this text, the first-order differential (or variation) δy of a function $y = y(x)$ may be obtained from the relationship

$$\delta y = \frac{dy}{dx} \, \delta x$$

For example, $\delta(x^2) = 2x \, \delta x$, $\delta(\sin x) = (\cos x) \, \delta x$, etc.

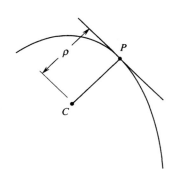

B5.4   Curvature κ and Radius of Curvature ρ

(a) In rectangular coordinates:

$$\kappa = \frac{\left|\dfrac{d^2 y}{dx^2}\right|}{\left[1 + \left(\dfrac{dy}{dx}\right)^2\right]^{3/2}} \qquad \rho = \frac{1}{\kappa}$$

(b) In polar coordinates:

$$\kappa = \frac{\left|r^2 + 2\left(\dfrac{dr}{d\theta}\right)^2 - r\dfrac{d^2 r}{d\theta^2}\right|}{\left[r^2 + \left(\dfrac{dr}{d\theta}\right)^2\right]^{3/2}} \qquad \rho = \frac{1}{\kappa}$$

B5.5   Expansion of $f(x)$ into Series

(a) Taylor's series:

$$f(x) = f(a) + (x - a)f'(a) + \frac{(x - a)^2}{2!}f''(a) + \cdots$$

$$= \sum_{n=0}^{\infty} \frac{(x - a)^n}{n!} f^{(n)}(a)$$

(b) Maclaurin's series:

$$f(x) = f(0) + x f'(0) + \frac{x^2}{2!}f''(0) + \cdots = \sum_{n=0}^{\infty} \frac{x^n}{n!} f^{(n)}(0)$$

B5.6   Hyperbolic Sine and Cosine Functions

(a) $\sinh x = \frac{1}{2}(e^x - e^{-x}) = x + \dfrac{x^3}{3!} + \dfrac{x^5}{5!} + \dfrac{x^7}{7!} + \cdots$

(b) $\cosh x = \frac{1}{2}(e^x + e^{-x}) = 1 + \dfrac{x^2}{2!} + \dfrac{x^4}{4!} + \dfrac{x^6}{6!} + \cdots$

(c) $\sinh^{-1} x = \ln\left(x + \sqrt{x^2 + 1}\right)$

(d) $\cosh^{-1} x = \ln\left(x + \sqrt{x^2 - 1}\right)$

(e) $\dfrac{d}{dx}(\sinh x) = \cosh x$

(f) $\dfrac{d}{dx}(\cosh x) = \sinh x$

(g) $\int \sinh x \, dx = \cosh x$

(h) $\int \cosh x \, dx = \sinh x$

B5.7   Some Hyperbolic Identities with Mnemonics

The mnemonics associated with the hexagon for hyperbolic functions shown consists of the following basic rules:

1. The product of the functions at the ends of each diagonal is equal to 1. We have

$$\sinh x \; \operatorname{csch} x = 1 \qquad \cosh x \; \operatorname{sech} x = 1 \qquad \tanh x \; \coth x = 1$$

2. Any function is equal to the product of its two neighboring functions.

$$\sinh x = \cosh x \; \tanh x \qquad\qquad \operatorname{csch} x = \operatorname{sech} x \; \coth x$$

$$\cosh x = \sinh x \; \coth x \qquad\qquad \operatorname{sech} x = \tanh x \; \operatorname{csch} x$$

$$\tanh x = \sinh x \; \operatorname{sech} x \qquad\qquad \coth x = \cosh x \; \operatorname{csch} x$$

3. In each of the three shaded triangles, the square of the quantity at the upper right side angle is equal to the sum of the squares of the quantities at the other two angles. We have

$$\cosh^2 x = 1 + \sinh^2 x \qquad 1 = \tanh^2 x + \operatorname{sech}^2 x$$
$$\coth^2 x = 1 + \operatorname{csch}^2 x$$

B5.8   Newton-Raphson Method

To extract a root of the equation $y(x) = 0$ by the Newton-Raphson method, we first assume a trial root $x = x_1$. Then the next improved trial root $x = x_2$ is given by $x_2 = x_1 - y(x_1)/y'(x_1)$ where $y(x_1) = y|_{x=x_1}$, and $y'(x_1) = dy/dx|_{x=x_1}$. The procedure is repeated until the correction term $y(x_i)/y'(x_i)$ for the $(i + 1)$th trial is a negligible fraction of the trial root $x = x_i$.

B5.9   Leibnitz's Rule

The taking of a derivative of a definite integral must be consistent with the rule that

$$\frac{d}{dx} \int_\alpha^\beta f(x,t) \, dt = \int_\alpha^\beta \frac{\partial}{\partial x} f(x,t) \, dt + f(x,\beta) \frac{d\beta}{dx} - f(x,\alpha) \frac{d\alpha}{dx}$$

where $\alpha = \alpha(x)$ and $\beta = \beta(x)$.

Appendix C   Infinitesimal Angular Displacements

Finite angular displacements (or finite rotations) are known to disobey the commutative law for the addition of vectors.[†] Thus, *finite* angular displacements do *not* qualify as vectors, despite the fact that they have magnitudes and directions given by the right-hand rule. However, we shall here show that *infinitesimal angular displacements are vectors*.

Let **r** be the position of the particle P of a rigid body, which initially occupies a region indicated by the dashed curve as shown in Fig. C.1. If the body undergoes successively an infinitesimal rotation $d\theta_1$ about the axis OA_1 and then an infinitesimal rotation $d\theta_2$ about the axis OA_2, the particle at P will go first to the point Q and then to the point P' as indicated, where $\overline{PQ} \approx \overline{C_1P}\, d\theta_1$ and $\overline{QP'} \approx \overline{C_2Q}\, d\theta_2$. If such successive infinitesimal rotations took place in reverse order, the particle at P would have gone first to the point R and then to the point P'' as indicated.

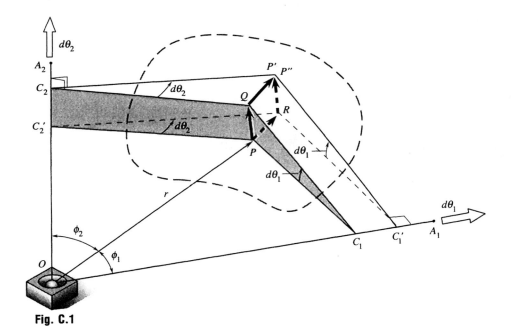

Fig. C.1

[†]Cf. D1.3.

Since the rotations $d\theta_1$ and $d\theta_2$ are infinitesimal, we infer that, within the framework of *first-order accuracy*, P'' must coincide with P', and

$$\overline{PQ} = \overline{RP'} = (r\sin\phi_1)d\theta_1$$

$$\overline{PR} = \overline{QP'} = (r\sin\phi_2)d\theta_2$$

Recalling that a rotation has a magnitude and a direction consistent with the right-hand rule, we may use the "mechanics" of cross products to write the differential displacements

$$\overrightarrow{PQ} = \overrightarrow{RP'} = \overrightarrow{d\theta}_1 \times \mathbf{r} \tag{C.1}$$

$$\overrightarrow{PR} = \overrightarrow{QP'} = \overrightarrow{d\theta}_2 \times \mathbf{r} \tag{C.2}$$

As the quadrilateral $PQP'R$ is a parallelogram, we have

$$\overrightarrow{PQ} + \overrightarrow{QP'} = \overrightarrow{PR} + \overrightarrow{RP'} \tag{C.3}$$

Substituting Eqs. (C.1) and (C.2) into Eq. (C.3), we write

$$\overrightarrow{d\theta}_1 \times \mathbf{r} + \overrightarrow{d\theta}_2 \times \mathbf{r} = \overrightarrow{d\theta}_2 \times \mathbf{r} + \overrightarrow{d\theta}_1 \times \mathbf{r}$$

$$(\overrightarrow{d\theta}_1 + \overrightarrow{d\theta}_2) \times \mathbf{r} = (\overrightarrow{d\theta}_2 + \overrightarrow{d\theta}_1) \times \mathbf{r} \tag{C.4}$$

Since Eq. (C.4) holds true for the position vector $\mathbf{r}$ of any point P, we must have

$$\overrightarrow{d\theta}_1 + \overrightarrow{d\theta}_2 = \overrightarrow{d\theta}_2 + \overrightarrow{d\theta}_1 \tag{C.5}$$

Thus, *infinitesimal angular displacements* (or infinitesimal rotations, such as virtual angular displacements) *are vectors* because they obey the commutative law for the addition of vectors as shown in Eq. (C.5) and are consistent with the parallelogram law. Since the time variable t is a scalar, we readily see that *angular velocities* $\boldsymbol{\omega}_1 = \overrightarrow{d\theta}_1/dt$, $\boldsymbol{\omega}_2 = \overrightarrow{d\theta}_2/dt$ *and angular accelerations* $\boldsymbol{\alpha}_1 = d\boldsymbol{\omega}_1/dt$, $\boldsymbol{\alpha}_2 = d\boldsymbol{\omega}_2/dt$ *are also vectors.*

Simultaneous Equations Solver

```
10 REM   SOLUTION OF SIMULTANEOUS EQUATIONS    (File name: SOSE)
12 REM
14 REM   A program to solve N linear simultaneous equations
16 REM   in N unknowns, [A][X] = [B], via matrix inversion
18 REM
20 REM      This is a BASIC program which runs on IBM PC and
22 REM   compatibles.  With 60K or more bytes free for BASIC,
24 REM   it can accommodate the solution and checking of at
26 REM   least 45 equations in 45 unknowns if running in double
28 REM   precision, and 65 equations in 65 unknowns if running
30 REM   in single precision.  The program automatically prints
32 REM   the DETERMINANT of the coefficient matrix [A] and does
34 REM   the back substitution of the solution for [X] into the
36 REM   original equations to see how well the product [A][X]
38 REM   CHECKs the constant vector [B].
40 REM
42 REM            D I R E C T I O N S :
44 REM   (a) If it is desired to divert the output to a
46 REM       printer, use, instead, the program SOSEP,
48 REM       which is also stored on the diskette.
50 REM   (b) Enter data as illustrated in Lines 1000-4000,
52 REM       then run the program.
54 REM   -------------------------------------------------
56 REM   |              C A U T I O N !                   |
58 REM   |    This program is intended to find unique solu- |
60 REM   | tions.  Sometimes, the rounding-off process with- |
62 REM   | in the computer may prevent the computer from    |
64 REM   | obtaining an exact value of zero for the determi- |
66 REM   | nant of a singular matrix.  Thus, in the output  |
68 REM   | of this program, if the product [A][X] in the    |
70 REM   | checking by back substitution differs greatly    |
72 REM   | from the original [B], then the solution for [X] |
74 REM   | should be discarded.  In this case, the matrix   |
76 REM   | [A] is possibly singular or the set of simulta-  |
78 REM   | neous equations is ill-conditioned.              |
80 REM   -------------------------------------------------
82 REM      Programmed by: I. C. Jong, University of Arkansas
84 REM
300 DEFDBL A-H,O-Z: REM Omit this line for single precision.
310 CLS: PRINT "There are N equations in [A][X] = [B]."
320 B$ = "Rowwise, the matrix [A] =": G1$ = "Subroutine "
330 C$ = "The constant vector [B] =": G2$="execution time ="
340 PRINT "N =";: READ N: D$ = "The determinant of [A] = "
350 E$ = "In checking, [A][X] gives [B] =": J$ = ") = "
360 F$ = "There is no unique solution for [A][X] = [B]."
370 IF N=0 OR N <> ABS(INT(N)) THEN PRINT " Illegal N!": END
380 PRINT N: ND=N+1: IF N>1 THEN ND=N: P$ = "*IN PROGRESS..."
390 DIM  A(ND,ND), AP(N,N), U(ND,ND), X(N), B(N): PRINT B$
400 FOR I=1 TO N: FOR J=1 TO N: READ A(I,J): AP(I,J)=A(I,J)
410 PRINT A(I,J);: NEXT J: PRINT: NEXT I: PRINT C$
420 FOR I = 1 TO N: READ B(I): PRINT B(I);: NEXT I: PRINT
430 PRINT: PRINT "*PRESS SPACE BAR TO COMPUTE..."
440 SP$ = INKEY$: ON -(SP$ <> CHR$(32)) GOTO 440: PRINT P$
450 PRINT D$;: GOSUB 900: C = T: GOSUB 700: GOSUB 900
460 E = T: PRINT D: IF D = 0 THEN PRINT F$: GOTO 540
470 PRINT "The solution yields [X] =": FOR I=1 TO N: X(I)=0
480 FOR J = 1 TO N: X(I) = X(I) + U(I,J)*B(J): NEXT J
490 PRINT SPC(8);"X ("; I; J$; X(I): NEXT I
500 ON -(N<10) GOTO 520: PRINT "*PRESS SPACE BAR TO CHECK.";
510 SP$ = INKEY$: ON -(SP$ <> CHR$(32)) GOTO 510: PRINT
520 PRINT E$: FOR I = 1 TO N: B(I) = 0: FOR J = 1 TO N
530 B(I)=B(I)+AP(I,J)*X(J):NEXT J:PRINT SPC(11);B(I):NEXT I
540 I$ = " second.": K$ = " seconds.": ET$ = " Less than one"
550 IF E >= C+1 THEN ET$ = STR$(E-C): IF E >= C+2 THEN I$=K$
```

```
560 PRINT G1$; G2$; ET$; I$: END
570 REM
700 REM .......... SUBROUTINE  DANDI  .......... (10-27-86)
702 REM 1. To let this subroutine use maximum pivots in find-
704 REM     ing the Determinant and Inverse of [A], do the
706 REM     following: (a) omit Lines 740 and 750, and (b)
708 REM     delete the word REM in each of Lines 745 and 746.
710 REM 2. Upon return: D = Determinant, [U] = Inverse of [A]
720 FOR I=1 TO N:FOR J=1 TO N:U(I,J)=0:IF I=J THEN U(I,J)=1
730 NEXT J: NEXT I: D = 1: N1 = N - 1: FOR I = 1 TO N1
740 ON -(A(I,I) <> 0) GOTO 790: K = I: FOR J = I TO N1
745 REM      BG = ABS(A(I,I)): K = I: FOR J = I TO N1: JP=J+1
746 REM      IF BG < ABS(A(JP,I)) THEN K=JP: BG = ABS(A(K,I))
750 JP = J + 1: IF A(JP,I) <> 0 THEN K = JP: GOTO 770
760 NEXT J: IF A(K,I) = 0 THEN D = 0: RETURN
770 ON -(K=I) GOTO 790: FOR J=1 TO N:P=A(I,J):A(I,J)=A(K,J)
780 A(K,J)=P:P=U(I,J):U(I,J)=U(K,J):U(K,J)=P:NEXT J: D = -D
790 AI = A(I,I): D = D*AI: FOR J=1 TO N: A(I,J) = A(I,J)/AI
800 U(I,J)=U(I,J)/AI: NEXT J: NI=N-I: FOR J=1 TO NI: IJ=I+J
810 AJ = A(IJ,I): FOR K=1 TO N: A(IJ,K) = A(IJ,K)-AJ*A(I,K)
820 U(IJ,K) = U(IJ,K) - AJ*U(I,K): NEXT K: NEXT J: NEXT I
830 D = D*A(N,N): IF D = 0 THEN RETURN
840 FOR I=1 TO N: U(N,I) = U(N,I)/A(N,N): NEXT I: A(N,N)=1
850 FOR I=1 TO N1: NI=N-I: FOR J=1 TO NI: IJ=I+J:AK=A(I,IJ)
860 FOR K = 1 TO N: A(I,K) = A(I,K) - AK*A(IJ,K)
870 U(I,K)=U(I,K)-AK*U(IJ,K): NEXT K: NEXT J: NEXT I:RETURN
900 REM .............. SUBROUTINE  TIME  ..............
910 T$=TIME$:H$=MID$(T$,1,2):M$=MID$(T$,4,2):S$=MID$(T$,7,2)
920 T = VAL(H$)*3600 + VAL(M$)*60 + VAL(S$): RETURN
930 REM
1000 DATA 6:      REM Enter N, the number of equations, here.
2000 REM Enter, rowwise, the coefficient matrix [A] below:
2010 DATA  .6, .96, -1, 0, 0, 0
2020 DATA  .8, -.28, 0, 0, 0, 0
2030 DATA  -.6, 0, 0, .384615, 0, 0
2040 DATA  -.8, 0, 0, -.923077, 1, 0
2050 DATA  0, -.96, 0, -.384615, 0, .882353
2060 DATA  0, .28, 0, .923077, 0, .470588
3000 REM Enter the constant vector [B] below:
4000 DATA 0, 0, 0, 0, 0, 98.1
```

The preceding program, named SOSE, is to be run under BASIC and is stored on the computer diskette labeled TUTORIAL, as described in App. F. Following the DIRECTIONS contained in the first REMark segment of the program, we show two sample runs as follows:[†]

☐ SOLUTION OF SIMULTANEOUS EQUATIONS IN EXAMPLE 3.8

```
There are N equations in [A][X] = [B].
N = 6
Rowwise, the matrix [A] =
 .6  .96 -1   0    0    0
 .8 -.28  0   0    0    0
-.6   0   0  .384615  0    0
-.8   0   0 -.923077  1    0
 0  -.96  0 -.384615  0  .882353
 0   .28  0  .923077  0  .470588
The constant vector [B] =
 0   0   0   0   0  98.1

The determinant of [A] =  .382262138896008
The solution yields [X] =
        X ( 1 ) =    24.38565007683738
        X ( 2 ) =    69.6732859338211
        X ( 3 ) =    81.51774454257069
        X ( 4 ) =    38.04165216151848
        X ( 5 ) =    54.6238942137679
        X ( 6 ) =    92.38677098912871
In checking, [A][X] gives [B] =
           1.77635683940025D-15
          -4.440892098500626D-16
          -1.332267629550188D-15
           0
          -1.77635683940025D-15
          98.1
Subroutine execution time = 3 seconds.
```

[†]Notice the first item in the DIRECTIONS in the program.

☐ CASE OF SINGULAR COEFFICIENT MATRIX [A]

```
There are N equations in [A][X] = [B].
N = 3
Rowwise, the matrix [A] =
  1   2   3
  4   5   6
  7   8   9
The constant vector [B] =
  1   2   3

The determinant of [A] =   0
There is no unique solution for [A][X] = [B].
Subroutine execution time = Less than one second.
```

Digital Root Finder[†]

```
10 REM    DIGITAL ROOT FINDER AND PLOTTER      (File name: DIGITAL)
12 REM
14 REM       A program to find the real roots of Y(X) = 0 to any
16 REM        desired digits of accuracy by plotting Y versus X
18 REM
20 REM       This is a BASIC program which runs on IBM PC and
22 REM    compatibles.  It plots in 80 columns from X = XI, with
24 REM    increment DX, to X = XF.
26 REM
28 REM                    D I R E C T I O N S:
30 REM    ---------------------------------------------------------------
32 REM    1. If it is desired to divert the output to a printer,
34 REM       use, instead, the program DIGITALP, which is also
36 REM       stored on the diskette.
38 REM    2. To find a root of a given nonlinear equation defined
40 REM       by Y(X) = 0, do the following: (a) define Y = Y(X) in
42 REM       BASIC in Lines 1000-1999; (b) run the program to ex-
44 REM       plore the locations of the roots by making trial runs
46 REM       until the graph crosses the X axis; (c) read the two
48 REM       neighboring digital values of X which bound the root
50 REM       at the crossing of the graph over the X axis; (d) use
52 REM       a smaller increment (e.g., one-tenth of the preceding
54 REM       increment) to plot between those two values of X; (e)
56 REM       repeat steps (c) and (d) until the root is determined
58 REM       to as many digits of accuracy as desired or permitted.
60 REM    ---------------------------------------------------------------
62 REM       Programmed by: I. C. Jong, University of Arkansas
64 REM
100 DEFDBL A-H, O-Z: REM Omit this line for single precision.
110 INPUT "XI = "; XI: INPUT "DX = "; DX: INPUT "XF = "; XF
120 N% = INT(ABS((XF - XI)/DX) + 1.5): DIM Y(N%): PRINT
130 PRINT "*IN PROGRESS": R=0: X = XI: FOR I=1 TO N%:    GOSUB 1000
140 Y(I) = Y: X = XI + I*DX: AY = ABS(Y(I)): IF AY>R THEN R = AY
150 NEXT I: SF=R/38: X = XI: FOR I = 1 TO 7: Y$ = Y$+":----!----"
160 NEXT I: Y$ = " !--" + Y$ + ":--!Y": I = 1: PRINT: PRINT
170 PRINT "XI =";XI; "    DX =";DX; "    XF =";XF; "      R =";R
180 PRINT: PRINT "-R"; SPC(37); "0"; SPC(37); "R": PRINT Y$
190 P% = INT(Y(I)/SF + 39 + .5): X% = P% - 40: LT = LEN(STR$(X))
200 Y% = INT(Y(I)/SF-LT-1.5): Z% = 38 - P%: IF P% >= 39 THEN 220
210 PRINT SPC(P%); "*"; SPC(Z%); "!"; X: GOTO 260
220 IF P% = 39 THEN PRINT SPC(P%); "*"; X: GOTO 260
230 IF 44>=P% AND P%>39 THEN PRINT SPC(39)"!"SPC(X%)"*":GOTO 260
240 IF Y% < 0 THEN PRINT SPC(39); "!"; X: GOTO 260
250 PRINT SPC(39); "!"; X; SPC(Y%); "*"
260 X = XI + I*DX: I = I + 1: IF X < XF + DX/3 THEN 190
270 PRINT SPC(39); "X": PRINT: LIST 1000-1999: END
1000 Y = 20 - X*((EXP(50/X)+EXP(-50/X))/2 - 1)◄────────── Y = Y(X)
2000 RETURN
```

[†]This program is provided for the convenience of those who may find it helpful in finding the root or roots of any nonlinear equation $y(x) = 0$, as well as plotting curves, with a computer. In the plot, the y axis is proportionally calibrated with the computed value of R; however, the x axis is digitally calibrated with any increment chosen by the user.

SOLUTION OF A TRANSCENDENTAL EQUATION (CF. EXAMPLE 8.24.)

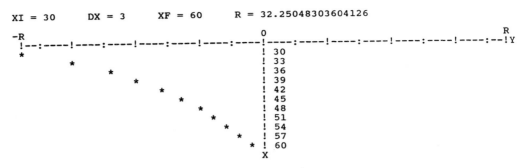

1000 Y = 20 - X*((EXP(50/X)+EXP(-50/X))/2 - 1)

(a) A trial run which indicates that the root is located at X > 60

1000 Y = 20 - X*((EXP(50/X)+EXP(-50/X))/2 - 1)

(b) Next trial run

1000 Y = 20 - X*((EXP(50/X)+EXP(-50/X))/2 - 1)

(c) Next run[†]

[†]Additional runs and outputs show that $Y = 0$ when $X \approx 65.586$. Since $\cosh(50/x) = [\exp(50/x) + \exp(-50/x)]/2$, we may take $c = 65.59$ as the root of the equation $20 = c(\cosh(50/c) - 1)$ encountered in Example 8.24.

Supplementary Software (Statics)

Appendix F

As alluded to in the Preface, diskettes containing supplementary software for this text may be obtained from bookstores or the publisher. The software is designed for an IBM PC® or compatible computer with at least 384K of random access memory, a CGA or compatible graphics display card, and DOS 2.0 or higher.

The BASIC programs CROSSPD, STRIPPD, CENTL, CENTA, AMOI, SOSE, and DIGITAL have been documented in Secs. 4.9, 4.15, 6.4, 6.8, 7.10, and Apps. D and E of the text. These seven BASIC programs plus a 40-column version of DIGITAL, named DIGIT40, and a compiled program named ISE40.EXE (which is an interactive version of SOSE for $N \le 40$), as well as the tutorial modules for all chapters, are stored on the diskette labeled TUTORIAL. Of course, the BASIC programs are to be run under a BASIC interpreter (e.g., BASICA.COM or GWBASIC.EXE). They all output to the screen of the monitor. If it is desired to output to a printer which is properly connected, the user may use, instead, the corresponding programs ending with a P (e.g., CROSSPDP, STRIPPDP, CENTLP, CENTAP, AMOIP, SOSEP, and DIGITALP) which are also stored on that diskette. To run the tutorial modules, the user may do the following: (a) set the computer in DOS, (b) insert the TUTORIAL diskette in drive A, (c) key in TUTOR at the prompt from the DOS (e.g., A:>TUTOR), and (d) make a selection from the menu as directed on the screen.

The four programs UNITS, VECTOR, AREA, and TRUSS are compiled programs designed to complement selected sections in the text. These compiled programs are stored on the diskette labeled STATSOFT. They are described in this appendix and may be run by doing the following: (a) set the computer in DOS, (b) insert the STATSOFT diskette in drive A, (c) key in STATSOFT at the prompt from the DOS (e.g., A:>STATSOFT), and (d) make a selection from the menu as directed on the screen.

Naturally, the principal sources of guidance in learning a subject are the *instructor* and the *textbook*. Students are advised to first study the text as assigned by the instructor. Then, as needed, they may use the software to strengthen their understanding of the concepts as covered in the following programs:

TUTOR: A Multiple-Choice Tutorial

This program contains ten modules covering Chaps. 1 through 10. It is designed to help students strengthen as well as assess their understanding of basic concepts in engineering mechanics by providing an interactive tutorial consisting of a series of multiple-choice questions related to one chapter at a time. As a user, you are to read each question carefully, go over the prospective answers thoughtfully, and key in the letter (e.g., A, B, C, D, or E) corresponding to the correct or best answer and then press the RETURN key. After your response, a window will appear to provide a comment or a hint. This process will continue until the correct answer is entered.

You receive a full credit for a question if you answer it correctly on the first try, half a credit if you answer it correctly on the second try, and no credit thereafter. Your score is continually updated on the screen. At the end of the quiz, a summary of credits earned for the individual questions, along with the total score, is shown. If you did not make 100% on the quiz, you may elect to retry those questions you missed on the first try. Note that the order of the prospective answers has been randomized and may not be the same when the question is posed again.

In responding to any question, you may enter END to terminate the quiz or SAVE to save the record of the quiz up to the current question for future restart. Entering HELP will recall the foregoing information.

UNITS: A Program to Convert Units

This program complements the material covered in Secs. 1.7 and 1.8 of the text. It is designed to convert one set of units to another. When it is run, a window appears on the top half of the screen giving a list of recognized units where upper and lower case modes are significant. Compound units may be formed by using the symbols * for multiplication, / for division, and ∧ for raising a unit to a power. Any prefixed or compound unit may be converted to any compatible unit(s). In handling a compound unit, note that this program regards all units to the right side of / as being divided into the units written to its left side. Furthermore, units raised to powers are assumed to have integer exponents.

At the prompt to enter the unit(s), one has several options. For example, one may enter a number with a unit or combination of units. If the number is omitted, it is defaulted to 1. If a question mark (?) is entered at the prompt, it will highlight a unit in the window, and a brief explanation of its meaning will appear. The left and right cursor keys may be used to move the highlight around the window. To leave the window and return to the prompt, press the ESC key. Entering nothing at the prompt (by simply pressing the RETURN key) will terminate the program.

VECTOR: A Program to Introduce Operations of Vectors in a Plane

This program complements the material covered in Secs. 2.1 through 2.5, 4.7 through 4.9, and 4.14 of the text. It is designed to introduce and demonstrate certain algebraic operations of vectors. It presents the addition, subtraction, dot product, and cross product of vectors in the xy plane. Initially, the screen shows a set of rectangular axes and prompts the user to interactively generate a vector **A** in either cartesian or polar form. By using the cursor keys, the initial vector shown on the screen can be incrementally elongated, shortened, or rotated to the liking of the user. This vector is then saved by pressing the RETURN key. A prompt to generate a second vector **B** will follow in a similar manner. Meanwhile, pressing the F1 key during the input process will call up a HELP screen.

After both vectors **A** and **B** have been saved into memory, the program will proceed to show a table which contains blanks to be completed by the user. However, the user may let the computer fill a blank by entering a question mark (?) in the blank. After the table is completed, the user will be asked to enter a choice which will demonstrate one of the algebraic operations as listed. Note that pressing the F2 key will recall the graphical display of the vectors **A** and **B** which were generated earlier.

AREA: A Program to Evaluate Areal Properties

This program complements the material covered in Secs. 6.7, 6.8, 7.2 through 7.8, and 7.10 of the text. It is designed to evaluate the areal properties (e.g., area, centroid, moments and products of inertia, principal moments and axes of inertia, and radii of gyration) of any plane area whose boundary is defined (or approximated) by N line segments connecting N nodes (or points), where $N \leq 500$. The program allows the following three ways to enter the data:

1. *Line input*: Enter interactively the numerical values of the coordinates of each node from the keyboard as prompted by the program.
2. *Graphical input*: Use the cursor keys to move the cross hair on the screen of the monitor to a desired position and then press the RETURN key to graphically enter the coordinates of a node. Repeat until the coordinates of all nodes on the boundary are entered.
3. *File input*: If the data for the coordinates of the various nodes on the boundary of an area have previously been saved by this program in a data file, those data may be recalled by choosing the file input mode and specifying the name of that file (e.g., ZBAR.DAT).

Once an area is defined, its properties may be evaluated and displayed. If desired, the area can be modified, translated, or rotated by using the Text Editor and/or the Graphics Editor in the program. Besides, the data can be saved to a disk file and recalled later. Thus, this program can give the user an experience in computer aided design (CAD).

TRUSS: A Program to Evaluate Forces in Truss Members

This program complements the material covered in Secs. 8.1 through 8.9 of the text. It is designed to evaluate the axial forces in the members of any statically determinate plane truss under the action of any forces at the joints. The truss may contain up to 26 joints (designated by the letters of the alphabet, A through Z), and it may be supported by an appropriate number of hinges, rollers, or links at its joints, as long as Eq. (8.2) is satisfied. The program allows the following two ways to enter the data:

1. *Line input*: Enter interactively the answers to a series of questions about the truss as prompted by the computer. A user is expected to specify the following: a name for identifying the given truss and for saving the data if desired, the names of the joints (A through Z) in the truss, the coordinates of the joints, the supports (e.g., hinge, roller, or link) existing at the joints, and the x and y components of the forces acting on the joints.

2. *File input*: If the data for the truss have previously been saved by this program in a data file, those data may be recalled by choosing the file input mode and specifying the name of that file (e.g., EX8P2.DAT).

Upon completing either the line input or the file input, the truss with its loads will first be graphically displayed. The user may then follow the prompt and the menu on the screen for calculating the axial forces and other editing and recalculating opportunities. Naturally, this program can give the user an experience in computer aided engineering (CAE).

Answers to Selected Developmental Exercises and Problems

Chapter 1

Developmental Exercises

D1.3 (i) No, (ii) no. **D1.4** No. **D1.5** F, a. **D1.6** (a)
Magnitude, (b) orientation, (c) sense. **D1.9** a, b, d. **D1.10** (a) (a,
b), (b) $\angle \theta$, (c) no. **D1.17** a, c, d. **D1.20** Yes. **D1.23**
Seven. **D1.28** (a) kg, m, s; (b) lb, ft, s. **D1.29** (a) 10 lb·s²/ft, (b)
5.24 kg·m/s², (c) 1.452 kg, (d) 30×10^{-6} s. **D1.30** (a) 5, (b) 6, (c)
4, (d) 5, (e) 4, (f) 7, (g) 4. **D1.31** (a) 41.5, (b) 41.5, (c) 42.5, (d)
42.5, (e) 0.00624, (f) 1.36×10^6, (g) 5.22×10^6. **D1.32** (a) 42, (b)
42, (c) 42, (d) 43, (e) 0.0062, (f) 1.4×10^6, (g) 5.2×10^6. **D1.33**
(a) 33.4, (b) 31.0, (c) 16.80, (d) 0.499, (e) 26.7×10^3, (f) 12.17 $\times$
10^{-6}, (g) 5.67, (h) 2.28, (i) 1.264, (j) 0.814, (k) -0.581. **D1.34**
No. **D1.35** (a) 0.0508 m, (b) 5.49 m, (c) 28.3 m³. **D1.37** Yes.
D1.39 (a) No, (b) no, (c) yes, (d) no. **D1.41** b. **D1.43** (i) a, c; (ii)
b, d. **D1.44** 34.4×10^{-9} ft³/(slug·s²). **D1.45** c.

Problems

1.1 (a) 55.9 mi/h, (b) 82.0 ft/s. **1.2** 238 L/s. **1.4** 76.0 ft/s. **1.5**
3.32 kg. **1.6** 17.39 psi. **1.7** 137.9 MPa. **1.9** (a) 53.5 lb, (b)
238 N. **1.10** 1.030 in. **1.12** 7.71 kg. **1.13** 24.1×10^3 mi/h.
1.15 b. **1.16** c. **1.17** b. **1.18** d. **1.19** a. **1.20** a.
1.21 b. **1.22** e. **1.23** e. **1.24** d.

Chapter 2

Developmental Exercises

D2.2

D2.3
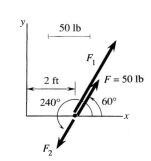

D2.4 $\theta_P = 36.9°$, $\theta_Q = 157.4°$.   **D2.5** c.   **D2.6** d.   **D2.8** Parallelogram law.   **D2.9** $P = 73.2$ lb, $Q = 51.8$ lb.   **D2.10** $Q = 29.8$ lb, $\theta_Q = 320.9°$.   **D2.13** (a) $G_x = 80\mathbf{i}$ N, $G_y = 60\mathbf{j}$ N, $H_x = -120\mathbf{i}$ N, $H_y = 50\mathbf{j}$ N; (b) $G_x = 80$ N, $G_y = 60$ N, $H_x = -120$ N, $H_y = 50$ N; (c) 120 N.   **D2.14** (a) $-240\mathbf{i} + 70\mathbf{j}$ N, (b) 163.7°.   **D2.15** $50\mathbf{i} - 480\mathbf{j}$ lb.   **D2.16** 71.8°.   **D2.17** $x = -2$, $y = 41$.   **D2.18** (a) $F_x = 3$ kN, $F_y = -4$ kN, $F_z = -12$ kN; (b) $r_x = 40$ m, $r_y = 0$, $r_z = -9$ m.   **D2.19** 9 kips.   **D2.20** (a) θ_y, θ_z; (b) no; (c) no; (d) yes, $\cos\theta_x$.   **D2.21** (a) $F_x = 0$, $F_y = 8$ lb, $F_z = -15$ lb; (b) $\theta_x = 90°$, $\theta_y = 61.9°$, $\theta_z = 151.9°$.   **D2.22** (a) $6\mathbf{i} - 4\mathbf{j} + 12\mathbf{k}$ m, (b) 14 m.   **D2.23** (a) $-6\mathbf{i} + 2\mathbf{j} + 3\mathbf{k}$ m, (b) 7 m, (c) $\frac{1}{7}(-6\mathbf{i} + 2\mathbf{j} + 3\mathbf{k})$.   **D2.24** (a) $0.707\mathbf{i} - 0.5\mathbf{j} + 0.5\mathbf{k}$, (b) $70.7\mathbf{i} - 50\mathbf{j} + 50\mathbf{k}$ lb.   **D2.25** (a) $\frac{1}{9}(4\mathbf{i} + 7\mathbf{j} - 4\mathbf{k})$, (b) $60\mathbf{i} - 240\mathbf{j} + 80\mathbf{k}$ lb.   **D2.26** $460\mathbf{i} + 460\mathbf{j} - 320\mathbf{k}$ lb at O.

Problems

2.1 $C = 184.4$ lb, $\theta_C = 192.5°$.   **2.3** $P = 544$ N, $Q = 254$ N, $\theta_P = 143.1°$, $\theta_Q = 53.1°$.   **2.4** $D = 680$ lb, $\theta_D = 298.5°$.   **2.6** $P = 135.9$ N, $Q = 63.4$ N, $\theta_P = 165°$, $\theta_Q = 255°$.   **2.7** $\theta_A = 223.9°$, $\theta_B = 115.9°$.   **2.9** $A = 273$ N, $B = 245$ N, $\theta_B = 315°$.   **2.11** 6.64 kips.   **2.12** 34.4°.   **2.14** $B = 1050$ N, $\theta_B = 150.9°$.   **2.16** $\theta_2 = 27.3°$, $R = 1544$ lb.   **2.17** $F = 170$ N, $S = 410$ N, $\theta_F = 61.9°$, $\theta_S = 282.7°$.   **2.19** $P = 1.44$ kips, $Q = 3.08$ kips.   **2.20** $18.5\mathbf{i} - 1.5\mathbf{j}$ kN.   **2.22** $\mathbf{R} = -3.5\mathbf{i} + 4.73\mathbf{j}$ kN.   **2.24** $200\mathbf{i} + 347\mathbf{j}$ N.   **2.25** 390 N.   **2.27** 20 lb.   **2.28** $F_{AC} = 250$ N, $F_{BC} = 520$ N.   **2.43** (a) $F_x = 3$ kN, $F_y = -12$ kN, $F_z = 4$ kN; (b) 5 kN; (c) $\frac{1}{13}(3\mathbf{i} - 12\mathbf{j} + 4\mathbf{k})$; (d) $\theta_x = 76.7°$, $\theta_y = 157.4°$, $\theta_z = 72.1°$.   **2.46** (a) $-0.388\mathbf{i} + 0.423\mathbf{j} + 0.819\mathbf{k}$, (b) 0.906.   **2.47** (a) $0.322\mathbf{i} + 0.928\mathbf{j} - 0.1857\mathbf{k}$, (b) 0.371.   **2.49** (a) $0.1664\mathbf{i} - 0.766\mathbf{j} + 0.621\mathbf{k}$, (b) 0.643.   **2.51** $-35\mathbf{i} + 360\mathbf{j} - 40\mathbf{k}$ lb.   **2.52** $F_{PA} = 125$ lb, $F_{PB} = 90$ lb, $F_{PC} = 195$ lb.   **2.53** $-16.8\mathbf{j}$ kN.   **2.54** $P = 10.77$ N, $Q = 6.32$ N, $D = 17.09$ N, $\theta_P = 158.2°$, $\theta_Q = 341.6°$, $\theta_D = 159.4°$.   **2.55** $\theta_A = 64.2°$, $\theta_B = 161.4°$.   **2.56** $S = 621$ N, $T = 312$ N.   **2.57** 2.12 kN.   **2.58** $-50\mathbf{i} + 320\mathbf{j} - 100\mathbf{k}$ lb.   **2.59** $F_{DA} = 130$ N, $F_{DB} = 125$ N, $F_{DC} = 175$ N.   **2.61** $T_{DA} = 9.9$ kN, $T_{DB} = 1.2$ kN, $T_{DC} = 15.6$ kN.

Chapter 3

Developmental Exercises

D3.3 True: a, b, d, false: c, e.   **D3.4** (a) Internal, (b) external. **D3.7** (a) From B to A, (b) from A to B.   **D3.8** Yes: a, b, c; no: d. **D3.9** a, b.   **D3.10** (a) 200 N, (b) 400 N.   **D3.14** As a body with a shape.   **D3.15** No.   **D3.16** b.

   D3.17

   D3.18

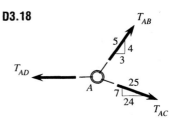

D3.19 431 N.   **D3.20** Yes.   **D3.23** 3.82 MN/m.   **D3.24** 162.6 lb
D3.25 Two.   **D3.29** $-600\mathbf{i} + 1800\mathbf{k}$ lb.

Problems

P3.1

P3.4

P3.9

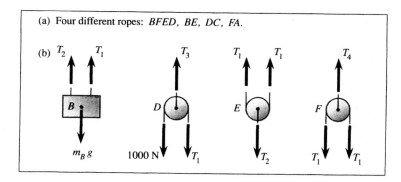

3.17 $T_{CA} = 863$ N, $T_{CB} = 640$ N.   **3.18** $T_{CA} = 1.766$ kN, $T_{CB} =$
2.35 kN.   **3.20** $T_{CA} = 750$ lb, $T_{CB} = 680$ lb.   **3.21** $T_{CA} = 1436$ N,
$T_{CB} = 1759$ N.   **3.23** 25 lb.   **3.25** 306 kg.   **3.26** 6 slugs.
3.28 $P = 577$ lb, $Q = 911$ lb.   **3.29** $S_1 = S_2 = 11.33$ N.   **3.30**
1.057 kN.   **3.31** 49.9°.   **3.32** (a) 294 N, (b) 196.2 N, (c) 196.2 N,
(d) 147.2 N.   **3.34** 2.45 kN.   **3.35** $P = 1.218$ kN, $\theta_P = 97.5°$.
3.37 $T_{CA} = 100$ lb, $T_{CB} = 340$ lb.   **3.38** 207 lb $< F <$ 430 lb.
3.39 26.5 kg.   **3.40** 312 kg.   **3.42** $m_A = 176.7$ kg, $m_B = 413$
kg.   **3.43** $T_{CG} = 29.4$ N, $T_{AB} = 20.6$ N, $T_{AC} = T_{BC} = 44.1$ N, T_{AD}
$= 58.9$ N, $T_{BE} = 58.9$ N, $T_{CF} = 76.5$ N.   **3.45** $T_{AB} = 4.47$ lb, T_{AD}
$= 11.31$ lb, $T_{AE} = 15.68$ lb, $T_{AJ} = 9$ lb, $T_{BC} = 3.48$ lb, $T_{BF} = 3.22$ lb,

P3.12

$T_{CD} = 3.14$ lb, $T_{CG} = 3.78$ lb, $T_{DH} = 13.35$ lb, $T_{DI} = 12$ lb. **3.46**
$(-\frac{2}{3} T_{AB} + \frac{3}{5} T_{AD})\mathbf{i} + (\frac{2}{3} T_{AB} + \frac{4}{5} T_{AC} + \frac{4}{5} T_{AD} - 450g)\mathbf{j} + (-\frac{1}{3} T_{AB} + \frac{3}{5} T_{AC})\mathbf{k}$ N. **3.48** $(-\frac{3}{7} T_{BA} + \frac{3}{5} T_{CA} - \frac{3}{7} T_{DA})\mathbf{i} + (\frac{6}{7} T_{BA} + \frac{4}{5} T_{CA} + \frac{6}{7} T_{DA} - 300)\mathbf{j} + (-\frac{2}{7} T_{BA} + \frac{2}{7} T_{DA})\mathbf{k}$ lb. **3.50** 2.21 kN. **3.51** 353
N. **3.52** 105 lb. **3.54** $50\mathbf{i} + 27.7\mathbf{j} - 15\mathbf{k}$ N. **3.55** 1099 N.
3.56 2.75 kN. **3.57** $T_{AP} = 88.3$ N, $T_{BQ} = 78.5$ N, $T_{OR} = 98.1$ N,
$T_{AB} = 33.7$ N, $T_{AC} = 66.4$ N, $T_{AD} = 52.5$ N, $T_{AO} = 26.2$ N, $T_{BG} = 57.2$ N, $T_{BH} = 72.8$ N, $T_{BO} = 47.0$ N, $T_{OE} = 58.8$ N, $T_{OF} = 68.1$ N.
3.58 196.2 N. **3.59** 578 lb. **3.60** (a) 156 lb/ft, (b) 1.675 ft.
3.63 $T_{DA} = 125$ N, $T_{DB} = 260$ N, $T_{DC} = 325$ N. **3.65** 12 lb.
3.66 6.87 kN.

Chapter 4

Developmental Exercises

D4.4 (a) Newton-meters (N · m), (b) pound-feet (lb · ft). **D4.6** (a)
Up. **D4.8** (a) 25 kN ⦨ 16.26°, (b) 361 m ⦨ 56.3°, (c) 4 kN · m ↻ ,
(d) 700 lb · ft ↺ . **D4.9** 1540 lb · ft ↻ . **D4.10** 560 N · m ↺ .
D4.14 (a) $M_A - 600$ N · m ↻ , (b) $M_A - 4A_y + 1000$ N · m ↻ .
D4.15 $M_x = -320$ lb · ft, $M_y = -800$ lb · ft, $M_z = 640$ lb · ft. **D4.16**
$M_{AB} = M_{Az} - \frac{18}{13} T_{CD} + 15$ kN · m, $M_{CB} = 12A_y - 30$ kN · m.
D4.18 a, b, d. **D4.19** 0. **D4.20** (a) 150°, (b) 750 lb · ft, (c) $\mathbf{k}$, (d)
$750\mathbf{k}$ lb · ft. **D4.23** $-3\mathbf{i} + 4\mathbf{j} - 12\mathbf{k}$. **D4.24** (a) $-\mathbf{k}$, (b) $\mathbf{0}$.
D4.26 (a) $-3\mathbf{i} - 6\mathbf{k}$, (b) $-3\mathbf{i} - 12\mathbf{j}$, (c) $\mathbf{i} + \mathbf{j} + \mathbf{k}$. **D4.27** (a)
78.2 ft², (b) 42.0°, (c) $-0.0767\mathbf{i} + 0.383\mathbf{j} + 0.920\mathbf{k}$. **D4.29** (a) $\mathbf{P} = -140\mathbf{i} + 20\mathbf{j} - 50\mathbf{k}$ N; (b) $\mathbf{r}_{AB} \times \mathbf{P}$; (c) $-10\mathbf{i} - 70\mathbf{j}$ N · m; (d) yes; (e)
0.471 m. **D4.30** $-10\mathbf{i} + 70\mathbf{j}$ N · m. **D4.32** $2\mathbf{i} - 6\mathbf{j} - 24\mathbf{k}$
kN · m. **D4.34** $400\mathbf{i} + 800\mathbf{j}$ lb · ft for (a) through (d). **D4.35** (a)
$-8\mathbf{i}$ kN · m, (b) $16\mathbf{i} - 12\mathbf{k}$ kN · m, (c) $8\mathbf{i} - 12\mathbf{k}$ kN · m. **D4.36** (a)
-4, (b) -4, (c) -10.39, (d) 0, (e) 16, (f) 0. **D4.37** (a) 173.2, (b)
8.66, (c) 17.32. **D4.38** 7. **D4.39** (a) $\frac{1}{5}(4\mathbf{i} + 3\mathbf{k})$, (b) $\frac{1}{7}(6\mathbf{i} + 3\mathbf{j} + 2\mathbf{k})$, (c) $\frac{6}{7}$, (d) $\frac{6}{7}$, (e) 31.0°, (f) 6 ft, (g) 6 ft. **D4.40** 125.7°. **D4.41**
(a) 6 ft³, (b) 6 ft³, (c) no. **D4.43** 5 ft. **D4.44** $450\mathbf{i} - 700\mathbf{j} + 500\mathbf{k}$
lb · ft. **D4.45** (a) $\frac{1}{7}(2\mathbf{i} - 3\mathbf{j} + 6\mathbf{k})$, (b) 2100 N · m, (c) $600\mathbf{i} - 900\mathbf{j} + 1800\mathbf{k}$ N · m. **D4.46** (a) $-420\mathbf{i} + 140\mathbf{k}$ N, (b) 443N, (c) 443 N, (d)
4.74 m. **D4.47** a, b, c, e.

Problems

4.1 20.1 N · m ↺ . **4.2** 620 lb · ft ↺ . **4.4** 8.11 lb · ft ↻ . **4.5** 40
lb · ft ↻ . **4.6** 53.6 N · m ↺ . **4.8** $3P - 30$ kN · m ↻ . **4.10** $8B_y - 2000$ lb · ft ↻ . **4.11** $24 - 6A_y$ kN · m ↻ . **4.12** $3A_x - 2A_y - 8$
kN · m ↻ . **4.14** (a) 66.6° or 246.6°, (b) 349 N · m. **4.16** 62.4° or
185.0°. **4.17** 220 lb · ft ↻ . **4.18** 300 N · m ↺ . **4.20** $200 - \frac{2}{5} P$
N · m ↻ . **4.21** $240 - \frac{6}{5} P$ N · m ↻ . **4.24** $2A_x - 3A_y - 240$ N · m
↻ . **4.25** $380 - \frac{38}{13} R$ lb · ft ↻ . **4.26** $M_x = -160$ N · m, $M_y = -220$ N · m, $M_z = 120$ N · m. **4.28** $M_x = -1200$ lb · ft, $M_y = 1800$
lb · ft, $M_z = 1200$ lb · ft. **4.29** $M_x^R = -4P + 8$ kN · m, $M_y^R = -6A_z + M_{Ay} - 3P + 2$ kN · m, $M_z^R = 6A_y - M_{Az}$ kN · m. **4.30** $M_x^R = -6A_y + 36$ kN · m, $M_y^R = 6A_x + \frac{5}{13} M_A - \frac{12}{7} T_{BC} - 11$ kN · m, $M_z^R =$

$\frac{12}{13} M_A + \frac{6}{7} T_{BC} - 24$ kN·m.   **4.32** $M_x^R = 870 - 9C_y$ lb·ft, $M_y^R = 4T_{EF} + 9B_x - 12C_z + M_{Cy}$ lb·ft, $M_z^R = 2T_{EF} + 12C_y + M_{Cz} - 1800$ lb·ft.   **4.33** (a) $5\mathbf{i} + \mathbf{j} - 3\mathbf{k}$, (b) $-5\mathbf{i} - \mathbf{j} + 3\mathbf{k}$, (c) $-5\mathbf{i} - \mathbf{j} + 3\mathbf{k}$, (d) 5.92, (e) 40.2°.   **4.35** 65.7 m².   **4.36** 122.2 ft².   **4.38** 42.4°.   **4.39** 53.0°.   **4.40** 19.46°.   **4.41** (a) 51.2 ft², (b) 34.7°.   **4.42** (a) 163.2 m², (b) 40.4°.   **4.44** 6.5k N·m.   **4.45** $-680\mathbf{k}$ lb·ft.   **4.46** $-800\mathbf{i} - 170\mathbf{j} - 200\mathbf{k}$ N·m.   **4.47** (a) $-640\mathbf{i} - 800\mathbf{j} - 320\mathbf{k}$ N·m, (b) $-740\mathbf{i} - 1000\mathbf{j} - 1120\mathbf{k}$ N·m.   **4.49** $(\frac{24}{7} T_{BF} - \frac{24}{7} T_{BG} + 300)\mathbf{i} + (\frac{36}{7} T_{BF} + \frac{36}{7} T_{BG} - 1200)\mathbf{k}$ kN·m.   **4.50** $(M_C + B_y + C_y + 110)\mathbf{i} + (-C_x - 2C_z - 20)\mathbf{j} + (2C_y + 100)\mathbf{k}$ lb·ft.   **4.51** $(-12A_z + 300)\mathbf{i} + (12A_x - 1200)\mathbf{k}$ kN·m.   **4.53** (a) $\mathbf{M}_1 = -6\mathbf{i} - 8\mathbf{k}$ kN·m, $\mathbf{M}_2 = -8\mathbf{k}$ kN·m; (b) and (c) $-6\mathbf{i} - 16\mathbf{k}$ kN·m.   **4.54** (a) $\mathbf{M}_1 = 6\mathbf{j} + 8\mathbf{k}$ kN·m, $\mathbf{M}_2 = 20\mathbf{k}$ kN·m; (b) and (c) $6\mathbf{j} + 28\mathbf{k}$ kN·m.   **4.55** (a) $\mathbf{M}_1 = 400\mathbf{i} + 600\mathbf{k}$ lb·ft, $\mathbf{M}_2 = -800\mathbf{i} + 1200\mathbf{k}$ lb·ft, $\mathbf{M}_3 = 2000\mathbf{j}$ lb·ft; (b) and (c) $-400\mathbf{i} + 2000\mathbf{j} + 1800\mathbf{k}$ lb·ft.   **4.57** (a) 76.8°, (b) 105.0°.   **4.60** (a) $-600\mathbf{i} + 400\mathbf{j} - 300\mathbf{k}$ N·m, (b) 500 N·m, (c) 500 N·m, (d) $400\mathbf{j} - 300\mathbf{k}$ N·m.   **4.61** 500 N·m.   **4.62** 62.4 N.   **4.64** (a) -18 N·m, (b) tighten.   **4.65** 600 mm.   **4.67** 196.1 m.   **4.68** $8B_y - 540$ lb·ft ↺.   **4.69** (a) $-126\mathbf{i} - 108\mathbf{j} - 36\mathbf{k}$ N·m, (b) -24 N·m, (c) tighten, (d) 629 mm, (e) 141.4 mm.   **4.70** (a) $1000\mathbf{i} + 400\mathbf{j} + 300\mathbf{k}$ N·m, (b) 840 N·m.   **4.71** (a) $(-4B_y + 2B_z)\mathbf{i} + (\frac{4}{3} T_{DE} + 4B_x - 4B_z + 1800)\mathbf{j} + (\frac{4}{3} T_{DE} + 2B_x - 4B_y)\mathbf{k}$ N·m, (b) $600 - \frac{4}{9} T_{DE}$ N·m.

Chapter 5

Developmental Exercises

D5.1 $\mathbf{M}_P' = \vec{PB} \times \mathbf{F} + \mathbf{M}_B = \vec{PB} \times \mathbf{F} + \vec{BA} \times \mathbf{F} = (\vec{PB} + \vec{BA}) \times \mathbf{F} = \vec{PA} \times \mathbf{F} = \mathbf{M}_P$. Thus, $\mathbf{M}_P = \mathbf{M}_P'$ is true.   **D5.2** (a) $\mathbf{F}_O = 100\mathbf{j}$ N, $\mathbf{M}_O = 1200\mathbf{k}$ N·m; (b) $\mathbf{F}_B = 100\mathbf{j}$ N, $\mathbf{M}_B = 500\mathbf{i} + 1200\mathbf{k}$ N·m; (c) yes.   **D5.3** (a) $4\mathbf{i} + 3\mathbf{j}$ kN, (b) $3\mathbf{i} + 6\mathbf{k}$ kN·m, (c) $4\mathbf{j}$ kN·m, (d) $3\mathbf{i} + 4\mathbf{j} + 6\mathbf{k}$ kN·m.   **D5.5** $-10\mathbf{i} + 5\mathbf{j} + 10\mathbf{k}$ lb·ft.   **D5.7** a, c, and f; d and e.   **D5.8** (a) $10\mathbf{j}$ kN, (b) $-20\mathbf{j} + 10\mathbf{k}$ kN·m, (c) $\mathbf{j}$, (d) $-20\mathbf{j}$ kN·m, (e) $10\mathbf{k}$ kN·m, (f) $-\mathbf{i}$ m, (g) (1, 0, 0) m.

D5.12

D5.15 b, c.   **D5.16** (a) Equilibrium problem of a *particle* in a plane, (b) equilibrium problem of a *rigid body* in a plane.   **D5.17** (a) $F_A = 80$ lb, $F_B = 60$ lb; (b) $F_A = 64$ lb, $F_B = 36$ lb.   **D5.18** (a) Unstable, (b) neutral, (c) stable.   **D5.20** Two-force: a, b; three-force: c.   **D5.21** Two-force: (a) *ABC*; three force: (a) *CDE*, (b) *CDE*, (c) *ABC*, *CDE*. **D5.22** (a) Third degree, (b) first degree, (c) second degree.

D5.23

(a)

(b) no,

(c) no,

(d) yes.

D5.24 Partially constrained: three prospective independent equilibrium equations to be applied, $\Sigma F_y \neq 0$, two unknowns.

D5.25

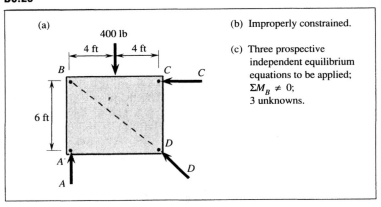

(a)

(b) Improperly constrained.

(c) Three prospective independent equilibrium equations to be applied; $\Sigma M_B \neq 0$; 3 unknowns.

D5.26 (a) 4, (b) 3, (c) B_y, (d) A_y, (e) no, (f) no, (g) no, (h) no.   **D5.27** $B_x = 980$ lb ←, $B_y = 2.35$ kips ↑.   **D5.29** $A_x = 50$ lb ←, $A_y = 30$ lb ↑.   **D5.32** No.   **D5.33** Wide.   **D5.34** (a) (i) 6, (ii) 6, (iii) no; (b) (i) 9, (ii) 6, (iii) statically indeterminate to the third degree.

Problems

5.1 $F_O = 100$ N ⬈ 53.1°, $M_O = 176$ N·m ↺.   **5.2** $F_O = -42\mathbf{i} + 24\mathbf{j} + 24\mathbf{k}$ lb, $M_O = -160\mathbf{i} - 280\mathbf{j}$ lb·ft.   **5.3** $F_C = 100$ N ⬈ 53.1°, $M_C = 112$ N·m ↺.   **5.5** $F_O = -50\mathbf{i} - 50\mathbf{j}$ lb, $M_O = -240\mathbf{k}$ lb·ft.   **5.6** $F_O = 30$ kN ↑, $M_O = 1440$ kN·m ↻.   **5.7**

$\mathbf{F}_O = -500\mathbf{j}$ lb, $\mathbf{M}_O = 2.8\mathbf{i} - 4.5\mathbf{k}$ kip·ft.   **5.9** $\mathbf{F}_A = -50\mathbf{i} - 50\mathbf{j}$ lb, $\mathbf{M}_A = -140\mathbf{k}$ lb·ft.   **5.10** $\mathbf{F}_A = 30$ kN ↑, $\mathbf{M}_A = 1560$ kN·m ↻.   **5.11** $\mathbf{F}_A = 500$ lb ↓, $\mathbf{M}_A = 2.8\mathbf{i} + 3\mathbf{k}$ kip·ft.   **5.13** (b) $P = 200$ N, $Q = 200$ N, $M = 100$ N·m; (c) $P = 100$ N, $Q = 100$ N, $M = 400$ N·m; (d) $P = 300$ N, $Q = 100$ N, $M = 200$ N·m.   **5.15** $P = 80$ lb, $Q = 35$ lb, $M = 17.5$ lb·ft.   **5.16** $P = 50$ N, $Q = 16$ N, $d = 0.2$ m.   **5.17** $A_x = 100$ N, $B = 50$ N, $P_y = 150$ N, $P_z = 225$ N, $Q = 100$ N, $S = 200$ N.   **5.19** (a) $P = 40$ N, $Q = 230$ N, $S = 200$ N; (b) $\mathbf{R} = 40\mathbf{i} - 30\mathbf{j}$ N at A, $\mathbf{M}_\| = 400\mathbf{i} - 300\mathbf{j}$ N·m.   **5.21** $-50\mathbf{i} - 50\mathbf{j}$ lb at $(4.8, 0, 0)$ ft.   **5.22** $30\mathbf{j}$ kN at $(48, 0, 0)$ m.   **5.23** $-500\mathbf{j}$ lb at $(9, 0, 5.6)$ ft.

P5.25

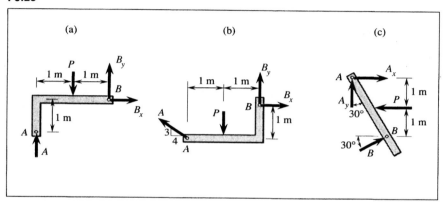

5.28 $\mathbf{A} = 110$ lb ⦣ 53.1°, $\mathbf{B} = 66\mathbf{i} + 32\mathbf{j}$ lb.   **5.29** $\mathbf{A} = 160$ lb ⦣ 53.1°, $\mathbf{B} = 216\mathbf{i} - 128\mathbf{j}$ lb.   **5.31** $\mathbf{A} = 6$ kN ∠ 53.1°, $\mathbf{C} = -3.6\mathbf{i} + 3.2\mathbf{j}$ kN.   **5.32** $\mathbf{A} = 6.4\mathbf{i} + 3.2\mathbf{j}$ kN, $\mathbf{C} = 8$ kN ⦣ 36.9°.   **5.34** (a) $\mathbf{A} = 3$ kN ↑, $\mathbf{B} = 3$ kN ↑; (b) $\mathbf{A} = 3$ kN ⦣ 36.9°, $\mathbf{B} = 2.4\mathbf{i} + 4.2\mathbf{j}$ kN; (c) $\mathbf{A} = 3.75\mathbf{i} - 1.299\mathbf{j}$ kN, $\mathbf{B} = 2.60$ kN ∠ 30°.   **5.35** (a) $\mathbf{A} = 8$ kN ∠ 53.1°, $\mathbf{B} = -4.8\mathbf{i} - 0.4\mathbf{j}$ kN; (b) $\mathbf{A} = 6$ kN ↑, $\mathbf{M}_A = 18$ kN·m ↻; (c) $\mathbf{A} = 1.6\mathbf{i} + 4.8\mathbf{j}$ kN, $\mathbf{B} = 2$ kN ⦣ 36.9°.   **5.36** (a) $\mathbf{A} = 500$ lb ⦣ 53.1°, $\mathbf{B} = 300\mathbf{i} + 200\mathbf{j}$ lb; (b) $\mathbf{A} = 2.6$ kips ∠ 53.1°, $\mathbf{B} = -1.56\mathbf{i} - 1.48\mathbf{j}$ kips; (c) $\mathbf{A} = 200$ lb ∠ 36.9°, $\mathbf{B} = -160\mathbf{i} + 480\mathbf{j}$ lb.   **5.38** $\mathbf{A} = 48\mathbf{i} + 86\mathbf{j}$ lb, $\mathbf{D} = 80$ lb ⦣ 53.1°.   **5.39** $\mathbf{A} = 58.9\mathbf{i} + 1109\mathbf{j}$ N, $\mathbf{D} = 490$ N ∠ 53.1°.   **5.40** 74.8 kN ∠ 87.7°.   **5.41** (a) 20.6 kN; (b) 5.21 kN ∠ 65.8°.   **5.42** 4 Mg.   **5.43** $\mathbf{A} = 900\mathbf{i} - 900\mathbf{j}$ lb.   **5.45** (a) 100.3 N; (b) $T_{AB} = 56.6$ N, $T_{AC} = 88.2$ N, $T_{AD} = 126.7$ N, $T_{BC} = 18.37$ N, $T_{BE} = 63.6$ N, $T_{CG} = 98.1$ N.   **5.46** (a) 32.5°, (b) 63.8 lb.   **5.47** (a) 4.71 m, (b) 72.5 N.   **5.48** Partially constrained.   **5.50** Improperly constrained.   **5.52** W.   **5.54** $T_A = 490$ N, $T_B = 981$ N, $T_C = 490$ N.   **5.56** $\mathbf{B} = 1600\mathbf{i} + 98.1\mathbf{j}$ N, $\mathbf{C} = -1200\mathbf{i} - 49.0\mathbf{j}$ N.   **5.57** $\mathbf{B} = 1500\mathbf{i} + 122.6\mathbf{j}$ N, $\mathbf{C} = -1200\mathbf{i} - 73.6\mathbf{j}$ N.   **5.58** $Wr^2\theta/L$.   **5.60** $T_{BF} = 85.8$ kN, $T_{BG} = 188.8$ kN, $\mathbf{A} = 117.7\mathbf{i} + 353\mathbf{j} + 29.4\mathbf{k}$ kN.   **5.61** $\mathbf{A} = 24.5\mathbf{i} + 98.1\mathbf{j}$ N, $\mathbf{B} = 24.5\mathbf{i}$ N, $\mathbf{C} = -49.0\mathbf{i}$ N.   **5.62** 4 kN.   **5.63** 1120 lb.   **5.64** 108 kN.   **5.65** 84 kN.   **5.68** $\mathbf{A} = 2.4\mathbf{i} + 8.8\mathbf{j}$ kN, $\mathbf{M}_A = 12\mathbf{j} - 27.2\mathbf{k}$ kN·m.   **5.69** $\mathbf{A} = 400\mathbf{j} + 360\mathbf{k}$ lb, $\mathbf{M}_A = -4.16\mathbf{k}$

kip·ft.   **5.70 A** = 120i − 12j − 8k kN.   **5.71 A** = 21i + 36j − 75k kN, M_A = 30j + 72k kN·m.   **5.74** T_{DF} = 1 kN, **A** = 6i + 2j + 6k kN, **B** = −3j − 10k kN.   **5.75** T_{EF} = 1800 lb, **A** = 900i − 450j lb, **B** = −300i − 750j + 1500k lb.   **5.78 A** = 500i + 600k N, M_A = 900j + 1200k N·m, **B** = −500i + 600j N.   **5.80 A** = 200i + 150k lb, **C** = −200i + 300j lb, **D** = 300j − 150k lb.   **5.82 B** = −25j − 30k N, **C** = 125j + 30k N, **D** = −60i − 100j N.   **5.83 A** = −175i + 100j + 100k lb, **B** = −75i + 200j − 100k lb, **D** = 250i lb.   **5.84** (a) (3.5, 6, 0) m; (b) (j + k)/$\sqrt{2}$; (c) **A** = −128.8i + 196.2j − 98.1k N, **B** = 128.8i N, **E** = 98.1j + 98.1k N.   **5.85** (a) 8, (b) 6, (c) yes, (d) C_y = 360j N.   **5.86** (a) 7, (b) 6, (c) yes, (d) B_y = 450j lb.   **5.87** (a) 8; (b) 6; (c) yes; (d) T_{EF} = 1080 N, C_y = 360j N, M_{Cz} = −720k N·m.   **5.89** T_{AD} = 54.6 N, T_{AE} = 38.7 N, T_{BF} = 78.6 N, T_{BG} = 48.0 N, T_{CI} = 34.0 N, T_{CH} = 75.1 N.   **5.90** (a) (i) A_x, A_y, M_A, and C; (ii) ΣF_x = 0, ΣF_y = 0, ΣM_A = 0; (iii) none; (iv) statically indeterminate. (b) (i) A, C, T_{EF}; (ii) ΣF_x = 0, ΣF_y = 0, ΣM_A = 0; (iii) ΣF_x = 0; (iv) improperly constrained. (c) (i) A_x, A_y, A_z, M_{Ay}, M_{Az}, T_{EF}; (ii) all 6 eqs., (iii) none; (iv) statically determinate.   **5.93** (a) **A** = −2i + j kN, M_A = 13 kN·m ↺ ; (b) **A** = 275 lb ⦨ 60°; (c) **A** = 12i + 37j lb.   **5.97** 18 kN.   **5.98 A** = −20i + 18j − 10k kN, M_A = −20j − 90k kN·m, **C** = 12j kN.   **5.99** (a) 7 unknowns (A:4, B:1, D:1, F:1), 6 equations; (b) A_x = 3i kips, A_z = 6k kips, **D** = 1j kip, T_{FG} = 18 kips.

Chapter 6

Developmental Exercises

D6.1 (a) $\frac{1}{4}$ kN/m³, (b) $\frac{1}{4}$ kN/m.   **D6.2** 1 ft.   **D6.3** 2 ft.   **D6.4** (a) (−4.25, 8.65) km; (b) (−1, 5.6) km.   **D6.7** (a) $d\theta$, (b) $\pi/2$ ft, (c) $\frac{1}{2}$ ft², (d) $1/\pi$ ft, (e) $\pi/4$ ft², (f) $\frac{1}{2}$ ft.   **D6.8** (0.652, 2.52) ft.   **D6.9** (2.49, −0.1195, −0.448) m.   **D6.10** T_A = 10 lb, T_B = 15 lb, T_C = 75 lb.   **D6.12** (a) $(4 − \frac{1}{4} y^2) \, dy$, (b) $2y^{1/2} \, dy$.   **D6.13** (a) $4x^{1/2} \, dx$ (b) $(4 − \frac{1}{4} x^2) \, dx$.   **D6.15** (a) $(2\pi R/h^2) (R^2 + h^2)^{1/2} x \, dx$, (b) $\pi R(R^2 + h^2)^{1/2}$, (c) $\frac{2}{3} \pi Rh(R^2 + h^2)^{1/2}$, (d) $\frac{2}{3} h$.   **D6.16** $\bar{x}$ = [4/(3π)] (a + b), $\bar{y}$, = 4b/(3π).   **D6.17** T_A = 610 N, T_B = 809 N.   **D6.18** 6.38 m.   **D6.19** 1 kip ↑.   **D6.20 B** = 2 kN ↑, M_B = 1 kN·m ↺ .   **D6.22** (a) R/2, (b) πR, (c) πRL.   **D6.24** (a) R/3, (b) $\frac{2}{3} \pi R$, (c) $\frac{1}{3} \pi R^2 h$.   **D6.25** (a) $(\pi/h^2) (R^2 x^2 \, dx)$, (b) $\frac{1}{3} \pi R^2 h$, (c) $\frac{1}{4} \pi R^2 h^2$, (d) $\frac{3}{4} h$.   **D6.26** No.   **D6.27** 9.42 kN ←.   **D6.28** ($\frac{11}{8}$ a, 0, 0).   **D6.29** (a) (2, 1, 0.5) m; (b) (1.6, 1.6, 0.4) m.   **D6.30** (0.568, 0, 0) m.

Problems

6.1 (a) c = 16 m, k = −64 m²; (b) c = 3 ft, k = 2.   **6.2** (a) 0.25 kN/m, (b) 0.707 kN/m.   **6.3** $\bar{x}$ = 1 m.   **6.4** $\bar{x}$ = 1.85 m, $\bar{z}$ = 0.6 m.   **6.5** $\bar{x}$ = 1.5 ft, $\bar{y}$ = 2.5 ft.   **6.6** 1/π m.   **6.7** R (sinα)/α.   **6.8** 2.82 ft.   **6.9** $\bar{x}$ = 0.5 m, $\bar{y}$ = 3 m.   **6.10** $\bar{x}$ = 0.569 m, $\bar{y}$ =

1.395 m.   **6.12** $\bar{x} = -0.0981$ m, $\bar{y} = 1.465$ m.   **6.13** $\bar{x} =$ $-0.200a$, $\bar{y} = 0.280a$.   **6.15** $\bar{x} = 2.55$ m, $\bar{y} = 0.8$ m, $\bar{z} = 0.75$ m.   **6.16** $\bar{x} = 0.553$ ft, $\bar{y} = 0.246$ ft, $\bar{z} = 1.510$ ft.   **6.18** 1.075 m.   **6.19** 0.931m.   **6.21 A** $= 71.9$N $\uparrow$, **B** $= 87.0$ N $\uparrow$.   **6.22** $T_A = 22.9$ N, $T_D = 128.4$ N.   **6.24 B**$_x = 35.7$ lb $\leftarrow$, **B**$_y = 186.7$ lb $\uparrow$, **D** $= 59.5$ lb $\measuredangle$ $53.1°$.   **6.25** $T_B = 134.3$ lb, $T_C = 51.3$ lb, $T_D = 80$ lb.   **6.27** $\bar{x} = 3$ m, $\bar{y} = 1.2$ m.   **6.28** $\bar{x} = 1.5$ ft, $\bar{y} = 1.2$ ft.   **6.30** $\bar{x} = 1.75$ ft, $\bar{y} = 1.7$ ft.   **6.31** $\bar{x} = 4a/(3\pi)$, $\bar{y} = 4b/(3\pi)$.   **6.33** $\bar{x} = 0.934$ ft, $\bar{y} = 0.75$ ft.   **6.34** $\bar{x} = 1.098$ m, $\bar{y} = 0.224$ m.   **6.36** $\bar{x} = 36$ mm, $\bar{y} = 14$ mm.   **6.37** $\bar{x} = 23.5$ in., $\bar{y} = 60.8$ in.   **6.40** $\bar{x} = 17.42$ in., $\bar{y} = 12.79$ in.   **6.42** $\bar{x} = 6.49$ in., $\bar{y} = 5.35$ in.   **6.43** $\bar{x} = 100$ mm, $\bar{y} = 65$ mm.   **6.45** $T_A = 14.72$ N, $T_B = 8.83$ N.   **6.46** $T_A = 20.4$ N, $T_B = 21.2$ N.   **6.48** $\bar{x} = 2.8$ ft, $\bar{y} = \bar{z} = 0$.   **6.49** $\bar{x} = 0.456$ m, $\bar{y} = \bar{z} = 0$.   **6.50** $\bar{x} = \pi R/4$, $\bar{y} = \pi h/8$, $\bar{z} = 0$.   **6.52** $\bar{x} = 0.446$ m, $\bar{y} = -0.1377$ m, $\bar{z} = 0.895$ m.   **6.53** $\bar{x} = 0.394$ m, $\bar{y} = -1.873$ m, $\bar{z} = -0.986$ m.   **6.54 A** $= 1.2$ kN $\uparrow$, **B** $= 1.5$ kN $\uparrow$.   **6.55 A** $= 9$ kips $\uparrow$, **M**$_A = 27$ kip·ft $\downarrow$.   **6.56 A** $= 2.08\mathbf{i} + 4.86\mathbf{j}$ kips, **B** $= 3.47$ kips $\measuredangle$ $53.1°$.   **6.57 A** $= 1100$ N $\uparrow$, **B** $= 1450$ N $\uparrow$.   **6.58 B** $= 27.2$ kN $\uparrow$, **C** $= 42.8$ kN $\uparrow$.   **6.60** 143.6 in².   **6.61** $(\pi/2)(8 + \pi)a^2$.   **6.63** 118.4 in³.   **6.64** $(\pi/12)(4 + 3\pi)a^3$.   **6.66** 4.95 kg.   **6.67** 39.0 L.   **6.68** $\bar{x} = \bar{y} = 0$, $\bar{z} = 1.5$ m.   **6.69** $\bar{x} = \bar{y} = 0$, $\bar{z} = 3h/8$.   **6.70** $\bar{x} = 0$, $\bar{y} = 2.5$ ft, $\bar{z} = -1$ ft.   **6.71** $\bar{x} = 0$, $\bar{y} = a(8R - 3a)/[2(3\pi R - 4a)]$, $\bar{z} = -(12\pi R^2 - 32Ra + 3\pi a^2)/[2\pi(3\pi R - 4a)]$.   **6.72** $\bar{x} = \bar{y} = 0$, $\bar{z} = 9$ in.   **6.74** 4.   **6.75** $\bar{x} = \bar{z} = 0$, $\bar{y} = 5.60$ ft.   **6.76** $\bar{x} = 5.93$ in., $\bar{y} = 0.414$ in., $\bar{z} = 0$.   **6.77** 3.07 m.   **6.78** 11.51 ft.   **6.80** 749 lb.   **6.82** 70.9 kN.   **6.83** 9.6 ft.   **6.84** 981 N.   **6.86** (a) $\pi R^3 \rho g$, (b) $3R/4$.   **6.87** $T_A = 988$ N, $T_B = 2.45$ kN.   **6.89** $\bar{x} = 4a/5$, $\bar{y} = 3b/4$, $\bar{z} = 2c/3$.   **6.90** $\bar{x} = 2.4$ ft, $\bar{y} = 1.2$ ft, $\bar{z} = 0.8$ ft.   **6.91** $\bar{x} = 20$ ft.   **6.92 D** $= 288$ lb $\measuredangle$ $67.4°$, **A** $= -110.7\mathbf{i} - 8.30\mathbf{j}$ lb.   **6.93** (a) 140 mm, (b) 138.3 mm.   **6.95 A** $= 9.68$ kN $\uparrow$, **B** $= 7.32$ kN $\uparrow$.   **6.97** $\bar{x} = \bar{y} = 0$, $\bar{z} = 6$ in.   **6.99** $\bar{x} = h/4$, $\bar{y} = R (\sin\alpha)/(2\alpha)$, $\bar{z} = 0$.   **6.100** $T_A = 0$, $T_B = 9.07$ kN.   **6.102** $\bar{x} = 2.79$ in., $\bar{y} = \bar{z} = 0$.

Chapter 7

Developmental Exercises

D7.4 $(bh/12)(b^2 + h^2)$.   **D7.5** $\frac{1}{4}\pi R^4$.   **D7.6** $I_x = \frac{1}{12} bh^3$, $I_y = \frac{1}{12} b^3 h$.   **D7.7** No.   **D7.8** No.   **D7.9** through **D7.11** No. The integrals do not conform to the definition in Sec. 7.2.   **D7.13** $\bar{k}_x = (\sqrt{3}/6)h$, $\bar{k}_y = (\sqrt{3}/6) b$.   **D7.14** $(\sqrt{2}/2) R$.   **D7.17** (a) $bh^3/3$, (b) $7b^3h/3$, (c) $(bh/3)(h^2 + 7b^2)$.   **D7.18** 50 ft⁴.   **D7.20** (a) 0, (b) 0, (c) 0, (d) 0.   **D7.21** $\frac{1}{24} b^2h^2$.   **D7.23** (a) $\frac{1}{3} x^2\, dx$, (b) $\frac{1}{18} x^5\, dx$, (c) 6.75 ft⁴.   **D7.24** (a) 4.32×10^3 mm⁴, (b) 5.48 mm.   **D7.25** (a) 27.6×10^3 mm⁴, (b) 13.86 mm.   **D7.26** (a) 779 mm⁴, (b) 4.27 mm.   **D7.27** $I_x = 114.6$ in⁴, $I_y = 427$ in⁴, $I_{xy} = -173.2$ in⁴.   **D7.30** Yes.   **D7.31** 300 mm⁴.   **D7.32** X: (I_x, I_{xy}); Y: $(I_y, -I_{xy})$.   **D7.34** $(I_{avg}, 0)$.

D7.35

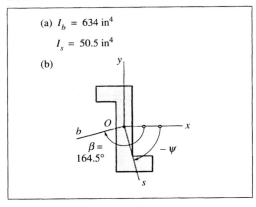

(a) $I_b = 634$ in^4

$I_s = 50.5$ in^4

(b)

$\beta = 164.5°$

$-\psi$

D7.39 $\frac{1}{12} mc^2$. **D7.40** $\frac{1}{2} mr^2$. **D7.41** No. **D7.42** 1088 slug·ft^2. **D7.44** 0.548r. **D7.45** 8 kg·m^2. **D7.46** No. **D7.49** $(m/20)(3r^2 + 2h)$. **D7.50** $I_x = 1.89$ kg·m^2, $k_x = 410$ mm. **D7.51** $I_z = 8.2$ kg·m^2, $k_z = 854$ mm. **D7.52** $I_x = 0.310$ slug·ft^2, $k_x = 7.89$ in. **D7.53** $I_z = 0.487$ slug·ft^2, $k_z = 9.89$ in. **D7.54** $I_x = 9.31 \times 10^{-3}$ kg·m^2, $k_x = 56.6$ mm. **D7.55** $I_y = 4.09 \times 10^{-3}$ kg·m^2, $k_y = 37.5$ mm.

Problems

7.1 $\frac{1}{12} bh^3$. **7.2** 3.15×10^3 mm^4. **7.3** 6.75×10^3 mm^4. **7.4** $\frac{3}{35} a^4$. **7.5** 0.305 ft^4. **7.6** $(\pi/16) ab^3$. **7.8** 1470 mm^4. **7.9** 686 mm^4. **7.10** $\frac{3}{35} a^4$. **7.11** 1.6 ft^4. **7.13** $\frac{1}{12} bh(3b^2 + h^2)$. **7.14** 4.62×10^3 mm^4. **7.16** $\frac{6}{35} a^4$. **7.17** 1.905 ft^4. **7.25** $I_x = 32.4$ m^4, $k_x = 1.342$ m. **7.26** $I_x = 43.2$ ft^4, $k_x = 1.897$ ft. **7.28** $I_y = 12.15$ m^4, $k_y = 1.643$ m. **7.29** $I_y = 10.8$ m^4, $k_y = 1.342$ m. **7.31** $I_x = \frac{1}{8}(2\alpha - \sin 2\alpha)R^4$, $k_x = [(2\alpha - \sin 2\alpha)/(8\alpha)]^{1/2}R$. **7.32** $I_x = 6.96 \times 10^{-3}a^4$, $k_x = 0.1332a$. **7.34** $I_y = \frac{1}{8}(2\alpha + \sin 2\alpha)R^4$, $k_y = [(2\alpha + \sin 2\alpha)/(8\alpha)]^{1/2}R$. **7.35** $I_y = 0.1403a^4$, $k_y = 0.598a$. **7.37** $J_O = \frac{1}{2} \alpha R^4$, $k_O = (\sqrt{2}/2) R$. **7.38** $J_O = 0.1473a^4$, $k_O = 0.612a$. **7.40** 19.50 m^4. **7.41** 0.914 ft^4. **7.43** 0.305 m^4. **7.46** 5.33 m^4. **7.47** 0.583 ft^4. **7.49** -0.533 m^4. **7.50** -1.333 m^4. **7.52** $\bar{I}_x = 1.102$ m^4, $\bar{k}_x = 0.714$ m. **7.53** $\bar{I}_x = 355$ in^4, $\bar{k}_x = 3.33$ in. **7.55** $\bar{I}_x = 13.6 \times 10^9$ mm^4, $\bar{k}_x = 238$ mm. **7.56** $\bar{I}_x = 35.1 \times 10^6$ mm^4, $\bar{k}_x = 62.4$ mm. **7.58** $\bar{I}_y = 0.324$ m^4, $\bar{k}_y = 0.387$ m. **7.59** $\bar{I}_y = 50.7$ in^4, $\bar{k}_y = 1.258$ in. **7.61** $\bar{I}_y = 6.4 \times 10^9$ mm^4, $\bar{k}_y = 163.3$ mm. **7.62** $\bar{I}_y = 8.78 \times 10^6$ mm^4, $\bar{k}_y = 31.2$ mm. **7.64** -4.8×10^9 mm^4. **7.65** -12.15×10^6 mm^4. **7.67** 83.5×10^3 in^4. **7.69** 1619 in^4. **7.71** 1.541×10^9 mm^4. **7.74** 1.019×10^9 mm^4. **7.75** 1670 in^4. **7.76** 237×10^3 in^4. **7.79** $\bar{I}_x = 127.5$ in^4, $\bar{I}_y = 27.2$ in^4, $\bar{J}_C = 154.7$ in^4. **7.80** $\bar{I}_x = 92.8$ in^4, $\bar{I}_y = 497$ in^4, $\bar{J}_C = 590$ in^4. **7.81** $\bar{I}_x = 267$ in^4, $\bar{I}_y = 305$ in^4, $\bar{J}_C = 572$ in^4. **7.82** 9.25 in. **7.84** $\bar{I}_u = 279$ in^4, $\bar{I}_v = 126.7$ in^4, $\bar{I}_{uv} = -131.6$ in^4. **7.85** $\bar{I}_u = 4.56 \times 10^6$ mm^4, $\bar{I}_v = 2.09 \times 10^6$ mm^4, $\bar{I}_{uv} = -1.036 \times 10^6$ mm^4. **7.86** $\bar{I}_u = 12.36 \times 10^9$ mm^4, $\bar{I}_v = 7.64 \times 10^9$ mm^4, $\bar{I}_{uv} = 5.52 \times 10^9$ mm^4. **7.88** $\bar{I}_u = 270$ in^4, $\bar{I}_v = 178.4$ in^4, $\bar{I}_{uv} = -90.3$

in^4. **7.98** $\bar{I}_u = 7.69 \times 10^6$ mm^4, $\bar{I}_v = 36.2 \times 10^6$ mm^4, $\bar{I}_{uv} = -10.85 \times 10^6$ mm^4. **7.99** $\bar{I}_u = 9.31 \times 10^6$ mm^4, $\bar{I}_v = 8.08 \times 10^6$ mm^4, $\bar{I}_{uv} = 4.82 \times 10^6$ mm^4.

P7.101

$\bar{I}_b = 16 \times 10^9$ mm^4

$\bar{I}_s = 4 \times 10^9$ mm^4

$\beta = 153.4°$

$\psi = -63.4°$

P7.103

$\bar{I}_b = 325$ in^4

$\bar{I}_s = 122.8$ in^4

$\beta = 125.8°$

$\psi = -35.8°$

P7.104

$\bar{I}_b = 39.9 \times 10^6$ mm^4

$\bar{I}_s = 4.02 \times 10^6$ mm^4

$\beta = 158.6°$

$\psi = -68.6°$

P7.106

$\bar{I}_b = 70.2 \times 10^9$ mm^4

$\bar{I}_s = 20.7 \times 10^9$ mm^4

$\beta = 79.98°$

$\psi = 10.02°$

P7.107

$I_{xy} = 4 \times 10^6$ mm^4

$I_{min} = 6 \times 10^6$ mm^4

$\beta = 26.6°$

$\psi = 63.4°$

P7.159

$\bar{I}_b = 1306$ in^4

$\bar{I}_s = 254$ in^4

$\beta = 117.6°$

$\psi = -27.6°$

7.110 $I_x = \frac{1}{3} mb^2$, $I_y = \frac{1}{3} ma^2$, $I_z = \frac{1}{3} m(a^2 + b^2)$. **7.117** $I_x = 21.5 \times 10^{-3}$ slug·ft^2, $k_x = 2.31$ in. **7.118** $I_x = 326$ kg·mm^2, $k_x = 18.97$ mm. **7.120** $\frac{1}{12} m(4b^2 + c^2)$. **7.122** $I_y = \frac{1}{12} m(4a^2 + c^2)$, $k_y = (\sqrt{3}/6)(4a^2 + c^2)^{1/2}$. **7.123** $I_y = \frac{1}{6} m(a^2 + c^2)$, $k_y = (\sqrt{6}/6)(a^2 + c^2)^{1/2}$. **7.124** $I_x = 19.44$ kg·m^2, $k_x = 367$ mm. **7.126** $I_x = 0.944$ kg·m^2, $I_y = 0.512$ kg·m^2. **7.128** $I_x = 60.9 \times 10^{-3}$ kg·m^2, $I_y = 143.2 \times 10^{-3}$ kg·m^2. **7.130** $I_z = 2.4$ slug·ft^2, $k_z = 1.414$ ft. **7.131** $I_z = 103.6 \times 10^{-3}$ kg·m^2, $k_z = 158.7$ mm. **7.132** $I_x = 55 \times 10^{-3}$ slug·ft^2, $I_y = 185 \times 10^{-3}$ slug·ft^2. **7.134** $I_x = 1.259$ slug·ft^2, $I_y = 2.32$ slug·ft^2. **7.136** $I_z = 748$ kg·mm^2, $k_z = 69.3$ mm. **7.137** $I_z = 1.179$ slug·ft^2, $k_z = 1.145$ ft. **7.138** $I_y = 110.8$ kg·mm^2, $k_y = 35.4$ mm. **7.140** $I_y = 419$ kg·mm^2, $k_y = 47.1$ mm. **7.142** $I_x = 0.933$ slug·ft^2, $k_x = 11.04$ in. **7.143** $I_x = 681$ kg·mm^2, $k_x = 60.1$ mm. **7.144** $I_y = 1.570 \times 10^{-3}$ kg·m^2, $k_y = 25.2$ mm. **7.146** $I_y = 7.46 \times 10^{-3}$ kg·m^2, $k_y = 34.5$ mm. **7.148** $I_x = 39.5$ slug·ft^2, $k_x = 10.36$ in. **7.149** $I_x = 20.2 \times 10^{-3}$ kg·m^2, $k_x = 56.7$ mm. **7.150** $I_z = 21.3 \times 10^{-3}$ slug·ft^2, $k_z = 2.46$ in. **7.151** $I_x = 4.25$ kg·m^2, $k_x = 204$ mm. **7.152** $I_y = 4.66 \times 10^{-3}$ kg·m^2, $I_z = 2.63 \times 10^{-3}$ kg·m^2. **7.156** $\bar{I}_x = 136$ in^4, $\bar{I}_y = 40$ in^4, $\bar{k}_x = 2.38$ in.,

$\bar{k}_y = 1.291$ in.   **7.157** $\bar{I}_x = 480$ in^4, $\bar{I}_y = 1080$ in^4, $\bar{I}_{xy} = -432$ in^4, $\bar{I}_u$ = 1004 in^4, $\bar{I}_v = 556$ in^4, $\bar{I}_{uv} = -476$ in^4.   **7.159** $\bar{I}_b = 1306$ in^4, $\bar{I}_s$ = 254 in^4, $\beta = 117.6°$, $\psi = -27.6°$.   **7.161** $I_x = 1.660 \times 10^9$ mm^4, $I_y = 2.31 \times 10^9$ mm^4, $I_{xy} = 1.720 \times 10^9$ mm^4.   **7.162** 2.60 N.   **7.166** $I_x = 26.6$ slug·ft^2, $k_x = 3.79$ ft.   **7.168** $I_x = 1.104 \times 10^{-3}$ kg·m^2, $k_x = 48.3$ mm.   **7.169** $I_y = 2.18 \times 10^{-3}$ kg·m^2, $k_y = 67.9$ mm.   **7.171** $I_y = 46.1 \times 10^{-3}$ kg·m^2, $k_y = 86.1$ mm.

Chapter 8

Developmental Exercises

D8.6 9 joints.   **D8.7** 6 members.   **D8.8** No.   **D8.10** (a) *ACF* and *BDE*; (b) *AB*, *CD*, and *EF*; (c) compound truss.   **D8.12** F_{CD}.   **D8.13** Yes.   **D8.15** *AB*, *BC*, *EF*, *EG*.   **D8.16** $F_{AB} = 6$ kips *C*, $F_{AD} = 4.8$ kips *T*, $F_{BC} = 8$ kips *C*, $F_{BD} = 10$ kips *T*, $F_{CD} = 4.8$ kips *T*.   **D8.17** $F_{CF} = -\frac{10}{13} F_{EF}$, $F_{CD} = \frac{25}{39} F_{EF}$.   **D8.18** 10 kips *T*.   **D8.19** $F_{EK} = 46.9$ kips *T*, $F_{DE} = 30$ kips *T*, $F_{HO} = 30$ kips *C*.   **D8.20** 15 kips *C*.
D8.21 c.   **D8.22** 17 kN *T*.   **D8.23** 3.9 kN *C*.   **D8.24** 5.2 kN *T*.   **D8.25** 28 kN *T*.   **D8.26** 17 kN *C*.   **D8.28** 24 members.
D8.29 (a) $\overline{AM} = \frac{1}{2}$ m, $\overline{OA} = (\sqrt{3}/3)$ m, $\overline{OD} = (\sqrt{6}/3)$ m; (b) $\frac{1}{3}$ kN *C*; (c) $F_{AB} = F_{AC} = F_{BC} = 136.1$ N *T*.   **D8.32** (a) A solid circular dot, (b) a small circular curve.

D8.33

D8.34

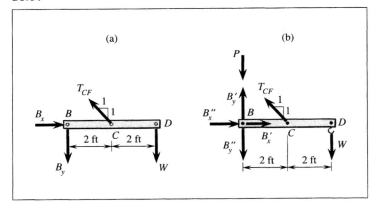

D8.36 A $= -360\mathbf{i}$ lb, **B** $= 480\mathbf{j}$ lb, **C'** $= 360\mathbf{i} - 480\mathbf{j}$ lb.　　**D8.37 A** $= -960\mathbf{i} - 1220\mathbf{j}$ lb, **B'** $= 960\mathbf{i} + 1340\mathbf{j}$ lb.　　**D8.38** 11.70 kips.
D8.39 (a) $(\sqrt{3}\mathbf{i} - \mathbf{j})/2$, (b) $16.35\mathbf{j}$ N, (c) $-7.08\mathbf{i} + 4.09\mathbf{j}$ N, (d) $7.08\mathbf{i} + 12.26\mathbf{j}$ N, (e) $13.83\mathbf{k}$ N·m, (f) **0**, (g) $13.83\mathbf{k}$ N·m.　　**D8.40** Horizontal.　　**D8.46** True.　　**D8.47** Yes.　　**D8.49** (a) -12 kN/m, (b) -18 kN·m/m, (c) 16 kN·m/m.　　**D9.50** Yes.　　**D8.51** (a) ↑, (b) ↓.　　**D8.54** 1082 lb.　　**D8.55** $T_{BC} = 1082$ lb, $T_{CD} = 949$ lb.
D8.56 d.　　**D8.57** No.　　**D8.58** Horizontal.　　**D8.60** 5.64 kN.
D8.61 (a) 30.7 m, (b) 51.9 m.　　**D8.62** (a) 126.6 m, (b) 1340 N.
D8.63 109.6 m.

D8.52

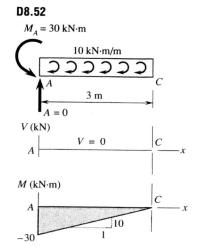

Problems

8.1 $F_{AB} = 12.75$ kN T, $F_{AC} = 15.70$ kN T, $F_{BC} = 19.62$ kN C.　　**8.2** $F_{AB} = 3.2$ kN C, $F_{AC} = 0$, $F_{BC} = 12.6$ kN C.　　**8.4** $F_{AB} = 260$ lb T, $F_{AC} = 170$ lb T, $F_{AD} = 0$, $F_{BC} = 240$ lb C, $F_{CD} = 320$ lb C.　　**8.5** $F_{AB} = 5.6$ kN T, $F_{AH} = 6$ kN T, $F_{BC} = 0$, $F_{BD} = 3.4$ kN T, $F_{BH} = 5$ kN C, $F_{CD} = 0$, $F_{DE} = 6$ kN T, $F_{DF} = 3.4$ kN C, $F_{DH} = 0$, $F_{EF} = 0$, $F_{FG} = 5.6$ kN C, $F_{FH} = 5$ kN T, $F_{GH} = 0$.　　**8.7** $F_{AB} = 75$ lb T, $F_{AF} = 45$ lb C, $F_{BC} = 90$ lb T, $F_{BE} = 75$ lb C, $F_{BF} = 0$, $F_{CD} = 90$ lb T, $F_{CE} = 0$, $F_{DE} = 150$ lb C, $F_{EF} = 45$ lb C.　　**8.8** $F_{AB} = 15$ kN C, $F_{AH} = 12$ kN T, $F_{BC} = 5$ kN C, $F_{BG} = 10$ kN C, $F_{BH} = 12$ kN T, $F_{CD} = 5$ kN C, $F_{CG} = 6$ kN T, $F_{DE} = 5$ kN T, $F_{DF} = 0$, $F_{DG} = 0$, $F_{EF} = 4$ kN T, $F_{FG} = 4$ kN T, $F_{GH} = 12$ kN T.　　**8.10** $F_{AB} = 157.0$ kN T, $F_{BC} = 78.5$ kN T, $F_{BE} = 58.9$ kN T, $F_{BF} = 98.1$ kN C, $F_{CD} = 58.9$ kN T, $F_{CE} = 98.1$ kN C, $F_{DE} = 0$, $F_{EF} = 78.5$ kN C.　　**8.11** $F_{AB} = 52$ kN C, $F_{AF} = 16$ kN C, $F_{BC} = 39$ kN C, $F_{BE} = 15$ kN C, $F_{BF} = 15.62$ kN T, $F_{CD} = 30$ kN C, $F_{CE} = 39$ kN T, $F_{DE} = 0$, $F_{EF} = 36$ kN T.　　**8.14** Simple truss.　　**8.15** Compound truss (LM's: *EI, EM, HI*).　　**8.18** *AN, DL, DM, EL, FK, FL, GJ, GL, IJ, JK, KL, LM, MN*.　　**8.21** $F_{AB} = 8$ kN C, $F_{AD} = 12$ kN C, $F_{AE} = 8$ kN C, $F_{BC} = 6$ kN C, $F_{BD} = 10$ kN T, $F_{CD} = 10$ kN T, $F_{CH} = 8$ kN T, $F_{DE} = 20$ kN T, $F_{EF} = 24$ kN T, $F_{EI} = 20$ kN C, $F_{FG} = 8$ kN T, $F_{FI} = 12$ kN T, $F_{GH} = 6$ kN T, $F_{GI} = 10$ kN C, $F_{HI} = 10$ kN C.　　**8.22** $F_{AB} = 3$ kN T, $F_{AD} = 12.5$ kN T, $F_{BC} = 2.5$ kN T, $F_{BE} = 2.5$ kN C, $F_{CD} = 2.5$ kN C, $F_{DE} = 9$ kN C.
8.24 $F_{AB} = 425$ lb T, $F_{AC} = 225$ lb C, $F_{AD} = 380$ lb T, $F_{BC} = 375$ lb C, $F_{BF} = 200$ lb C, $F_{CD} = 300$ lb C, $F_{DE} = 20$ lb T, $F_{DG} = 300$ lb T, $F_{EF} = 425$ lb C, $F_{EG} = 225$ lb T, $F_{FG} = 375$ lb T.　　**8.26** $F_{BK} = 40$ kN T, $F_{CJ} = 20$ kN C.　　**8.27** $F_{EI} = 6$ kN C, $F_{DJ} = 72$ kN T.
8.29 $F_{AF} = 2.55$ kips T, $F_{BE} = 1$ kip T.　　**8.30** $F_{BL} = 1.5$ kN C, $F_{CL} = 22.4$ kN T.　　**8.32** $F_{CO} = 27.5$ kips T, $F_{EP} = 22.5$ kips T.　　**8.34** $F_{AB} = 39$ kN C, $F_{FG} = 42$ kN C.　　**8.35** $F_{GI} = 0$, $F_{DL} = 5$ kN T.
8.37 $F_{FG} = 2.5$ kN C, $F_{CJ} = 12.5$ kN C.　　**8.38** $F_{LM} = 1250$ lb T, $F_{EN} = 966$ lb T, $F_{GK} = 0$.　　**8.39** $F_{BC} = 150$ kN T, $F_{BI} = 120$ kN C, $F_{DH} = 30$ kN C.　　**8.41** $F_{AB} = P/3$ C, $F_{CD} = P$ C, $F_{EF} = 2P/3$ C.
8.42 $F_{AB} = 0.471P$ C, $F_{CD} = 0.770P$ C, $F_{EF} = 0.345P$ T.　　**8.44** $F_{BH} = 89.3$ kN T, $F_{AI} = 233$ kN C.　　**8.46** $F_{IJ} = 50$ kN C, $F_{DG} = 25$ kN T.　　**8.48** *AF, BF, EF, FO, FG*.　　**8.49** $F_{BA} = 5$ kN C, $F_{BC} = 43.2$ kN C, $F_{BD} = 33.8$ kN T, $F_{BE} = 13.93$ kN T.

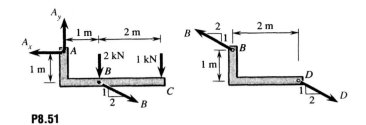

P8.51

8.55 A $= -10\mathbf{i} + 8\mathbf{j}$ kN, **B** $= 10\mathbf{i} - 7\mathbf{j}$ kN.   **8.57 A** $= 12\mathbf{i} - 6\mathbf{j}$ kN, **B** $= -21\mathbf{i} + 8\mathbf{j}$ kN, **C** $= 9\mathbf{i} - 2\mathbf{j}$ kN.   **8.58 A** $= 400\mathbf{i} + 1000\mathbf{j}$ lb, **B** $= -2.4\mathbf{i} - \mathbf{j}$ kips, **C** $= 2\mathbf{i}$ kips.   **8.59 A** $= -900\mathbf{i} - 435\mathbf{j}$ N, **F** $= 900\mathbf{i} + 1155\mathbf{j}$ N.   **8.60 A** $= -200\mathbf{i} - 1200\mathbf{j}$ lb, **B** $= 400\mathbf{i} + 600\mathbf{j}$ lb, **C** $= -200\mathbf{i} + 600\mathbf{j}$ lb.   **8.61 A** $= -16\mathbf{i} + 3.5\mathbf{j}$ kN, $M_A = 10.5$ kN·m ↻, **G** $= 8\mathbf{j}$ kN.   **8.64 A** $= 57.7$ lb →, **B** $= 200$ lb ↑, **C** $= 200$ lb ↑.   **8.65** 300 N.   **8.67 C** $= 171$ lb ←, **D** $= 75\mathbf{i} - 40\mathbf{j}$ lb.   **8.68** (a) 500 N·m; (b) **B** $= 375\mathbf{j}$ N, **C** $= -375\mathbf{j}$ N, **E** $= \mathbf{0}$.
8.70 8.51 lb.   **8.71** 173.2 N·m   **8.72** 720 lb.   **8.74** 260 N.
8.75 5.5 kN.   **8.77** 26P.   **8.78** 506 N.   **8.79** $F_{DE} = 10.61$ kN C, $F_{CF} = 17.30$ kN C.   **8.80** (a) $F_{BJ} = 493$ N C, $F_{CE} = 4$ kN C, $F_{FG} = 4.68$ kN C; (b) **A** $= -1132\mathbf{i} - 40.5\mathbf{j}$ N, **D** $= 4.28\mathbf{i} + 1.95\mathbf{j}$ kN.
8.81 (a) $-14\mathbf{k}$ kN, (b) $8\mathbf{i} - 8\mathbf{j}$ kN, (c) $-24\mathbf{k}$ kN·m, (d) $56\mathbf{i} + 98\mathbf{j}$ kN·m.

8.99 $V = \dfrac{w_o L}{12} - \dfrac{w_o}{L^2}\left(\dfrac{x^2 L}{2} - \dfrac{x^3}{3}\right)$, $M = \dfrac{w_o L x}{12} - \dfrac{w_o}{L^2}\left(\dfrac{x^3 L}{6} - \dfrac{x^4}{12}\right)$.

8.102 $h_B = 15$ ft, $h_C = 20$ ft, $h_D = 12$ ft.   **8.103** (a) 108 N, (b) 1020 N.   **8.104** 5.3 m.   **8.106** 1345 m.   **8.108** $T_{min} = 2.5$ kips, $T_{max} = 2.69$ kips.   **8.109** $T_{min} = 9.91$ kN, $T_{max} = 11.29$ kN.   **8.110** 102.6 ft.   **8.111** 104.0 m.   **8.113** 4.73 m.   **8.115** 22.6 ft.
8.116 (a) 11.35°, (b) 16.88°, (c) 2.53 kips.   **8.118** 22.5 ft.   **8.119** (a) 11.31°, (b) 16.70°, (c) 2.53 kips.   **8.123** $F_{GH} = 62.5$ kN C, $F_{EI} =$

18.23 kN T, F_{BK} = 42.4 kN T, F_{KJ} = 73.7 kN C, F_{CL} = 36.1 kN T.
8.124 B = $-7.5\mathbf{i} - 0.4\mathbf{j}$ kips, E = $7.5\mathbf{i} + 1.2\mathbf{j}$ kips.   **8.125** 2.51
kN.   **8.127** (a) 8.00 kips, (b) 54.8 ft.   **8.128** 2.37 m.

Chapter 9

Developmental Exercises

D9.3

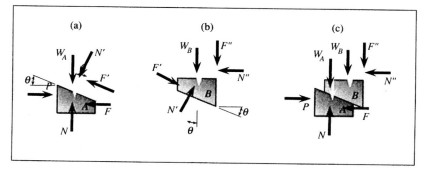

D9.5 No.   **D9.6** (a) 0.4, (b) 0.35, (c) 25 lb →, (d) 28 lb →.   **D9.9**
(a) 20°, (b) 0.364.   **D9.10** (a) 30°, (b) 0.577.   **D9.12** (a) 110 lb, (b)
24 lb, (c) 24.2 lb, (d) not move.   **D9.13** (a) Yes (sliding), (b) no, (c)
yes (tipping).   **D9.14** 346 lb.   **D9.15** (a) 14.60°, (b) 942 N.
D9.16 290 lb.   **D9.17** 21.4 N.   **D9.18** (a) 25.2 lb, (b) through the
corner at D, (c) 25 lb, (d) 25 lb.   **D9.19** 0.268.   **D9.22** P = 5.62
kN, M' = 5.04 N·m.   **D9.23** 51 lb ↑.   **D9.24** (a) 1.875 lb·ft, (b)
0.075 in.   **D9.27** (a) 2π rad, (b) 0.366.   **D9.28** $2\pi/3$ rad, (b) P_{min}
= 58.1 N, P_{max} = 165.6 N.   **D9.30** (a) 19.92, (b) 12.99.

Problems

9.1 $\frac{1}{2}$.   **9.2** $4/(3\pi)$.   **9.4** 30°.   **9.6** 26.1°.   **9.7** Yes.   **9.8**
Yes.   **9.10** No.   **9.11** Yes.   **9.13** $\frac{1}{2}$.   **9.15** 0.686.   **9.16**
0.646.   **9.18** 0.611.   **9.20** (a) $\frac{1}{3}$, (b) 20 lb.   **9.21** 17.27 N.
9.23 (a) Crates will move, (b) F_A = 193.6 N, F_B = 92.2 N.   **9.25**
73.3°.   **9.26** 132 lb.   **9.27** (a) Yes, (b) no, (c) no.   **9.28** 28.4
lb.   **9.29** 32.9 N $\leq$ P $\leq$ 1.265 kN.   **9.31** 26.6 mm.   **9.32**
36.9°.   **9.34** 0.48.   **9.35** 0.204.   **9.37** 0.478.   **9.39** No.
9.41 5.66 kg $\leq m_A$ = 9.43 kg.   **9.42** 65.4° $\leq \theta \leq$ 69.7°.   **9.43** 9.5
kg $\leq m_B \leq$ 188 kg.   **9.45** 10 lb.   **9.46** 10.8 lb.   **9.48** 10.44
lb.   **9.49** 10.8 lb.   **9.51** 56.3°.   **9.52** 17.40 lb.   **9.53** 59.5
N.   **9.54** 172.4 N.   **9.56** $\frac{1}{3}$.   **9.57** 36.9°.   **9.60** 126.1 N.
9.61 157.6 N.   **9.62** 53.6 lb.   **9.64** 8.81 lb·ft.   **9.65** 28.1.
9.67 9.03 N·m.   **9.69** (a) 20.4 N·m, (b) 11.74 N·m.   **9.70** (a)
114.3 lb, (b) 109.9 lb.   **9.71** 536 N.   **9.73** 0.215.   **9.75** (a)
$3P/(2\pi R^2)$, (b) $\mu_s PR$.   **9.76** 15.18 N.   **9.77** 114.8 N.   **9.79** 4 full
turns.   **9.80** 3.62 N.   **9.82** 0.350.   **9.83** 0.1166.   **9.85**
0.348.   **9.87** 0.348.   **9.88** 336 N.   **9.89** 1.033 kN.   **9.91** 68.5
lb.   **9.93** 306 N.   **9.94** 139.7 N.   **9.95** 86.0 lb·ft ↩.

9.97 285 lb.　**9.98** 31.9 N.　**9.100** 26.3 kg.　**9.102** 0.221.
9.104 58.9 N · m $\curvearrowleft$.　**9.106** 1.382 kip · ft.　**9.108** 13 lb.　**9.109**
74.3 N (rotation of A and tipping of B).　**9.110** (a) 496 N, (b) 0.497
N · m.　**9.113** 216 mm.　**9.115** 3.42 kg $\leq m_B \leq$ 181.5 kg.
9.116 $\theta_A = 49.1°$, $\mu_s = 0.235$.　**9.117** 5.53.

Chapter 10

Developmental Exercises

D10.2 (a) 3 ft →, (b) 5 ft.　**D10.3** (a) $\pi/4$ rad $\curvearrowright$, (b) $5\pi/4$ rad.
D10.5 (a) − 200 J, (b) 200 J, (c) − 100 J.　**D10.6** (a) Equal to, (b) less
than.　**D10.7** (a) 98.1 J, (b) through (e) 0.　**D10.8** No.　**D10.10**
($\frac{4}{5} F - A_x$) dx.　**D10.11** ($\frac{3}{5} F - A_y$) dy.　**D10.13** (a) $L(\delta\theta)^2$, (b) yes,
(c) $L\ \delta\theta$ ↑.　**D10.14** $\overrightarrow{\delta y_C} = 4\ \delta\theta$ ft ↓, $\overrightarrow{\delta x_D} = 10\ \delta\theta$ ft →, $\overrightarrow{\delta y_E} =$
$8\ \delta\theta$ ft ↓, $\overrightarrow{\delta x_A} = 0$, $\overrightarrow{\delta y_A} = 4\ \delta\theta$ ft ↓.　**D10.15** $\overrightarrow{\delta x_E} = 6\ \delta\theta$ ft →,
$\overrightarrow{\delta y_E} = 4\ \delta\theta$ ft ↓.　**D10.16** (1) $\sin\theta = 1.7\ \sin\phi$; (2) (a) $0.4\ \delta\theta \curvearrowright$,
(b) $\overrightarrow{\delta x_E} = 1.12\ \delta\theta$ m ←, $\overrightarrow{\delta y_C} = 2.1\ \delta\theta$ m ↑.　**D10.19** (a) **P** =
$(P/5)(-3\mathbf{i} + 4\mathbf{j})$, **Q** = $(Q/17)(15\mathbf{i} + 8\mathbf{j})$; (b) ($\frac{84}{85} Q - 168)\delta r_1$; (c) 170 N;
(d) 250 N.　**D10.20** (a) $\overrightarrow{\delta x_C} = 3\ \delta\theta$ m →, $\overrightarrow{\delta y_C} = 2\ \delta\theta$ m ↑, $\delta\phi =$
$\delta\theta \curvearrowright$; (b) 4 kN·m $\curvearrowright$.　**D10.21** 400 lb →.　**D10.22** D_x = 1050 lb
→, D_y = 1100 lb ↑.　**D10.23** **G** = 7 kN ↑, **K** = 1 kN ↑.
D10.24 9 kN ←.　**D10.25** 200 N←.　**D10.26** 60 lb.　**D10.33**
No.　**D10.35** (a) $-(WL/2)(4 \cos\theta + \sin\theta)$, (b) 14.04°.　**D10.37**
Stable.　**D10.38** Unstable.　**D10.39** Unstable.

Problems

10.1 (a) 240 ft · lb, (b) 240 ft · lb.　**10.2** (a) 288 ft · lb, (b) 100.8
ft · lb.　**10.3** 90 ft · lb.　**10.5** 95.1 J.　**10.6** 185.7 J.　**10.7**
2.5 J.　**10.9** 28.0 kg.　**10.10** 1.453 slugs.　**10.12** (A_y − 300) dy
ft · lb.　**10.15** 300 N ↑.　**10.16** 400 lb ↑.　**10.18** 4.2 ft.
10.19 510 N.　**10.21** 3 kN.　**10.22** 1.2 kips.　**10.24** 0.477 m and
1.566 m.　**10.25** 33.4° and 81.2°.　**10.27 A** = 2 kN ↑, **H** = 13
kN ↑.　**10.29** A_x = 900 lb ←, A_y = 500 lb ↑, M_A = 2000 lb · ft $\curvearrowright$;
G = 650 lb ↑.　**10.31** A_x = 9.5 kN ←, A_y = 47 kN ↑, B_x = 9.5
kN →, B_y = 27 kN ↓.　**10.32** 119.6 kN.　**10.33** 111.4 lb.
10.35 1.6 kips.　**10.37** 34.2 kN.　**10.38** 20 lb.　**10.45** Neutral.
10.46 Unstable.　**10.48** Unstable.　**10.50** Unstable.　**10.54** $a/b <$
$4/(3\pi - 4)$.　**10.56** $a/b < \sqrt{2}/2$.　**10.57** $a/b < 1$.　**10.58** $h <$
1.019 m.　**10.60** $k > mg/h$.　**10.61** $\theta = \tan^{-1} mg/(2kL)$, $dV/d\theta =$
0, $d^2V/d\theta^2 = L[(kL)^2 + (\frac{1}{2} mg)^2]^{1/2} > 0$.　**10.62** θ = 0, unstable; θ =
75.5°, stable; θ = 180°, unstable.　**10.64** θ = 0, unstable; θ = 180°,
stable.　**10.65** θ = 0, unstable; θ = 77.8°, stable.　**10.66** 8 kg.
10.68 $m > 8$ kg.　**10.69** θ = 30°, stable; θ = 90°, unstable.　**10.71**
Unstable.　**10.72** 90 lb, not dependent on a and b.　**10.73 A** = −6**i**
+ 4.5**j** kN, M_A = 13.5 kN · m $\curvearrowright$.　**10.74 A** = 250**i** + 120**j** lb, M_A
= 900 lb · ft $\curvearrowright$.　**10.75** 464 lb · ft $\curvearrowright$.　**10.79 A** = 22**i** + 6**j** kN,
M_A = 66**k** kN · m; **B** = −2**i** + 6**j** kN.　**10.80** $P < ka^2/(a + b)$.
10.81 (a) a/b, (b) $Q > WLa/(2b^2)$.　**10.82** $P < 0.293\ K/L$.

Appendix A

1. d	**2.** a	**3.** d	**4.** c	**5.** e	**6.** c	**7.** a	**8.** c
9. e	**10.** a	**11.** d	**12.** e	**13.** c	**14.** b	**15.** a	
16. b	**17.** b	**18.** c	**19.** e	**20.** b	**21.** a	**22.** a	
23. d	**24.** b	**25.** e	**26.** c	**27.** b	**28.** c	**29.** d	
30. e	**31.** d	**32.** c	**33.** b	**34.** e	**35.** e	**36.** a	
37. d	**38.** d	**39.** b	**40.** a	**41.** d	**42.** c	**43.** b	
44. c	**45.** e	**46.** d	**47.** b	**48.** a	**49.** e	**50.** d	

INDEX

I-2

CENTROIDS OF LINES AND AREAS OF COMMON SHAPES

Quarter-circular arc	**Semicircular arc**	**Segment of circular arc**
$\bar{x} = \bar{y} = \dfrac{2R}{\pi}$ $L = \dfrac{\pi R}{2}$	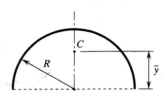 $\bar{y} = \dfrac{2R}{\pi}$ $L = \pi R$	$\bar{x} = \dfrac{R \sin \alpha}{\alpha}$ $L = 2\alpha R$
Triangular area	**Circular sector**	**Quarter-circular area**
$\bar{y} = \dfrac{1}{3}h$ $A = \dfrac{1}{2}bh$	$\bar{x} = \dfrac{2R \sin \alpha}{3\alpha}$ $A = \alpha R^2$	$\bar{x} = \bar{y} = \dfrac{4R}{3\pi}$ $A = \dfrac{1}{4}\pi R^2$
Semicircular area	**Parabolic area**	**Semiparabolic area**
$\bar{y} = \dfrac{4R}{3\pi}$ $A = \dfrac{1}{2}\pi R^2$	$\bar{y} = \dfrac{3}{5}h$ $A = \dfrac{4}{3}ah$	$\bar{x} = \dfrac{3}{8}a$ $\bar{y} = \dfrac{3}{5}h$ $A = \dfrac{2}{3}ah$
Hemispherical shell	**Circular conic shell**	**Half conic shell**
$\bar{x} = \dfrac{R}{2}$ $A = 2\pi R^2$	$\bar{x} = \dfrac{h}{3}$ $A = \pi R (h^2 + R^2)^{1/2}$	$\bar{x} = \dfrac{h}{3}$ $\bar{y} = -\dfrac{4R}{3\pi}$ $A = \dfrac{1}{2}\pi R (h^2 + R^2)^{1/2}$

CENTROIDS OF VOLUMES OF COMMON SHAPES

Hemisphere	**Paraboloid of revolution**	**Semiellipsoid of revolution**
$\bar{x} = \dfrac{3}{8}R$ $V = \dfrac{2}{3}\pi R^3$	$\bar{x} = \dfrac{h}{3}$ $V = \dfrac{1}{2}\pi R^2 h$	$\bar{x} = \dfrac{3}{8}h$ $V = \dfrac{2}{3}\pi R^2 h$
Circular cone	**Half circular cone**	**Pyramid**
$\bar{x} = \dfrac{h}{4}$ $V = \dfrac{1}{3}\pi R^2 h$	$\bar{x} = \dfrac{h}{4}$ $\bar{y} = -\dfrac{R}{\pi}$ $V = \dfrac{1}{6}\pi R^2 h$	$\bar{x} = \dfrac{h}{4}$ $V = \dfrac{1}{3}abh$